Lecture Notes in Mathematics

Edited by A. Dold and B. Eckmann

1117

D. J. Aldous
I. A. Ibragimov
J. Jacod

École d'Été de Probabilités de Saint-Flour XIII – 1983

Édité par P.L. Hennequin

Springer-Verlag
Berlin Heidelberg New York Tokyo

Auteurs

David J. Aldous
University of California
Department of Statistics
Berkeley, CA 94720, USA

Illdar A. Ibragimov
Math. Institute Ac.Sci.
Fontanka 27
191011 Leningrad, USSR

Jean Jacod
Laboratoire de Probabilités
Tour 56 (3ème étage)
4, Place Jussieu
75230 Paris Cedex 05, France

Editeur

P.L. Hennequin
Université de Clermont II, Complexe Scientifique des Cézeaux
Département de Mathématiques Appliquées
B.P. 45, 63170 Aubière, France

AMS Subject Classifications (1980): 60-02, 60F05, 60G05, 60G09, 60G46, 60G50, 62-02, 62A05, 60D05

ISBN 3-540-15203-2 Springer-Verlag Berlin Heidelberg New York Tokyo
ISBN 0-387-15203-2 Springer-Verlag New York Heidelberg Berlin Tokyo

Printing and binding: Beltz Offsetdruck, Hemsbach/Bergstr.
2146/3140-543210

INTRODUCTION

La Treizième Ecole d'Eté de Calcul des Probabilités de Saint-Flour s'est tenue du 3 au 20 Juillet 1983 et a rassemblé, outre les conférenciers, une quarantaine de participants dans les locaux acceuillants du Foyer des Planchettes.

Les trois conférenciers, Messieurs ALDOUS, IBRAGIMOV et JACOD ont entièrement repris la rédaction de leurs cours qui constitue maintenant un texte de référence et ceci justifie le nombre d'années mis pour les publier.

En outre plusieurs exposés ont été faits par les participants durant leur séjour à Saint-Flour :

A. BADRIKIAN	"Approximation de la mécanique quantique"
J. DESHAYES	"Ruptures de modèles pour des processus de Poisson"
A. EHRHARD	"L'inégalité isopérimétrique de Borell et l'opérateur $-\Delta + x\nabla$ "
L. GALLARDO	"Une formule de Lévy-Khintchine sur les hypergroupes (au sens de Jewett ou de Spector) commutatifs dénombrables"
M. LEDOUX	"Théorèmes limite central dans les espaces $\ell_p(B)$ $(1 \leqslant p < \infty)$"
D. LOTI VIAUD	"Modélisation dans l'asymptotique des grandes déviations de processus de branchements spatiaux multitypes non homogènes saturés par le vecteur des proportions des types"
A. MILLET	"Processus à deux indices : problèmes de convergence et d'arrêt optimal"
P. MORALES	"La propriété de Bochner dans les espaces vectoriels topologiques localement convexes"
J.L. PALACIOS	"The exchangeable sigma-field of a markov chain"
J. PICARD	"Un problème de filtrage avec un petit terme non linéaire"
P. SPREIJ	"Filtering and parameter estimation for software retrability models"

A. TOUATI "Grandes déviations pour les mesures aléatoires et applications à la statistique des processus"

M. WEBER "Mesures de Hausdorff et points multiples du mouvement brownien fractionnaire dans $\mathbb{R}^n$ "

M. WSCHEBOR "Ensembles de niveau des surfaces aléatoires"

La frappe du manuscrit a été assurée par les départements de l'Université de Californie, de Clermont-Ferrand et par l'auteur et nous remercions pour leur soin et leur efficacité les secrétaires qui se sont chargées de ce travail délicat.

Nous exprimons enfin notre gratitude à la Société Springer-Verlag qui permet d'accroître l'audience internationale de notre Ecole en accueillant une nouvelle fois ces textes dans la collection Lecture Notes in Mathematics.

P.L. HENNEQUIN
Professeur à l'Université de Clermont II
B.P. n° 45
F-63170 AUBIERE

LISTE DES AUDITEURS

Mr. AZEMA J.	Université de Paris V
Mr. BADRIKIAN A.	Université de Clermont-Ferrand II
Mr. BRETAGNOLLE J.	Université de Paris XIII
Mle CHEVET S.	Université de Clermont-Ferrand II
Mr. COMETS F.	Université de Paris XI
Mr. DESHAYES J.	Ecole Nationale Supérieure des Télécommunications à Paris
Mr. DOUKHAN P.	Université de Paris XI
Mr. EHRHARD A.	Université de Clermont-Ferrand II
Mr. FERNIQUE X.	Université de Strasbourg I
Mr. FOURT G.	Université de Clermont-Ferrand II
Mr. GALLARDO L.	Université de Nancy I
Mr. HENNEQUIN P.L.	Université de Clermont-Ferrand II
Mr. HEBUTERNE G.	C.N.E.T. à Lannion
Mme HUBER C.	Université de Paris XIII
Mle KADI Y.	Université de Paris XI
Mr. KREE P.	Université de Paris VI
Mr. LEDOUX	Université de Strasbourg I
Mr. LEONARD C.	Université de Paris XI
Mr. LETAC G.	Université de Toulouse III
Mle MILLET A.	Université d'Angers
Mr. MORALES P.	Université de Sherbrooke (Canada)
Mr. NGUYEN Van HO	Université de Paris XI
Mr. PALACIOS J.	Université Simon Bolivar à Caracas (Vénézuela)
Mr. PARDOUX E.	Université d'Aix-Marseille I
Mle PICARD D.	Université de Paris XI
Mr. PICARD J.	I.N.R.I.A. à Valbonne
Mr. PORTAL F.	Université de Paris XI
Mr. ROUX D.	Université de Clermont-Ferrand II
Mr. SANTIBANEZ J.	Université de Mexico (Mexique)
Mr. SCHOTT R.	Université de Nancy I
Mr. SPREIJ P.	Mathematisch Centrum à Amsterdam (Pays-Bas)
Mr. SUBIA CEPEDA N.	Escuela Politècnica Nacional à Quito (Equateur)

Mr. TALAY D.	I.N.R.I.A. à Valbonne
Mr. TOUATI A.	Ecole Normale Supérieure de Bizerte (Tunisie)
Mr. TU LY HOANG	Université de Paris XI
Mr. WEBER M.	Université de Strasbourg I
Mr. WONG H.	Université d'Ottawa (Canada)
Mr. WSCHEBOR M.	Université de Paris XI

TABLE DES MATIERES

D.J. ALDOUS : "EXCHANGEABILITY AND RELATED TOPICS"

INTRODUCTION 2

PART I

1. Definitions and immediate consequences 6
2. Mixtures of i.i.d. sequences 10
3. de Finetti's theorem 20
4. Exchangeable sequences and their directing random measures 28
5. Finite exchangeable sequences 37

PART II

6. Properties equivalent to exchangeability 43
7. Abstract spaces 50
8. The subsequence principle 59
9. Other discrete structures 66
10. Continuous-time processes 71
11. Exchangeable random partitions 85

PART III

12. Abstract results 98
13. The infinitary tree 109
14. Partial exchangeability for arrays : the basic structure results 122
15. Partial exchangeability for arrays : complements 134
16. The infinite-dimensional cube 146

PART IV

17. Exchangeable random sets 156
18. Sufficient statistics and mixtures 158
19. Exchangeability in population genetics 166
20. Sampling processes and weak convergences 170
21. Other results and open problems 179

APPENDIX 183

NOTATION 185

REFERENCES 186

I.A. IBRAGIMOV : "THEOREMES LIMITES POUR LES MARCHES ALEATOIRES"

INTRODUCTION 200

CHAPITRE I

1. Lois stables. Conditions de convergence vers une loi stable 202
2. Processus stables. Conditions de convergence vers un processus stable 204
3. Problème sur la loi limite de fonctionnelles définies sur une marche aléatoire 205
4. Temps local de processus stables 211
5. Convergence en probabilité et convergence en loi 214
6. Une propriété caractéristique du processus de Wiener 217
7. Commentaire 218

CHAPITRE II

1. Introduction 219
2. Théorèmes limites du premier type 222
3. Théorèmes limites du premier type, suite 234
4. Convergence des processus engendrés par $\sum_{k \leqq nt} f_n(Sn_k)$ 240
5. Théorèmes limites du deuxième type 247
6. Commentaire 260

CHAPITRE III

1. Introduction 262
2. Théorèmes limites dans le cas $A_n = 0$ 262
3. Théorèmes limites pour des fonctions sommables 268
4. Cas $\gamma > \frac{\alpha-1}{2}$ 270
5. Marche aléatoire de Cauchy 274
6. Marche aléatoire dans R^2 278
7. Le cas des fonctions périodiques 280
8. Commentaire 291

NOTATIONS 293

BIBLIOGRAPHIE 294

J. JACOD : "THEOREMES LIMITE POUR LES PROCESSUS"

INTRODUCTION 299

CHAPITRE I - TOPOLOGIE DE SKOROKHOD ET CONVERGENCE EN LOI DE PROCESSUS

1. L'espace de Skorokhod 301
2. Convergence en loi de processus 306
3. Un critère de compacité adapté aux processus asymptotiquement quasi-continus à gauche 311

CHAPITRE II - CONVERGENCE DES PROCESSUS A ACCROISSEMENTS INDEPENDANTS

1. Les caractéristiques d'un processus à accroissements indépendants 318
2. Condition nécessaire et suffisante de convergence vers un pai sans discontinuités fixes 329
3. Application aux sommes de variables indépendantes 341

CHAPITRE III - CONVERGENCE DE SEMIMARTINGALES VERS UN PROCESSUS A ACCROISSEMENTS INDEPENDANTS

1. Semimartingales et caractéristiques locales 343
2. Convergence de semimartingales vers un pai 358
3. Deux exemples 367

CHAPITRE IV - CONVERGENCE VERS UNE SEMIMARTINGALE

1. Un théorème général de convergence 373
2. Théorème de convergence : une condition plus faible 384
3. Convergence de processus de Markov 388

CHAPITRE V - CONDITIONS NECESSAIRES DE CONVERGENCE

1. Convergence et variation quadratique 395
2. Conditions nécessaires de convergence vers un processus continu 401

BIBLIOGRAPHIE 407

EXCHANGEABILITY AND RELATED TOPICS

PAR David J. ALDOUS

0. Introduction

If you had asked a probabilist in 1970 what was known about exchangeability, you would likely have received the answer "There's de Finetti's theorem: what else is there to say?" The purpose of these notes is to dispel this (still prevalent) attitude by presenting, in Parts II-IV, a variety of mostly post-1970 results relating to exchangeability. The selection of topics is biased toward my own interests, and away from those areas for which survey articles already exist. Any student who has taken a standard first year graduate course in measure-theoretic probability theory (e.g. Breiman (1968)) should be able to follow most of this article; some sections require knowledge of weak convergence.

In Bayesian language, de Finetti's theorem says that the general infinite exchangeable sequence (Z_i) is obtained by first picking a distribution θ at random from some prior, and then taking (Z_i) to be i.i.d. with distribution θ. Rephrasing in the language of probability theory, the theorem says that with (Z_i) we can associate a random distribution $\alpha(\omega,\cdot)$ such that, conditional on $\alpha = \theta$, the variables (Z_i) are i.i.d. with distribution θ. This formulation is the central fact in the circle of ideas surrounding de Finetti's theorem, which occupies most of Part I. No previous knowledge of exchangeability is assumed, though the reader who finds my proofs overly concise should take time out to read the more carefully detailed account in Chow and Teicher (1978), Section 7.3.

Part II contains results complementary to de Finetti's theorem. Dacunha-Castelle's "spreading-invariance" property and Kallenberg's stopping time property give conditions on an infinite sequence which turn out to be equivalent to exchangeability. Kingman's "paintbox" description of exchangeable random partitions leads to Cauchy's formula for the distribution of cycle lengths in a uniform random permutation, and to results about components of random functions. Continuous-time processes with interchangeable increments are discussed; a notable result is that any continuous-path process on $[0,\infty)$ (resp. $[0,1]$) with interchangeable increments is a mixture of processes which are linear transformations of Brownian motion (resp. Brownian bridge). The subsequence principle reveals exchangeable-like sequences lurking unsuspectedly within arbitrary sequences of random variables. And we discuss exchangeability in abstract spaces, and weak convergence issues.

The class of exchangeable sequences is the class of processes whose distributions are invariant under a certain group of transformations; in Part III related invariance concepts are described. After giving the abstract result on ergodic decompositions of measures invariant under a group of transformations, we specialize to the setting of <u>partial exchangeability</u>, where we study the class of processes $(X_i : i \in I)$ invariant under the action of some group of transformations of the index set I. Whether anything can be proved about partially exchangeable classes in general is a challenging open problem; we can only discuss three particular instances. The most-studied instance, investigated by Hoover and by myself, is partial exchangeability for arrays of random variables, where the picture is fairly complete. We also discuss partial exchangeability on trees of infinite degree, where the basic examples are reversible Markov chains; and on infinite-dimensional cubes, where it appears that the basic examples are random walks, though here the picture remains fragmentary.

Part IV outlines other topics of current research. A now-classical result on convergence of partial sum processes from sampling without replacement to Brownian bridge leads to general questions of convergence for triangular arrays of finite exchangeable sequences, where the present picture is unsatisfactory for applications. Kingman's uses of exchangeability in mathematical genetics will be sketched. The theory of sufficient statistics and mixtures of processes of a specified form will also be sketched--actually, this topic is perhaps the most widely studied relative of exchangeability, but in view of the existing accounts in Lauritzen (1982) and Diaconis and Freedman (1982), I have not emphasized it in these notes. Kallenberg's stopping time approach to continuous-time exchangeability is illustrated by the study of exchangeable subsets of $[0,\infty)$. A final section provides references to work related to exchangeability not elsewhere discussed: I apologize in advance to those colleagues whose favorite theorems I have overlooked.

General references. Chow and Teicher (1978) is the only textbook (known to me) to give more than a cursory mention to exchangeability. A short but elegant survey of exchangeability, whose influence can be seen in these notes, has been given by Kingman (1978a). In 1981 a conference on "Exchangeability in Probability and Statistics" was held in Rome to honor Professor Bruno de Finetti: the conference proceedings (EPS in the References) form a sample of the current interests of workers in exchangeability. Dynkin (1978) gives a concise abstract treatment of the "sufficient statistics" approach in several areas of probability including exchangeability.

The material in Sections 13 and 16 is new, and perhaps a couple of proofs elsewhere may be new; otherwise no novelty is claimed.

Notation and terminology. The mathematical notation is intended to be standard, so the reader should seldom find it necessary to consult the list of notation at the end. As for terminology, "exchangeable" is more popular and shorter than the synonyms "symmetrically dependent" and "interchangeable". I have introduced "directing random measure" in place of Kallenberg's "canonical random measure", partly as a more vivid metaphor and partly for more grammatical flexibility, so one can say "directed by ...". I use "partial exchangeability" in the narrow sense of Section 12 (processes with certain types of invariance) rather than in the wider context of Section 18 (processes with specified sufficient statistics). "Problem" means "unsolved problem" rather than "exercise": if you can solve one, please let me know.

Acknowledgements. My thanks to Persi Diaconis for innumerable invaluable discussions over the last several years; and to the members of the audiences at St. Flour and the Berkeley preview who detected errors and contributed to the presentation. Research supported by National Science Foundation Grant MCS80-02698.

PART I

The purpose of Part I is to give an account of de Finetti's theorem and some straightforward consequences, using the language and techniques of modern probability theory. I have not attempted to assign attributions to these results: historical accounts of de Finetti's work on exchangeability and the subsequent development of the subject can be found in EPS (Foreword, and Furst's article) and in Hewitt and Savage (1955).

1. Definitions and immediate consequences

A finite sequence $(Z_1,\ldots,Z_N)$ of random variables is called exchangeable (or N-exchangeable, to indicate the number of random variables) if

$$(Z_1,\ldots,Z_N) \stackrel{\mathcal{D}}{=} (Z_{\pi(1)},\ldots,Z_{\pi(N)}) \;; \tag{1.1}$$

each permutation π of $\{1,\ldots,N\}$. An infinite sequence $(Z_1,Z_2,\ldots)$ is called exchangeable if

$$(Z_1,Z_2,\ldots) \stackrel{\mathcal{D}}{=} (Z_{\pi(1)},Z_{\pi(2)},\ldots) \tag{1.2}$$

for each finite permutation π of $\{1,2,\ldots\}$, that is each permutation for which $\#\{i\colon \pi(i)\neq i\} < \infty$. Throughout Part I we shall regard random variables Z_i as real-valued; but we shall see in Section 7 that most results remain true whenever the Z_i have any "non-pathological" range space.

There are several obvious reformulations of these definitions. Any finite permutation can be obtained by composing permutations which transpose 1 and $n > 1$; so (1.2) is equivalent to the at first sight weaker condition

(1.3) $(Z_1,\dots,Z_{n-1},Z_n,Z_{n+1},\dots) \overset{\mathcal{D}}{=} (Z_n,Z_2,\dots,Z_{n-1},Z_1,Z_{n+1},\dots)$;

each $n > 1$. In the other direction, (1.2) implies the at first sight stronger condition

(1.4) $$(Z_1,Z_2,Z_3,\dots) \overset{\mathcal{D}}{=} (Z_{n_1},Z_{n_2},Z_{n_3},\dots) ;$$

each sequence (n_i) with distinct elements. In Section 6 we shall see some non-trivially equivalent conditions.

Sampling variables. The most elementary examples of exchangeability arise in sampling. Suppose an urn contains N balls labelled $x_1,\dots,x_N$. The results $Z_1,Z_2,\dots$ of an infinite sequence of draws with replacement form an infinite exchangeable sequence; the results $Z_1,\dots,Z_N$ of N draws without replacement form a N-exchangeable sequence (sequences of this latter type we call urn sequences).

Both ideas generalize. In the first case, (Z_i) is i.i.d. uniform on $\{x_1,\dots,x_N\}$; obviously any i.i.d. sequence is exchangeable. In the second case we can write

(1.5) $$(Z_1,\dots,Z_N) = (x_{\pi^*(1)},\dots,x_{\pi^*(N)})$$

where π^* denotes the uniform random permutation on $\{1,\dots,N\}$, that is $P(\pi^* = \pi) = 1/N!$ for each π. More generally, let $(Y_1,\dots,Y_N)$ be arbitrary random variables, take π^* independent of (Y_i), and then

(1.6) $$(Z_1,\dots,Z_N) = (Y_{\pi^*(1)},\dots,Y_{\pi^*(N)})$$

defines a N-exchangeable sequence. This doesn't work for infinite sequences, since we cannot have a uniform permutation of a countable infinite set (without abandoning countable additivity--see (13.27)). However we can define a uniform random ordering on a countable infinite set: simply define $i \overset{\omega}{\leq} j$ to mean $\xi_i(\omega) \leq \xi_j(\omega)$, for i.i.d. continuous $(\xi_1, \xi_2, \ldots)$. This trick is useful in several contexts--see (11.9), (17.4), (19.8).

Correlation structure. Exchangeability restricts the possible correlation structure for square-integrable sequences. Let (Z_i) be N-exchangeable. Then there is a correlation $\rho = \rho(Z_i, Z_j)$, $i \neq j$. We assert

(1.7) $\rho \geq \frac{-1}{N-1}$, with equality iff $\sum Z_i$ is a.s. constant.

In particular, $\rho = -1/(N-1)$ for sampling without replacement from an N-element urn. To prove (1.7), linearly scale to make $EZ_i = 0$, $EZ_i^2 = 1$, and then

$$0 \leq E(\sum Z_i)^2 = EZ_i^2 + \sum_{1 \leq i \neq j \leq N} \sum EZ_i Z_j = N + N(N-1)\rho .$$

So $\rho \geq -1/(N-1)$, with equality iff $\sum Z_i = 0$ a.s.

Observe that (1.7) implies

(1.8) $\rho \geq 0$ for an infinite exchangeable sequence.

Conversely, every $\rho \leq 1$ satisfying (1.7) (resp. (1.8)) occurs as the correlation in some N-exchangeable (resp. infinite exchangeable) sequence. To prove this, let (ξ_i) be i.i.d., $E\xi_i = 0$, $E\xi_i^2 = 1$. Define

$$Z_i = \xi_i + c \sum_{j=1}^{N} \xi_j \; ; \quad 1 \leq i \leq N ,$$

for some constant c. Then (Z_i) is N-exchangeable, and a simple computation gives $\rho = 1 - (Nc^2 + 2c + 1)^{-1}$. As c varies we get all values $1 > \rho \geq -(N-1)^{-1}$. (The case $c = -1/N$ which gives $\rho = -1/(N-1)$ will be familiar from statistics!). Of course we can get $\rho = 1$ by setting $Z_1 = \cdots = Z_N$. In the infinite case, take $(\xi_i : i \geq 0)$ i.i.d. and set

$$Z_i = c\xi_0 + \xi_i \; ; \quad i \geq 1$$

for some constant c. Then (Z_i) is an infinite exchangeable sequence with $\rho = c^2/(c^2+1)$, and as c varies we get all values $0 \leq \rho < 1$.

Gaussian exchangeable sequences. Since the distribution of a Gaussian sequence is determined by the covariance structure, the results above enable us to describe explicitly the Gaussian exchangeable sequences. Let $X_0, X_1, X_2, \ldots$ be independent $N(0,1)$. The general N-exchangeable Gaussian sequence is, in distribution, of the form

$$(1.9) \qquad Z_i = a + bX_i + c\sum_{j=1}^{N} X_j \, , \quad 1 \leq i \leq N$$

for some constants a, b, c. The general infinite exchangeable sequence is, in distribution, of the form

$$(1.10) \qquad Z_i = a + bX_0 + cX_i \, , \quad i \geq 1$$

for some constants a, b, c.

Extendibility. An N-exchangeable sequence (Z_i) is M-extendible $(M > N)$ if $(Z_1,\ldots,Z_N) \stackrel{\mathcal{D}}{=} (\hat{Z}_1,\ldots,\hat{Z}_N)$ for some M-exchangeable sequence $(\hat{Z}_i)$. By (1.7) there exist N-exchangeable sequences which are not (N+1)-extendible; for example, sampling without replacement from an urn with N elements. This suggests several problems, which we state rather vaguely.

(1.11) Problem. Find effective criteria for deciding whether a given N-exchangeable sequence is M-extendible.

(1.12) Problem. What proportion of N-exchangeable sequences are M-extendible?

Such problems seem difficult. Some results can be found in Diaconis (1977), Crisma (1982) and Spizzichino (1982).

Combinatorial arguments. Many identities and inequalities for i.i.d. sequences are proved by combinatorial arguments which remain valid for exchangeable sequences. Such results are scattered in the literature; for a selection, see Kingman (1978) Section 1 and Marshall and Olkin (1979).

2. Mixtures of i.i.d. sequences

Everyone agrees on how to say de Finetti's theorem in words:

"An infinite exchangeable sequence is a mixture of i.i.d. sequences."

But there are several mathematical formalizations (at first sight different, though in fact equivalent) of the theorem in the literature, because the concept of "a mixture of i.i.d. sequences" can be defined in several ways. Our strategy is to discuss this concept in detail in this section, and defer discussion of exchangeability and de Finetti's theorem until the next section.

Let $\theta_1,\ldots,\theta_k$ be probability distributions on R, and let $p_1,\ldots,p_k > 0$, $\sum p_i = 1$. Then we can describe a sequence (Y_i) by the two-stage procedure:

(2.1) (i) Pick θ at random from $\{\theta_1,\ldots,\theta_k\}$, $P(\theta = \theta_i) = p_i$;

(ii) then let (Y_i) be i.i.d. with distribution θ.

More generally, write $\mathcal{P}$ for the set of probability measure on R, let Θ be a distribution on $\mathcal{P}$, and replace (i) by

(i') Pick θ at random from distribution Θ.

Here we are merely giving the familiar Bayesian idea that (Y_i) is i.i.d. (θ), where θ has a prior distribution Θ. The easiest way to formalize this verbal description is to say

$$P(\underset{\sim}{Y} \in A) = \int_P \theta^\infty(A)\Theta(d\theta) \; ; \quad A \subset R^\infty \tag{2.2}$$

where $\underset{\sim}{Y} = (Y_1, Y_2, \ldots)$, regarded as a random variable with values in R^∞, and $\theta^\infty = \theta \times \theta \times \cdots$ is the distribution on R^∞ of an i.i.d. (θ) sequence. This describes the distribution of a sequence which is a mixture of i.i.d. sequences. This is a special case of a general idea. Given a family $\{\mu_\gamma : \gamma \in \Gamma\}$ of distributions on a space S, call a distribution ν a mixture of (μ_γ)'s if

$$\nu(\cdot) = \int_\Gamma \mu_\gamma(\cdot)\Theta(d\gamma) \quad \text{for some distribution } \Theta \text{ on } \Gamma \, . \tag{2.3}$$

But in practice it is much more convenient to use a definition of "(Y_i) is a mixture of i.i.d. sequences" which involves the random variables Y_i explicitly. To do so, we need a brief digression to discuss random measures and regular conditional distributions.

A random measure α is simply a P-valued random variable. So for each ω there is a probability measure $\alpha(\omega)$ and this assigns probability $\alpha(\omega, A)$ to subsets $A \subset R$. To make this definition precise we need to specify a σ-field on P: the natural σ-field is that generated by the maps

$$\theta \longrightarrow \theta(A) \; ; \quad \text{measurable } A \subset R \, .$$

The technicalities about measurability in P that we need are straightforward and will be omitted. We may equivalently define a random measure as a function $\alpha(\omega, A)$, $\omega \in \Omega$, $A \subset R$, such that

$$\alpha(\omega,\cdot) \text{ is a probability measure; each } \omega \in \Omega .$$
$$\alpha(\cdot,A) \text{ is a random variable; each } A \subset R .$$

Say $\alpha_1 = \alpha_2$ a.s. if they are a.s. equal as random variables in P; or equivalently, if $\alpha_1(\cdot,A) = \alpha_2(\cdot,A)$ a.s. for each $A \subset R$.

Given a real-valued random variable Y and a σ-field F, a <u>regular conditional distribution</u> (r.c.d.) for Y given F is a random measure α such that

$$\alpha(\cdot,A) = P(Y \in A | F) \text{ a.s., each } A \subset R .$$

It is well known that r.c.d.'s exist, are a.s. unique, and satisfy the <u>fundamental property</u>

(2.4) $$E(g(X,Y)|F) = \int g(X,y)\alpha(\omega,dy) \text{ a.s.; } X \in F, \; g(X,Y) \text{ integrable.}$$

We now come to the key idea of this section. Given a random measure α, it is possible to construct (Y_i) such that conditional on $\alpha = \theta$ (where θ denotes a generic probability distribution), the sequence (Y_i) is i.i.d. with distribution θ. One way of doing so is to formalize the required properties of (Y_i) in an abstract way (2.6) and appeal to abstract existence theorems to show that random variables with the required properties exist. We prefer to give a concrete construction first.

Let $F(\theta,t) = \theta(-\infty,t]$ be the distribution function of θ, and let $F^{-1}(\theta,x) = \inf\{t: F(\theta,t) \geq x\}$ be the inverse distribution function. It is well known that if ξ is uniform on $(0,1)$ ("ξ is $U(0,1)$") then $F^{-1}(\theta,\xi)$ is a random variable with distribution θ. So if (ξ_i) is an i.i.d. $U(0,1)$ sequence then $(F^{-1}(\theta,\xi_i))$ is an i.i.d. (θ) sequence. Now given a random measure α, take (ξ_i) as above, independent of α, and let

(2.5) $$\hat{Y}_i = F^{-1}(\alpha,\xi_i) .$$

This construction captures the intuitive idea that, conditional on $\alpha = \theta$, the variables $\hat{Y}_i$ are i.i.d. (θ). The abstract properties of $(\hat{Y}_i, i \geq 1; \alpha)$ are given in

(2.6) Definition. Let α be a random measure and let $\underset{\sim}{Y} = (Y_i)$ be a sequence of random varibles. Say $\underset{\sim}{Y}$ is a mixture of i.i.d.'s directed by α if

$$(\alpha(\omega))^\infty \text{ is a r.c.d. for } \underset{\sim}{Y} \text{ given } \sigma(\alpha) .$$

Plainly this implies that the distribution of $\underset{\sim}{Y}$ is of the form (2.2), where Θ is the distribution of α. We remark that this idea can be abstracted to the general setting of (2.3); X is a mixture of $(\mu_\gamma: \gamma \in \Gamma)$ directed by a random element $\beta: \Omega \to \Gamma$ if $\mu_{\beta(\omega)}$ is a r.c.d. for X given $\sigma(\beta)$. Think of this as the "strong" notion of mixture corresponding to the "weak" notion (2.3).

The condition in (2.6) is equivalent to

(2.6a) $$P(Y_i \in A_i, 1 \leq i \leq n | \alpha) = \prod_i \alpha(\omega, A_i) \; ; \quad \text{all } A_1, \ldots, A_n, \quad n \geq 1 .$$

And this splits into two conditions, as follows.

(2.7) Lemma. Write $F = \sigma(\alpha)$. Then $\underset{\sim}{Y}$ is a mixture of i.i.d.'s directed by α iff

(2.8) $(Y_i: i \geq 1)$ are conditionally independent given F, that is
$P(Y_i \in A_i, 1 \leq i \leq n | F) = \prod_i P(Y_i \in A_i | F)$.

(2.9) the conditional distribution of Y_i given F is α; that is,
$P(Y_i \in A_i | F) = \alpha(\omega, A_i)$.

Readers unfamiliar with the concept of conditional independence defined in (2.8) should consult the Appendix, which lists properties (A1)-(A9) and references.

Lemma 2.7 suggests a definition of "conditionally i.i.d." without explicit refernece to a random measure.

(2.10) <u>Definition</u>. Let (Y_i) be random variables and let F be a σ-field. Say (Y_i) <u>is conditionally i.i.d. given</u> F if (2.8) holds and if

$$(2.11) \qquad P(Y_i \in A|F) = P(Y_j \in A|F) \text{ a.s., each } A, \; i \neq j.$$

Here is a useful technical lemma.

(2.12) <u>Lemma</u>. <u>Suppose</u> (Y_i) <u>are conditionally i.i.d. given</u> F. <u>Let</u> α <u>be a r.c.d. for</u> Y_1 <u>given</u> F. <u>Then</u>

(a) (Y_i) <u>is a mixture of i.i.d.'s directed by</u> α.

(b) $\underset{\sim}{Y}$ <u>and</u> F <u>are conditionally independent given</u> α.

<u>Proof</u>. By (2.11), for each i we have that α is a r.c.d. for Y_i given F. So by (2.8), $P(Y_i \in A_i, 1 \leq i \leq n|F) = \prod_i \alpha(\cdot, A_i)$. Now α is F-measurable, so conditioning on α gives $P(Y_i \in A_i, 1 \leq i \leq n|\alpha) = \prod_i \alpha(\cdot, A_i)$. This implies (a) by (2.6a). And it also implies

$$P(\underset{\sim}{Y} \in A|F) = P(\underset{\sim}{Y} \in A|\alpha) \text{ a.s.,} \quad A \subset R^\infty,$$

which gives (b) by a standard property of conditional independence (A3).

We have been stating results for infinite sequences (Y_i), but the results so far are true for finite sequences also. Suppose now we are told that a sequence (Y_i) is a mixture of i.i.d.'s for some unspecified random measure α. Can we determine α from (Y_i)? For finite sequences the answer is no, in general (4.7). But for infinite sequences we can. Define $\Lambda_n: R^n \to P$ and $\Lambda: R^\infty \to P$ by

(2.13) $\Lambda_n(x_1,\ldots,x_n) = n^{-1}\sum_i \delta_{x_i}$, the empirical distribution of (x_i);

(2.14) $$\Lambda(\underset{\sim}{x}) = \underset{n\to\infty}{\text{weak-limit}}\ \Lambda_n(x_1,\ldots,x_n)$$
$$= \delta_0 \text{ , say, if the limit does not exist,}$$

(δ_x denotes the degenerate distribution $\delta_x(A) = 1_{(x\in A)}$). If $\underset{\sim}{X} = (X_i)$ is an infinite i.i.d. (θ) sequence, then the Glivenko-Cantelli theorem says that $\Lambda(\underset{\sim}{X}) = \theta$ a.s. Thus for a mixture $(\hat{Y}_i)$ of i.i.d.'s directed by α we have $\Lambda(\hat{\underset{\sim}{Y}}) = \alpha$ a.s. by (2.6) and the fundamental property of r.c.d.'s. Hence

(2.15) Lemma. *If the infinite sequence $\underset{\sim}{Y}$ is a mixture of i.i.d.'s, then it is directed by $\alpha = \Lambda(\underset{\sim}{Y})$, and this directing random measure is a.s. unique.*

So we can talk about "the" directing random measure. Since $\Lambda(\underset{\sim}{x})$ is unchanged by changing a finite number of coordinates of $\underset{\sim}{x}$, we have

(2.16) Lemma. *Let the infinite sequence $\underset{\sim}{Y}$ be a mixture of i.i.d.'s, let α be the directing random measure. Then α is essentially T-measurable, where T is the tail σ-field of $\underset{\sim}{Y}$.*

Keeping the notation of Lemma 2.16, we see that the following are a.s. equal.

(2.17) (a) $P(Y_i \in A_i,\ 1 \le i \le n \mid Y_m, Y_{m+1},\ldots)$, $m > n$

(b) $P(Y_i \in A_i,\ 1 \le i \le n \mid Y_m, Y_{m+1},\ldots;\alpha)$, $m > n$

(c) $P(Y_i \in A_i,\ 1 \le i \le n \mid T)$

(d) $P(Y_i \in A_i,\ 1 \leq i \leq n | \alpha)$

(e) $\prod_i \alpha(\cdot, A_i)$.

Indeed, (a) = (b) since the conditioning is the same by Lemma 2.16; (b) = (d) by conditional independence; (d) = (e) by Lemma 2.7; and (b) = (c) = (d) since $\sigma(\alpha) \subset \mathcal{T} \subset \sigma(Y_m, Y_{m+1}, \ldots; \alpha)$ a.s. These identities establish the next lemmas.

(2.18) Lemma. _(Y_i) is a mixture of i.i.d.'s if and only if (Y_i) is conditionally i.i.d. given $\mathcal{T}$._

(2.19) Lemma. _If the infinite sequence $\underset{\sim}{Y}$ is a mixture of i.i.d.'s then it is directed by_

(a) _a r.c.d. for Y_1 given $\mathcal{T}$;_

(b) _a r.c.d. for Y_1 given $(Y_{m+1}, Y_{m+2}, \ldots)$, $m \geq 1$._

Observe that Lemmas 2.15 and 2.19 provide three ways to obtain (in principle, at least) the directing random measure. Each of these ways is useful in some circumstances.

Facts about mixtures of i.i.d.'s are almost always easiest to prove by conditioning on the directing measure. Let us spell this out in detail. For bounded $g: R \rightarrow R$ define $\bar{g}: \mathcal{P} \rightarrow R$ by

$$\bar{g}(\theta) = \int g \, d\theta \,. \tag{2.20}$$

By (2.9) and the fundamental property of r.c.d.'s

$$E(g(Y_i)|\alpha) = \bar{g}(\alpha) \text{ a.s., and hence } Eg(Y_i) = E\bar{g}(\alpha) \,. \tag{2.21}$$

And using the conditional independence (2.8),

$$(2.22)\qquad E(g_1(Y_i)g_2(Y_j)|\alpha) = \bar{g}_1(\alpha)\bar{g}_2(\alpha) \text{ a.s.}, \quad i \neq j, \quad \text{and hence}$$
$$Eg_1(Y_i)g_2(Y_j) = E\bar{g}_1(\alpha)\bar{g}_2(\alpha) .$$

These extend to unbounded g provided $|g(Y_1)|$, $|g_1(Y_i)g_2(Y_j)|$ are integrable. Let us use these to record some technical facts about moments of Y_i and α. For a distribution θ let $\text{mean}(\theta)$, $\text{var}(\theta)$, $\text{abs}_r(\theta)$ denote the mean, variance and r^{th} absolute moment of θ. Let $V(Y)$ denote the variance of a random variable Y. The next lemma gives properties which follow immediately from (2.21) and (2.22).

(2.23) Lemma.

(a) $E|Y_i|^r = E\,\text{abs}_r(\alpha)$.

(b) *If* $E|Y_i|$ *is finite, then* $EY_i = E\,\text{mean}(\alpha)$.

(c) *If* EY_i^2 *is finite, then*

(i) $EY_iY_j = E(\text{mean}(\alpha))^2$, $i \neq j$;

(ii) $V(Y_i) = E\,\text{abs}_2(\alpha) - (E\,\text{mean}(\alpha))^2 = E\,\text{var}(\alpha) + V(\text{mean}(\alpha))$.

Most classical limit theorems for i.i.d. sequences extend immediately to mixtures of i.i.d. sequences. For instance, writing $S_n = Y_1 + \cdots + Y_n$,

$$(2.24)\qquad \lim_{n\to\infty} n^{-1}S_n = \text{mean}(\alpha) \text{ a.s.}, \quad \text{provided } E|Y_1| < \infty .$$

$$(2.25)\qquad \limsup_{n\to\infty} \frac{S_n - n\cdot\text{mean}(\alpha)}{\{\text{var}(\alpha)\cdot 2n\log\log(n)\}^{1/2}} = 1 \text{ a.s.}, \quad \text{provided } E\,Y_1^2 < \infty .$$

$$(2.26)\qquad \frac{S_n - n\cdot\text{mean}(\alpha)}{\{n\,\text{var}(\alpha)\}^{1/2}} \xrightarrow{\mathcal{D}} \text{Normal}(0,1) \text{ as } n\to\infty, \quad \text{provided } E\,Y_1^2 < \infty .$$

Let us spell out some details. The fundamental property of r.c.d.'s says $P(\lim n^{-1}S_n = \text{mean}(\alpha)|\alpha) = h(\alpha)$, where $h(\theta) = P(\lim n^{-1}(X_1+\cdots+X_n) = \text{mean}(\theta))$ for (X_i) i.i.d. (θ). The strong law of large numbers says $h(\theta) = 1$

provided $abs_1(\theta) < \infty$. So (2.24) holds provided $abs_1(\alpha) < \infty$ a.s., and by Lemma 2.23 this is a consequence of $E|Y_1| < \infty$. The same argument works for (2.25), and for any a.s. convergence theorem for i.i.d. variables. For weak convergence theorems like (2.26), one more step is needed. Let $g: R \to R$ be bounded continuous. Then

$$E(g(\frac{S_n - n\cdot mean(\alpha)}{\{n\ var(\alpha)\}^{1/2}})|\alpha) = g_n(\alpha)$$

where

$$g_n(\theta) = E\ g(\frac{\sum_1^n X_i - n\cdot mean(\theta)}{\{n\ var(\theta)\}^{1/2}}) \quad \text{for } (X_i) \text{ i.i.d. } (\theta).$$

The central limit theorem says $g_n(\theta) \to Eg(V)$ when $abs_2(\theta) < \infty$, where V indicates a $N(0,1)$ variable. Hence

$$E\ g(\frac{S_n - n\cdot mean(\alpha)}{\{n\ var(\alpha)\}^{1/2}}) \to Eg(V)$$

provided $abs_2(\alpha) < \infty$ a.s., which by Lemma 2.23 is a consequence of $E\ Y_1^2 < \infty$. The same technique works for any weak convergence theorem for i.i.d. variables.

The form of results obtained in this way is slightly unusual, in that random normalization is involved, but they can easily be translated into a more familiar form. To ease notation, suppose $EY_1^2 < \infty$ and $mean(\alpha) = 0$ a.s. (which by Lemma 2.23 is equivalent to assuming $EY_i = 0$ and (Y_i) uncorrelated). Then (2.25) translates immediately to

$$\limsup_{n\to\infty} \frac{S_n}{\{2n \log\log(n)\}^{1/2}} = \{var(\alpha)\}^{1/2} \text{ a.s.}$$

And (2.26) translates to

(2.27) $$n^{-1/2}S_n \xrightarrow{\mathcal{D}} V\cdot\{\mathrm{var}(\alpha)\}^{1/2} ;$$

where V is Normal(0,1) independent of α. To see this, observe that the argument for (2.26) gives

$$(\mathrm{var}(\alpha), \frac{S_n - n\cdot\mathrm{mean}(\alpha)}{\{n\ \mathrm{var}(\alpha)\}^{1/2}}) \xrightarrow{\mathcal{D}} (\mathrm{var}(\alpha), V) ,$$

and then the continuous mapping theorem gives (2.27). Finally, keeping the assumptions above, (2.22) shows that $\mathrm{var}(\alpha) = \sigma^2$, constant, iff $EY_i^2 = \sigma^2$ and $EY_i^2Y_j^2 = \sigma^4$, so that these extra assumptions are what is needed to obtain $n^{-1/2}S_n \xrightarrow{\mathcal{D}} N(0,\sigma^2)$. Weak convergence theorems for mixtures of i.i.d. sequences are a simple class of <u>stable</u> convergence theorems, described further in Section 7.

Finally, we should point out there are occasional subtleties in extending results from i.i.d. sequences to mixtures. Consider the weak law of large numbers. The set $\mathcal{S}$ of distributions θ such that

(2.28) $$n^{-1}\sum_1^n X_i \xrightarrow{p} 0 \quad \text{for } (X_i) \text{ i.i.d. } (\theta)$$

is known. For a mixture of i.i.d. sequences to satisfy this weak law, it is certainly <u>sufficient</u> that the directing random measure satisfies $\alpha(\omega) \in \mathcal{S}$ a.s., by the usual conditioning argument. But this is not necessary, because (informally speaking) we can arrange the mixture to satisfy the weak law although α takes values θ such that convergence in (2.28) holds as $n \to \infty$ through some set of integers with asymptotic density one but not when $n \to \infty$ through all integers. (My thanks to Mike Klass for this observation.)

Another instance where the results for mixtures are more complicated is the estimation of L^p norms for weighted sums $\sum a_i X_i$. Such estimates are given by Dacunha-Castelle and Schreiber (1974) in connection with Banach space questions; see also (7.21).

We end with an intriguing open problem.

(2.29) Problem. Let $S_n = \sum_{i=1}^{n} X_i$, $\hat{S}_n = \sum_{i=1}^{n} \hat{X}_i$, where each of the sequences (X_i), $(\hat{X}_i)$ is a mixture of i.i.d. sequences. Suppose $S_n \overset{\mathcal{D}}{=} \hat{S}_n$ for each n. Does this imply $(X_i) \overset{\mathcal{D}}{=} (\hat{X}_i)$?

3. de Finetti's theorem

Our verbal description of de Finetti's theorem is now a precise assertion, which we restate as

(3.1) de Finetti's Theorem. *An infinite exchangeable sequence (Z_i) is a mixture of i.i.d. sequences.*

Remarks

(a) The converse, that a mixture of i.i.d.'s is exchangeable, is plain.

(b) As noted in Section 1, a finite exchangeable sequence may not be to an infinite sequence, and so a finite exchangeable sequence may not be a mixture of i.i.d.'s.

(c) The directing random measure can in principle be obtained from Lemma 2.15 or 2.19.

Most modern proofs of de Finetti's theorem rely on martingale convergence. We shall present both the standard proof (which goes back at least to Loève (1960)) and also a more sophisticated variant. Both proofs contain useful techniques which will be used later in other settings.

First proof of Theorem 3.1. Let

$$G_n = \sigma\{f_n(Z_1,\dots,Z_n)\colon f_n \text{ symmetric}\}$$
$$H_n = \sigma\{G_n, Z_{n+1}, Z_{n+2}, \dots\} .$$

Then $H_n \supset H_{n+1}$, $n \geq 1$. Exchangeability implies

$$(Z_i, Y) \overset{\mathcal{D}}{=} (Z_1, Y) , \quad 1 \leq i \leq n ,$$

for Y of the form $(f_n(Z_1,\dots,Z_n), Z_{n+1}, Z_{n+2}, \dots)$, f_n symmetric, and hence for all $Y \in H_n$. So for bounded ϕ,

$$E(\phi(Z_1)|H_n) = E(\phi(Z_i)|H_n) , \quad 1 \leq i \leq n$$
$$= E(n^{-1} \sum_{i=1}^{n} \phi(Z_i)|H_n)$$

(3.2) $$= n^{-1} \sum_{i=1}^{n} \phi(Z_i) , \quad \text{since this is } G_n\text{-measurable.}$$

The reversed martingale convergence theorem implies

(3.3) $$n^{-1} \sum_{i=1}^{n} \phi(Z_i) \to E(\phi(Z_1)|H) \text{ a.s., where } H = \bigcap_n H_n .$$

For bounded $\phi(x_1,\dots,x_k)$, the argument for (3.2) shows that for $n > k$

$$E(\phi(Z_1,\dots,Z_k)|H_n) = \frac{1}{n(n-1)\cdots(n-k+1)} \sum_{j_1=1}^{n} \cdots \sum_{j_k=1}^{n} \phi(Z_{j_1},\dots,Z_{j_k}) 1_{D_{k,n}}$$

where $D_{k,n} = \{(j_1,\dots,j_k)\colon 1 \leq j_r \leq n, (j_r) \text{ distinct}\}$. Using martingale convergence, and the fact that $\#D_{k,n}$ is $O(n^{k-1})$ as $n \to \infty$ for fixed k,

$$n^{-k} \sum_{j_1=1}^{n} \cdots \sum_{j_k=1}^{n} \phi(Z_{j_1},\dots,Z_{j_k}) \to E(\phi(Z_1,\dots,Z_k)|H) \text{ a.s.}$$

By considering $\phi(x_1,\dots,x_k)$ of the form $\phi_1(x_1)\phi_2(x_2)\cdots\phi_k(x_k)$ and using (3.3),

$$E\left(\prod_{r=1}^{k} \phi_r(Z_r) \mid \mathcal{H}\right) = \prod_{r=1}^{k} E(\phi_r(Z_1) \mid \mathcal{H}) .$$

This says that (Z_i) is conditionally i.i.d. given $\mathcal{H}$, and as discussed in Section 2 this is one of several equivalent formalizations of "mixture of i.i.d. sequences".

For the second proof we need an easy lemma.

(3.4) Lemma. *Let Y be a bounded real-valued random variable, and let $\mathcal{F} \subset \mathcal{G}$ be σ-fields. If $E(E(Y|\mathcal{G}))^2 = E(E(Y|\mathcal{F}))^2$, in particular if $E(Y|\mathcal{F}) \stackrel{\mathcal{D}}{=} E(Y|\mathcal{G})$, then $E(Y|\mathcal{G}) = E(Y|\mathcal{F})$ a.s.*

Proof. This is immediate from the identity

$$E(E(Y|\mathcal{G}) - E(Y|\mathcal{F}))^2 = E(E(Y|\mathcal{G}))^2 - E(E(Y|\mathcal{F}))^2 .$$

Second Proof of Theorem 3.1. Write $\mathcal{F}_n = \sigma(Z_{n+1}, Z_{n+2}, \dots)$, so $\bigcap_n \mathcal{F}_n = \mathcal{T}$, the tail σ-field. We shall show that (Z_i) is conditionally i.i.d. given $\mathcal{T}$. By exchangeability, $(Z_1, Z_2, Z_3, \dots) \stackrel{\mathcal{D}}{=} (Z_1, Z_{n+1}, Z_{n+2}, \dots)$ and so $E(\phi(Z_1)|\mathcal{F}_2) \stackrel{\mathcal{D}}{=} E(\phi(Z_1)|\mathcal{F}_n)$ for each bounded $\phi: R \to R$. The reversed martingale convergence theorem says $E(\phi(Z_1)|\mathcal{F}_n) \to E(\phi(Z_1)|\mathcal{T})$ a.s. as $n \to \infty$, and so $E(\phi(Z_1)|\mathcal{F}_2) \stackrel{\mathcal{D}}{=} E(\phi(Z_1)|\mathcal{T})$. But Lemma 3.4 now implies there is a.s. equality, and this means (A4)

Z_1 and $\mathcal{F}_2$ are conditionally independent given $\mathcal{T}$.

The same argument applied to $(Z_m, Z_{m+1}, \dots)$ gives

Z_m and $\mathcal{F}_{m+1}$ are conditionally independent given $\mathcal{T}$; $m \geq 1$.

These imply that the whole sequence $(Z_i; i \geq 1)$ is conditionally independent given $\mathcal{T}$.

For each $n \geq 1$, exchangeability says $(Z_1, Z_{n+1}, Z_{n+2}, \ldots) \stackrel{\mathcal{D}}{=} (Z_n, Z_{n+1}, Z_{n+2}, \ldots)$, and so $E(\phi(Z_1)|F_n) = E(\phi(Z_n)|F_n)$ a.s. for each bounded ϕ. Conditioning on T gives $E(\phi(Z_1)|T) = E(\phi(Z_n)|T)$ a.s. This is (2.11), and so $(Z_i;\ i \geq 1)$ is indeed conditionally i.i.d. given T.

<u>Spherically symmetric sequences</u>. Another classical result can be regarded as a specialization of de Finetti's theorem. Call a random vector $Y^n = (Y_1, \ldots, Y_n)$ <u>spherically symmetric</u> if $UY^n \stackrel{\mathcal{D}}{=} Y^n$ for each orthogonal $n \times n$ matrix U. Call an infinite sequence $\underset{\sim}{Y}$ spherically symmetric if Y^n is spherically symmetric for each n. It is easy to check that an i.i.d. Normal $N(0,v)$ sequence is spherically symmetric; and hence so is a mixture (over v) of i.i.d. $N(0,v)$ sequences. On the other hand, computations with characteristic functions (Feller (1971) Section III.4) give

(3.5) <u>Maxwell's Theorem</u>. <u>An independent spherically symmetric sequence has $N(0,\sigma^2)$ distribution, for some $\sigma^2 \geq 0$.</u>

Now a spherically symmetric sequence is exchangeable, since for any permutation π of $\{1, \ldots, n\}$, the map $(y_1, \ldots, y_n) \rightarrow (y_{\pi(1)}, \ldots, y_{\pi(n)})$ is a rotation. We shall show that Maxwell's theorem and de Finetti's theorem imply

(3.6) <u>Schoenberg's Theorem</u>. <u>An infinite spherically symmetric sequence $\underset{\sim}{Y}$ is a mixture of i.i.d. $N(0,\sigma^2)$ sequences.</u>

This is apparently due to Schoenberg (1938), and has been rediscovered many times. See Eaton (1981), Letac (1981a) for variations and references. This result also fits naturally into the "sufficient statistics" setting of Section 18.

Proof. Let U be a $n \times n$ orthogonal matrix, let $Y^n = (Y_1, \ldots, Y_n)$, $\hat{Y}^{n,m} = (Y_{n+1}, \ldots, Y_{n+m})$. By considering $\begin{pmatrix} U & 0 \\ 0 & I_m \end{pmatrix}$, an orthogonal $(m+n) \times (m+n)$ matrix, we see $(UY^n, \hat{Y}^{n,m}) \stackrel{\mathcal{D}}{=} (Y^n, \hat{Y}^{n,m})$. Letting $m \to \infty$ gives $(UY^n, \tilde{Y}^n) \stackrel{\mathcal{D}}{=} (Y^n, \tilde{Y}^n)$, where $\tilde{Y}^n = (Y_{n+1}, Y_{n+2}, \ldots)$. By conditioning on the tail σ-field $\mathcal{T} \subset \sigma(\tilde{Y}^n)$, we see that the conditional distribution of Y^n and of UY^n given $\mathcal{T}$ are a.s. identical. Applying this to a countable dense set of orthogonal $n \times n$ matrices and to each $n \geq 1$, we see that the conditional distribution of $\underset{\sim}{Y}$ given $\mathcal{T}$ is a.s. spherically symmetric. But de Finetti's theorem says that the conditional distribution of $\underset{\sim}{Y}$ given $\mathcal{T}$ is a.s. an i.i.d. sequence, so the result follows from Maxwell's theorem.

Here is a slight variation on de Finetti's theorem. Given an exchangeable sequence Z, say Z is exchangeable over V if

$$(3.7) \quad (V, Z_1, Z_2, \ldots) \stackrel{\mathcal{D}}{=} (V, Z_{\pi(1)}, Z_{\pi(2)}, \ldots), \text{ all finite permutations } \pi.$$

Similarly, say Z is exchangeable over a σ-field $\mathcal{G}$ if (3.7) holds for each $V \in \mathcal{G}$.

(3.8) Proposition. Let $\underset{\sim}{Z}$ be an infinite sequence, exchangeable over V. Then (a) (Z_i) is conditionally i.i.d. given (V, α), where α is the directing random measure for $\underset{\sim}{Z}$.

(b) $\underset{\sim}{Z}$ and V are conditionally independent given α.

Proof. Let $\hat{Z}_i = (V, Z_i)$. Then $\underset{\sim}{\hat{Z}}$ is exchangeable, so de Finetti's theorem implies

$$(\hat{Z}_i) \text{ is conditionally i.i.d. given } \hat{\alpha},$$

where $\hat{\alpha}$ is the directing random measure for $\underset{\sim}{\hat{Z}}$. So in particular,

$$(Z_i) \text{ is conditionally i.i.d. given } \hat{\alpha} .$$

But applying Lemma 2.15 to $\hat{\underset{\sim}{Z}}$, we see $\hat{\alpha} = \delta_V \times \alpha$, and so $\sigma(\hat{\alpha}) = \sigma(V,\alpha)$. This gives (a). And (b) follows from (a) and Lemma 2.12(b).

We remark that the conditional independence assertion of (b) is a special case of a very general result, Proposition 12.12. Proposition 3.8 plays an important role in the study of partial exchangeability in Part III. As a simple example, here is a version of de Finetti's theorem for a family of sequences, where each is "internally exchangeable".

(3.9) _Corollary_. _For_ $1 \leq j \leq k$ _let_ $Z^j = (Z_i^j, i \geq 1)$ _be exchangeable_. _Suppose further that for each_ j_0 _and each finite permutation_ π _we have_ $(Z_i^j) \overset{\mathcal{D}}{=} (\hat{Z}_i^j)$, _where_

$$\begin{aligned} \hat{Z}_i^j &= Z_i^j \quad , \quad j \neq j_0 \\ &= \hat{Z}_{\pi(i)}^j \quad , \quad j = j_0 \, . \end{aligned}$$

Let α_j _be the directing measure for_ Z^j, _and let_ $\mathcal{F} = \sigma(\alpha_j, 1 \leq j \leq k)$. _Then_ (a) $(Z_i^j: 1 \leq j \leq k, i \geq 1)$ _are conditionally independent given_ $\mathcal{F}$.

(b) α_j _is a r.c.d. for_ Z_i^j _given_ $\mathcal{F}$.

This result goes back to de Finetti, and has been rediscovered many times. In Bayesian language, the family is obtained as follows.

(i) Pick a k-tuple $(\theta_1,\ldots,\theta_k)$ of distributions according to a prior Θ on $\mathcal{P}^k$;

(ii) then for each j let the sequence $(Z_i^j, i \geq 1)$ be i.i.d. (θ_j), independent for different j.

Proof. Fix j. Proposition 3.8 shows

Z^j and $\sigma(Z^m: m \neq j)$ are conditionally independent given α_j.

Then de Finetti's theorem for Z^j yields

$Z_1^j, Z_2^j, \ldots;\ \sigma(Z^m: m \neq j)$ are conditionally independent given α_j;

α_j is a r.c.d. for Z_i^j given α_j.

Since $\sigma(\alpha_j) \subset \mathcal{F} \subset \sigma(\alpha_j, Z^m: m \neq j)$ this is equivalent to

$Z_1^j, Z_2^j, \ldots;\ \sigma(Z^m: m \neq j)$ are conditionally independent given $\mathcal{F}$;

α_j is a r.c.d. for Z_i^j given $\mathcal{F}$.

Since j is arbitrary, this establishes the result.

Here is another application of Proposition 3.8. Call a subset B of R^∞ exchangeable if

$$(x_1, x_2, \ldots) \in B \quad \text{implies} \quad (x_{\pi(1)}, x_{\pi(2)}, \ldots) \in B$$

for each finite permutation π. Given an infinite sequence $\underset{\sim}{X} = (X_i)$, call events $\{\underset{\sim}{X} \in B\}$, B exchangeable, exchangeable events, and call the set of exchangeable events the exchangeable σ-field $\mathcal{E}_X$. It is easy to check that $\mathcal{E}_X \supset \mathcal{T}_X$ a.s., where $\mathcal{T}_X$ is the tail σ-field of $\underset{\sim}{X}$.

(3.10) Corollary. *If $\underset{\sim}{Z}$ is exchangeable then $\mathcal{E}_Z = \mathcal{T}_Z = \sigma(\alpha)$ a.s.*

Proof. For $A \in \mathcal{E}_Z$ the random variable $V = 1_A$ satisfies (3.7), and so by Proposition 3.4, V and $\underset{\sim}{Z}$ are conditionally independent given α. Hence $\mathcal{E}_Z$ and $\underset{\sim}{Z}$ are conditionally independent given α. But $\mathcal{E}_Z \subset \sigma(\underset{\sim}{Z})$ and so $\mathcal{E}_Z$ and $\mathcal{E}_Z$ are conditionally independent given α, which implies

(A6) that $\mathcal{E}_Z \subset \sigma(\alpha)$ a.s. But $\sigma(\alpha) \subset \mathcal{T}_Z$ a.s. by Lemma 2.16, and $\mathcal{T}_Z \subset \mathcal{E}_Z$ a.s.

In particular, Corollary 3.10 gives the well-known Hewitt-Savage 0-1 law:

(3.11) Corollary. *If* $\underset{\sim}{X}$ *is i.i.d. then* $\mathcal{E}_X$ *is trivial.*

This can be proved by more elementary methods (Breiman (1968), Section 3.9).

There are several other equivalences possible in Corollary 3.10; let us state two.

(3.12) Corollary. *If* $\underset{\sim}{Z}$ *is exchangeable then* $\sigma(\alpha)$ *coincides a.s. with*

(a) *the invariant* σ-*field of* $\underset{\sim}{Z}$

(b) *the tail* σ-*field of* $(Z_{n_1}, Z_{n_2}, \ldots)$, *for any distinct* (n_i).

In particular, for an exchangeable process $(Z_i : -\infty < i < \infty)$ the "left tail" and "right tail" σ-fields each coincides a.s. with $\sigma(\alpha)$. In the partial exchangeability setting of Part III, we shall see that subprocesses may generate different σ-fields.

Corollary 3.10 is one generalization of the Hewitt-Savage law. Another type of generalization is to consider which random sequences X have the property that $\mathcal{E}_X$ is trivial. For independent sequences, the condition

$$(3.13) \qquad \sum \text{variance}(\phi(X_i)) = 0 \text{ or } \infty \ ; \quad \text{each bounded } \phi$$

is necessary in order that $\mathcal{E}_X$ be trivial, since if (3.13) fails for ϕ then $\sum(\phi(X_i) - E\phi(X_i))$ defines a non-degenerate $\mathcal{E}_X$-measurable random variable. For finite state space sequences, condition (3.13) is sufficient (see Aldous and Pitman (1979) for this and equivalent conditions). But still open is

(3.14) Problem. For an independent sequence (X_i) taking values in a countable set, is (3.13) sufficient for E_X to be trivial?

Results about E_X for Markovian sequences have been given by Blackwell and Freedman (1964), Grigorenko (1979), and Palacios (1982).

4. Exchangeable sequences and their directing random measures

Convention. In this section $\underset{\sim}{Z} = (Z_i)$ is a real-valued exchangeable infinite sequence directed by some random measure α.

The purpose of this section is to point out some concrete ways of constructing exchangeable sequences, and to investigate how particular properties of $\underset{\sim}{Z}$ correspond to particular properties of α. The results are mostly straightforward consequences of de Finetti's theorem, but will give the reader some experience in manipulating random measures.

There are two ways in which an exchangeable sequence may be considered "degenerate". First, if it is i.i.d. This corresponds to $\alpha = \theta$ a.s., for some fixed distribution θ. Second, if $Z_1 = Z_2 = \cdots$ a.s. This corresponds to $\alpha = \delta_X$ a.s. for some random variable X.

The "simple" exchangeable sequences described at (2.1)(i) are those with $\alpha = \theta_I$ a.s., where $\{\theta_1,\ldots,\theta_k\}$ are distributions and I is a random variable taking values in $\{1,\ldots,k\}$.

One natural way to construct an exchangeable sequence is to take a parametric family of distributions, choose the parameter randomly, and take an i.i.d. sequence whose distribution has this random parameter. For example, let $\mu_{\theta,s}$ denote the Normal(θ,s^2) distribution. For random (Θ,S), we can define an exchangeable sequence (Z_i) which is i.i.d. Normal(θ,s^2) conditional on $(\Theta = \theta,\ S = s)$. This is the exchangeable sequence directed by $\alpha = \mu_{\Theta,S}$. Because the Normal family is a location-scale family, we can construct (Z_i) very simply by putting $Z_i = \Theta + SX_i$, for (X_i) i.i.d. Normal(0,1).

In general, $\underset{\sim}{Z}$ is a <u>mixture</u> of a parametric family (μ_θ) if $\alpha \in (\mu_\theta)$ a.s. It is natural to ask for intrinsic conditions on $\underset{\sim}{Z}$ (rather than α) which determine whether $\underset{\sim}{Z}$ is a mixture of a specified family; results of this kind are given in Section 18. Such processes arise naturally in the Bayesian analysis of parametric statistical problems. For the Bayesian analysis of non-parametric problems, one needs tractable random measures whose values are not restricted to small subsets of P; the most popular are the Dirichlet random measures (Ferguson (1974)), described briefly in Section 10.

Here are two more ways of producing exchangeable sequences.

(4.1)
Let $(Y_1, Y_2, \ldots)$ be arbitrary;
let $(X_1, X_2, \ldots)$ be i.i.d., independent of $\underset{\sim}{Y}$, taking values $\{1,2,\ldots\}$;
let $Z_i = Y_{X_i}$.

Then $\underset{\sim}{Z}$ is exchangeable, and using Lemma 2.15 we see $\alpha = \sum_i p_i \delta_{Y_i(\omega)}$, where $p_i = P(X_1 = i)$.

(4.2)
Let $X_1, X_2, \ldots$ be i.i.d., distribution θ;
let Y be independent of $\underset{\sim}{X}$, with distribution ϕ;
let $Z_i = f(Y, X_i)$, for some function f.

Then $\underset{\sim}{Z}$ is exchangeable. Indeed, from the canonical construction (2.5) and de Finetti's theorem, every exchangeable sequence is of this form (in distribution). However, exchangeable sequences arising in practice can often be put into the form (4.2) where θ, ϕ, f have some simple form with intuitive significance (e.g. the representation (1.10) for Gaussian exchangeable sequences). To describe the directing random measure for such a sequence, we need some notation.

(4.3) Definition. Given $f: R \to R$ define the induced map $\tilde{f}: \mathcal{P} \to \mathcal{P}$ by

$$\tilde{f}(\mathcal{L}(Y)) = \mathcal{L}(f(Y)) .$$

Given $f: R \times R \to R$ define the induced map $\tilde{f}: R \times \mathcal{P} \to \mathcal{P}$ by

$$\tilde{f}(x,\mathcal{L}(Y)) = \mathcal{L}(f(x,Y)) .$$

This definition and Lemma 2.15 give the next lemma.

(4.4) Lemma. (a) *Let* $\underset{\sim}{Z}$ *be exchangeable, directed by* α, *let* $f: R \to R$ *have induced map* $\tilde{f}$, *and let* $\hat{Z}_i = f(Z_i)$. *Then* $\underset{\sim}{\hat{Z}}$ *is exchangeable and is directed by* $\tilde{f}(\alpha)$.

(b) *Let* $\underset{\sim}{Z}$ *be of the form* (4.2) *for some* $f: R \times R \to R$, *and let* $\tilde{f}: R \times \mathcal{P} \to \mathcal{P}$ *be the induced map. Then* $\underset{\sim}{Z}$ *is exchangeable and is directed by* $\tilde{f}(Y,\theta)$.

In particular, for the addition function $f(x,y) = x + y$ we have $\tilde{f}(x,\theta) = \delta_x * \theta$, where $*$ denotes convolution. So (1.10) implies:

(4.5) $\underset{\sim}{Z}$ is Gaussian if and only if $\alpha(\omega,\cdot) = \delta_{X(\omega)} * \theta$, where θ and $\mathcal{L}(X)$ are Normal.

Another simple special case is 0-1 valued exchangeable sequences. Call events $(A_i, i \geq 1)$ exchangeable if the indicator random variables $Z_i = 1_{A_i}$ are exchangeable. In this case α must have the form

$X(\omega)\delta_1 + (1 - X(\omega))\delta_0$, for some random variable $0 \leq X \leq 1$. Informally, conditional on $\{X = p\}$ the events (A_i) are independent and have probability p.

There is a curious connection between exchangeable events and a classical moment problem. Let $p_n = P(A_1 \cap A_2 \cap \cdots \cap A_n)$. Since $P(A_1 \cap \cdots \cap A_n | X) = X^n$ a.s. we have

(a) $p_n = EX^n$.

Now the distribution of the sequence (A_i) is, by exchangeability, determined by the numbers $q_{n,k} = P(A_1 \cap \cdots \cap A_k \cap A_{k+1}^c \cap \cdots \cap A_n^c)$. But the relation $q_{n,k} = q_{n-1,k} - q_{n,k+1}$ shows that the numbers $(q_{n,k})$ are determined by (p_n) $(= q_{n,n})$. But the distribution of (A_i) determines the distribution of X, so

(b) the numbers (p_n) determine $\mathcal{L}(X)$.

Since any distribution θ on $[0,1]$ is possible for X, facts (a) and (b) imply

(4.6) a distribution θ on $[0,1]$ is determined by its moments

$$p_n = \int x^n \theta(dx), \quad n \geq 1.$$

This is the classical "Hausdorff moment problem". The sequence (p_n) is completely monotone; for further discussion of monotonicity and exchangeability, see Kallenberg (1976) Chapter 9; Daboni (1982); Kimberling (1973).

(4.7) Remark. Given N we can find different distributions θ_1, θ_2 on $[0,1]$ such that $\int x^n \theta_1(dx) = \int x^n \theta_2(dx)$, $n \leq N$. Consider the corresponding finite exchangeable sequences $(1_{A_1}, \ldots, 1_{A_N})$. These have the same distribution, by the argument above. Thus a finite exchangeable sequence may be extendible to more than one infinite exchangeable sequence.

We now consider covariance properties. Let $\underset{\sim}{Z}$ be exchangeable, and suppose $EZ_1^2 < \infty$. Let $\rho = E(Z_i - EZ_i)(EZ_j - EZ_j)$, $i \neq j$, be the covariance. In (1.8) it was proved directly that $\rho \geq 0$. But de Finetti's theorem gives more information: by Lemma 2.23,

(4.8) $\quad \rho = 0$ if and only if $\operatorname{mean}(\alpha) = c$ a.s., some constant c.

Of course, from de Finetti's theorem

$$\operatorname{mean}(\alpha) = E(Z_1 | \alpha) = E(Z_1 | T) \text{ a.s.} \tag{4.9}$$

In particular, an exchangeable $\underset{\sim}{Z}$ with $EZ_1 = 0$ is uncorrelated if and only if the random variable in (4.9) is a.s. zero. Curiously, this implies the (generally stronger) property that $\underset{\sim}{Z}$ is a martingale difference sequence.

(4.10) *Lemma. Suppose* $\underset{\sim}{Z}$ *is exchangeable*, $EZ_i = 0$. *Then* $\underset{\sim}{Z}$ *is a martingale difference sequence if and only if* $\operatorname{mean}(\alpha) = 0$ a.s.

Proof. By conditional independence,

$$E(Z_n | Z_1, \dots, Z_{n-1}, \alpha) = E(Z_n | \alpha) = \operatorname{mean}(\alpha) \text{ a.s.}$$

So if $\operatorname{mean}(\alpha) = 0$ a.s. then $E(Z_n | Z_1, \dots, Z_{n-1}) = 0$ a.s., and so $\underset{\sim}{Z}$ is a martingale difference sequence. Conversely, if $\underset{\sim}{Z}$ is a martingale difference sequence then $E(Z_n | Z_1, \dots, Z_{n-1}) = 0$ a.s., so by exchangeability $E(Z_1 | Z_2, \dots, Z_n) = 0$ a.s. The martingale convergence theorem now implies $E(Z_1 | Z_i, i > 1) = 0$ a.s., and since $\sigma(\alpha) \subset \sigma(Z_i, i > 1)$ we have $\operatorname{mean}(\alpha) = E(Z_1 | \alpha) = 0$ a.s.

Here is another instance where for exchangeable sequences one property implies a generally stronger property.

(4.11) Lemma. *If an infinite exchangeable sequence* $(Z_1, Z_2, \ldots)$ *is pairwise independent then it is i.i.d.*

Proof. Fix some bounded function $f: R \to R$. The sequence $\hat{Z}_i = f(Z_i)$ is pairwise independent, and hence uncorrelated. By Lemma 4.4(a), $\underset{\sim}{\hat{Z}}$ is directed by $\tilde{f}(\alpha)$, and now (4.8) implies $\text{mean}(\tilde{f}(\alpha))$ is a.s. constant, c_f say. In other words:

$$\int f(x)\alpha(\cdot, dx) = c_f \text{ a.s.; each bounded } f.$$

Standard arguments (7.12) show this implies $\alpha = \theta$ a.s., where θ is a distribution with $\int f(x)\theta(dx) \equiv c_f$. So $\underset{\sim}{Z}$ is i.i.d. (θ).

(4.12) Example. Fix $N > 2$. Let $(Y_1, \ldots, Y_N)$ be uniform on the set of sequences $(y_1, \ldots, y_N)$ of 1's and 0's satisfying $\sum y_i = 0 \bmod (2)$. Then

(a) $(Y_1, \ldots, Y_N)$ is N-exchangeable;

(b) $(Y_1, \ldots, Y_{N-1})$ are independent.

So Lemma 4.11 is not true for finite exchangeable sequences. And by considering $X_i = 2Y_i - 1$, we see that a finite exchangeable sequence may be uncorrelated but not a martingale difference sequence.

Our next topic is the Markov property.

(4.13) Lemma. *For an infinite exchangeable sequence* $\underset{\sim}{Z}$ *the following are equivalent.*

(a) $\underset{\sim}{Z}$ *is Markov.*

(b) $\sigma(\alpha) \subset \sigma(Z_1)$ a.s.

(c) $\sigma(\alpha) = \bigcap_i \sigma(Z_i)$ a.s.

Remark. When the support of α is some countable set (θ_j) of distributions, these conditions are equivalent to

(d) the distributions (θ_j) are mutually singular.

It seems hard to formalize this in the general case.

Proof. $\underset{\sim}{Z}$ is Markov if and only if for each bounded $\phi: R \rightarrow R$ and each $n \geq 2$,

(a') $E(\phi(Z_n)|Z_1,\ldots,Z_{n-1}) = E(\phi(Z_n)|Z_{n-1})$ a.s.

Suppose this holds. Then by exchangeability,

$$E(\phi(Z_2)|Z_1,Z_3,\ldots,Z_{n-1}) = E(\phi(Z_2)|Z_1) \text{ a.s.}$$

So by martingale convergence

$$E(\phi(Z_2)|Z_1) = E(\phi(Z_2)|Z_1,Z_3,Z_4,\ldots) = E(\phi(Z_2)|\alpha) .$$

In particular, $\alpha(\cdot,A) = P(Z_2 \in A|\alpha)$ is essentially $\sigma(Z_1)$-measurable, for each A. This gives (b).

If (b) holds then by symmetry $\sigma(\alpha) \subset \sigma(Z_i)$ a.s. for each i, and so $\sigma(\alpha) \subset \cap \sigma(Z_i)$ a.s. And Corollary 3.10 says $\sigma(\alpha) = \mathcal{T} \supset \cap \sigma(Z_i)$ a.s., which gives (c).

If (c) holds then

$$\begin{aligned} E(\phi(Z_n)|Z_1,\ldots,Z_{n-1}) &= E(\phi(Z_n)|Z_1,\ldots,Z_{n-1},\alpha) \quad \text{by (c)} \\ &= E(\phi(Z_n)|Z_{n-1},\alpha) \quad \text{by conditional independence} \\ &= E(\phi(Z_n)|Z_{n-1}) \quad \text{by (c)} \end{aligned}$$

and this is (a').

This is one situation where the behavior of finite exchangeable sequences is the same as for the infinite case, by the next result.

(4.14) Lemma. Any finite Markov exchangeable sequence $(Z_1,\ldots,Z_N)$ extends to an infinite Markov exchangeable sequence $\underset{\sim}{Z}$, provided $N \geq 3$.

This cannot be true for $N = 2$, since a 2-exchangeable sequence is vacuously Markov.

Proof. The given $(Z_1,\ldots,Z_N)$ extends to an infinite sequence $\underset{\sim}{Z}$ whose distribution is specified by the conditions

(4.15) (i) $\underset{\sim}{Z}$ is Markov;

(ii) $(Z_i,Z_{i+1}) \overset{\mathcal{D}}{=} (Z_1,Z_2)$.

We must prove $\underset{\sim}{Z}$ is exchangeable. Suppose, inductively, that $(Z_1,\ldots,Z_m)$ is m-exchangeable for some $m \geq N$. Then $(Z_2,\ldots,Z_m) \overset{\mathcal{D}}{=} (Z_1,\ldots,Z_{m-1})$, so by (i) and (ii) we get

$$(Z_2,\ldots,Z_{m+1}) \overset{\mathcal{D}}{=} (Z_1,\ldots,Z_m) . \tag{4.16}$$

So these vectors are m-exchangeable. Hence $(Z_2,\ldots,Z_{m+1}) \overset{\mathcal{D}}{=} (Z_2,\ldots,Z_{m-1},Z_{m+1},Z_m)$ and so by the Markov property (at time 2)

$$(Z_1,\ldots,Z_{m+1}) \overset{\mathcal{D}}{=} (Z_1,\ldots,Z_{m-1},Z_{m+1},Z_m) . \tag{4.17a}$$

Similarly, using the Markov property at time m,

$$(Z_1,\ldots,Z_{m+1}) \overset{\mathcal{D}}{=} (Z_2,Z_1,Z_3,\ldots,Z_{m+1}) . \tag{4.17b}$$

Finally, for any permutation π of $(2,\ldots,m)$ we assert

$$(Z_1,Z_{\pi(2)},\ldots,Z_{\pi(m)},Z_{m+1}) \overset{\mathcal{D}}{=} (Z_1,\ldots,Z_{m+1}) . \tag{4.17c}$$

For let $Y = (Z_2,\ldots,Z_m)$, $\hat{Y} = (Z_{\pi(2)},\ldots,Z_{\pi(m)})$. Then the triples (Z_1,Y,Z_{m+1}) and $(Z_1,\hat{Y},Z_{m+1})$ are Markov, and (4.16) shows $(Z_1,Y) \overset{\mathcal{D}}{=} (Z_1,\hat{Y})$

and $(Y,Z_{m+1}) \overset{\mathcal{D}}{=} (\hat{Y},Z_{m+1})$, which gives (4.17c). Now (4.17a-c) establish the (m+1)-exchangeability of $(Z_1,\ldots,Z_{m+1})$.

Let us digress slightly to present the following two results, due to Carnal (1980). Informally, these can be regarded as extensions of Lemmas 4.13 and 4.14 to non-exchangeable sequences.

(4.18) Lemma. *Let X_1, X_2, X_3 be random variables such that for any ordering (i,j,k) of $(1,2,3)$, X_i and X_j are conditionally independent given X_k. Then X_1, X_2, X_3 are conditionally independent given* $\mathcal{F} = \sigma(X_1)\cap\sigma(X_2)\cap\sigma(X_3)$.

(4.19) Lemma. *For an infinite sequence $\underset{\sim}{X}$, the following are equivalent:*

(a) $(X_{n_1},X_{n_2},\ldots)$ *is Markov, for any distinct* $n_1,n_2,\ldots$

(b) $(X_i;\ i\geq 1)$ *are conditionally independent given* $\mathcal{G} = \bigcap_i \sigma(X_i)$.

Proof of Lemma 4.18. We shall prove that for any ordering (i,j,k),

(a) $\sigma(X_i)\cap\sigma(X_j) = \mathcal{F}$ a.s.;

(b) X_i and $\sigma(X_j,X_k)$ are conditionally independent given $\sigma(X_j)\cap\sigma(X_k)$.

These imply that X_i and $\sigma(X_j,X_k)$ are conditionally independent given $\mathcal{F}$, and the lemma follows.

For $A\in\sigma(X_i)$ and $B\in\sigma(X_j)$, conditional independence given $P(A\cap B|X_k) = P(A|X_k)\cdot P(B|X_k)$. So for $A\in\sigma(X_i)\cap\sigma(X_j)$ we have $P(A|X_k) = \{P(A|X_k)\}^2$, so $A\in\sigma(X_k)$ a.s., giving (a).

For bounded $\phi\colon R\to R$, conditional independence gives

$$E(\phi(X_i)|X_j,X_k) = E(\phi(X_i)|X_j) = E(\phi(X_i)|X_k)\ .$$

So $E(\phi(X_i)|X_j,X_k)$ is essentially $\sigma(X_j)\cap\sigma(X_k)$-measurable, proving (b).

Proof of Lemma 4.19. Suppose (a) holds. For distinct i, j, k, the hypothesis of Lemma 4.18 holds, so by its conclusion

$$E(\phi(X_i)|X_j) = E(\phi(X_i)|\sigma(X_j) \cap \sigma(X_j) \cap \sigma(X_k)) .$$

Since this holds for each $k \neq i, j$,

$$E(\phi(X_i)|X_j) \in \bigcap_{m=1} \sigma(X_m) = G \text{ a.s.}$$

In other words, X_i and X_j are conditionally independent given G. But the Markov hypothesis implies that X_i and $\sigma(X_k;\ k \neq i)$ are conditionally independent given X_j. These last two facts imply (A7) that X_i and $\sigma(X_k;\ k \neq i)$ are conditionally independent given G, and the result follows.

5. Finite exchangeable sequences

As mentioned in Section 1, the basic way to obtain an N-exchangeable sequence is as an urn process: take N constants $y_1,\ldots,y_N$, not necessarily distinct, and put them in random order.

$$(5.1) \qquad \underset{\sim}{Y} = (Y_1,\ldots,Y_N) = (y_{\hat{\pi}(1)},\ldots,y_{\hat{\pi}(N)}) ;$$

$\hat{\pi}$ the uniform random permutation on $\{1,\ldots,N\}$. In the notation of (2.13), $\underset{\sim}{Y}$ has empirical distribution

$$(5.2) \qquad \Lambda_N(\underset{\sim}{Y}) = \frac{1}{N}\sum \delta_{y_i} = \Lambda_N(\underset{\sim}{y}) .$$

Conversely, it is clear that:

(5.3) if $\underset{\sim}{Y}$ is N-exchangeable and satisfies (5.2) then $\underset{\sim}{Y}$ has distribution (5.1).

Let U_N denote the set of distributions $L(\underset{\sim}{Y})$ for urn processes $\underset{\sim}{Y}$.

Let $\mathcal{U}_N^*$ denote the set of empirical distributions $\frac{1}{N}\sum \delta_{y_i}$. Let $\Phi: \mathcal{U}_N \to \mathcal{U}_N^*$ be the natural bijection $\Phi(L(\underset{\sim}{Y})) = \Lambda_N(\underset{\sim}{Y})$. The following simple result is a partial analogue of de Finetti's theorem in the finite case.

(5.4) <u>Lemma</u>. <u>Let</u> $\underset{\sim}{Z} = (Z_1,\ldots,Z_N)$ <u>be</u> <u>N-exchangeable</u>. <u>Then</u> $\Phi^{-1}(\Lambda_N(\underset{\sim}{Z}))$ <u>is</u> <u>a</u> <u>regular</u> <u>conditional</u> <u>distribution</u> <u>for</u> $\underset{\sim}{Z}$ <u>given</u> $\Lambda_N(\underset{\sim}{Z})$.

In words: conditional on the empirical distribution, the N (possibly repeated) values comprising the empirical distribution occur in random order. In the real-valued case we can replace "empirical distribution" by "order statistics", which convey the same information.

<u>Proof</u>. For any permutation π of $\{1,\ldots,N\}$,

$$(Z_1,\ldots,Z_N,\Lambda_N(\underset{\sim}{Z})) \stackrel{\mathcal{D}}{=} (Z_{\pi(1)},\ldots,Z_{\pi(N)},\Lambda_N(\underset{\sim}{Z})) .$$

So if $\beta(\omega,\cdot)$ is a regular conditional distribution for $\underset{\sim}{Z}$ given $\Lambda_N(\underset{\sim}{Z})$ then (a.s. ω)

(a) the distribution $\beta(\omega,\cdot)$ is N-exchangeable.

But from the fundamental property of r.c.d.'s, for a.s. ω we have

(b) the N-vector with distribution $\beta(\omega,\cdot)$ has empirical distribution $\Lambda_N(\underset{\sim}{Z}(\omega))$.

And (5.3) says that (a) and (b) imply $\beta(\omega,\cdot) = \Phi^{-1}(\Lambda_N(\underset{\sim}{Z}(\omega)))$ a.s.

Thus for some purposes the study of finite exchangeable sequences reduces to the study of "sampling without replacement" sequences of the form (5.1). Unlike de Finetti's theorem, this idea is not always useful: for example, it does not seem to help with the weak convergence problems discussed in Section 20.

From an abstract viewpoint, the difference between finite and infinite exchangeability is that the group of permutations on a finite set is compact, whereas on an infinite set it is non-compact. Lemma 5.4 has an analogue for distributions invariant under a specified compact group of transformations; see (12.15).

We know that an N-exchangeable sequence need not be a mixture of i.i.d.'s, that is to say it need not extend to an infinite exchangeable sequence. But we can ask how "close" it is to some mixture of i.i.d.'s. Let us measure closeness of two distributions μ, ν by the total variation distance

$$(5.5)\qquad \|\mu-\nu\| = \sup_A |\mu(A)-\nu(A)| \ .$$

The next result implies that an M-exchangeable sequence which can be extended to an N-exchangeable sequence, where N is large compared to M^2, is close to a mixture of i.i.d.'s.

(5.6) Proposition. Let $\underset{\sim}{Y}$ be N-exchangeable. Then there exists an infinite exchangeable sequence $\underset{\sim}{Z}$ such that, for $1 \leq M \leq N$,

$$\|L(Y_1,\ldots,Y_M) - L(Z_1,\ldots,Z_M)\| \leq 1 - \prod_{i=1}^{M-1}(1 - i/N) \leq \frac{M(M-1)}{2N} \ .$$

This is one formalization of the familiar fact that sampling with replacement and without replacement are almost equivalent when the sample size is small compared to the population size. See Proposition 20.6 for another formalization.

Proof. We quote the straightforward estimate

$$(5.7)\qquad \|L(V) - L(V|B)\| \leq 1 - P(B) \ ; \quad \text{all random variables } V, \text{ events } B$$

where $L(V|B)$ is the conditional distribution. Let $(I_1, I_2, \ldots)$ be i.i.d. uniform on $\{1,2,\ldots,N\}$. Let $B_{M,N}$ be the event $\{I_1,\ldots,I_M$ all distinct$\}$. Then

(5.8) $$P(B_{M,N}) = \prod_{i=1}^{M-1} (1 - i/N) .$$

Let $Z_i = Y_{I_i}$, $i \geq 1$. Then (Z_i) is an infinite exchangeable sequence. Since, by exchangeability of Y,

$$(Y_{j_1},\ldots,Y_{j_M}) \stackrel{\mathcal{D}}{=} (Y_1,\ldots,Y_M) \ ; \quad \text{any distinct } (j_k) ,$$

we see that

$$L((Z_1,\ldots,Z_M)|B_{M,N}) = L(Y_1,\ldots,Y_M) .$$

Now (5.7) and (5.8) establish the first inequality of the Proposition; the second is calculus.

Better bounds can be obtained if there are bounds on the number of possible values of Y_i, but more delicate arguments are required. We quote a result from Diaconis and Freedman (1980a), which also contains Proposition 5.6 and further discussion.

(5.9) Proposition. *Let $(Y_1,\ldots,Y_N)$ be N-exchangeable, taking values in a set of cardinality c. Then there exists an infinite exchangeable sequence $\underset{\sim}{Z}$ such that for $1 \leq M \leq N$*

$$\| L(Y_1,\ldots,Y_M) - L(Z_1,\ldots,Z_M)\| \leq cM/N .$$

Diaconis and Freedman (unpublished) also have a similar result relating to Schoenberg's theorem.

(5.10) <u>Proposition</u>. <u>Let</u> $(Y_1,\ldots,Y_N)$ <u>be spherically symmetric</u>. <u>Then there exists a sequence</u> $\underset{\sim}{Z}$ <u>which is a mixture of i.i.d.</u> $N(0,\sigma^2)$ <u>sequences such that</u>

$$\|L(Y_1,\ldots,Y_M) - L(Z_1,\ldots,Z_M)\| \leq bM/N , \quad 1 \leq M \leq N ,$$

<u>where</u> b <u>is a constant not depending on</u> (Y_i).

The obvious way of getting M-exchangeable sequences from N-exchangeable sequences (N > M) is by taking the first M variables; Proposition 5.6 and the discussion of extendibility in Section 1 show that the M-exchangeable sequences obtainable in this way are restricted. Here is another way of getting new exchangeable sequences from old. Let $N, K \geq 1$. Let $Z = (Z_i: 1 \leq i \leq KN)$ be exchangeable, and let $f: R^N \to R$ be a function. Define

$$\hat{Y}_j = f(Z_{(j-1)N+1},\ldots,Z_{jN}) , \quad 1 \leq j \leq K . \tag{5.11}$$

Then $\hat{Y} = (\hat{Y}_j: 1 \leq j \leq K)$ is exchangeable. Now any given K-exchangeable sequence Y may or may not have the property

(5.12) for each N there exists a NK-exchangeable Z and a function f such that $Y \overset{\mathcal{D}}{=} \hat{Y}$, for $\hat{Y}$ defined at (5.11).

This can be regarded as a non-linear analogue of "infinite divisibility".

(5.13) <u>Problem</u>. Give an intrinsic characterization of K-exchangeable sequences Y with property (5.12).

(5.14) <u>Example</u>. Let (Y_1,Y_2,Y_3) be the urn sequence from urn $\{a,b,c\}$. Then (5.12) fails for $N = 2$. Here is an outline of the argument. Suppose (5.12) held, so some $(f(Z_1,Z_2),f(Z_3,Z_4),f(Z_5,Z_6))$ is a random ordering of $\{a,b,c\}$. Using Lemma 5.4, we may suppose (Z_i) is an urn process, with urn (z_i) say. Since $f(Z_5,Z_6)$ is determined by (Z_1,Z_2,Z_3,Z_4), we may take f symmetric. Now picture a, b, c as colors; consider the complete graph on the 6 points (z_i) and paint the edge (z_i,z_j) with color $f(z_i,z_j)$. Then edges with distinct endpoints must have different colors, and it is easy to verify this is impossible.

(5.15) Example. The Gaussian exchangeable sequences (1.9) do have property (5.12).

Finally, let us mention a curious result given in Dellacherie and Meyer (1980), V.51: any finite exchangeable sequence is a "mixture" of i.i.d. sequences, if we allow the mixing measure to be a signed measure.

PART II

In Part II we present those extensions and analogues of de Finetti's theorem which are close in spirit to the theorem itself; subsequent parts will take us further afield.

6. Properties equivalent to exchangeability

In Section 1 we pointed out some conditions which were trivially equivalent to exchangeability. The next result collects together some remarkable, non-trivial equivalences.

(6.1) Theorem. *For an infinite sequence of random variables* $Z = (Z_i)$, *each of the following conditions is equivalent to exchangeability:*

(A) $Z \stackrel{\mathcal{D}}{=} (Z_{n_1}, Z_{n_2}, \ldots)$ *for each increasing sequence* $1 \le n_1 < n_2 < \cdots$ *of constants.*

(B) $Z \stackrel{\mathcal{D}}{=} (Z_{T_1+1}, Z_{T_2+1}, Z_{T_3+1}, \ldots)$ *for each increasing sequence* $0 \le T_1 < T_2 < \cdots$ *of stopping times.*

(C) $Z \stackrel{\mathcal{D}}{=} (Z_{T+1}, Z_{T+2}, Z_{T+3}, \ldots)$ *for each stopping time* $T \ge 0$

(*Stopping times are relative to the filtration* $\mathcal{F}_n = \sigma(Z_1, \ldots, Z_n)$, $\mathcal{F}_0$ *trivial.*)

Remarks

(a) Ryll-Nardzewski (1957) proved (A) implies exchangeability. Property (A), under the name "spreading-invariance", arises naturally in the work of Dacunha-Castelle and others who have studied certain Banach space problems using probabilistic techniques. A good survey of this area is Dacunha-Castelle (1982).

(b) The fact that (B) and (C) are equivalent to exchangeability is due to Kallenberg (1982a), who calls property (C) "strong stationarity". The idea of expressing exchangeability-type properties in terms of stopping times seems a promising technique for the study of exchangeability concepts for continuous-time processes, where there is a well-developed technical machinery involving stopping times. See Section 17 for one such study.

(c) Stopping times of the form $T+1$ are _predictable_ stopping times.

(d) The difficult part is proving these conditions imply exchangeability. Let us state the (vague) question

(6.2) _Problem_. What hypotheses _prima facie_ weaker than exchangeability do in fact imply exchangeability?

The best result known seems to be that obtained by combining Lemma 6.5 and Proposition 6.4 below, which are taken from Aldous (1982b).

For the proof of Theorem 6.1 we need the following extension of Lemma 3.4.

(6.3) _Lemma_. _Let_ $(\mathcal{G}_n)$ _be an increasing sequence of_ σ_-fields, let_ $\mathcal{G} = \bigvee_n \mathcal{G}_n$, _and let_ $\mathcal{F} \subset \mathcal{G}$. _Let_ Y _be a bounded random variable such that for each_ n _there exists_ $\mathcal{F}_n \subset \mathcal{F}$ _such that_ $E(Y|\mathcal{F}_n) \stackrel{\mathcal{D}}{=} E(Y|\mathcal{G}_n)$. _Then_ $E(Y|\mathcal{F}) = E(Y|\mathcal{G})$ a.s.

Proof. Write $\|U\| = EU^2$. Then $\|E(Y|\mathcal{G}_n)\| = \|E(Y|\mathcal{F}_n)\| \leq \|E(Y|\mathcal{F})\|$. Since $E(Y|\mathcal{G}_n) \rightarrow E(Y|\mathcal{G})$ in L^2 by martingale convergence, we obtain $\|E(Y|\mathcal{G})\| \leq \|E(Y|\mathcal{F})\|$. But $\mathcal{F} \subset \mathcal{G}$ implies $\|E(Y|\mathcal{F})\| \leq \|E(Y|\mathcal{G})\|$. So $\|E(Y|\mathcal{G})\| = \|E(Y|\mathcal{F})\|$, and now Lemma 3.4 establishes the result.

(6.4) _Proposition_. _Let_ $\underset{\sim}{X}$ _be an infinite sequence with tail_ σ_-field_ $\mathcal{T}$. _Suppose that for each_ $j, k \geq 1$ _there exist_ $n_1, \ldots, n_k$ _such that_ $n_i > i$

and $(X_j, X_{j+1}, \ldots, X_{j+k}) \overset{\mathcal{D}}{=} (X_j, X_{j+n_1}, \ldots, X_{j+n_k})$. *Then* $(X_i;\ i \geq 1)$ *are conditionally independent given* $\mathcal{T}$.

Proof. Fix $m, n \geq 1$, let $\mathcal{F} = \sigma(X_m, X_{m+1}, \ldots)$ and let $\mathcal{G}_n = \sigma(X_2, \ldots, X_n)$. By repeatedly applying the hypothesis, there exist $q_2, \ldots, q_n$ such that $q_i > m$ and $(X_1, \ldots, X_n) \overset{\mathcal{D}}{=} (X_1, X_{q_2}, \ldots, X_{q_n})$. So for bounded $\phi: R \to R$ we have $E(\phi(X_1)|\mathcal{G}_n) \overset{\mathcal{D}}{=} E(\phi(X_1)|\mathcal{F}_n)$, where $\mathcal{F}_n = \sigma(X_{q_2}, \ldots, X_{q_n}) \subset \mathcal{F}$. Applying Lemma 6.3,

$$\begin{aligned} E(\phi(X_1)|X_2, X_3, \ldots) &= E(\phi(X_1)|X_m, X_{m+1}, \ldots) \text{ a.s.} \\ &= E(\phi(X_1)|\mathcal{T}) \text{ a.s. by martingale convergence.} \end{aligned}$$

This says that X_1 and $\sigma(X_2, X_3, \ldots)$ are conditionally independent given $\mathcal{T}$. But for each j the sequence $(X_j, X_{j+1}, \ldots)$ satisfies the hypotheses of the Proposition, so X_j and $\sigma(X_{j+1}, X_{j+2}, \ldots)$ are conditionally independent given $\mathcal{T}$. This establishes the result.

(6.5) *Lemma. Let* X *be an infinite sequence with tail* σ*-field* $\mathcal{T}$. *Suppose*

$$(X_1, X_{n+1}, X_{n+2}, X_{n+3}, \ldots) \overset{\mathcal{D}}{=} (X_n, X_{n+1}, X_{n+2}, X_{n+3}, \ldots);\ \text{each } n \geq 1.$$

Then the random variables X_i *are conditionally identically distributed given* $\mathcal{T}$, *that is* $E(\phi(X_1)|\mathcal{T}) = E(\phi(X_n)|\mathcal{T})$, *each* $n \geq 1$, ϕ *bounded.*

Proof. By hypothesis $E(\phi(X_1)|X_{n+1}, X_{n+2}, \ldots) = E(\phi(X_n)|X_{n+1}, X_{n+2}, \ldots)$. Condition on $\mathcal{T}$.

Proof of Theorem 6.1. It is well known (and easy) that an i.i.d. sequence has property (B). It is also easy to check that any stopping time T on $(\mathcal{F}_n)$ can be taken to have the form $T = t(Z_1, Z_2, \ldots)$, where the function $t(\underset{\sim}{x})$ satisfies the condition

(*) if $t(\underset{\sim}{x}) = n$ and $x_i' = x_i$, $i \leq n$, then $t(\underset{\sim}{x}') = n$.

Let Z be exchangeable, directed by α say, and let $T_r = t_r(Z)$ be an increasing sequence of stopping times. Conditional on α, the variables Z_i are i.i.d. and the times (T_r) are stopping times, since the property (*) is unaffected by conditioning. Thus we can apply the i.i.d. result to see that conditional on α the distributions of Z and $(Z_{T_1+1}, Z_{T_2+1}, \ldots)$ are identical; hence the unconditional distributions are identical. Thus exchangeability implies (B).

Plainly property (B) implies (A) and (C). It remains to show that each of the properties (A), (C) implies both the hypotheses of (6.4) and (6.5), so that (Z_i) are conditionally i.i.d. given $\mathcal{T}$, and hence exchangeable. For (A) these implications are obvious. So suppose (C) holds. Let $j, k, n \geq 1$. Applying (C) to the stopping times S, T, where $S = j$ and

$$T = j \quad \text{on } \{Z_j \in F\}, \qquad = j + n \quad \text{on } \{Z_j \notin F\},$$

we have $(Z_{j+1}, Z_{j+2}, \ldots) \overset{\mathcal{D}}{=} (Z_{T+1}, Z_{T+2}, \ldots)$. Since these vectors are identical on $\{Z_j \in F\}$, the conditional distributions given $\{Z_j \notin F\}$ must be the same. That is, conditional on $\{Z_j \notin F\}$ the distributions of $(Z_{j+1}, Z_{j+2}, \ldots)$ and of $(Z_{j+n+1}, Z_{j+n+2}, \ldots)$ are the same. Since F is arbitrary, we obtain

$$(6.6) \qquad (Z_j, Z_{j+1}, Z_{j+2}, \ldots, Z_{j+k}) \overset{\mathcal{D}}{=} (Z_j, Z_{j+n+1}, \ldots, Z_{j+n+k}) ,$$

and this implies the hypothesis of (6.4). Finally, property (C) with $T = 1$ shows Z is stationary, so

$$\begin{aligned}(Z_{j+n}, Z_{j+n+1}, \ldots, Z_{j+n+k}) &\overset{\mathcal{D}}{=} (Z_j, Z_{j+1}, \ldots, Z_{j+k}) \\ &\overset{\mathcal{D}}{=} (Z_j, Z_{j+n+1}, \ldots, Z_{j+n+k}) \quad \text{by (6.6)}\end{aligned}$$

and this gives the hypothesis of (6.5).

<u>Remark</u>. Here we have used Proposition 6.3 for sequences which eventually turn out to be exchangeable. However, it can also give information for sequences which are in fact not exchangeable; see (14.7).

<u>Finite exchangeable sequences</u>. For a finite sequence $(Z_1,\ldots,Z_N)$ conditions (A)-(C) do not imply exchangeability. For instance, when $N = 2$ they merely imply $Z_1 \stackrel{\mathcal{D}}{=} Z_2$. On the other hand an exchangeable sequence obviously satisfies (A); what is less obvious is that (B) (and hence (C)) holds in the finite case, where the argument used in the infinite case based on de Finetti's theorem cannot be used.

(6.7) <u>Proposition</u>. <u>Let</u> $(Z_1,\ldots,Z_N)$ <u>be exchangeable and let</u> $0 \leq T_1 < T_2 < \cdots < T_k < N$ <u>be stopping times</u>. <u>Then</u> $(Z_{T_1+1},\ldots,Z_{T_k+1}) \stackrel{\mathcal{D}}{=} (Z_1,\ldots,Z_k)$.

This result is related to a gambling game called "play red". I shuffle a standard deck of cards (52 cards; 26 red and 26 black) and slowly deal them out, face up so you can see them. At some time, you bet that the next card will be red; and you must make this bet sometime before all the cards are dealt. What is your best strategy for deciding when to make the bet? One strategy is to bet on the first card, which gives you chance 1/2 of winning: is there a better strategy? By counting the colors of the cards already dealt, you can know the proportion of red cards remaining in the deck, so a natural strategy is to wait until this proportion is greater than 1/2 and then bet; intuitively, this should give you a chance greater than 1/2 of winning. However, this intuitive argument is wrong. Let Z_i be the ith card dealt, and let T be the time you decide to bet; then Proposition 6.7 says that the next card Z_{T+1} has the same distribution as Z_1, that is uniform over the deck of 52 cards.

Proposition 6.7 is due to Kallenberg (1982a). The fact that $Z_{T_1+1} \stackrel{\mathcal{D}}{=} Z_1$ is easy. For the process

$$P(Z_{i+1} \in A | Z_1, \ldots, Z_i) = P(Z_N \in A | Z_1, \ldots, Z_i)$$

is a martingale, so for a stopping time $T < N$

$$\begin{aligned} P(Z_{T+1} \in A) &= E\, P(Z_{T+1} \in A | Z_1, \ldots, Z_T) \\ &= E\, P(Z_N \in A | Z_1, \ldots, Z_T) \\ &= P(Z_N \in A) \quad \text{by the optional sampling theorem.} \end{aligned}$$

However, making an honest proof of the k-stopping-times result requires some effort; we take a slightly different approach. Recall that (Z_i) is exchangeable over V if $(V, Z_1, \ldots, Z_N) \stackrel{\mathcal{D}}{=} (V, Z_{\pi(1)}, \ldots, Z_{\pi(N)})$ for all permutations π. The following lemma is immediate.

(6.8) Lemma. *Let* $(Z_1, \ldots, Z_N)$ *be exchangeable over* V, *let* $0 \leq i \leq N$, *let* $A \in \sigma(V, Z_1, \ldots, Z_i)$ *and let* $V^i = (V, Z_1, \ldots, Z_i)$, $Z^i_j = Z_{i+j}$, $1 \leq j \leq N-i$. *Then conditional on* A, $(Z^i_1, \ldots, Z^i_{N-1})$ *is exchangeable over* V^i.

We now establish Proposition 6.7 by proving, by induction on $k \geq 1$, the following more general fact.

(6.9)(k) Assertion. *Whenever* (Z_i) *is exchangeable over* V *and* $0 \leq T_1 < \cdots < T_k$ *are stopping times relative to* $\mathcal{G}_n = \sigma(V, Z_1, \ldots, Z_n)$, *then*

$$(V, Z_{T_1+1}, \ldots, Z_{T_k+1}) \stackrel{\mathcal{D}}{=} (V, Z_{N-k+1}, \ldots, Z_N) .$$

Proof. $P(V \in A, Z_{T_1+1} \in B) = \sum_i P(V \in A, Z_{i+1} \in B, T_1 = i)$

$$= \sum_i P(V \in A, Z_1^i \in B | T_1 = i) P(T_1 = i)$$

$$= \sum P(V \in A, Z_{N-i}^i \in B | T_1 = i) P(T_1 = i) \quad \text{by (6.8)}$$

$$= \sum_i P(V \in A, Z_N \in B, T_1 = i)$$

$$= P(V \in A, Z_N \in B),$$

establishing (6.9) for $k = 1$. Suppose (6.9) holds for some k. We shall prove it for $k+1$. Consider first the special case where $T_1 = 0$. Then the sequence $(Z_2, \ldots, Z_N)$ is exchangeable over (V, Z_1), so

$$(V, Z_1, Z_{T_2+1}, \ldots, Z_{T_{k+1}+1}) \stackrel{\mathcal{D}}{=} (V, Z_1, Z_{N-k+1}, \ldots, Z_N) \quad \text{by (6.9) for } k$$
$$\stackrel{\mathcal{D}}{=} (V, Z_{N-k}, \ldots, Z_N) \quad \text{by exchangeability over } V,$$

establishing (6.9) for $k+1$ in the special case $T_1 = 0$. In the general case, fix i, and define V^i, Z_j^i as at (6.8). On the set $\{T_1 = i\}$ we have $T_j = i + \hat{T}_j$, where $\hat{T}_1 = 0$ and $\hat{T}_j$ is a stopping time with respect to $\mathcal{G}_n^i = \sigma(V^i, Z_1^i, \ldots, Z_n^i) = \mathcal{G}_{i+n}$. So by (6.8) and the special case,

$$(V^i, Z^i_{\hat{T}_1+1}, \ldots, Z^i_{\hat{T}_{k+1}+1}) \stackrel{\mathcal{D}}{=} (V^i, Z^i_{N-i-k}, \ldots, Z^i_{N-i}), \quad \text{conditional on } \{T_1 = i\}.$$

This implies

$$(V, Z_{T_1+1}, \ldots, Z_{T_{k+1}+1}) \stackrel{\mathcal{D}}{=} (V, Z_{N-k}, \ldots, Z_N), \quad \text{conditional on } \{T_1 = i\}.$$

Since this holds for each i, it holds unconditionally, establishing (6.9) for $k+1$.

7. Abstract spaces

Let S be an arbitrary measurable space. For a sequence $\underset{\sim}{Z} = (Z_1, Z_2, \ldots)$ of S-valued random variables the definition (1.2) of "exchangeable" and the definition (2.6) of "mixture of i.i.d.'s" make sense. So we can ask whether de Finetti's theorem is true for S-valued sequences, i.e. whether these definitions are equivalent. Dubins and Freedman (1979) give an example to show that for general S de Finetti's theorem is false: an exchangeable sequence need not be a mixture of i.i.d. sequences. See also Freedman (1980). But, loosely speaking, de Finetti's theorem *is* true for "non-pathological" spaces. One way to try to prove this would be to examine the proof of the theorem for R and consider what abstract properties of the range space S were needed to make the proof work for S. However, there is a much simpler technique which enables results for real-valued processes to be extended without effort to a large class of abstract spaces. We now describe this technique.

(7.1) *Definition*. Spaces S_1, S_2 are *Borel-isomorphic* if there exists a bijection $\phi: S_1 \to S_2$ such that ϕ and ϕ^{-1} are measurable. A space S is a *Borel* (or *standard*) space if it is Borel-isomorphic to some Borel-measurable subset of R.

It is well known (see e.g. Breiman (1968) A7) that any Polish (i.e. complete separate metric) space is Borel; in particular R^n, R^∞ and the familiar function spaces C(0,1) and D(0,1) are Borel. Restricting attention to Borel spaces costs us some generality; for instance, the general compact Hausdorf space is not Borel, and it is known (Diaconis and Freedman (1980a)) that de Finetti's theorem is true for compact Hausdorf spaces, but has the great advantage that results extend automatically from the real-valued setting to the Borel space-valued setting.

We need some notation. Let $P(S)$ denote the set of probability measures on S. As at (4.3), for functions $f: S_1 \to S_2$ or $g: S_1 \times S_2 \to S_3$ define the induced maps $\tilde{f}: P(S_1) \to P(S_2)$, $\tilde{g}: S_1 \times P(S_2) \to P(S_3)$ by

(7.3) $$\tilde{f}(L(Y)) = L(f(Y)) , \quad \tilde{g}(x, L(Y)) = L(g(x,Y)) .$$

(7.4) Proposition. Let $\underset{\sim}{Z}$ be an infinite exchangeable sequence, taking values in a Borel space S. Then $\underset{\sim}{Z}$ is a mixture of i.i.d. sequences.

Proof. Let $\phi: S \to B$ be an isomorphism as in (7.1) between S and a Borel subset B of R. Let $\hat{\underset{\sim}{Z}}$ be the real-valued sequence $(\phi(Z_i))$. Then $\hat{\underset{\sim}{Z}}$ is exchangeable, so by the result (3.1) for the real-valued case, $\hat{\underset{\sim}{Z}}$ is a mixture of i.i.d. sequences, directed by a random measure $\hat{\alpha}$, say. Since $\hat{Z}_i \in B$ we have $\hat{\alpha}(\cdot, B) = 1$ a.s., so we may regard $\hat{\alpha}$ as $P(B)$-valued. The map $\psi = \phi^{-1}: B \to S$ induces a map $\hat{\psi}: P(B) \to P(S)$, and $\alpha = \hat{\psi}(\hat{\alpha})$ defines a random measure on S. It is straightforward to check that $\underset{\sim}{Z}$ is a mixture of i.i.d.'s directed by α.

Exactly the same arguments show that all our results for real-valued exchangeable sequences which involve only "measure-theoretic" properties of R can be extended to S-valued sequences. We shall not write them all out explicitly. Let us just mention two facts.

(7.5) There exists a regular conditional distribution for any S-valued random variable given any σ-field.

(7.6) Let ξ be $U(0,1)$. For any distribution μ on S there exists $f: (0,1) \to S$ such that $f(\xi)$ has distribution μ.

<u>Topological spaces</u>. To discuss convergence results we need a topology on the range space S. We shall simply make the

<u>Convention</u>. All abstract spaces S mentioned are assumed to be Polish.

Roughly speaking, convergence results for real-valued exchangeable processes extend to the Polish space setting.

Let us record some notation and basic facts about weak convergence in a Polish space S. We assume the reader has some familiarity with this topic (see e.g. Billingsley (1968); Parthasarathy (1967)).

For bounded $f\colon S \to R$ write

$$\bar{f}(\theta) = \int f(x)\theta(dx) . \tag{7.7}$$

Let $C(S)$ be the set of bounded continuous functions $f\colon S \to R$. Give $\mathcal{P}(S)$ the topology of weak convergence:

$$\theta_n \to \theta \quad \text{iff} \quad \bar{f}(\theta_n) \to \bar{f}(\theta); \quad \text{each} \quad f \in C(S) .$$

The space $\mathcal{P}(S)$ itself is Polish: if d is a bounded complete metric on S then

$$\tilde{d}(\mu,\nu) = \inf\{Ed(X,Y): \mathcal{L}(X) = \mu,\ \mathcal{L}(Y) = \nu\} \tag{7.8}$$

defines a complete metrization of $\mathcal{P}(S)$.

(7.9) <u>Skorohod Representation Theorem</u>. <u>Given</u> $\theta_n \to \theta$, <u>we can construct random variables</u> $X_n \to X$ <u>a.s. and such that</u> $\mathcal{L}(X_n) = \theta_n$, $\mathcal{L}(X) = \theta$.

A sequence (θ_n) is relatively compact iff it is <u>tight</u>, that is for each $\varepsilon > 0$ there exists a compact $K_\varepsilon \subset S$ such that $\inf_n \theta_n(K_\varepsilon) \geq 1 - \varepsilon$. There exists a countable subset $\mathcal{H}$ of $C(S)$ which is <u>convergence-determining</u>:

(7.10) $\quad$ if $\lim_n \bar{h}(\theta_n) = \bar{h}(\theta)$, $h \in H$, then $\theta_n \to \theta$.

In particular H is <u>determining</u>:

(7.11) $\quad$ if $\bar{h}(\theta) = \bar{h}(\mu)$, $h \in H$, then $\theta = \mu$.

For a random measure α on S, that is to say a $P(S)$-valued random variable, and for bounded $h: S \to R$, the expression $\bar{h}(\alpha)$ gives the real-valued random variable $\int h(x)\alpha(\cdot,dx)$. By (7.11), if

(7.12) $\quad$ $\bar{h}(\alpha_1) = \bar{h}(\alpha_2)$ a.s., $h \in H$,

where H is a countable determining class, then $\alpha_1 = \alpha_2$ a.s.

For a random measure α on S define

(7.13) $\quad$ $\bar{\alpha}(A) = E\alpha(\cdot,A)$, $A \subset S$,

so $\bar{\alpha}$ is a distribution on S.

Here is a technical lemma.

(7.14) <u>Lemma</u>. <u>Let</u> (α_n) <u>be random measures on</u> S.

(a) <u>If</u> $(\bar{\alpha}_n)$ <u>is tight on</u> $P(S)$ <u>then</u> $(L(\alpha_n))$ <u>is tight in</u> $P(P(S))$.

(b) <u>If</u> (α_n) <u>is a martingale, in the sense that</u>

$$E(\alpha_{n+1}(\cdot,B)|F_n) = \alpha_n(\cdot,B) \text{ a.s.}; \quad B \subset S, \quad n \geq 1,$$

<u>for some increasing</u> σ-<u>fields</u> (F_n), <u>then there exists a random measure</u> β <u>such that</u> $\alpha_n \to \beta$ <u>a.s., that is</u>

$$P(\omega: \alpha_n(\omega,\cdot) \to \beta(\omega,\cdot) \text{ in } P(S)) = 1 .$$

<u>Proof</u>. (a) Fix $\varepsilon > 0$. By hypothesis there exist compact $K_j \subset S$ such that

$$(7.15) \qquad \bar{\alpha}_n(K_j^c) \le \varepsilon 2^{-2j}; \quad j, n \ge 1.$$

So by Markov's inequality

$$(7.16) \qquad P(\alpha_n(\cdot, K_j^c) > 2^{-j}) \le \varepsilon 2^{-2j}/2^{-j} = \varepsilon 2^{-j}; \quad j, n \ge 1.$$

So, setting

$$(7.17) \qquad \Theta = \{\theta \colon \theta(K_j^c) \le 2^{-j}, \text{ all } j \ge 1\},$$

we have from (7.16)

$$P(\alpha_n \in \Theta) \ge 1 - \varepsilon; \quad n \ge 1.$$

Since Θ is a compact subset of $P(S)$, this establishes (a).

(b) For each $h \in C(S)$ the sequence $\bar{h}(\alpha_n)$ is a real-valued martingale. So for a countable convergence-determining class H we have (a.s.)

$$\lim_{n\to\infty} \bar{h}(\alpha_n(\omega)) \text{ exists, each } h \in H.$$

Thus it suffices to prove that a.s.

$$(7.18) \qquad \text{the sequence of distributions } \alpha_n(\omega, \cdot) \text{ is tight.}$$

By the martingale property, $\bar{\alpha}_n$ does not depend on n. Take (K_j) as at (7.15). Using the maximal inequality for the martingale $\alpha_n(\cdot, K_j^c)$ gives

$$P(\alpha_n(\cdot, K_j^c) > 2^{-j} \text{ for some } n) \le \varepsilon 2^{-j}.$$

So for Θ as at (7.17),

$$P(\omega \colon \alpha_n(\omega, \cdot) \in \Theta \text{ for all } n) \ge 1 - \varepsilon .$$

This establishes (7.18).

Weak convergence of exchangeable processes. First observe that the class of exchangeable processes is closed under weak convergence. To say this precisely, suppose that for each $k \geq 1$ we have an infinite exchangeable (resp. N-exchangeable) sequence $\underset{\sim}{Z}^k = (Z_1^k)$. Think of $\underset{\sim}{Z}^k$ as a random element of S^∞ (resp. S^N), where this product space has the product technology. If $\underset{\sim}{Z}^k \xrightarrow{\mathcal{D}} \underset{\sim}{X}$, which in the infinite case is equivalent to

$$(7.19) \qquad (Z_1^k,\dots,Z_m^k) \xrightarrow{\mathcal{D}} (X_1,\dots,X_m) \quad \text{as } k\to\infty; \text{ each } m \geq 1,$$

then plainly $\underset{\sim}{X}$ is exchangeable. Note that by using interpretation (7.19) we can also talk about $\underset{\sim}{Z}^k \xrightarrow{\mathcal{D}} \underset{\sim}{X}$ where $\underset{\sim}{Z}^k$ is N^k-exchangeable, $N_k \to \infty$, and $\underset{\sim}{X}$ is infinite exchangeable.

Note also that tightness of a family $(\underset{\sim}{Z}^k)$ of exchangeable processes is equivalent to tightness of (Z_1^k). Given some class of exchangeable processes, one can consider the "weak closure" of the class, i.e. the (necessarily exchangeable) processes which are weak limits of processes from the given class.

We know that the distribution of an infinite (resp. finite) exchangeable process $\underset{\sim}{Z}$ is determined by the distribution of the directing random measure (resp. empirical distribution) α. The next result shows that weak convergence of exchangeable processes is equivalent to weak convergence of these associated random measures. Kallenberg (1973) gives this and more general results.

(7.20) Proposition. Let Z be an infinite exchangeable sequence directed by α. For $k \geq 1$ let $\underset{\sim}{Z}^k$ be exchangeable, and suppose either

(a) each $\underset{\sim}{Z}^k$ is infinite, directed by α_k, say; or

(b) $\underset{\sim}{Z}^k$ is N_k-exchangeable, with empirical distribution α_k, and $N_k \to \infty$

Then $\underset{\sim}{Z}^k \xrightarrow{\mathcal{D}} \underset{\sim}{Z}$ if and only if $\alpha_k \xrightarrow{\mathcal{D}} \alpha$, that is to say $L(\alpha_k) \rightarrow L(\alpha)$ in $P(P(S))$.

Proof. (a) Recall the definition (7.8) of $\tilde{d}$. It is easy to check that the infimum in (7.8) is attained by some distribution $L(X,Y)$ which may be taken to have the form $g(\theta,\mu)$ for some measurable $g: P(S) \times P(S) \rightarrow P(S \times S)$. To prove the "if" assertion, we may suppose $\alpha_k \rightarrow \alpha$ a.s., by the Skorohod representation (7.9). Then $\tilde{d}(\alpha_k,\alpha) \rightarrow 0$ a.s. For each k let $(\underset{\sim}{V}^k,\underset{\sim}{W}^k) = ((V_i^k,W_i^k);\ i \geq 1)$ be the S^2-valued infinite exchangeable sequence directed by $g(\alpha_k,\alpha)$. Then

(i) $\underset{\sim}{V}^k \overset{\mathcal{D}}{=} \underset{\sim}{Z}^k$; $\underset{\sim}{W}^k \overset{\mathcal{D}}{=} \underset{\sim}{Z}$; each $k \geq 1$.

Also $E(d(V_1^k,W_1^k) | g(\alpha_k,\alpha)) = \tilde{d}(\alpha_k,\alpha)$, and so

(ii) $Ed(V_1^k,W_1^k) \rightarrow 0$ as $k \rightarrow \infty$.

Properties (i) and (ii) imply $\underset{\sim}{Z}^k \xrightarrow{\mathcal{D}} \underset{\sim}{Z}$.

Conversely, suppose $Z_k^k \xrightarrow{\mathcal{D}} \underset{\sim}{Z}$. Since $\bar{\alpha}_k = L(Z_1^k)$, Lemma 7.14 shows that (α_k) is tight. If $\hat{\alpha}$ is a weak limit, the "if" assertion of the Proposition implies $\hat{\alpha} \overset{\mathcal{D}}{=} \alpha$. So $\alpha_k \xrightarrow{\mathcal{D}} \alpha$ as required.

(b) Let $\underset{\sim}{\hat{Z}}^k$ be the infinite exchangeable sequence directed by α_k. By Proposition 5.6, for fixed $m \geq 1$ the total variation distance $\| L(\hat{Z}_1^k,\ldots,\hat{Z}_m^k) - L(Z_1^k,\ldots,Z_m^k) \|$ tends to 0 as $k \rightarrow \infty$. So $\underset{\sim}{Z}^k \xrightarrow{\mathcal{D}} \underset{\sim}{Z}$ iff $\underset{\sim}{\hat{Z}}^k \xrightarrow{\mathcal{D}} \underset{\sim}{Z}$, and part (b) follows from part (a).

Proposition 7.20 is of little practical use in the finite case, e.g. in proving central limit theorems for triangular arrays of exchangeable variables, because generally finite exchangeable sequences are presented in such a way that the distribution of their empirical distribution is not manifest. Section 20 presents more practical results. Even in the infinite case, there are open problems, such as the following.

Let $\underset{\sim}{Z}$ be an infinite exchangeable real-valued sequence directed by α. For constants $(a_1,\dots,a_m)$ we can define an exchangeable sequence $\underset{\sim}{Y}$ by taking weighted sums of blocks of $\underset{\sim}{Z}$:

$$Y_i = \sum_{j=1}^{m} a_j Z_{j+(i-1)m} \, .$$

By varying $(m;a_1,\dots,a_m)$ we obtain a class of exchangeable sequences; let $C(\underset{\sim}{Z})$ be the weak closure of this class.

(7.21) <u>Problem</u>. Describe explicitly which exchangeable processes are in $C(\underset{\sim}{Z})$.

This problem arises in the study of the Banach space structure of subspaces of L^1; see Aldous (1981b). There it was shown that, under a uniform integrability hypothesis, $C(\underset{\sim}{Z})$ must contain a sequence of the special form (VY_i), where (Y_i) is i.i.d. symmetric stable, and V is independent of (Y_i). This implies that every infinite-dimensional linear subspace of L^1 contains a subspace linearly isomorphic to some ℓ_p space. Further information about Problem 7.21 might yield further information about isomorphisms between subspaces of L^1.

(7.22) <u>Stable convergence</u>. For random variables $X_1,X_2,\dots$ defined on the same probability space, say X_n converges <u>stably</u> if for each non-null event A the conditional distributions $\mathcal{L}(X_n|A)$ converge in distribution to some limit, μ_A say. Plainly stable convergence is stronger than convergence in distribution and weaker than convergence in probability. This concept is apparently due to Rényi (1963), but has been rediscovered by many authors; a recent survey of stability and its applications is in Aldous and

Eagleson (1978). Rényi and Révész (1963) observed that exchangeable processes provide an example of stable convergence. Let us briefly outline this idea.

Copying the usual proof of existence of regular conditional distributions, one readily obtains

(7.23) _Lemma_. _Suppose_ (X_n) _converges stably_. _Then there exists a random measure_ $\beta(\omega,\cdot)$ _which represents the limit distributions_ μ_A _via_

$$P(A)\mu_A(B) = \int 1_A(\omega)\beta(\omega,B)P(d\omega); \quad A \subset \Omega, \quad B \subset S.$$

Let us prove

(7.24) _Lemma_. _Suppose_ (Z_n) _is exchangeable, directed by_ α. _Then_ (Z_n) _converges stably, and the representing random measure_ $\beta = \alpha$.

Proof. Let $f \in C(S)$ and $A \in \sigma(Z_1,\ldots,Z_m)$. Then for $n > m$

$$\begin{aligned} P(A)E(f(Z_n)|A) &= E\,1_A\,E(f(Z_n)|Z_1,\ldots,Z_m,\alpha) \\ &= E\,1_A\,E(f(Z_n)|\alpha) \quad \text{by conditional independence} \\ &= E\,1_A \int f(x)\alpha(\omega,dx)\ . \end{aligned}$$

Thus $P(A)E(f(Z_n|A)) \to E\,1_A \int f(x)\alpha(\omega,dx)$ as $n \to \infty$ for $A \in \sigma(Z_1,\ldots,Z_m)$, and this easily extends to all A. Thus $L(Z_n|A) \to \mu_A$, where

$$P(A)\mu_A(\cdot) = E\,1_A\,\alpha(\omega,\cdot)$$

as required.

Note that our proof of Lemma (7.24) did not use the general result (7.23). It is actually possible to first prove the general result (7.23) and then use the type of argument above to give another proof of de Finetti's theorem; see Rényi and Révész (1963).

If X_n, defined on (Ω,F,P), converges stably, then we can extend the space to construct a "limit" variable X^* such that the representing measure β is a regular conditional distribution for X^* given F. Then (see e.g. Aldous and Eagleson (1978))

$$(Y,X_n) \xrightarrow{\mathcal{D}} (Y,X^*); \quad \text{all} \quad Y \in F. \tag{7.25}$$

Classical weak convergence theorems for exchangeable processes are stable. For instance, let (Z_i) be a square-integrable exchangeable sequence directed by α. Let $S_n = n^{-1/2} \sum_1^n (Z_i - \text{mean}(\alpha))$. Then S_n converges stably, and its representing measure $\beta(\omega,\cdot)$ is the Normal $N(0,\text{var}(\alpha))$ distribution. If we construct a $N(0,1)$ variable W independent of the original probability space, then not only do we have $S_n \xrightarrow{\mathcal{D}} S^* = \{\text{var}(\alpha)\}^{1/2}W$ as at (2.27), but also by (7.25)

$$(Y,S_n) \xrightarrow{\mathcal{D}} (Y,S^*); \quad \text{each } Y \text{ in the original space.}$$

8. The subsequence principle

Suppose we are given a sequence (X_i) of random variables whose distributions are tight. Then we know we can pick out a subsequence $Y_i = X_{n_i}$ which converges in distribution. Can we say more, e.g. can we pick (Y_i) to have some tractable kind of dependence structure? It turns out that we can: informally,

(A) we can find a subsequence (Y_i) which is similar to some exchangeable sequence $\underset{\sim}{Z}$.

Now we know from de Finetti's theorem that infinite exchangeable sequences are mixtures of i.i.d. sequences, and so satisfy analogues of the classical limit theorems for i.i.d. sequences. So (A) suggests the equally informal assertion

(B) we can find a subsequence (Y_i) which satisfies an analogue of any prescribed limit theorem for i.i.d. sequences.

Historically, the prototype for (B) was the following result of Komlós (1967).

(8.1) _Proposition_. _If_ $\sup_i E|X_i| < \infty$ _then there exists a subsequence_ (Y_i) _such that_ $N^{-1}\sum_1^N Y_i \to V$ _a.s., for some random variable_ V.

This is (B) for the strong law of large numbers. Chatterji (1974) formulated (B) as the _subsequence principle_ and established several other instances of it. A weak form of (A), in which (Y_i) is _asymptotically exchangeable_ in the sense

$$(Y_{j+1}, Y_{j+2}, \ldots) \xrightarrow{\mathcal{D}} (Z_1, Z_2, \ldots) \quad \text{as} \quad j \to \infty,$$

arose independently from several sources: Dacunha-Castelle (1974), Figiel and Sucheston (1976), and Kingman (unpublished), who was perhaps the first to note the connection between (A) and (B). We shall prove this weak form of (A) as Theorem 8.9. Unfortunately this form is not strong enough to imply (B); we shall discuss stronger results later.

The key idea in our proof is in (b) below. An infinite exchangeable sequence $\underset{\sim}{Z}$ has the property (stronger than the property of stable convergence) that the conditional distribution of Z_{n+1} given $(Z_1, \ldots, Z_n)$ converges to the directing random measure; the key idea is a kind of converse, that any sequence with this property is asymptotically exchangeable. Our arguments are rather pedestrian; the proof of Dacunha-Castelle (1974) uses ultrafilters to obtain limits, while Figiel and Sucheston (1976) use Ramsey's combinatorial theorem to prove a result for general Banach spaces which is readily adaptable to our setting.

Suppose random variables take values in a Polish space S.

(8.2) Lemma. *Let $\underset{\sim}{Z}$ be an infinite exchangeable sequence directed by* α.

(a) *Let α_n be a regular conditional distribution for Z_{n+1} given $(Z_1,\ldots,Z_n)$. Then $\alpha_n \to \alpha$ a.s.*

(b) *Let $\underset{\sim}{X}$ be an infinite sequence, let α_n be a regular conditional distribution for X_{n+1} given $(X_1,\ldots,X_n)$, and suppose $\alpha_n \to \alpha$ a.s. Then*

$$(X_{n+1},X_{n+2},\ldots) \xrightarrow{\mathcal{D}} (Z_1,Z_2,\ldots) \text{ as } n\to\infty. \tag{8.3}$$

Proof. (a) Construct Z_0 so that $(Z_i;\ i\geq 0)$ is exchangeable. Let $h \in C(S)$, and define $\bar{h}$ as at (7.7). Then

$$\begin{aligned}
\bar{h}(\alpha_n) &= E(h(Z_{n+1})|Z_1,\ldots,Z_n) \\
&= E(h(Z_0)|Z_1,\ldots,Z_n) && \text{by exchangeability} \\
&\to E(h(Z_0)|Z_i;\ i\geq 1) \text{ a.s.} && \text{by martingale convergence} \\
&= E(h(Z_0)|\alpha) \\
&= \bar{h}(\alpha).
\end{aligned}$$

Apply (7.10).

(b) Given $\underset{\sim}{X}$ and α, let $\mathcal{F}_m = \sigma(X_1,\ldots,X_m)$, $\mathcal{F} = \sigma(X_i;\ i\geq 1)$ and construct $\underset{\sim}{Z}$ such that $\underset{\sim}{Z}$ is an infinite exchangeable sequence directed by α and also

$$\underset{\sim}{Z} \text{ and } \mathcal{F} \text{ are conditionally independent given } \alpha. \tag{8.4}$$

We shall prove, by induction on k, that

$$(V,X_{n+1},\ldots,X_{n+k}) \xrightarrow{\mathcal{D}} (V,Z_1,\ldots,Z_k) \text{ as } n\to\infty;\ \text{each } V\in\mathcal{F}; \tag{8.5}$$

for each k. This will establish (b).

Suppose (8.5) holds for fixed $k \geq 0$. Let $f: S^k \times S \to R$ be bounded continuous. Define $\bar{f}: S^k \times P(S) \to R$ by

$$\bar{f}(x_1,\dots,x_k,L(Y)) = Ef(x_1,\dots,x_k,Y) .$$

Note $\bar{f}$ is continuous. By the fundamental property of conditional distributions,

$$(8.6) \qquad E(f(X_{n+1},\dots,X_{n+k},X_{n+k+1})|F_{n+k}) = \bar{f}(X_{n+1},\dots,X_{n+k},\alpha_{n+k})$$

$$(8.7) \qquad E(f(Z_1,\dots,Z_k,Z_{k+1})|F,Z_1,\dots,Z_k) = \bar{f}(Z_1,\dots,Z_k,\alpha), \text{ using (8.4).}$$

Fix $m \geq 1$ and $A \in F_m$. By inductive hypothesis

$$(\alpha,1_A,X_{n+1},\dots,X_{n+k}) \xrightarrow{\mathcal{D}} (\alpha,1_A,Z_1,\dots,Z_k) \quad \text{as} \quad n \to \infty .$$

Since $\alpha_{n+k} \to \alpha$ a.s.,

$$(8.8) \qquad (\alpha,1_A,X_{n+1},\dots,X_{n+k},\alpha_{n+k}) \xrightarrow{\mathcal{D}} (\alpha,1_A,Z_1,\dots,Z_k,\alpha) \quad \text{as} \quad n \to \infty.$$

Now

$$\begin{aligned} Ef(X_{n+1},\dots,X_{n+k+1})1_A &= E\bar{f}(X_{n+1},\dots,X_{n+k},\alpha_{n+k})1_A, \quad n \geq m, \text{ by (8.6)} \\ &\to E\bar{f}(Z_1,\dots,Z_k,\alpha)1_A \quad \text{as } n \to \infty, \text{ by (8.8) and continuity of } \bar{f}; \\ &= Ef(Z_1,\dots,Z_{k+1})1_A \quad \text{by (8.7).} \end{aligned}$$

Since this convergence holds for all f, we see that the inductive assertion (8.5) holds for $k+1$ when $V = 1_A$, $A \in F_m$. But m is arbitrary, so this extends to all $V \in F$.

(8.9) Theorem. _Let $\underset{\sim}{X}$ be a sequence of random variables such that $L(X_i)$ is tight. Then there exists a subsequence $Y_i = X_{n_i}$ such that_

$$(Y_{j+1}, Y_{j+2}, \ldots) \xrightarrow{\mathcal{D}} (Z_1, Z_2, \ldots) \quad \text{as} \quad j \to \infty$$

for some exchangeable $\underset{\sim}{Z}$.

We need one preliminary. A standard fact from functional analysis is that the unit ball of a Hilbert space is compact in the weak topology (i.e. the topology generated by the dual space): applying this fact to the space L^2 of random variables gives

(8.10) Lemma. _Let (V_i) be a uniformly bounded sequence of real-valued random variables. Then there exists a subsequence (V_{n_i}) and a random variable V such that $EV_{n_i}1_A \to EV1_A$ for all events A._

Proof of Theorem 8.9. By approximating, we may suppose each X_i takes values in some finite set S_i. Let (h_j) be a convergence-determining class. By Lemma 8.10 and a diagonal argument, we can pick a subsequence (X_n) such that as $n \to \infty$

(8.13) $$Eh_j(X_n)1_A \to EV_j1_A; \quad \text{each } A, j .$$

We can now pass to a further subsequence in which

(8.14) $$\left|E(h_j(X_{n+1})|A) - E(V_j|A)\right| \le 2^{-n}$$

for each $n \ge 1$, each $1 \le j \le n$ and each atom A of the finite σ-field $F_n = \sigma(X_1, \ldots, X_n)$ with $P(A) > 0$. Let α_n be a regular conditional distribution for X_{n+1} given F_n. We shall prove $\alpha_n \to \beta$ a.s. for some random measure β, and then Lemma 8.2(b) establishes the theorem. Note

(8.15) $$E(h_j(X_{n+1}|F_n) = \bar{h}_j(\alpha_n) .$$

Fix $m \geq 1$ and an atom A of F_m. By (8.13),

(8.16) $$L(X_n|A) \to \mu_A, \text{ say, where } \bar{h}_j(\mu_A) = E(V_j|A) .$$

Let β_m be the random measure such that $\beta_m(\omega,\cdot) = \mu_A(\cdot)$ for $\omega \in A$. So $\bar{h}_j(\beta_m) = E(V_j|F_m)$, and so by (8.14) and (8.15)

(8.17) $$|h_j(\alpha_n)-h_j(\beta_n)| \leq 2^{-n} ; \quad 1 \leq j \leq n .$$

We assert that (β_m) forms a martingale, in the sense of Lemma 7.14. For an atom A of F_m is a finite union of atoms A_k of F_{m+1}, and by (8.16) $\mu_A(B) = \sum P(A_k|A)\mu_{A_k}(B)$, $B \subset S$, which implies $E(\beta_{m+1}(\cdot,B)|F_m) = \beta_m(\cdot,B)$. Now by Lemma 7.14 we have $\beta_m \to \beta$ a.s., for some random measure β. And (8.17) implies $\bar{h}_j(\alpha_n) \to \bar{h}_j(\beta)$ a.s. for each j, and so $\alpha_n \to \beta$ a.s. as required.

Let us return to discussion of the subsequence principle. Call (Y_i) almost exchangeable if we can construct exchangeable (Z_i) such that $\sum|Y_i-Z_i| < \infty$ a.s. (we are now taking real-valued sequences). Plainly such a (Y_i) will inherit from (Z_i) the property of satisfying analogues of classical limit theorems. So if we can prove

(8.18) Every tight sequence (X_i) has an almost exchangeable subsequence (Y_i)

then we would have established a solid form of the subsequence principle (B). Unfortunately (8.18) is false. See Kingman (1978) for a counterexample, and Berkes and Rosenthal (1983) for more counterexamples and discussion of which sequences (X_i) do satisfy (8.18).

Thus we need a property weaker than "almost exchangeability" but stronger than "asymptotically exchangeable". Let $\varepsilon_k \downarrow 0$. Let (X_n) be such that for each k we can construct exchangeable $(Z_j^k, j \geq k)$ such that $P(|X_j - Z_j^k| > \varepsilon_k) \leq \varepsilon_k$ for each $j \geq k$. This property (actually, a slightly stronger but more complicated version) was introduced by Berkes and Péter (1983), who call such (X_n) <u>strongly exchangeable at infinity</u> with rate (ε_k). They prove

(8.19) <u>Theorem</u>. <u>Let</u> (X_i) <u>be tight, and let</u> $\varepsilon_k \downarrow 0$. <u>Then there exists a subsequence</u> (Y_i) <u>which is strongly exchangeable at infinity with rate</u> (ε_k).

(Again, they actually prove a slightly stronger result). From this can be deduced results of type (B), such as Proposition 8.1 and, to give another example, the analogue of the law of the iterated logarithm:

(8.20) <u>Proposition</u>. <u>If</u> $\sup EX_i^2 < \infty$ <u>then there exists a subsequence</u> (Y_i) <u>and random variables</u> V, S <u>such that</u> $\limsup_{n\to\infty} (2N \log\log(N))^{-1/2} \sum_{i=1}^N (Y_i - V) = S$ a.s.

A different approach to the subsequence principle is to abstract the idea of a "limit theorem". Let $A \subset P(R) \times R^\infty$ be the set

$$\{(\theta, \underset{\sim}{x}): \text{mean}(\theta) = \infty \text{ or } \lim N^{-1} \sum_1^N x_i = \text{mean}(\theta)\} .$$

Then the strong law of large numbers is the assertion

$$\text{(8.21)} \qquad P((\theta, X_1, X_2, \ldots) \in A) = 1 \quad \text{for } (X_i) \text{ i.i.d. } (\theta) .$$

Similarly, any a.s. limit theorem for i.i.d. variables can be put in the form of (8.21) for some set A, which we call a <u>statute</u>. Call A a <u>limit</u> statute if also

$$\text{if} \quad (\theta,\underset{\sim}{x}) \in A \quad \text{and if} \quad \sum|\hat{x}_i - x_i| < \infty \quad \text{then} \quad (\theta,\underset{\sim}{\hat{x}}) \in A \ .$$

Then Aldous (1977) shows

(8.22) _Theorem_. _Let_ A _be a limit statute and_ (X_i) _a tight sequence. Then there exists a subsequence_ (Y_i) _and a random measure_ α _such that_

$$(\alpha, Y_1, Y_2, \ldots) \in A \quad \text{a.s.}$$

Applying this to the statutes describing the strong law of large numbers or the law of the iterated logarithm, we recover Propositions 8.1 and 8.20. To appreciate (8.22), observe that for an exchangeable sequence (Z_i) directed by α we have $(\alpha, Z_1, Z_2, \ldots) \in A$ a.s. for each statute A, by (8.21). So for an almost exchangeable sequence (Y_i) and a limit statute A we have $(\alpha, Y_1, Y_2, \ldots) \in A$ a.s. Thus (8.22) is a consequence of (8.18), when (8.18) holds; what is important is that (8.22) holds in general while (8.18) does not.

The proofs of Theorems 8.19 and 8.22 are too technical to be described here: interested readers should consult the original papers.

9. _Other discrete structures_

In Part III we shall discuss processes $(X_i: i \in I)$ invariant under specified transformations of the index set I. As an introduction to this subject, we now treat some simple cases where the structure of the invariant processes can be deduced from de Finetti's theorem. We have already seen one result of this type, Corollary 3.9.

Two exchangeable sequences. Consider two infinite S-valued sequences (X_i), (Y_i) such that

(9.1) the sequence (X_i,Y_i), $i \geq 1$, of pairs is exchangeable.

Then this sequence of pairs is a mixture of i.i.d. bivariate sequences, directed by some random measure α on $S \times S$, and the marginals $\alpha_X(\omega)$, $\alpha_Y(\omega)$ are the directing measures for (X_i) and for (Y_i). Corollary 3.9 says that the stronger condition

$$(9.2) \quad (X_1,X_2,\ldots;Y_1,Y_2,\ldots) \stackrel{\mathcal{D}}{=} (X_{\pi(1)},X_{\pi(2)},\ldots;Y_{\sigma(1)},Y_{\sigma(2)},\ldots) \quad \text{for all finite permutations } \pi, \sigma$$

holds iff $\alpha(\omega) = \alpha_X(\omega) \times \alpha_Y(\omega)$.

If we wish to allow switching X's and Y's, consider the following possible conditions:

$$(9.3) \quad (X_1,X_2,X_3,\ldots;Y_1,Y_2,Y_3,\ldots) \stackrel{\mathcal{D}}{=} (Y_1,Y_2,\ldots;X_1,X_2,\ldots),$$

$$(9.4) \quad (X_1,X_2,X_3,\ldots;Y_1,Y_2,Y_3,\ldots) \stackrel{\mathcal{D}}{=} (Y_1,X_2,X_3,\ldots;X_1,Y_2,Y_3,\ldots) .$$

Let $h(x,y) = (y,x)$; let $\tilde{h}\colon \mathcal{P}(S \times S) \to \mathcal{P}(S \times S)$ be the induced map, and let $\mathcal{S}$ be the set of symmetric (i.e. $\tilde{h}$-invariant) measures on $S \times S$.

(9.5) Proposition.

(a) Both (9.1) and (9.3) hold iff $\alpha \stackrel{\mathcal{D}}{=} \tilde{h}(\alpha)$.

(b) Both (9.1) and (9.4) hold iff $\alpha(\omega) \in \mathcal{S}$ a.s.

(c) Both (9.2) and (9.3) hold iff $\alpha(\omega) = \alpha_X(\omega) \times \alpha_Y(\omega)$ a.s., where $(\alpha_X,\alpha_Y) \stackrel{\mathcal{D}}{=} (\alpha_Y,\alpha_X)$.

(d) Both (9.2) and (9.4) hold iff $\alpha(\omega) = \alpha_X(\omega) \times \alpha_Y(\omega)$ a.s., where $\alpha_X = \alpha_Y$ a.s., that is iff the whole family $(X_1,X_2,\ldots;Y_1,Y_2,\ldots)$ is exchangeable.

This is immediate from the remarks above and the following lemma, applied to $Z_i = (X_i, Y_i)$.

(9.6) Lemma. Let $h: S \to S$ be measurable, let $\tilde{h}: \mathcal{P}(S) \to \mathcal{P}(S)$ be the induced map, and let $\mathcal{P}_h$ be the set of distributions μ which are h-invariant: $\tilde{h}(\mu) = \mu$. Let $\underset{\sim}{Z}$ be an infinite exchangeable S-valued sequence directed by α.

(i) $\underset{\sim}{Z} \overset{\mathcal{D}}{=} (h(Z_1), h(Z_2), h(Z_3), \ldots)$ iff $\alpha \overset{\mathcal{D}}{=} \tilde{h}(\alpha)$.

(ii) $\underset{\sim}{Z} \overset{\mathcal{D}}{=} (h(Z_1), Z_2, Z_3, Z_4, \ldots)$ iff $\alpha = \tilde{h}(\alpha)$ a.s., that is $\alpha \in \mathcal{P}_h$ a.s.

Proof. Lemma 4.4(a) says that $(h(Z_i))$ is an exchangeable sequence directed by $\tilde{h}(\alpha)$, and this gives (i). For (ii), note first

$$\alpha \text{ is a r.c.d. for } Z_1 \text{ given } \alpha;$$
$$\tilde{h}(\alpha) \text{ is a r.c.d. for } h(Z_1) \text{ given } \alpha\,.$$

Writing $W = (Z_2, Z_3, \ldots)$, we have by Lemma 2.19

$$\alpha \text{ is a r.c.d. for } Z_1 \text{ given } W;$$
$$\tilde{h}(\alpha) \text{ is a r.c.d. for } h(Z_1) \text{ given } W.$$

Now $(Z_1, W) \overset{\mathcal{D}}{=} (h(Z_1), W)$ iff the conditional distribution for Z_1 and $h(Z_1)$ given W are a.s. equal: this is (ii).

It is convenient to record here a technical result we need in Section 13.

(9.7) Lemma. Let (X_i), (Y_i) be exchangeable. Suppose that for each subset A of $\{1,2,\ldots\}$ the sequence Z defined by

$$Z_i = X_i, \quad i \in A; \quad Z_i = Y_i, \quad i \notin A$$

satisfies $Z \stackrel{\mathcal{D}}{=} X$. _Then the directing random measures_ α_Z, α_X _satisfy_ $\alpha_Z = \alpha_X$ _a.s. for each_ Z.

Remark. This says that conditional on $\alpha = \theta$, the vectors (X_i, Y_i) are indepedent as i varies and have marginal distributions θ.

Proof. In the notation of Lemma 2.15, $\alpha_X = \Lambda(X_1, X_2, \ldots)$. Now a function of infinitely many variables may be approximated by functions of finitely many variables, so there exist functions g_k such that

$$E\, \tilde{d}(\alpha_X, g_k(X_1, \ldots, X_k)) = \delta_k\,, \tag{9.8}$$

where $\tilde{d}$ is a bounded metrisation of $P(S)$, and $\delta_k \to 0$ as $k \to \infty$. Fix $\underset{\sim}{Z}$ and define $\underset{\sim}{Z}^k$ by

$$\begin{aligned} Z_i^k &= X_i, \quad i \leq k \\ &= Z_i, \quad i > k. \end{aligned}$$

By hypotheses $\underset{\sim}{Z}^k \stackrel{\mathcal{D}}{=} \underset{\sim}{X}$, so by (9.8) $E\, \tilde{d}(\alpha_{Z^k}, g_k(Z_1^k, \ldots, Z_k^k)) = \delta_k$. But $\alpha_{Z^k} = \alpha_Z$ a.s. because α is tail-measurable; and $Z_i^k = X_i$ for $i \leq k$; so by (9.8)

$$E\, d(\alpha_Z, \alpha_X) \leq 2\delta_k\,.$$

Since k is arbitrary, $\alpha_Z = \alpha_X$ a.s.

A stratified tree. We now discuss a quite different structure, a type of stratified tree. For each $n \in \mathbf{Z}$ let $I_n = \{j2^n : j \geq 0\}$, and let $I = \{(n,i) : n \in \mathbf{Z},\, i \in I_n\}$. The set I has a natural tree structure--see the diagram. A point (n,i) has a set of "descendants", the points (m,i') such that $m \leq n$ and $i \leq i' < i + 2^{-n}$. Given n and i_1, $i_2 \in I_n$ we can define a map $\gamma: I \to I$ which switches the descendants of (n,i_1) with those of (n,i_2):

$$\begin{aligned}\gamma(m,i) &= (m,i) && \text{if } (m,i) \text{ is not a descendant of } (n,i_1) \text{ or } (n,i_2)\\ &= (m,i+i_2-i_1) && \text{if } (m,i) \text{ is a descendant of } (n,i_1)\\ &= (m,i+i_1-i_2) && \text{if } (m,i) \text{ is a descendant of } (n,i_2).\end{aligned}$$

Let Γ be the set of maps γ of this form. We want to consider processes $\underset{\sim}{X} = (X_i : i \in I)$ invariant under Γ; that is

(9.9) $$\underset{\sim}{X} \overset{\mathcal{D}}{=} (X_{\gamma(i)}, i \in I), \quad \text{each } \gamma \in \Gamma.$$

Suppose also that each $X_{n,i}$ is a function of its immediate descendants:

(9.10) $$X_{n,i} = f_n(X_{n-1,i}, X_{n-1,i+2^{n-1}}) .$$

(9.11) Lemma. *Under hypotheses (9.10) and (9.10), there is a σ-field F such that for each n the family $(X_{n,i};\ i \in I_n)$ is conditionally i.i.d. given F.*

Proof. For fixed n the family $(X_{n,i};\ i \in I_n)$ is exchangeable, and so has directing random measure α_n, say. Now consider $k < n$. The variables $(X_{n,i};\ i \in I_n)$ are functions of the variables $(X_{k,i};\ i \in I_k)$ which are conditionally i.i.d. given α_k, and hence $(X_{n,i};\ i \in I_n)$ are conditionally i.i.d. given α_k. Appealing to Lemma 2.12 we see

$$(X_{n,i};\ i \in I_n) \text{ and } \alpha_k \text{ are c.i. given } \alpha_n$$
$$\alpha_n \in \sigma(\alpha_k) \text{ a.s.}$$

Setting $F = \bigvee_{k<-\infty} \sigma(\alpha_k)$, we obtain

$$(X_{n,i};\ i \in I_n) \text{ and } F \text{ are c.i. given } \alpha_n.$$

Since the family $(X_{n,i};\ i \in I_n)$ is conditionally i.i.d. given α_n, the result follows.

(9.12) <u>Problem</u>. What is the analogue of Lemma 9.11 in the finite case, where we set $I_n = \{j2^n : 0 \leq j \leq 2^{-n}, n \leq 0\}$?

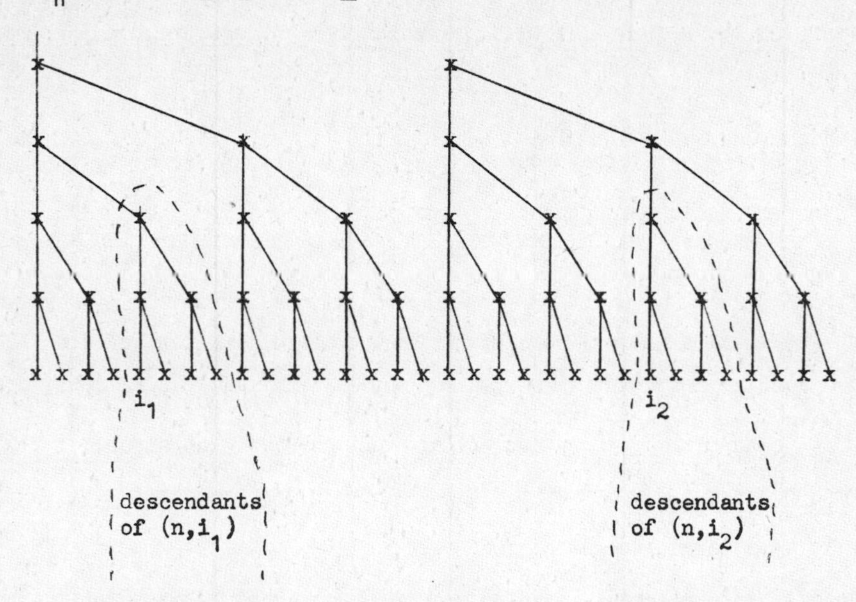

10. Continuous-time processes

de Finetti's theorem can be extended to uncountable families of random variables. But a process $(X_t: t \geq 0)$ with i.i.d. values has sample paths which are either constant or non-measurable, and this makes the concept of an exchangeable process $(X_t: t \geq 0)$ rather uninteresting. However, by considering instead the continuous analogue of processes of partial <u>sums</u> of exchangeable sequences, we get the concept of processes with interchangeable increments, and this leads to the simplest continuous-time analogues of discrete-time results. This theory has been developed in detail by Kallenberg (1973, 1974, 1975), who gives the results through (10.19) and many further results.

Consider real-valued processes $X = (X_t)$, $0 \leq t < \infty$ or $0 \leq t \leq 1$, with sample paths in the space D $(= D(0,\infty)$ or $D(0,1))$ of functions which are right-continuous with left limits. Say X has <u>interchangeable</u> <u>increments</u> if

(10.1a) for each $\delta > 0$ the sequence $(Z_i) = (X_{i\delta} - X_{(i-1)\delta})$ of increments is exchangeable;

(10.1b) $X_0 = 0$.

Assumption (b) entails no real loss of generality, since we can replace X_t by $X_t - X_0$ and preserve (a). Informally, think of interchangeable increments processes as integrals $X_t = \int_0^t Z_s \, ds$ of some underlying exchangeable "generalized process" Z.

Here is an alternative definition.

Given disjoint intervals $(a_1,b_1]$, $(a_2,b_2]$ of equal length, let $T\colon \mathbb{R}^+ \to \mathbb{R}^+$ be the natural map which switches these intervals:

(10.2)
$$T(t) = t; \quad t \notin \bigcup_i (a_i,b_i]$$
$$T(a_1+t) = a_2 + t, \quad T(a_2+t) = a_1 + t; \quad 0 < t \leq b_i - a_i .$$

Let $\mathcal{T}$ be the set of such maps T. Let $\mathcal{B}$ be the set of finite unions of disjoint intervals. With any real-valued point function $f = (f(t)\colon t \geq 0)$ we can associate a set function $(f(B)\colon B \in \mathcal{B})$ defined by

$$f(B) = \sum_i (f(t_i) - f(s_i)); \quad B = \cup (s_i,t_i], \text{ disjoint.}$$

Given a function f and a map T, let $\hat{T}f$ be the function whose associated set function is

$$(\hat{T}f)(B) = f(T(B)) .$$

The diagram is more informative than the formulas: $\hat{T}f$ is obtained from f by switching the increment of f over $(a_1, b_1]$ with the increment over $(a_2, b_2]$. Now $\hat{T}$ maps D to D. It is easy to see that definition (10.1a) is equivalent to

(10.3) $$\hat{T}(X) \stackrel{\mathcal{D}}{=} X; \quad \text{each } T \in \mathcal{T} .$$

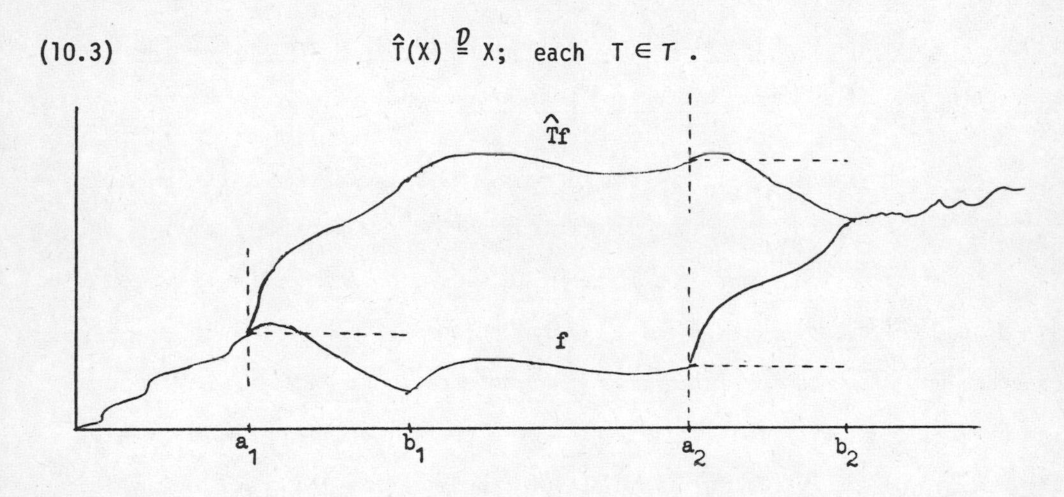

Recall the theory of <u>Lévy processes</u>, i.e. processes with stationary independent increments. For a Lévy process X, the distribution of X_1 is infinitely divisible; conversely, to any infinitely divisible distribution there corresponds a Lévy process for which X_1 has the given distribution. A continuous-path Lévy process is of the form

(10.4a) $$X_t = at + bB_t; \quad B_t \text{ Brownian motion; } a, b \text{ constants.}$$

A Lévy process which is a counting process is of the form

(10.4b) $$X_t = N_{\lambda t}; \quad N_t \text{ Poisson process of rate 1; } \lambda \text{ constant.}$$

It turns out that processes with interchangeable increments on the time-interval $[0,\infty)$ (resp. $[0,1]$) are analogous to infinite (resp. finite) exchangeable sequences. Here is the analogue of de Finetti's theorem, first noted by Bühlmann (1960).

(10.5) Proposition. *A process X with interchangeable increments on the time-interval $[0,\infty)$ is a mixture of Lévy processes.*

Proof. For $n \in \mathbb{Z}$, $j \geq 0$ set $i = j2^n$ and set $X_{n,i} = (X(i+t)-X(i): 0 \leq t \leq 2^n)$. So $X_{n,i}$ is a random element of $D[0,2^n]$. We assert that $(X_{n,i})$ satisfies the hypotheses of Lemma (9.11). In fact (9.10) is immediate, and assertion (9.9) is a reformulation of the property of interchangeable increments for intervals with dyadic rational endpoints. Now Lemma (9.11) concludes that, conditional on a certain σ-field F, the family $(X_{n,j2^n}; j \geq 0)$ is i.i.d. for each n. By approximating arbitrary intervals by intervals with dyadic rational endpoints, we deduce that conditional on F the process X is Lévy.

Using (10.4a,b) we get

(10.6) Corollary. *Let X be a process with interchangeable increments on the time-interval $[0,\infty)$. If X has continuous paths then*

(a) $X_t = \alpha t + \beta B_t$; *$B_t$ Brownian motion; α, β r.v.'s independent of B.*

If X is a counting process, then

(b) $X_t = N_{\Lambda t}$; *N_t the Poisson process of rate 1; Λ a r.v. independent of N.*

We should also mention the analogue of Theorem 6.1. For a process X write X^t for the process $X_u^t = X_{t+u} - X_t$. The strong Markov property shows that X is a Lévy process iff

$$X^T \text{ is independent of } \sigma(X_s, s \leq T);\ X^T \stackrel{\mathcal{D}}{=} X; \text{ for each stopping time } T.$$

Kallenberg (1982a,b) defines a process X to have <u>strongly stationary increments</u> if

$$(10.7) \qquad X^T \stackrel{\mathcal{D}}{=} X; \quad \text{each stopping time } T,$$

and shows that this property is equivalent to the interchangeable increments property. The proof requires only Theorem 6.1 and the arguments of Proposition 10.5.

Now consider processes on the time-interval $[0,1]$. Two processes with interchangeable increments are

(a) the counting process N^m of m draws from the uniform distribution:

$$(10.8) \qquad N_t^m = \sum_{i=1}^m 1_{(t \leq \xi_i)}, \quad (\xi_i) \text{ i.i.d. } U(0,1).$$

(b) the Brownian bridge B^0.

It turns out that these are the basic examples of counting processes (resp. continuous-path processes) with interchangeable increments on $[0,1]$.

(10.9) <u>Lemma</u>. <u>Let</u> X <u>be a counting process on</u> $[0,1]$ <u>with interchangeable increments, and suppose</u> $X_1 = m$. <u>Then</u> $X \stackrel{\mathcal{D}}{=} N^m$, <u>for</u> N^m <u>as at</u> (10.7).

<u>Proof</u>. Let $D_k = \{i2^{-k}: 0 \leq i < 2^k\}$ and let $\phi_k(x) = \max\{t \leq x: t \in D_k\}$. Let $0 \leq \theta_1 < \cdots < \theta_m \leq 1$ be the jump times of X, and let $\xi_{(1)}, \ldots, \xi_{(m)}$ be the increasing ordering of (ξ_i). Let A_k be the event that the random variables $\phi_k(\theta_i)$, $1 \leq i \leq m$, are distinct; let $\hat{A}_k$ be the corresponding event for (ξ_i). Then

(i) $P(A_k) \to 1$ and $P(\hat{A}_k) \to 1$ as $k \to \infty$;

(ii) $\phi_k(\theta_i) \to \theta_i$ a.s. and $\phi_k(\xi_i) \to \xi_i$ a.s. as $k \to \infty$;

(iii) the distribution of $(\phi_k(\theta_1),\ldots,\phi_k(\theta_m))$ conditional on A_k is the same as the distribution of $(\phi_k(\xi_{(1)}),\ldots,\phi_k(\xi_{(m)}))$ conditional on $\hat{A}_k$, because each is the distribution of the order statistics of m draws without replacement from D_k.

Properties (i)-(iii) imply $(\theta_1,\ldots,\theta_m) \stackrel{\mathcal{D}}{=} (\xi_{(1)},\ldots,\xi_{(m)})$, so $X \stackrel{\mathcal{D}}{=} N^m$.

From Lemma 10.9 we see the form of the general counting process with interchangeable increments on $[0,1]$: let M have an arbitrary distribution on $\{0,1,2,\ldots\}$ and then, conditional on $\{M=m\}$, let X have distribution (10.8).

We now consider continuous-path processes. Let $(B_t^0\colon 0 \le t \le 1)$ be the Brownian bridge, that is the Gaussian process with $B_0^0 = B_1^0 = 0$ and

$$EB_s^0 B_t^0 = s(1-t), \quad 0 \le s \le t \le 1 . \tag{10.10}$$

For fixed k put $Z_i = B_{i/k}^0 - B_{(i-1)/k}^0$. Then $(Z_1,\ldots,Z_k)$ is Gaussian, and using (10.10) we compute $EZ_i^2 = (k-1)k^{-2}$, $EZ_iZ_j = -k^{-2}$ $(i \ne j)$, and so (Z_i) is exchangeable. It follows that B^0 has interchangeable increments. From this we can construct more continuous-path interchangeable increments processes by putting

$$X_t = \alpha B_t^0 + \beta t; \quad \text{where } (\alpha,\beta) \text{ is independent of } B^0. \tag{10.11}$$

Note that Brownian motion on $[0,1]$ is of this form, for $\alpha = 1$ and β having $N(0,1)$ distribution.

(10.12) <u>Theorem</u>. <u>Every continuous-path interchangeable increments process</u> $(X_t\colon 0 \le t \le 1)$ <u>is of the form</u> (10.11).

This is a non-trivial result. The natural method of proof (Kallenberg (1973)) is via weak convergence of discrete approximations. However, we shall later (Section 20) use Theorem 10.12 as a starting point for proving weak convergence results, and so for aesthetic reasons we would like a proof of Theorem 10.12 independent of weak convergence ideas. The following proof uses martingale techniques, and assumes

(10.13) $$EX_t^2 < \infty, \quad \text{each } t.$$

We shall later indicate how to remove this restriction.

(10.14) Lemma. Let $(X_t: 0 \le t \le 1)$ be a continuous-path process adapted to a filtration $(\mathcal{G}_t)$ and satisfying

(i) $X_0 = X_1 = 0$

(ii) $E(X_u - X_t | \mathcal{G}_t) = -\frac{u-t}{1-t} X_t, \quad t \le u < 1$

(iii) $\mathrm{var}(X_u - X_t | \mathcal{G}_t) = \frac{(u-t)(1-u)}{1-t}, \quad t \le u < 1.$

Then $X \overset{\mathcal{D}}{=} B^0$ and X is independent of $\mathcal{G}_0^+ = \bigcap_{t>0} \mathcal{G}_t$.

Proof. Write

(10.15) $$Y_t = (1+t)X_{t/(1+t)}; \quad \mathcal{F}_t = \mathcal{G}_{t/(1+t)}; \quad 0 \le t < \infty.$$

Then $Y_0 = 0$, Y has continuous paths and by (ii) and (iii),

$$E(Y_u - Y_t | \mathcal{F}_t) = 0, \quad \mathrm{var}(Y_u - Y_t | \mathcal{F}_t) = u - t; \quad t \le u.$$

So by Lévy's Theorem (see e.g. Doob (1953) p. 78) Y is Brownian motion. Inverting (10.15) gives $X_t = (1-t)Y_{t/(1-t)}$, which implies X is Brownian bridge. Finally, for any event A in $\mathcal{G}_0^+$ the process X conditioned on A satisfies the hypotheses of the lemma, so the conditioned process is also Brownian bridge; this proves independence.

Proof of Theorem 10.12. Consider first the special case

$$X_1 = 0 .$$

For $m \geq 1$ let $D_m = \{i2^{-m}: 0 \leq i \leq 2^m\}$. Let $V_{m,j} = (X(j2^{-m}+u) - X(j2^{-m}):$ $u \leq 2^{-m})$, considered as a random element of $D(0,2^{-m})$. For $t \in D_m$ let G_t^m be the σ-field generated by $(X_s: s \leq t)$ and the empirical distribution of $(V_{m,j}: j2^{-m} \geq t)$. Then conditional on G_t^m the variables $(X_{(j+1)2^{-m}} - X_{j2^{-m}}: j2^{-m} \geq t)$ are an urn process, in the language of section 5. The elementary formulas for means and variances when sampling without replacement (20.1) show that for $u \in D_m$, $u \geq t$,

$$\text{(10.16)} \qquad \begin{aligned} E(X_u - X_t | G_t^m) &= -\frac{u-t}{1-t} X_t \\ \text{var}(X_u - X_t | G_t^m) &= \frac{(u-t)(1-u)}{(1-t)(1-t-2^{-m})} \{Q_1^m - Q_t^m + \frac{2^{-m}X_t^2}{1-t}\} \end{aligned}$$

where

$$\text{(10.17)} \qquad Q_t^m = \sum_{i=0}^{t2^m-1} (X_{(i+1)2^{-m}} - X_{i2^{-m}})^2; \quad t \in D_m .$$

For fixed t, the σ-fields G_t^m are decreasing as m increases; for $t \in D = \bigcup_m D_m$ let $G_t = \bigcap_m G_t^m$, and for general t let $G_t = \bigcap_{\substack{u>t \\ u \in D}} G_u$. Suppose we can prove there exists a random variable $\alpha \geq 0$ such that

$$\text{(10.18)} \qquad Q_t^m \xrightarrow{p} \alpha t; \quad \text{each } t \in D.$$

Then reverse martingale convergence in (10.16) shows that for $t \leq u$ $(t, u \in D)$

$$\begin{aligned} E(X_u - X_t | G_t) &= -\frac{u-t}{1-t} X_t \\ \text{var}(X_u - X_t | G_t) &= \frac{(u-t)(1-u)}{1-t} \alpha . \end{aligned}$$

These extend to all $t \le u$, by approximating from above. Note $\alpha \in G_0^+$. On $\{\alpha > 0\}$ set $V_t = \alpha^{-1/2} X_t$. Then V satisfies the hypotheses of Lemma 10.13, and so V is Brownian bridge, independent of α. Since $X_t = \alpha^{1/2} V_t$, this establishes the theorem in the special case $X_1 = 0$. For the general case, set $\hat{X}_t = X_t - tX_1$, define $\hat{G}_t^m$ using $\hat{X}$ as G_t^m was defined using X, and include X_1 in the σ-field. The previous argument gives that $V_t = \alpha^{-1}\hat{X}_t$ is Brownian bridge independent of $G_0^+ \supset \sigma(\alpha, X_1)$, and then writing $X_t = \alpha V_t + X_1 t$ establishes the theorem in the general case.

To prove (10.18), we quote the following lemma, which can be regarded as a consequence of maximal inequalities for sampling without replacement (20.5) or as the degenerate weak convergence result (20.10).

(10.19) Lemma. *For each* $m \ge 1$ *let* $(Z_{m,1}, \ldots, Z_{m,k_m})$ *be exchangeable. If*

(a) $\sum_i Z_{m,i} = 0$ *for each* m,

(b) $\sum_i Z_{m,i}^2 \xrightarrow[p]{} 0$ *as* $m \to \infty$,

then

(c) $\max_j \left| \sum_{i=1}^{j} Z_{m,i} \right| \xrightarrow[p]{} 0$ *as* $m \to \infty$.

Proof of (10.18). Set $Z_{m,i} = (X_{(i+1)2^{-m}} - X_{i2^{-m}})^2 - 2^{-m} Q_1^m$. Then (a) is immediate. For (b),

$$\begin{aligned} \sum_i Z_{m,i}^2 &\le \delta_m \sum_i |Z_{m,i}|; \quad \delta_m = \max_i |Z_{m,i}| \\ &\le \delta_m \cdot 2Q_1^m \\ &\xrightarrow[p]{} 0 \text{ since } \delta_m \xrightarrow[p]{} 0 \text{ by continuity and } Q_1^m \text{ converges a.s.} \\ &\qquad \text{by reverse martingale convergence in (10.16).} \end{aligned}$$

So conclusion (c) says

$$\max_{t \in D_m} |Q_t^m - tQ_1^m| \xrightarrow[p]{} 0 \quad \text{as} \quad m \to \infty ,$$

and this is (10.18).

Remark. To remove the integrability hypothesis (10.13), note first that for non-integrable variables we can define conditional expectations "locally": $E(U|F) = V$ means that for every $A \in F$ for which $V1_A$ is integrable, we have that $U1_A$ is integrable and $E(U1_A|F) = V1_A$. In the non-integrable case, (10.16) remains true with this interpretation. To establish convergence of the "local" martingale Q_1^m it is necessary to show that $(Q_1^m: m \geq 1)$ is tight, and for this we can appeal to results on sampling without replacement in the spirit of (10.19). However, there must be some simpler way of making a direct proof of Theorem 10.12.

Let us describe briefly the general case of processes $(X_t: 0 \leq t \leq 1)$ with interchangeable increments, and refer the reader to Kallenberg (1973) for the precise result. Given a finite set $J = (x_i)$, there is one process with interchangeable increments, with jump sizes (x_i), and constant between jumps

$$X_t^J = \sum_i x_i 1_{(t \geq \xi_i)} \ ; \quad \text{where} \quad (\xi_i) \ \text{are independent} \ U(0,1).$$

This sum can also be given an interpretation for certain infinite sets J, as a L^2-limit. The resulting processes X^J are the "pure jump" processes; taking constants a, b and taking a Brownian bridge B^0 independent of (ξ_i), and putting

$$X_t = X_t^J + aB_t^0 + bt$$

gives a process X with interchangeable increments. These are the "ergodic" processes in the sense of Section 12; the general process with interchangeable increments is obtained by first choosing (J,a,b) at random from some prior distribution.

(10.20) The Dirichlet process. An interesting and useful instance of a process with interchangeable increments and increasing discontinuous paths is the family of Dirichlet processes, which we now describe.

Fix $a > 0$. Recall that the Gamma(b,1) distribution has density $\frac{1}{\Gamma(b)} x^{b-1}e^{-x}$ on $\{x \geq 0\}$. Since this distribution is infinitely divisible, there exists a Lévy process (X_t) such that X_1 has Gamma(a,1) distribution, and hence X_t has Gamma(at,1) distribution. Call X the Gamma(a) process. Here is an alternative description. Let ν be the measure on $(0,\infty)$ with density

$$\nu(dx) = ax^{-1}e^{-x}dx \ . \tag{10.21}$$

Then $\nu(\varepsilon,\infty) < \infty$ for $\varepsilon > 0$, but $\nu(0,\infty) = \infty$. Let $\nu \times \lambda$ be the product of ν and Lebesgue measure on $Q = \{(x,t): x > 0, t \geq 0\}$. Let N be a Poisson point process on Q with intensity $\nu \times \lambda$. Then N is distributed as the times and sizes of the jumps of X:

$$N \overset{\mathcal{D}}{=} \{(x,t): X_t - X_{t-} = x\} \ .$$

So we can construct X from N by adding up jumps:

$$X_t = \sum x \, 1_{\{(x,s) \in N, \ s \leq t\}} \ .$$

The Dirichlet(a) process is $Y_t = X_t/X_1$, $0 \leq t \leq 1$. Thus Y has increasing paths, interchangeable increments, $Y_0 = 0$, $Y_1 = 1$. (The relation between the Dirichlet process and the Gamma process is reminiscent of the relation between Brownian bridge and Brownian motion). The marginal distribution of Y_t is Beta(at,a(1-t)). As $a \to \infty$ the distribution of (Y_t) converges to the deterministic process t; as $a \to 0$ it converges to the single jump process $1_{(t \geq \xi)}$, ξ uniform on $[0,1]$. For $0 < t_1 < t_2 < \cdots < t_k = 1$ the increments $(Y_{t_1}, Y_{t_2} - Y_{t_1}, \ldots, Y_{t_k} - Y_{t_{k-1}})$ have the distribution $\mathcal{D}(at_1, a(t_2 - t_1), \ldots, a(t_k - t_{k-1}))$, where $\mathcal{D}(\alpha_1, \ldots, \alpha_k)$ is the distribution on the simplex $\{(y_1, \ldots, y_k): y_i \geq 0, \sum y_i = 1\}$ with density

$$\frac{\Gamma(\sum \alpha_i)}{\Pi \Gamma(\alpha_i)} \Pi y_i^{\alpha_i - 1}$$

with respect to Lebesgue measure on the simplex. In particular, the increments

$$(10.22) \qquad (Z_1, Z_2, \ldots, Z_k) = (Y_{1/k}, Y_{2/k} - Y_{1/k}, \ldots, Y_1 - Y_{(k-1)/k})$$

have density

$$\frac{\Gamma(a)}{\{\Gamma(a/k)\}^k} (\Pi z_i)^{-1+a/k}$$

on the simplex, and this is k-exchangeable.

Dirichlet processes have been extensively studied as prior distributions in Bayesian statistics. For we can view $Y_t(\omega)$ as the distribution function of a random measure $\beta(\omega, \cdot)$ on $[0,1]$ which is specified by the requirement

(10.23) $(\beta(\cdot, A_1), \ldots, \beta(\cdot, A_k))$ has distribution $\mathcal{D}(a\lambda(A_1), \ldots, a\lambda(A_k))$ for each partition (A_i).

This is the Dirichlet random measure associated with the measure $a\lambda(\cdot)$ on $[0,1]$. Given any finite measure α on a Borel space S, we can construct a Dirichlet random measure β satisfying (10.23) with α in place of $a\lambda(\cdot)$, by simply applying a function $f: [0,1] \to S$ which takes $a\lambda(\cdot)$ into α.

The advantage of using random measures of this form as priors is that posterior distributions can be handled analytically. For a survey of the statistical results, see Ferguson (1973, 1974).

By construction the Dirichlet random measure takes values in the set of purely atomic distributions. It is interesting to note that a random measure on $[0,1]$ with the property

(10.24) $(\beta(\cdot,A_1),\dots,\beta(\cdot,A_k))$ is exchangeable, for partitions (A_i) with $\lambda(A_i) = 1/k$

and which takes values in the set of continuous distributions, must be of the trivial form $\beta(\omega,\cdot) = \lambda(\cdot)$. For consider the distribution function $F_t(\omega) = \beta(\omega,[0,t])$. By (10.24), (F_t) has interchangeable increments, so if it is continuous then by Theorem 10.12 we must have $F_t = \alpha B_t + \beta t$ for some (α,β). But $F_1 = 1$ forces $\beta = 1$, and for F to be increasing we must have $\alpha = 0$.

Our uses of Dirichlet processes arise from entirely different considerations, the Kingman-Watterson treatment of the infinite allele model and the Ewens sampling formula, to be described in Sections 19 and 11. We need one more concept. The countable set of jump sizes $(Y_t - Y_{t-})$ of a sample path $Y(\omega)$ of a Dirichlet(a) process can be arranged in decreasing order as $(D_1,D_2,\dots)$. This describes a distribution on the infinite-dimensional simplex $\{(x_1,x_2,\dots): x_i > 0, \sum x_i = 1\}$ which is called the <u>Poisson-Dirichlet(a)</u> distribution. The name is explained by an alternative description using the measure ν of (10.21). Take a Poisson point process of intensity ν on $(0,\infty)$, and let $(V_1,V_2,\dots)$ be the points of this process, arranged in decreasing order. Then $S = \sum V_i < \infty$ a.s., because $ES = \int_{0+}^{\infty} x\,\nu(dx) = a$;

and $(D_1,D_2,\dots) = (V_1/S,V_2/S,\dots)$ has the Poisson-Dirichlet distribution. Explicit though complicated expressions for the marginals are given in Watterson (1976); see also Kingman (1980), Section 3.6.

As a brief indication of the kind of calculations which can be done, consider the "intensity" of the points (D_i):

$$\phi(y)dy = P(\text{some } D_i \in (y,y+dy)) .$$

We shall show

$$(10.25) \qquad \phi(y) = ay^{-1}(1-y)^{a-1}, \quad 0 < y < 1 .$$

Consider first the intensity

$$\psi(v,s)dv\,ds = P(\text{some } V_i \in (v,v+dv),\ S = \textstyle\sum V_i \in (s,s+ds)) .$$

Conditional on some V being in $(v,v+dv)$, the remaining (V_i) still form a Poisson process of intensity ν, so their sum still has the Gamma(a,1) density f_S, say. Thus

$$\begin{aligned} \psi(v,s)dv\,ds &= \nu(dv)f_S(s-v)ds \\ (10.26) \qquad &= av^{-1}e^{-v}(s-v)^{a-1}e^{-(s-x)}dvds/\Gamma(a) . \end{aligned}$$

Now $D_i = V_i/S$, so the intensity

$$\begin{aligned} P(\text{some } D_i \in (y,y+dy),\ S \in (s,s+ds)) &= \psi(sy,s)\,s\,dy\,ds, \quad \text{putting } y = v/s \\ &= ay^{-1}(1-y)^{a-1}dy \cdot s^{a-1}e^{-s}/\Gamma(a) \quad \text{by (10.26).} \end{aligned}$$

This product form shows that the intensity $\phi(y)$ satisfies (10.25), independent of S.

Observing now that at most one of the D_i can be greater than 1/2 (since $\sum D_i = 1$), we see from (10.25) that the density of D_1 is

$$f_{D_1}(y) = ay^{-1}(1-y)^{a-1} \quad \text{on} \quad 1/2 < y < 1 \, .$$

However, the expression for the density over $0 < y < 1/2$ is more complicated.

11. Exchangeable random partitions

In this section we describe recent work of Kingman and others.

Let $\mathcal{S}_N$ be the (finite) set of all partitions $A = (A_i)$ of $\{1,2,\ldots,N\}$. Call the sets A_i the components of A. By a random partition $\mathcal{R}$ we simply mean a random element of $\mathcal{S}_N$. Each permutation π of $\{1,2,\ldots,N\}$ acts on subsets $B \subset \{1,2,\ldots,N\}$ by $\pi(B) = \{\pi(i): i \in B\}$ and so acts on partitions by $\pi(A_1,A_2,\ldots) = (\pi(A_1),\pi(A_2),\ldots)$. Thus we call $\mathcal{R}$ exchangeable if $\pi(\mathcal{R}) \stackrel{\mathcal{D}}{=} \mathcal{R}$ for each π.

It is easy to describe the general exchangeable partition of a finite set (this is analogous to the description of the general finite exchangeable sequence in Section 5). Let $I_N = \{(n_i): n_1 \geq n_2 \geq \cdots \geq 0, \sum n_i = N\}$. Define $L_N: \mathcal{S}_N \rightarrow I_N$ by

(11.1) $\qquad L_N((A_i))$ is the decreasing rearrangement of $(\#A_i)$.

Let π^* be the uniform random permutation. For any partition A, it is easy to see that $\pi^*(A)$ is uniform on $\{B: L_N(B) = L_N(A)\}$. Thus the general exchangeable partition $\mathcal{R}$ is obtained as follows:

(i) take a random element (η_i) of I_N;

(ii) conditional on $(\eta_i) = (n_i)$, let $\mathcal{R}$ be uniform on $\{A: L_N(A) = (n_i)\}$.

In particular, the distribution of $\mathcal{R}$ is determined by the distribution of $L_N(\mathcal{R})$.

An alternative description of partitions is sometimes useful. Given a partition $\mathcal{R}$, define

$$R_{i,j} = \{\omega\colon \text{i and j in the same component of } \mathcal{R}(\omega)\} \quad , \tag{11.2}$$

and then

$$R_{i,i} = \Omega; \quad R_{i,j} = R_{j,i}; \quad R_{i,k} \supset R_{i,j} \cap R_{j,k}; \quad 1 \le i,j,k \le N \ . \tag{11.3}$$

Conversely any family of events $(R_{i,j}\colon 1 \le i,j \le N)$ satisfying (11.3) defines a random partition $\mathcal{R}$, in which $\mathcal{R}(\omega)$ is the partition into equivalence classes of the equivalence relation

$$i \overset{\omega}{\sim} j \quad \text{iff} \quad \omega \in R_{i,j} \ .$$

Furthermore, $\mathcal{R}$ is exchangeable iff

$$(R_{i,j}\colon 1 \le i,j \le N) \overset{\mathcal{D}}{=} (R_{\pi(i),\pi(j)}\colon 1 \le i,j \le N); \quad \text{each permutation } \pi. \tag{11.4}$$

Consider now partitions of $\{1,2,3,\ldots\}$. The set S_∞ of all partitions is uncountable; we define a random partition to be a map $\mathcal{R}\colon \Omega \to S_\infty$ such that the sets $R_{i,j}$ defined at (11.2) are measurable. As before, a random partition of $\{1,2,\ldots\}$ can be described as a family of events $\{R_{i,j}\colon 1 \le i,j < \infty\}$ satisfying (11.3) for each N. Note that one way to construct random partitions of $\{1,2,3,\ldots\}$ is by appealing to the Kolmogorov extension theorem: if for each N we have an exchangeable partition $\mathcal{R}^N$ satisfying the consistency conditions

$$(R^{N+1}_{i,j}\colon 1 \le i,j \le N) \overset{\mathcal{D}}{=} (R^N_{i,j}\colon 1 \le i,j \le N); \quad \text{each } N \ge 1 \ , \tag{11.5}$$

then there exists an exchangeable random partition $\mathcal{R}$ of $\{1,2,3,\ldots\}$ such that

$$(R_{i,j}\colon 1 \le i,j \le N) \overset{\mathcal{D}}{=} (R^N_{i,j}\colon 1 \le i,j \le N); \quad \text{each } N \ge 1. \tag{11.6}$$

Our aim is to prove an analogue of de Finetti's theorem for exchangeable partitions of $\{1,2,3,\ldots\}$. The role of i.i.d. sequences is played by the "paintbox processes" which we now describe. Let μ be a distribution on $[0,1]$; think of μ as partly discrete and partly continuous. Let (X_i) be i.i.d. (μ). Let $\mathcal{R}(\omega)$ be the partition with components $\{i: X_i(\omega) = x\}$, $0 \leq x \leq 1$. In other words, $\mathcal{R}_{i,j} = \{\omega: X_i(\omega) = X_j(\omega)\}$. Clearly $\mathcal{R}$ is exchangeable. Kingman suggests the following mental picture: think of real numbers $0 \leq x \leq 1$ as labelling the colours of the spectrum; imagine colouring objects 1,2,3,... at random by painting object i with colour X_i; then we obtain a partition into sets of identically-coloured objects.

Clearly the distribution of $\mathcal{R}$ in this construction depends only on the sizes of the atoms of μ. Let $p_j = \mu(x_j)$, where (x_j) are the atoms of μ arranged so that (p_j) is decreasing (and put $p_j = 0$ if there are less than j atoms). This defines a map

(11.7) $$L(\mu) = (p_j)$$

from $\mathcal{P}[0,1]$ into the set of possible sequences

$$\nabla = \{(p_1,p_2,\ldots): p_1 \geq p_2 \geq \cdots \geq 0, \sum p_j \leq 1\} .$$

When $L(\mu) = \underset{\sim}{p}$, call the exchangeable partition $\mathcal{R}$ above the <u>paintbox</u>($\underset{\sim}{p}$) <u>process</u>, and denote its distribution by ψ_p. The following facts are easy consequences of the strong law $N^{-1}\#(i \leq N: X_i = x_j) \rightarrow p_j$ a.s.

(11.8) <u>Lemma</u>. <u>Let</u> $\mathcal{R}$ <u>be a paintbox</u>($\underset{\sim}{p}$) <u>process</u>.

(a) $N^{-1}L_N(\mathcal{R}^N) \rightarrow (p_1,p_2,\ldots)$ <u>in</u> ∇ <u>a.s.</u>, <u>where</u> $\mathcal{R}^N$ <u>is the restriction of</u> $\mathcal{R}$ <u>to</u> $\{1,2,\ldots,N\}$.

(b) <u>Let</u> C_1 <u>be the component of</u> $\mathcal{R}(\omega)$ <u>containing</u> 1. <u>Then</u>

$$N^{-1}\#(C_1\cap\{1,\dots,N\}) \xrightarrow{\mathcal{D}} p_J 1_{(J>0)} ,$$

<u>where</u> $P(J=j) = p_j$, $P(J=0) = 1-\sum p_j$.

(c) $P(1,2,\dots,r$ <u>in same component</u>$) = \sum_{j\geq 1} p_j^r$; $r \geq 2$.

Here is the analogue of de Finetti's theorem, due to Kingman (1978b, 1982a).

(11.9) <u>Proposition</u>. <u>Let</u> $\mathcal{R}$ <u>be an exchangeable partition of</u> $\{1,2,\dots\}$, <u>and let</u> $\mathcal{R}^N$ <u>be its restriction to</u> $\{1,2,\dots,N\}$. <u>Then</u>

(a) $N^{-1}L_N(\mathcal{R}^N) \xrightarrow{a.s.} (D_1,D_2,\dots) = D$ <u>for some random element</u> D <u>of</u> ∇;

(b) ψ_D <u>is a regular conditional distribution for</u> $\mathcal{R}$ <u>given</u> $\sigma(\psi_D)$.

So (b) says that conditional on $D = \underset{\sim}{p}$, the partition $\mathcal{R}$ has the paintbox($\underset{\sim}{p}$) distribution ψ_p. As discussed in Section 2, this is the "strong" notion of $\mathcal{R}$ being a mixture of paintbox processes.

<u>Proof</u>. Let (ξ_i) be i.i.d. uniform on $(0,1)$, independent of $\mathcal{R}$. Throwing out a null set, we can assume the values $(\xi_i(\omega): i\geq 1)$ are distinct. Define

$$F_i(\omega) = \min\{j: i \text{ and } j \text{ in same component of } \mathcal{R}(\omega)\} \leq i$$

$$Z_i = \xi_{F_i} .$$

So for each ω the partition $\mathcal{R}(\omega)$ is precisely the partition with components $\{i: Z_i(\omega)=z\}$, $0 \leq z \leq 1$. We assert (Z_i) is exchangeable. For $(Z_i) = g((\xi_i),\mathcal{R})$ for a certain function g, and $(Z_{\pi(i)}) = g((\xi_{\pi(i)}),\pi(\mathcal{R}))$, and $((\xi_{\pi(i)}),\pi(\mathcal{R})) \overset{\mathcal{D}}{=} ((\xi_i),\mathcal{R})$ by exchangeability and by independence of $\mathcal{R}$ and (ξ_i).

Let α be the directing random measure for (Z_i). Then conditional on $\alpha = \mu$ the sequence (X_i) is i.i.d. (μ) and so $\mathcal{R}$ has the paintbox distribution $\psi_{L(\mu)}$. In other words $\psi_{L(\alpha)}$ is a regular conditional distribution for $\mathcal{R}$ given α, and this establishes (b) for $D = L(\alpha)$. And then (a) follows from Lemma 11.8(a) by conditioning on D.

Remarks. Kingman used a direct martingale argument, in the spirit of the first proof of de Finetti's theorem in Section 3. Our trick of labelling components by external randomization enables us to apply de Finetti's theorem. Despite the external randomization, (a) shows that D is a function of $\mathcal{R}$.

Yet another proof of Proposition 11.9 can be obtained from deeper results on partial exchangeability--see (15.23).

The Ewens sampling formula. Proposition 11.5 is a recent result, so perhaps some presently unexpected applications will be found in future. The known applications involve situations where certain extra structure is present. An exchangeable random partition $\mathcal{R}$ on $\{1,2,3,\ldots\}$ can be regarded as a sequence of exchangeable random partitions $\mathcal{R}^N$ on $\{1,2,\ldots,N\}$ satisfying the consistency condition (11.5). Let us state informally another "consistency" condition which may hold. Fix N and $r < N$. Pick a partition $\mathcal{R}^N = (A_i)$, and suppose $1 \in A_j$, where $\#A_j = r$. The remaining sets $(A_i\colon i \neq j)$ form a partition of $\{1,2,\ldots,N\} \backslash A_j$; the new condition is that this partition $(A_i\colon i \neq j)$ should be distributed as $\mathcal{R}^{N-r}$.

To formalize this, we introduce more notation. For a partition $A = (A_i)$ of $\{1,2,\ldots,N\}$ define

(11.10) $a(A) = (a_1,\ldots,a_N)$, where a_j is the number of sets A_i for which $\#A_i = j$.

Of course $a(A)$ gives precisely the same information about A as does $L_N(A)$ used earlier. Given an exchangeable random partition $\mathcal{R}^N$, let

$$(11.11) \qquad Q_N(a_1,\ldots,a_N) = P(a(\mathcal{R}^N) = (a_1,\ldots,a_N)) .$$

By exchangeability, Q_N determines the distribution of $\mathcal{R}^N$. Let $B \subset \{1,2,\ldots,N\}$ be such that $1 \in B$, $\#B = r$. Let $\hat{\mathcal{R}}$ denote the partition of $\mathcal{R}^N$ with the set containing 1 removed. The condition we want is

$$(11.12) \qquad P\big(a(\hat{\mathcal{R}}) = (a_1,\ldots,a_{r-1},a_r-1,a_{r+1},\ldots) \,|\, B \in \mathcal{R}^N\big)$$
$$= Q_{N-r}(a_1,\ldots,a_{r-1},a_r-1,a_{r+1},\ldots) .$$

To rephrase this condition, observe that the left side equals

$$P\big(a(\mathcal{R}^N) = a \,|\, B \in \mathcal{R}^N\big), \quad \text{for } a = (a_1,\ldots,a_r,\ldots)$$
$$= P\big(a(\mathcal{R}^N) = a\big) \cdot P\big(B \in \mathcal{R}^N \,|\, a(\mathcal{R}^N) = a\big) \cdot 1/P(B \in \mathcal{R}^N)$$
$$= Q_N(a) \cdot a_r \Big/ \binom{N}{r} \cdot 1/P\big(\{1,2,\ldots,r\} \in \mathcal{R}^N\big) .$$

Thus condition (11.12) implies

$$\frac{a_r Q_N(a_1,\ldots,a_r,\ldots)}{Q_{N-r}(a_1,\ldots,a_r-1,\ldots)} \text{ depends only on } (N,r) .$$

This is the basis for the proof of (11.16) below.

Now consider, for some exchangeable $\mathcal{R}$, the chance that 1 and 2 belong to the same set in the partition:

$$(11.13) \qquad P\big(\mathcal{R}^2 = \{1,2\}\big) = Q_2(0,1) = 1/(1+\theta), \text{ say, for some } 0 \le \theta \le \infty .$$

There are two extreme cases: if $\theta = 0$ then $\mathcal{R}$ is a.s. the trivial partition $\{1,2,\ldots\}$; if $\theta = \infty$ then $\mathcal{R}$ is a.s. the discrete partition $(\{1\},\{2\},\{3\},\ldots)$. The interesting case is $0 < \theta < \infty$. It is a remarkable fact that if (11.12) holds then θ determines the distribution of $\mathcal{R}$, and some explicit formulas can be obtained. We quote the following result (see Kingman (1980), Sections 3.5-3.7).

(11.14) Theorem. *Let* R *be an exchangeable random partition on* $\{1,2,\ldots\}$ *satisfying* (11.12). *Define* θ *by* (11.13), *and suppose* $0 < \theta < \infty$. *Let* $D = (D_1, D_2, \ldots)$ *be as in Proposition* 11.9. *Then*

(11.15) $(D_1, D_2, \ldots)$ *has the Poisson-Dirichlet*(θ) *distribution,*

$$(11.16) \qquad Q_N(a_1,\ldots,a_N) = \frac{N!}{\theta(\theta+1)\cdots(\theta+N-1)} \prod_{r=1}^{N} \frac{\theta^{a_r}}{r^{a_r} a_r!}$$

Equation (11.16) arose in genetics as in the *Ewens sampling formula*--see section 19.

As an example of the calculations which are possible, let R be an exchangeable random partition and let C_1 be the component of R containing 1. By Proposition 11.9 and (11.8b),

$$(11.17a) \qquad N^{-1}\#(C_1 \cap \{1,2,\ldots,N\}) \xrightarrow{\mathcal{D}} T = D_J 1_{(J \geq 1)} ,$$

where $P(J = j \mid D) = D_j$. The limit distribution T can be described by

$$P(T \in (t, t+dt)) = t \sum_{j \geq 1} P(D_j \in (t, t+dt)) .$$

So under the hypotheses of Theorem 11.14 we can apply (10.24) to obtain the density

$$(11.17b) \qquad P(T \in (t, t+dt)) = \theta(1-t)^{\theta-1} dt , \quad 0 < t < 1 .$$

Equation (11.16) readily yields various special cases:

$$(11.18a) \qquad P(1,2,\ldots,N \text{ in same component}) = \prod_{i=2}^{N} \frac{i-1}{\theta+i-1}$$

$$(11.18b) \qquad P(1,2,\ldots,N \text{ in distinct components}) = \prod_{i=1}^{N} \frac{\theta}{\theta+i-1}$$

Finally, we remark that there is a "sequential" description of the partitions $\mathcal{R}^N$ of Theorem 11.14; following Jim Pitman, we call this

(11.19) The Chinese restaurant process. Imagine people $1,2,\ldots,N$ arriving sequentially at an initially empty restaurant with a large number of large tables. Person j either sits at the same table as person i (with probability $1/(j-1+\theta)$, for each $i < j$), or else sits at an empty table (with probability $\theta/(j-1+\theta)$). Then the partition "people at each table" has the distribution of Theorem 11.14.

Although Theorem 11.14 was motivated by a problem in genetics, it has some purely mathematical applications. The first was been noted by Kingman and others.

Cycle length in random permutations. Let π_N^* be the uniform random permutation on $\{1,2,\ldots,N\}$, and let $\mathcal{R}^N$ be the partition into cycles of π_N^*. We claim that the exchangeable random partitions $\mathcal{R}^N$, $N \geq 1$ are consistent in the sense of (11.5). To see this, for a permutation σ of $\{1,\ldots,N+1\}$ define a permutation $\hat{\sigma} = g(\sigma)$ of $\{1,\ldots,N\}$ by deleting $N+1$ from the cycle representation of σ:

$$\begin{aligned}\hat{\sigma}(i) &= \sigma(i) &&\text{if } \sigma(i) \neq N+1\\ &= \sigma(N+1) &&\text{if } \sigma(i) = N+1 .\end{aligned}$$

Then

$$g(\pi_{N+1}^*) \stackrel{\mathcal{D}}{=} \pi_N^* ;$$

a pair $i,j \leq N$ are in the same cycle of $g(\pi_{N+1}^*(\omega))$ iff they are in the same cycle of $\pi_{N+1}^*(\omega)$; and this implies consistency.

Next, consider $1 \in B \subset \{1,2,\ldots,N\}$. Then conditional on B being a cycle of π_N^*, the restriction of π_N^* to $\{1,\ldots,N\}\backslash B$ is distributed uniformly on the set of all permutations of $\{1,\ldots,N\}\backslash B$. This establishes the consistency property (11.12).

So Theorem 11.14 is applicable. To evaluate θ, let $X_0 = 1$, $X_i = \pi_N^*(X_{i-1})$. Then $P(X_i = 1 | X_1,\ldots,X_{i-1} \notin \{1,2\}) = P(X_i = 2 | X_1,\ldots,X_{i-1} \notin \{1,2\}) = (N-i+1)^{-1}$, and so $P(1 \text{ and } 2 \text{ are in same cycle of } \pi_N^*) = \frac{1}{2}$. Thus $\theta = 1$.

From (11.16) we obtain a classical theorem of Cauchy.

(11.20) Corollary. Let $a_r \geq 0$, $\sum r a_r = N$. The number of permutations of $\{1,\ldots,N\}$ with exactly a_r cycles of length r (each $r \geq 1$) is

$$N! \prod_{r \geq 1} \frac{1}{r^{a_r} a_r!}.$$

And if $(M_{N,1}, M_{N,2},\ldots)$ are the lengths of the cycles of a uniform random permutation of $\{1,\ldots,N\}$, arranged in decreasing order, then (11.15) says that $N^{-1}(M_{N,1}, M_{N,2},\ldots)$ converges in distribution to the Poisson-Dirichlet(1) process.

Remarks. There is a large literature on random permutations; we mention only a few recent papers related to the discussion above. Vershik and Schmidt (1977) give an interesting "process" description of the Poisson-Dirichlet(1) limit of $N^{-1}(M_{N,1}, M_{N,2},\ldots)$. Ignatov (1982) extends their ideas to general θ. Kerov and Vershik (1982) use the ideas of Theorem 11.14 in the analysis of some deeper structure of random permutations (e.g. Young's tableaux).

Components of random functions. A function $f: \{1,\ldots,N\} \to \{1,\ldots,N\}$ defines a directed graph with edges $i \to f(i)$ for each i. Thus f induces a partition of $\{1,\ldots,N\}$ into the components of this graph. The partition can be described by saying that i and j are in the same component iff $f^{k*}(i) = f^{m*}(j)$ for some $k,m \geq 1$; where $f^{k*}(i)$ denotes the k-fold iteration $f(f(\ldots f(i)\ldots))$.

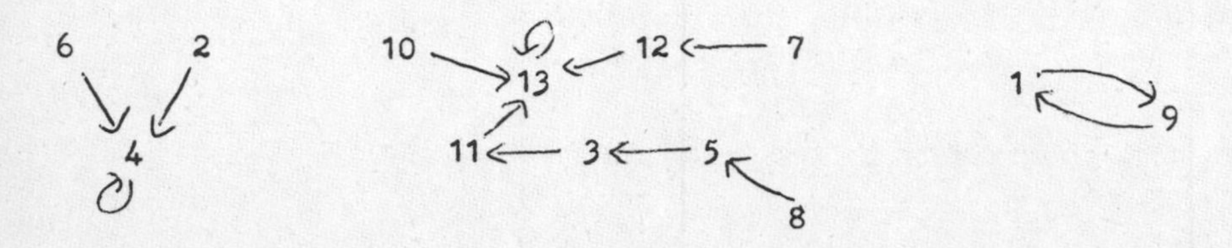

If we now let F_N be a random function, uniform over the set of all N^N possible functions, then we get a random partition $\mathcal{R}^N$ of $\{1,\ldots,N\}$ into the components of F_N. Clearly $\mathcal{R}^N$ is exchangeable. We shall outline how Theorem 11.14 can be used to get information about the asymptotic (as $N \to \infty$) sizes of the components.

Remark. Many different questions can be asked about iterating random functions. For results and references see e.g. Knuth (1981), pp. 8, 518-520; Pavlov (1982); Pittel (1983). I do not know references to the results outlined below, though I presume they are known.

Given that a specified set $B \subset \{1,\ldots,N\}$ is a component of F_N, it is clear that F_N restricted to $\{1,\ldots,N\}\backslash B$ is uniform on the set of functions from $\{1,\ldots,N\}\backslash B$ into itself. Thus

$$\mathcal{R}^N, \quad N \geq 1, \quad \text{satisfies the consistency condition (11.12).}$$

However, the consistency condition (11.5) is not satisfied exactly, but rather holds in an asymptotic sense. To make this precise, for $K < N$ let

$\mathcal{R}^{N,K}$ be the restriction of $\mathcal{R}^N$ to $\{1,\ldots,K\}$. Then

(11.21) <u>Lemma</u>. $\mathcal{R}^{N,K} \xrightarrow{\mathcal{D}} \hat{\mathcal{R}}^K$, <u>say</u>, <u>as</u> $N \to \infty$.

For each N the family $(\mathcal{R}^{N,K};\ K < N)$ is consistent in the sense of (11.5), and so Lemma 11.21 implies that $(\hat{\mathcal{R}}^K;\ K \geq 1)$ is consistent in that sense. It is then not hard to show

(11.22) <u>Lemma</u>. $(\hat{\mathcal{R}}^K;\ K \geq 1)$ <u>is consistent in sense</u> (11.12).

Then Theorem 11.14 is applicable to $(\hat{\mathcal{R}}^K)$. To identify θ, we need

(11.23) <u>Lemma</u>. $P\big(1$ <u>and</u> 2 <u>in same component of</u> $\mathcal{R}^N\big) \to 2/3$ <u>as</u> $N \to \infty$.

Then Lemma 11.21 implies $P\big(\hat{\mathcal{R}}^2 = \{1,2\}\big) = 2/3$, and so (11.13) identifies

$$\theta = \frac{1}{2}\,.$$

Now Theorem 11.14 gives information about $(\hat{\mathcal{R}}^K)$. For instance, writing $\hat{C}_1^K$ for the component of $\hat{\mathcal{R}}^K$ containing 1, (11.17) says

(11.24) $K^{-1}\,\#\hat{C}_1^K \xrightarrow{\mathcal{D}} T$, where T has density $f(t) = \frac{1}{2}(1-t)^{-1/2}$, $0 < t < 1$.

We are really interested in the corresponding assertion

(11.25) $$N^{-1}\,\#C_1^N \xrightarrow{\mathcal{D}} T$$

where C_1^N is the component of the random function F_N containing 1. Let us indicate how to pass from (11.24) to (11.25). Lemma 11.21 shows

(a) $\|\mathcal{L}(\#\hat{C}_1^K) - \mathcal{L}(\#(C_1^N \cap \{1,\ldots,K\}))\|_0 \to 0$ as $N \to \infty$; K fixed,

where $\|\ \|_0$ is total variation distance (5.5). By exchangeability, $\#(C_1^N \cap \{2,\ldots,K\})$ is distributed as the sum of $K-1$ draws without replacement from an urn with $\#C_1^N - 1$ "1"s and $N - \#C_1^N$ "0"s. Let $V_{N,K}$ be the corresponding sum with replacement. As $N \to \infty$, sampling with or without replacement become equivalent, so

(b) $\| L(\#(C_1^N \cap \{2,\dots,K\})) - L(V_{N,K}) \|_0 \to 0$ as $N \to \infty$; K fixed.

Now $V_{N,K}$ is conditionally Binomial$(K-1,(N-1)^{-1}(\#C_1^N - 1))$ given $\#C_1^N$, so by the weak law of large numbers

(c) $\lim_{K\to\infty} \limsup_{N\to\infty} E|K^{-1}V_{N,K} - N^{-1}\#C_1^N| = 0$.

Properties (a)-(c) lead from (11.24) to (11.25).

The same argument establishes the result corresponding to (11.15); if $(M_1^N, M_2^N, \dots)$ are the sizes of the components of F_N arranged in decreasing order, then $N^{-1}(M_1^N, M_2^N, \dots)$ converges in distribution to the Poisson-Dirichlet$(\frac{1}{2})$ process.

Let us outline the proofs of Lemmas 11.21 and 11.23. Let $X_0 = 1$, $X_n = F_N(X_{n-1})$, $S_1 = \min\{n: X_n(\omega) \in \{X_0(\omega),\dots,X_{n-1}(\omega)\}\}$. Then (X_n) is i.i.d. uniform until time S_1, and we get the simple formula

$$P(S_1 \geq n) = \prod_{m=0}^{n-1} (1 - m/N) .$$

Of course this is just "the birthday problem." Calculus gives

$$N^{-1/2} S_1 \xrightarrow{\mathcal{D}} \hat{S}_1, \text{ where } \hat{S}_1 \text{ has density } f(s) = s \cdot \exp(-s^2/2) .$$

Now let $Y_0 = 2$, $Y_n = F_N(Y_{n-1})$, $S_2 = \min\{n: Y_n(\omega) \in \{X_0(\omega),\dots,X_{S_1-1}(\omega), Y_0(\omega),\dots,Y_{n-1}(\omega)\}\}$, and let A_N be the event $\{Y_{S_2} \in \{X_0,\dots,X_{S_1-1}\}\} =$ $\{1$ and 2 in same component of $F_N\}$. Again (Y_n) is i.i.d. uniform until time S_2, there is a simple formula for $P(A_N, S_2 = n | S_1 = q)$, and calculus gives $(A_N, S_1, S_2) \xrightarrow{\mathcal{D}} (\hat{A}, \hat{S}_1, \hat{S}_2)$, where the limit has density

$$P(\hat{A},\ \hat{S}_2 \in (s_2, s_2+ds_2),\ \hat{S}_1 \in (s_1, s_1+ds_1)) = s_1^2 \cdot \exp(-\tfrac{1}{2}(s_1+s_2)^2)ds_1 ds_2 \ .$$

Integrating this density gives $P(\hat{A}) = 2/3$, and this is Lemma 11.23.

Call the process $(X_n,\ n < S_1;\ Y_n,\ n < S_2)$ above the <u>N-process</u>. Let B_N be the event "some element X_n or Y_n of the N-process equals N, and $F_N(N) = N$". Construct a process $(X_n^*,\ n < S_1^*;\ Y_n^*,\ n < S_2^*)$ by deleting any terms of the N-process which equal N. Then conditional on B_N^c the process (X_n^*, Y_n^*) is distributed as the (N-1)-process. So

$$|P(1 \text{ and } 2 \text{ in same component of } F_N) - P(1 \text{ and } 2 \text{ in same component of } F_{N-1})| \\ \leq P(B_N)\ .$$

But $P(B_N) \leq c_2 N^{-3/2}$, since S_1 and S_2 are of order $N^{1/2}$ and $P(F_N(N) = N) = N^{-1}$. A similar argument considering iterates of F_N starting from $1,2,\ldots,K$ shows

$$|P(\mathcal{R}^{N,K} = A) - P(\mathcal{R}^{N-1,K} = A)| \leq c_K N^{-3/2} \ ; \quad A \in \mathcal{S}_K \ .$$

Since $\sum N^{-3/2} < \infty$ this gives Lemma 11.21.

(11.26) <u>Random graphs</u>. Another way to construct random graphs on N vertices is to have each edge present with probability λ/N, independent for different possible edges. In this case the component containing "1", C_1^N, satisfies

$$N^{-1}\ \#C_1^N \xrightarrow{\mathcal{D}} T \quad \text{as} \quad N \to \infty\ ,$$

where $T = 0$ for $\lambda \leq 1$; $P(T = c(\lambda)) = c(\lambda)$, $P(T = 0) = 1 - c(\lambda)$ for $\lambda > 1$, where $c(\lambda) > 0$ for $\lambda > 1$. Thus (c.f. 11.17) the partition into components of these random graphs cannot be fitted into the framework of Theorem 11.14. It would be interesting to know which classes of random graphs had components following the Poisson-Dirichlet distribution predicted by Theorem 11.14.

PART III

The class of exchangeable sequences can be viewed as the class of distributions invariant under certain transformations. So a natural generalization of exchangeability is to consider distributions invariant under families of transformations. Section 12 describes some abstract theory; Sections 13-16 some particular cases.

12. Abstract results

In the general setting of distributions invariant under a group of transformations, there is one classical result: that each invariant measure is a mixture of ergodic invariant measures. We shall discuss this result (Theorem 12.10) without giving detailed proof; we then specialize to the "partial exchangeability" setting, and pose some hard general problems.

Until further notice, we work in the following general setting:

(12.1) S is a Polish space; $\mathcal{T}$ is a countable group (under composition) of measurable maps $T: S \to S$.

Call a random element X of S <u>invariant</u> if

$$T(X) \stackrel{\mathcal{D}}{=} X \; ; \quad T \in \mathcal{T} . \tag{12.2}$$

Call a distribution μ on S <u>invariant</u> if it is the distribution of an invariant random element, i.e. if

$$\tilde{T}(\mu) = \mu \; ; \quad T \in \mathcal{T} \tag{12.3}$$

where $\tilde{T}$ is the induced map (7.3). Let $\mathcal{M}$ denote the set of invariant distributions, and suppose $\mathcal{M}$ is non-empty. Call a subset A of S <u>invariant</u> if

(12.4) $$T(A) = A\ ; \quad T \in \mathcal{T}\ .$$

The family of invariant subsets forms the <u>invariant σ-field</u> $\mathcal{J}$. Call an invariant distribution μ <u>ergodic</u> if

(12.5) $$\mu(A) = 0 \text{ or } 1; \quad \text{each } A \in \mathcal{J}\ .$$

We quote two straightforward results:

(12.6) If μ is invariant and if A is an invariant set with $\mu(A) > 0$ then the conditional distribution $\mu(\cdot|A)$ is invariant.

(12.7) If μ is ergodic, and if a subset B of S is almost invariant, in the sense that $T(B) = B$ μ-a.s., each $T \in \mathcal{T}$, then $\mu(B) = 0$ or 1.

The two obvious examples (to a probabilist!) are the classes of stationary and of exchangeable sequences. To obtain stationarity, take $S = \mathbb{R}^{\mathbb{Z}}$, $\mathcal{T} = (T_k;\ k\in\mathbb{Z})$, where T_k is the "shift by k" map taking (x_i) to (x_{i-k}). Then a sequence $X = (X_i:\ i\in\mathbb{Z})$ is stationary iff it is invariant under $\mathcal{T}$, and the definitions of "invariant σ-field", "ergodic" given above are just the definitions of these concepts for stationary sequences.

To obtain exchangeability, take $S = \mathbb{R}^{\infty}$, $\mathcal{T} = (T_\pi)$, where for a finite permutation π the map T_π takes (x_i) to $(x_{\pi(i)})$. Then a sequence $X = (X_i)$ is exchangeable iff it is invariant under $\mathcal{T}$. The invariant σ-field $\mathcal{J}$ here is the exchangeable σ-field of Section 3; the ergodic processes are those with trivial exchangeable σ-fields, which by Corollary 3.10 are precisely the i.i.d. sequences.

Returning to the abstract setting, the set M of invariant distributions is <u>convex</u>:

(12.8) $\mu_1, \mu_2 \in M$ implies $\mu = c\mu_1 + (1-c)\mu_2 \in M$; $0 < c < 1$.

So we can define an extreme point: μ is extreme if the only representation of μ as $c\mu_1 + (1-c)\mu_2$, $\mu_i \in M$, has $\mu_1 = \mu_2 = \mu$.

(12.9) Lemma. An invariant distribution is extreme in M iff it is ergodic.

Proof. Let μ be invariant but not ergodic. Then $0 < \mu(A) < 1$ for some invariant A. And $\mu = P(A)\mu(\cdot|A) + P(A^c)\mu(\cdot|A^c)$ represents μ as a linear combination of invariant distributions, by (12.6), so μ is not extreme.

Conversely, suppose μ is ergodic, and suppose $\mu = c\mu_1 + (1-c)\mu_2$ for invariant μ_i, $0 < c < 1$. Let f be the Radon-Nikodym derivative $d\mu_1/d\mu$. Then by invariance, $f = f \circ T$ μ-a.s., each $T \in T$. So sets of the form $\{f \geq a\}$, a constants, are almost invariant in the sense of (12.7), which implies f is μ-a.s. constant. This implies $\mu_1 = \mu$, and hence μ is extreme.

Write E for the set of ergodic (= extreme) distributions. For an invariant random element X, write J_X for the σ-field of sets $\{X \in A\}$, $A \in J$. We can now state the abstract result: the reader is referred to Dynkin (1978), Theorems 3.1 and 6.1 for a proof (in somewhat different notation) and further results.

(12.10) Theorem. (a) E is a measurable subset of $P(S)$.

(b) Let X be an invariant random element. Let α be a r.c.d. for X given J_X. Then $\alpha(\omega) \in E$ a.s.

(c) To each invariant distribution μ there corresponds a distribution Λ_μ on E such that $\mu(\cdot) = \int_E \nu(\cdot)\Lambda_\mu(d\nu)$.

(d) The distribution Λ_μ in (c) is unique.

(12.11) Remarks. (i) Assertions (b) and (c) are different ways of saying that an invariant distribution is a mixture of ergodic distributions, corresponding to the "strong" and "weak" notions of mixture discussed in Section 2.

(ii) Dynkin (1978) proves the theorem directly. Maitra (1977) gives another direct proof. Parts (a), (c), (d) may alternatively be deduced from general results on the representation of elements of convex sets as means of distributions on the extreme points. See Choquet (1969), Theorem 31.3, for a version of Theorem 12.10 under somewhat different hypotheses; see also Phelps (1966).

(iii) In the usual case where $\mathcal{T}$ consists of continuous maps T, the set $\mathcal{M}$ is closed (in the weak topology). But $\mathcal{E}$ need not be closed; for instance, in the case of stationary $\{0,1\}$-valued sequences $\mathcal{E}$ is a dense G_δ.

(iv) It is easy to see informally why (b) should be true. Fix X and α. Property (12.6) implies that α is also a r.c.d. for T(X) given $\mathcal{J}_X$, and so

$$\tilde{T}(\alpha(\omega)) = \alpha(\omega) \text{ a.s.; each } T \in \mathcal{T} .$$

Since $\mathcal{T}$ is countable, this implies

$$\alpha(\omega) \in \mathcal{M} \text{ a.s.}$$

Now for $A \in \mathcal{J}$ we have $\alpha(\omega,A) = P(X\in A|\mathcal{J}_X) = 1_{(X\in A)}$ a.s. Then for a sub-σ-field $\mathcal{A}$ of $\mathcal{J}$ generated by a countable sequence

$$\mathcal{A} = \sigma(A_1,A_2,A_3,\ldots) \subset \mathcal{J} \tag{12.11a}$$

we have

$$P(\omega: \alpha(\omega,A) = 0 \text{ or } 1 \text{ for each } A\in\mathcal{A}) = 1 .$$

Unfortunately we cannot conclude

$$P(\omega: \alpha(\omega,A) = 0 \text{ or } 1 \text{ for each } A \in \mathcal{J}) = 1$$

because in general the invariant σ-field $\mathcal{J}$ itself cannot be expressed in the form (12.11a); this technical difficulty forces proofs of Theorem 12.10 to take a less direct approach.

Proposition 3.8 generalizes to our abstract setting in the following way (implicit in the "sufficiency" discussion in Dynkin (1978)).

(12.12) Proposition. *Let* X, V *be random elements of* S, S' *such that* $(X,V) \stackrel{\mathcal{D}}{=} (T(X),V)$ *for each* $T \in \mathcal{T}$ *(so in particular* X *is invariant). Then* X *and* V *are conditionally independent given* $\mathcal{J}_X$.

Proof. Consider first the special case where X is ergodic. If $0 < P(V \in B) < 1$ then X is a mixture of the conditional distributions $P(X \in \cdot | V \in B)$ and $P(X \in \cdot | V \in B^c)$ which are invariant by hypothesis; so by extremality $\mathcal{L}(X) = \mathcal{L}(X | V \in B)$, and so X and V are independent.

For the general case, let $\mathcal{M}^2$ be the class of distributions of (X^*,V^*) on $S \times S'$ such that $(X^*,V^*) \stackrel{\mathcal{D}}{=} (T(X^*),V^*)$ for each T. Let ψ be a r.c.d. for (X,V) given $\mathcal{J}_X$.

We assert

$$\psi(\omega) \in \mathcal{M}^2 \text{ a.s.} \tag{12.13}$$

Observe first that by the countability of $\mathcal{T}$ and (7.11), there exists a countable subset (h_i) of $C(S \times S')$ such that

$$\theta \in \mathcal{M}^2 \quad \text{iff} \quad \int h_i \, d\theta = 0, \quad i \geq 1 . \tag{12.14}$$

Next, for $A \in \mathcal{J}$ with $P(X \in A) > 0$ the hypothesis implies that the conditional distribution of (X,V) given $\{X \in A\}$ is in $\mathcal{M}^2$. Thus if ψ_n is a r.c.d. for (V,X) given a finite sub-σ-field F_n of J_X, then $\psi_n(\omega) \in M^2$ a.s. Because J_X is a sub-σ-field of the separable σ-field $\sigma(X)$, it is essentially separable, that is $J_X = F_\infty$ a.s. for some separable F_∞. Taking finite σ-fields $F_n \uparrow F_\infty$, Lemma 7.14(b) says $\psi_n \to \psi$ a.s. Since $\psi_n \in M^2$ a.s., (12.14) establishes (12.13).

For $\theta \in P(S \times S')$ let $\theta_1 \in P(S)$ be the marginal distribution. Then $\psi_1(\omega)$ is a r.c.d. for X given J_X, and so by Theorem 12.10 $\psi_1(\omega) \in E$ a.s. But now for each ω we can use the result in the special case to show that $\psi(\omega)$ is a product measure; this establishes the Proposition.

(12.15) Remarks. (i) In the case where T is a compact group (in particular, when T is finite), let T^* be a random element of T with Haar distribution (i.e. uniform, when T is finite). Then for fixed $s \in S$ the random element $T^*(s)$ is invariant; and it is not hard to show that the set of distributions of $T^*(s)$, as s varies, is the set of ergodic distributions. This is an abstract version of Lemma 5.4 for finite exchangeable sequences.

(ii) In the setting (12.1), call a distribution μ quasi-invariant if for each $T \in T$ the distributions μ and $\tilde{T}(\mu)$ are mutually absolutely continuous. Much of the general theory extends to this setting: any quasi-invariant distribution is a mixture of ergodic quasi-invariant distributions. See e.g. Blum (1982); Brown and Dooley (1983).

So far we have been working under assumptions (12.1). We now specialize. Suppose

(12.16) I is a countable set, Γ is a countable group (under convolution) of maps $\gamma: I \to I$, and S is a Polish space.

By a process X we mean a family $(X_i: i \in I)$ with X_i taking values in S.

The process is <u>invariant</u> if

(12.17) $$X \overset{\mathcal{D}}{=} (X_{\gamma(i)}: i \in I), \quad \text{each } \gamma \in \Gamma.$$

To see that this is a particular case of (12.1), take $S^* = S^I$, and take T to be the family $(\gamma^*: \gamma \in \Gamma)$ of maps $S^* \to S^*$, where γ^* maps (x_i) to $(x_{\gamma(i)})$. Note that γ^* is continuous (in the product topology on S^I), and so the class of invariant processes is closed under weak convergence.

Obviously any exchangeable process X is invariant. We use the phrase <u>partially exchangeable</u> rather loosely to indicate the invariant processes when (I,Γ) is given. Our main interest is in

(12.18) <u>The Characterization Problem</u>. Given (I,Γ), can we say explicitly what the partially exchangeable processes are?

<u>Remarks</u>. (a) To the reader who answers "the ones satisfying (12.17)" we offer the following analogy: the definition of a "simple" finite group tells us when a particular group is simple; the problem of saying explicitly what are the finite simple groups is harder!

(b) In view of Borel-isomorphism (Section 7), the nature of S is essentially unimportant for the characterization problem: one can assume $S = [0,1]$.

Theorem 12.10 gives some information: any partially exchangeable process is a mixture of ergodic partially exchangeable processes. This is all that is presently known in general. But there seem possibilities for better results, as we now describe.

Suppose we are given a collection M_0 of invariant processes. How can we construct from these more invariant processes? In the general setting (12.1), the only way apparent is to take mixtures. Thus it is natural to ask

(12.19) What is the minimal class M_0 from which we can obtain all invariant distributions by taking mixtures?

Of course Theorem 12.10 tells us that the minimal class is E, the ergodic distributions. However, in the partial exchangeability setting (12.17) there are other ways to get new processes from old. Let X be an (ergodic) partially exchangeable process, and let $f: S \to S'$ be a function; then $Y_i = f(X_i)$ defines an (ergodic) partially exchangeable process Y. Let X' be another (ergodic) partially exchangeable process, independent of X; then $Z_i = (X_i, X_i')$ defines an (ergodic) partially exchangeable process Z. Combining these ideas, we get

> Let M_0 be a class of (ergodic) partially exchangeable processes. Let $\hat{M}_0$ be the class of processes of the form $Y_i = f(X_i^1, X_i^2, X_i^3, \ldots)$, where each process X^k is in M_0, the family (X^k) are independent, and f is any function. Then $\hat{M}_0$ consists of (ergodic) partially exchangeable processes.

In view of this observation, it is natural to pose the problem analogous to (12.19).

(12.20) Problem. Is there a minimal class M_0 such that $\hat{M}_0 = E$, the class of all ergodic processes? Can this minimal class be specified abstractly?

Nothing is known about this problem in general. Let us discuss Problems 12.18 and 12.20 in connection with the examples already mentioned; in the next sections we will discuss further examples.

For the class of exchangeable sequences, de Finetti's theorem answers Problem 12.18, and the ergodic processes are the i.i.d. sequences. Now recall (7.6) that for any distribution μ on a Polish space S there exists a function $f: [0,1] \to S$ such that $f(\xi)$ has distribution for ξ with $U(0,1)$ distribution. Thus for exchangeable sequences we can take the class $\mathcal{M}_0$ in (12.20) to consist of a single element, the i.i.d. $U(0,1)$ sequence.

For the simple generalizations of exchangeability described in Section 9, the results in Section 9 answer Problem 12.18. For each of those hypotheses it is not difficult to identify the ergodic processes and to describe a class $\mathcal{M}_0$ in Problem 12.20 consisting of a single element. Let us describe the hardest case, and leave the others to the interested reader.

<u>Hypotheses (9.1) and (9.3)</u>. As in Section 9 let $h(x,y) = (y,x)$, so that for a distribution μ on $S \times S$ the distribution $\tilde{h}(\mu)$ is the measure obtained by interchanging the coordinates. Let $Z_i = (X_i, Y_i)$ satisfy (A) and (C). Then Z is ergodic iff its directing random measure α is of the form

$$(12.21) \qquad P(\alpha = \mu) = P(\alpha = \tilde{h}(\mu)) = \tfrac{1}{2}; \quad \text{some} \quad \mu \in P(S \times S) .$$

The existence of a single-element class M_0 follows from the next lemma, since we can use the process satisfying (12.21) for the particular distribution μ_0 below.

(12.22) <u>Lemma</u>. <u>There exists a distribution</u> μ_0 <u>on</u> $[0,1] \times [0,1]$ <u>such that for any distribution</u> μ <u>on</u> $S \times S$ <u>there exists a function</u> $f: [0,1] \to S$ <u>such that</u> $(f(U_1), f(U_2))$ <u>has distribution</u> μ <u>when</u> (U_1, U_2) <u>has distribution</u> μ_0.

<u>Proof</u>. Let $\Omega = [0,1]$. By considering the set of random elements $\underset{\sim}{V} = (V_1, V_2): \Omega \to [0,1]^2$ as a subset of Hilbert space $L^2(\Omega; [0,1]^2)$, we see that there exists an L^2-dense sequence $\underset{\sim}{V}^1, \underset{\sim}{V}^2, \ldots$. Set

$V_j = (V_j^1, V_j^2, V_j^3, \ldots)$, $j = 1,2$. Now given μ, there exists a random element $\underset{\sim}{W}: \Omega \to [0,1]^2$ with distribution μ. By L^2-denseness, there exists a subsequence $\underset{\sim}{V}^{n_k} \to \underset{\sim}{W}$ in L^2, and by passing to a further subsequence we may take $\underset{\sim}{V}^{\tilde{n}_k} \to \underset{\sim}{W}$ a.s. Setting $g(x) = \limsup x_{n_k}$ gives $L(g(V_1), g(V_2)) = \mu$. Finally, let $\phi: [0,1]^\infty \to A \subset [0,1]$ be a Borel-isomorphism, and set $\mu_0 = L(\phi(V_1), \phi(V_2))$. Then $f = g \circ \phi^{-1}$ satisfies the assertion of the lemma.

For the class of stationary sequences, these types of problems are much harder. Informally, the class seems too big to allow any simple explicit description of the general stationary ergodic process. Let us just describe one result in this area. Let S be finite. Let $(\xi_i)_{i \in \mathbb{Z}}$ be i.i.d. $U(0,1)$. Let $f: [0,1]^{\mathbb{Z}} \to S$ and let

$$Y_k = f((\xi_{i+k})_{i \in \mathbb{Z}}); \quad \underset{\sim}{Y} = (Y_k)_{k \in \mathbb{Z}} \tag{12.23}$$

Then $\underset{\sim}{Y}$ is a stationary ergodic S-valued sequence. Ornstein showed that not all stationary ergodic processes can be expressed in this form: in fact, $\underset{\sim}{Y}$ can be represented in form (12.23) iff $\underset{\sim}{Y}$ satisfies a condition called "finitely determined." See Shields (1979) for an account of these results.

(12.24) Second-order structure. When a partially exchangeable process $X = (X_i : i \in I)$ is real-valued square-integrable, it has a correlation function $\rho(i,j) = \rho(X_i, X_j)$ which plainly satisfies

$$\begin{aligned} &\rho \text{ is non-negative definite;} \\ &\rho \text{ is invariant, that is } \rho(i,j) = \rho(\gamma(i), \gamma(j)), \quad \gamma \in \Gamma. \end{aligned} \tag{12.25}$$

Conversely, any ρ satisfying (12.25) is the correlation function of some (Gaussian) partially exchangeable process. Thus the study of the second-order structure of partially exchangeable processes reduces to the study of functions satisfying (12.25), and this can be regarded as a problem in abstract harmonic analysis. In the classical case of stationary sequences, ρ satisfies (12.25) iff $\rho(i,j) = \rho(j-i)$, where $\rho(n) = \int_0^\pi \cos(n\lambda)\mu(d\lambda)$ for some distribution μ on $[0,\pi]$. For exchangeable sequences, the possible correlations were described in Section 1. For the examples in the following sections, we shall obtain the correlation structure as a corollary of the (much deeper) probabilistic structure; and give references to alternative derivations using harmonic analysis. Letac (1981b) is a good introduction to the analytic methods.

(12.26) Topologies on $\mathcal{M}$. Finally, we mention some abstract topological questions. Consider the general setting (12.1), and suppose the maps T are continuous. The set $\mathcal{M}$ of invariant distributions inherits the weak topology from $\mathcal{P}(S)$, but there is another topology which can be defined on $\mathcal{M}$. Call a pair (X,Y) of S-valued random elements jointly invariant if $(X,Y) \stackrel{\mathcal{D}}{=} (T(X),T(Y))$, $T \in \mathcal{T}$. Let d be a bounded metric on S. Then

$$(12.27) \quad \bar{d}(\mu,\nu) = \inf\{Ed(X,Y): (X,Y) \text{ jointly invariant}, \mathcal{L}(X)=\mu, \mathcal{L}(Y)=\nu\}$$

defines a metric on $\mathcal{M}$ which is stronger than (or equivalent to) the weak topology. Now specialize to the partial exchangeability setting (12.16), and suppose there exists $i_0 \in I$ such that $\{\gamma(i_0): \gamma \in \Gamma\} = I$. Then processes (X,Y) are jointly invariant if $((X_i,Y_i): i \in I) \stackrel{\mathcal{D}}{=} ((X_{\gamma(i)},Y_{\gamma(i)}): i \in I)$ for $\gamma \in \Gamma$; and

$$(12.28) \quad \bar{d}(\mu,\nu) = \inf\{Ed(X_{i_0},Y_{i_0}): (X,Y) \text{ jointly invariant}, \mathcal{L}(X)=\mu, \mathcal{L}(Y)=\nu\}$$

defines a metric equivalent to that of (12.27). This $\bar{d}$-topology was introduced by Ornstein for stationary sequences; see Ornstein (1973) for a survey. A discussion of its possible uses in partial exchangeability is given in Aldous (1982a). Informally, the $\bar{d}$-topology seems more compatible with the characterization problem than is the weak topology (for instance, E is always $\bar{d}$-closed). A natural open problem is:

(12.29) Problem. Under what conditions are the $\bar{d}$-topology and the weak topology equivalent?

This is true for exchangeable sequences, and in the general setting when T is a compact group. It is not true for stationary sequences, or for the examples in Sections 13 and 14.

13. The infinitary tree

In this section we analyse a particular example of partial exchangeability. Though this example is somewhat artificial, we shall see some connections with naturally-arising problems.

A tree is a connected undirected graph without loops. Two vertices are neighbors if they are joined by an edge. For any two distinct vertices i, j there is a unique path of distinct vertices $i = i_1, i_2, \ldots, i_n = j$ such that i_r and i_{r+1} are neighbors, $1 \le r < n$. A simple example is the infinite binary tree, in which there are a countable infinite number of vertices, and each vertex has exactly 3 neighbors. Similarly there is the infinite k-ary tree, where each vertex has exactly $k+1$ neighbors. We shall consider the infinite ∞-ary tree (the infinitary tree) where each vertex has infinitely many neighbors: for our purposes this is simpler than the finite case (see Remark 13.11). Let T denote the set of vertices of this tree. Let D be the set of finite sequences $d = (d_1, d_2, \ldots, d_n)$ of positive integers, and include in D the empty sequence $\emptyset$. Pick a vertex of T arbitrarily, and label it $j_\emptyset$. Then put the neighbors of $j_\emptyset$ in arbitrary order and label them $j_1, j_2, j_3, \ldots$. Then for each n put the neighbors of j_n (other than $j_\emptyset$) in arbitrary order and label them $j_{n1}, j_{n2}, j_{n3}, \ldots$. Continuing in the obvious manner, we obtain a labelling scheme $\{j_d: d \in D\}$ for all T.

An automorphism of T is a bijection $\gamma: T \to T$ such that $\gamma(i)$ and $\gamma(j)$ are neighbors iff i and j are neighbors. The easiest way to think of automorphisms is via labelling schemes. Given two labelling schemes $\{j_d: d \in D\}$, $\{\hat{j}_d: d \in D\}$, the map $\gamma(j_d) = \hat{j}_d$ is an automorphism; conversely, every automorphism is of this form.

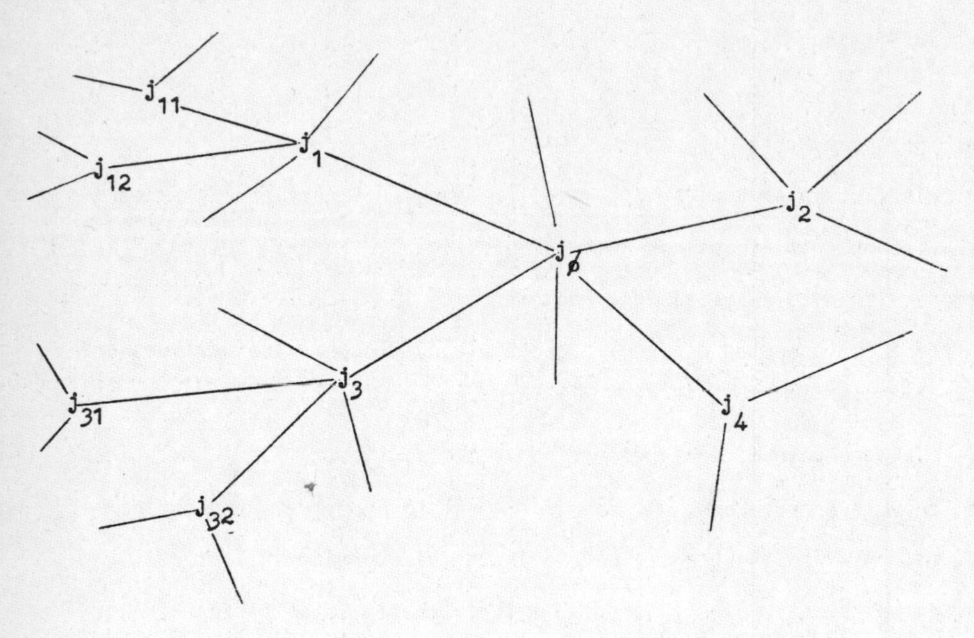

A tree-process X is a family $(X_i: i \in T)$ of random variables indexed by T and taking values in some space S. Given a tree-process and a labelling scheme, we can define a process indexed by D:

(13.1) $$X_d^* = X_{j_d} .$$

A tree-process X is <u>invariant</u> if

(13.2) $$X \stackrel{\mathcal{D}}{=} (X_{\gamma(i)}: i \in T), \text{ each automorphism } \gamma.$$

Equivalently, X is invariant iff the distribution of $X^* = (X_d^*: d \in D)$ defined at (13.1) does not depend on the particular labelling scheme $(j_d: d \in D)$ used. Informally, this says the distribution of any $(X_{i_1},\ldots,X_{i_n})$ depends only on the graph structure on $(i_1,\ldots,i_n)$. The set Γ of automorphisms is uncountable, but it is easy to see there exists a countable subset Γ_0 such that a process is Γ-invariant iff it is Γ_0-invariant. Thus invariant tree-processes do indeed fit into the "partial exchangeability" set-up of Section 12.

Our purpose is to describe explicitly the general invariant tree-process (i.e. to solve Problem 12.18 for this particular instance of partial exchangeability).

First we describe the well-known special case of Markov tree-processes. Informally, a tree-process Y is Markov if, conditional on the value Y_i at a vertex i, the values of the process along different branches from i are independent. To make this precise, for vertex i let N_i denote the set of vertices neighboring i. For $j \in N_i$ let the <u>branch</u> $B_{i,j}$ be the set of vertices k for which the path $i = i_1, i_2, \ldots, i_n = k$ of distinct neighboring vertices has $i_2 = j$. Let $Y(B_{i,j})$ be the array $(Y_k: k \in B_{i,j})$. Then Y is <u>Markov</u> if, for each i,

(13.3) $(Y(B_{i,j}): j \in N_i)$ are conditionally independent given Y_i.

<u>Warning</u>. There are two definitions of "Markov" for processes on trees. In the language of Spitzer (1975) we mean "Markov chain", not "Markov random

field"--see Remark 13.13.

If Y is an invariant tree-process then the distribution

$$\theta = L(Y_i, Y_j); \quad i, j \text{ neighbors} \tag{13.4}$$

is a symmetric distribution on $S \times S$ which does not depend on the particular pair i,j of neighbors. It is easy to verify (c.f. Spitzer (1975), Theorem 2) that for any symmetric θ there exists a unique (in distribution) invariant Markov tree-process satisfying (13.4). Call this the _invariant Markov tree-process associated with_ θ. Given such a process Y and a map $f\colon S \to S'$, the process $(f(Y_i)\colon i \in T)$ is a tree-process with range space S'. This process inherits the invariance property from Y, but in general will not inherit the Markov property. The next result says that every invariant process can be obtained this way.

(13.5) _Theorem. Let_ X _be an invariant tree-process with range space_ S. _Then there exists an invariant Markov tree-process_ Y _with range space_ $[0,1]$ _and a function_ $f\colon [0,1] \to S$ _such that_ $X \stackrel{\mathcal{D}}{=} (f(Y_i)\colon i \in T)$.

Proof. The idea of the proof is simple. The family of sub-processes on different branches from i is exchangeable; let Y_i be the directing random measure; then it turns out that (X_i, Y_i) is Markov.

For the formalities, we need some notation. For $n \geq 1$ and $d = (d_1, \ldots, d_m) \in D$ let $nd = (n, d_1, \ldots, d_m)$, and let $\emptyset d = d$. Fix a vertex i, and set up a labelling scheme $(j_d\colon d \in D)$ with $j_\emptyset = i$. For $n \geq 1$, let $C_n = (X_{nd}\colon d \in D)$ be the values of the process X on the branch B_{i,j_n}. So C_n takes values in S^D.

(13.6) <u>Lemma</u>. (a) $(C_1,C_2,C_3,\ldots)$ <u>is exchangeable over</u> X_i.

(b) <u>The distribution of</u> $(X_i;C_1,C_2,C_3,\ldots)$ <u>does not depend on</u> i <u>or on the labelling scheme</u>.

<u>Proof</u>. Assertion (b) is immediate from invariance. To prove (a), let π be a finite permutation of $\{1,2,3,\ldots\}$, and let $\gamma_\pi(i) = i$, $\gamma_\pi(j_{nd}) = j_{\pi(n)d}$. Then γ_π is an automorphism, and the invariance of X under γ_π implies $(X_i;C_1,C_2,\ldots) \stackrel{\mathcal{D}}{=} (X_i;C_{\pi(1)},C_{\pi(2)},\ldots)$.

Let Y_i be the directing random measure for (C_n). So Y_i takes values in $P(S^D)$. Keep i fixed. It is clear from (13.6)(b) that the <u>distribution</u> of Y_i does not depend on the labelling scheme used to define (C_n); what is crucial is that the actual <u>random variable</u> Y_i does not depend (a.s.) on the labelling scheme. To prove this, let Y_i, $\hat{Y}_i$ be derived from schemes (j_d), $(\hat{j}_d)$ with $j_\emptyset = \hat{j}_\emptyset = i$. Now the directing random measure for an infinite permutation $(Z_{\pi(i)})$ of an exchangeable sequence Z is a.s. equal to the directing measure for Z; so by re-ordering the neighbors of i we may suppose $j_n = \hat{j}_n$, $n \geq 1$. Now let A be a subset of $\{1,2,3,\ldots\}$, and let j^A be the labelling scheme $j^A_\emptyset = i$, $j^A_{nd} = j_{nd}$, $n \in A$, $j^A_{nd} = \hat{j}_{nd}$, $n \notin A$. By (13.6)(b) the distribution of $(X_{j^A_d}: d \in D)$ does not depend on A, and then Lemma 9.7 implies $Y_i = \hat{Y}_i$ a.s.

Now set $Y^*_i = (X_i,Y_i)$. We shall prove that the process $Y^* = (Y_i: i \in T)$ is invariant. Let γ be an automorphism. The invariance of X under γ implies

$$(X_i,\ C^i_n: n \geq 1,\ i \in T) \stackrel{\mathcal{D}}{=} (X_{\gamma(i)},\ \hat{C}^{\gamma(i)}_n: n \geq 1,\ i \in T)$$

where for each i the sequence $(\hat{C}^i_n: n \geq 1)$ describes the values of X on the branches $B_{i,j}$, $j \in N_i$, using some labelling scheme perhaps different from that used to define (C_n^i). Now the fact that Y_i does not depend on the labelling scheme shows

$$(X_i, Y_i : i \in T) \stackrel{\mathcal{D}}{=} (X_{\gamma(i)}, Y_{\gamma(i)} : i \in T) .$$

This gives invariance of Y^*.

Since X_i is a function of Y_i^*, and since the range space of Y_i^* is Borel-isomorphic to some subset of $[0,1]$, the proof of Theorem 13.5 will be complete when we show Y^* is Markov. Write $C_n^i = (X_{j_{nd}} : d \in D)$, for some labelling scheme (j_d) with $j_\emptyset = i$. So $\sigma(C_n^i) = \sigma(X_v : v \in B_{i,j_n}) = F_j^i$, say, where $j = j_n$.

(13.7) <u>Lemma</u>. <u>Let</u> i,j <u>be neighbors, and let</u> $k \in B_{i,j}$. <u>Then</u> $Y_k \in F_j^i$ a.s.

<u>Proof</u>. Let $i = i_1, i_2, \ldots, i_{r+1} = k$ be the path of distinct vertices linking i with k, so $i_2 = j$. Clearly for any neighbor v of k, $v \neq i_r$, we have

$$B_{k,v} \subset B_{i,j} . \tag{13.8}$$

Consider a labelling scheme $(\hat{j}_d)$ with $\hat{j}_\emptyset = k$, $\hat{j}_1 = i_r$. Now Y_k is the canonical random measure for the exchangeable family $(C_n^k : n \geq 1)$ defined using $(\hat{j}_d)$. So Y_k is some function of $(C_n^k : n \geq 2)$. But by (13.8) this array is contained in the array $(X_q : q \in B_{i,j})$, which generates the σ-field F_j^i.

To prove Y^* is Markov, fix i. In the notation of (13.3), we must prove $(X(B_{i,j}), Y(B_{i,j}))$, $j \in N_i$ are conditionally independent given (X_i, Y_i). Using Lemma 13.7, this reduces to proving

(13.9) $\quad (F_j^i : j \in N_i)$ are conditionally independent given (X_i, Y_i) .

But Lemma 13.6(a) and Proposition 3.8 show

$$\{X_i, C_1^i, C_2^i, \ldots\} \text{ are conditionally independent given } Y_i,$$

and this implies (13.9).

The remainder of this section is devoted to various remarks suggested by Theorem 13.5.

Theorem 13.5 answers the "characterization problem" (12.18). In view of the discussion in Section 12 it is natural to ask which are the ergodic processes. It is easy to see that an invariant Markov tree-process X is ergodic iff there do not exist sets A_1, A_2 such that for neighbors i, j

$$\{X_i \in A_1\} = \{X_j \in A_2\} \text{ a.s.}, \quad 0 < P(X_i \in A_1) < 1.$$

And a general invariant tree-process X is ergodic iff it has a representation $X_i = f(Y_i)$ for some ergodic Markov Y.

Problem 12.20 and the subsequent discussion suggest

(13.10) <u>Problem</u>. Does there exist an invariant Markov tree-process Y which is "universal," in that for any invariant X there exists f such that $X \stackrel{\mathcal{D}}{=} (f(Y_i))$?

In connection with this problem, we remark that there exists (c.f. 12.22) a symmetric distribution θ on $[0,1]^2$ which is universal, in that for $\theta = L(V_1, V_2)$ the distributions $L(f(V_1), f(V_2))$ as f varies give all symmetric distributions. But this is not sufficient to answer Problem 13.10.

(13.11) <u>The finite case</u>. The conclusion of Theorem 13.5 is false for T_k, the infinite k-ary tree. For it is easy to construct an invariant process X on T_k such that $\{X_i, X_j : j \in N_i\}$ is distributed as a random permutation of $\{1,\ldots,k+2\}$. Then $\{X_j: j \in N_i\}$ cannot be extended to an infinite exchangeable sequence, so the conclusion of Theorem 13.5 fails.

One way to construct invariant processes on T_k is to take $(\xi_i: i \in T_k)$ i.i.d. $U(0,1)$, and take $g: [0,1]^{k+2} \to S$ such that $g(x,y_1,\ldots,y_{k+1})$ is symmetric in (y_i) for each x. Then the process

(13.12) $$X_i = g(\xi_i,\xi_{j_1},\ldots,\xi_{j_{k+1}}); \quad j_r \in N_i$$

is invariant. More generally, we can take $X_i = g(\xi_{j_d} : d \in D)$ for a labelling scheme (j_d) with $j_\emptyset = i$ and for g satisfying the natural symmetry conditions. Presumably, as in the case (12.23) of ordinary stationary sequences, this construction gives all "finitely determined" invariant processes--but this looks hard to prove.

(13.13) Markov random fields. A tree-process X is a Markov random field if, for each vertex i,

$\sigma(X_j: j \neq i)$ and X_i are conditionally independent given $\sigma(X_j: j \in N_i)$.

This is weaker than our definition (13.3) of "Markov": see Kindermann and Snell (1980). It is natural to ask

(13.14) Problem. Is there a simple explicit description of the invariant Markov random fields on the infinitary tree T?

In the notation of Theorem 13.5, the Markov random field property for X is equivalent to some messy conditional independence properties for the triple $(Y_i,Y_j,f(Y_j))$, i,j neighbors, but it seems hard to say when these properties hold.

Spitzer (1975) (see also Zachary (1983)) discusses $\{0,1\}$-valued Markov random fields on the tree T_k. For such a field, let $V_i = \sum_{j \in N_i} X_j$. Then

(a) there is an explicit description of the possible conditional distribution of X_i given V_i in an invariant Markov random field;

(b) any such conditional distribution is attained by some Markov tree-process.

For the tree T we can set $V_i = \lim_n n^{-1} \sum_1^n X_{j_r}$, where $N_i = \{j_1, j_2, \ldots\}$, and ask the same questions. It can be shown that (b) is not true on T; the analog of (a) is unclear.

(13.15) Stationary reversible processes. A stationary process $X = (X_n : n \in \mathbb{Z})$ is called reversible if $X \stackrel{\mathcal{D}}{=} (X_{-n} : n \in Z)$. So a stationary Markov process Y is reversible iff the distribution $L(Y_0, Y_1)$ is symmetric. Reversible Markov processes occur often in applied probability models--see e.g. Kelly (1979)--and since we are frequently interested in functions of the underlying process, the following theoretical definition and problem are suggested.

(13.16) Definition. Given a space S, let H_0 be the class of S-valued processes of the form $(f(Y_n) : n \in \mathbb{Z})$, where Y is a stationary reversible Markov process on some space S' ($= [0,1]$, without loss of generality) and $f : S' \to S$ is some function.

(13.17) Problem. Give intrinsic conditions for a stationary reversible process X to belong to H_0.

We shall see later (13.24) that not every stationary reversible sequence is in H_0.

To see the connection between Problem 13.17 and tree-processes, consider

(13.18) Definitions. A line in the infinitary tree T is a sequence $(i_n : n \in \mathbb{Z})$ of distinct vertices with i_n, i_{n+1} neighbors for each n.

Given a space S, let H_1 be the class of processes $(X_{i_n} : n \in \mathbb{Z})$, where X is some S-valued invariant tree-process, and (i_n) is a line in T.

Now Theorem 13.5 has the immediate

(13.19) Corollary. $H_1 = H_0$.

Though this hardly solves Problem 13.17, it does yield some information. For instance, H_1 is easily seen to be closed under weak convergence, so

(13.20) Corollary. H_0 *is closed under weak convergence*.

This fact is not apparent from the definition of H_0.

(13.21) Correlation structure. On the tree T there is the natural distance $d(i,j)$, the number of edges on the minimal path from i to j. For an invariant tree-process X, where X_i is real-valued and square-integrable, the correlations must be of the form $\rho(X_i, X_j) = \rho_{d(i,j)}$ for some *correlation function* $(\rho_n : n \geq 0)$.

(13.22) Proposition. *A sequence* $(\rho_n : n \geq 0)$ *is the correlation function of some invariant tree-process iff* $\rho_n = \int x^n \lambda(dx)$ *for some probability measure* λ *on* $[-1,1]$.

Remark. This result can be deduced from harmonic analysis results for the trees T_k. Cartier (1973) shows that correlation functions on T_k are mixtures of functions of the form

$$\rho(n) = \frac{k(\lambda^{n+1} - \lambda^{-n-1}) - (\lambda^{n-1} - \lambda^{-n+1})}{k^{n/2}(k+1)(\lambda - \lambda^{-1})}$$

where λ is either real with $|\lambda| \leq k^{1/2}$, or complex with $|\lambda| = 1$. And ρ is a correlation function on T iff it is a correlation function on each T_k. See also Arnaud (1980).

Instead, we shall see how to deduce Proposition 13.22 from Theorem 13.5.

Proof. Let C denote the set of sequences of the form $c_n = \int x^n \lambda(dx)$, $\lambda(\cdot)$ some probability distribution on $[-1,1]$.

Let Y be the Markov tree-process associated with θ, where

$$\theta(0,0) = \theta(1,1) = \tfrac{1}{4}(1+\lambda); \quad \theta(0,1) = \theta(1,0) = \tfrac{1}{4}(1-\lambda)$$

for some $-1 \leq \lambda \leq 1$. Then Y has correlation function $\rho(n) = \lambda^n$. By taking mixtures over λ, we see that every sequence in C is indeed the correlation function of some invariant process.

By Theorem 13.5 the converse reduces to determining the correlation function for sequences $Y_n = f(X_n)$, where $X = (X_n: n \geq 0)$ is stationary reversible Markov. Consider first the case where X has a finite state space S. Let P be the matrix of transition probabilities, and let π be the vector $(\pi(s))$ of the stationary distribution. Then $\rho_n = c_n/c_0$, where

$$(13.23) \qquad c_n = \sum_s \sum_t f(s)\pi(s)P^n(s,t)f(t) .$$

Reversibility implies $\pi(s)P(s,t) = \pi(t)P(t,s)$, and so the matrix $Q(s,t) = \pi^{1/2}(s)P(s,t)\pi^{-1/2}(t)$ is symmetric. The spectral theorem for symmetric matrices says we can write $Q = U\Lambda U^T$, where U is orthogonal and Λ is diagonal with real eigenvalues (λ_i); since these are also the eigenvalues of P we have $|\lambda_i| \leq 1$. Substituting into (13.23), putting $v(s) = f(s)\pi^{1/2}(s)$,

$$c_n = vU\Lambda^n U^T v^T = \sum a_i^2 \lambda_i^n , \quad \text{say.}$$

So ρ_n is indeed in C.

The general case, where X has state space $[0,1]$, can presumably be obtained by appealing to a more sophisticated spectral theorem. Let us instead give a probabilistic argument. For $N \geq 1$ let $\phi_N(x) = i/2^N$ on $i/2^N \leq x < (i+1)/2^N$. Let $(X_0^N, X_1^N, X_2^N, \ldots)$ be the stationary reversible Markov chain for which $(X_0^N, X_1^N) \stackrel{\mathcal{D}}{=} (\phi_N(X_0), \phi_N(X_1))$. Obviously $(X_0^N, X_1^N) \xrightarrow{\mathcal{D}} (X_0, X_1)$, and it can be shown (we leave this to the reader as a hard exercise) that

(13.24) $\quad (X_0^N, X_1^N, \ldots, X_k^N) \xrightarrow{\mathcal{D}} (X_0, X_1, \ldots, X_k)$ as $N \to \infty$, k fixed.

Observe also that by weak compactness of $\mathcal{P}[-1,1]$,

(13.25) $\quad C$ is closed under pointwise convergence.

Now consider $Y_n = f(X_n)$, where f is bounded continuous, and let (ρ_n) be the correlation function for Y. Using the result for finite state spaces, (13.24) and (13.25), we deduce that (ρ_n) is in C. Measure theory extends this to general f.

(13.26) Remarks. Take X_0 and Y independent, X_0 uniform on $\{0,1,2,3\}$, Y uniform on $\{-1,1\}$, and define $X_n = X_0 + nY$ modulo 4. Then X is a stationary process which is reversible, and $\rho(X_0, X_2) = -1$, so by Proposition 13.22 X cannot be represented as a function of any stationary reversible Markov process.

We also remark that the analog of Problem 13.17 without reversibility (which stationary processes are functions of stationary Markov processes?) is uninteresting as it stands. For given a stationary process (X_n), let (Y_n) be the stationary Markov process given by $Y_n = (X_n, X_{n+1}, X_{n+2}, \ldots)$, and then $X_n = f(Y_n)$ for $f(x_0, x_1, x_2, \ldots) = x_0$. However, the problem of which finite state stationary processes are functions of finite state stationary Markov processes is non-trivial--see Heller (1965).

(13.27) A curious argument. Readers unwilling to consider the possibility of picking uniformly at random from a countable infinite set should skip this section. For the others, we present a curious argument for de Finetti's theorem and for Corollary 13.19 (that a line in an invariant tree-process yields a function of a stationary reversible Markov process).

Argument for de Finetti's theorem. Let $Z = (Z_n: n \in \mathbb{N})$ be exchangeable, so

(a) $Z \overset{\mathcal{D}}{=} (Z_{i_1}, Z_{i_2}, \ldots)$, any distinct $(i_1, i_2, \ldots)$.

Take $I_1, I_2, \ldots$ independent of each other and Z, uniform on $\mathbb{N}$. Then

(b) the values $I_1, I_2, \ldots$ are distinct;

(c) for any function z defined on $\mathbb{N}$, the sequence $(z(I_1), z(I_2), \ldots)$ is i.i.d.

Now consider the sequence Z_{I_n}. By conditioning on $(I_1, I_2, \ldots)$ and using (a) and (b), we see

(d) $Z \overset{\mathcal{D}}{=} (Z_{I_n})$.

For any sequence $z = (z_n)$ of constants, the distribution of (Z_{I_n}) given $Z = z$ is i.i.d., by (c). So

(e) (Z_{I_n}) is a mixture of i.i.d. sequences.

And (d) and (e) give de Finetti's theorem.

Argument for Corollary 13.19. Let X be an invariant tree-process, let $(i_n: n \geq 0)$ be a line in T, so

(a) $(X_{i_n}) \overset{\mathcal{D}}{=} (X_{j_n})$ for any line (j_n).

Take I_0 uniform on T, I_1 uniform on the neighbors of I_0, I_2 uniform on the neighbors of I_1, and so on. Then

(b) the values $I_0, I_1, I_2, \ldots$ form a line in T;

(c) (I_n) is a stationary reversible Markov process on T.

Now consider the sequence (X_{I_n}). By conditioning on $(I_1, I_2, \ldots)$ and using (a) and (b),

(d) $(X_{i_n}) \overset{\mathcal{D}}{=} (X_{I_n})$.

Recall definition (13.16) of H_0. For any function x on T, the sequence (x_{I_n}) is in H_0, by (c). So by conditioning on X,

(e) (X_{I_n}) is a mixture of processes in H_0.

But it is not hard to see H_0 is closed under mixtures, so (d) and (e) give Corollary 13.19.

(13.28) Problem. Is it possible to formalize the arguments above (e.g. by using finite additivity)?

14. Partial exchangeability for arrays: the basic structure results

Several closely related concepts of partial exchangeability for arrays of random variables have been studied recently. In this section we describe the analog of de Finetti's theorem for arrays.

Consider an array $X = (X_{i,j}: i,j \geq 1)$ of S-valued random variables such that

(14.1) $$X \stackrel{\mathcal{D}}{=} (X_{\pi_1(i),\pi_2(j)})$$

for all finite permutations π_1, π_2 of $\mathbb{N}$. In terms of the general description (12.16) of partial exchangeability, $I = \mathbb{N}^2$ and Γ is the product of the finite permutation groups. Writing $R_i = (X_{i,j}: j \geq 1)$ for the ith row of the array X, and $C_j = (X_{i,j}: i \geq 1)$ for the jth column, condition (14.1) is equivalent to the conditions

(14.2)(a) $(R_1, R_2, R_3, \ldots)$ is exchangeable;

(b) $(C_1, C_2, C_3, \ldots)$ is exchangeable.

If these conditions hold, call X a row-and-column-exchangeable (RCE) array. Here are three obvious examples of such arrays.

(14.3)(a) $(X_{i,j})$ i.i.d.

(b) $X_{i,j} = \xi_i$, where $(\xi_1, \xi_2, \ldots)$ are i.i.d. Here the entries within a column are i.i.d., but different columns are identical; or equivalently, entries within a row are identical, but different rows are i.i.d.

(c) $X_{i,j} = \eta_j$, where $(\eta_1, \eta_2, \ldots)$ are i.i.d. Similar to (b), interchanging rows with columns.

Here is a more interesting class of examples. Let $\phi: [0,1]^2 \to [0,1]$ be an arbitrary measurable function. Let $(\xi_1, \xi_2, \ldots; \eta_1, \eta_2, \ldots)$ be i.i.d. uniform on $[0,1]$. Then we can define a $\{0,1\}$-valued array X by

(14.4) conditional on $(\xi_1, \xi_2, \ldots; \eta_1, \eta_2, \ldots)$, $P(X_{i,j} = 1) = \phi(\xi_i, \eta_j)$, $P(X_{i,j} = 0) = 1 - \phi(\xi_i, \eta_j)$, and the $X_{i,j}$ are independent.

Such processes, called <u>ϕ-processes</u>, can be simulated on a computer and the realizations drawn as a pattern of black and white squares. Diaconis and Freedman (1981a) present such simulations and use them to discuss hypotheses about human visual perception.

What is the analog of de Finetti's theorem for RCE arrays? There are several easy but rather superficial results which we describe below; the harder and deeper result is Theorem 14.11, and the reader may well skip to the statement of that theorem.

One method of analysing RCE arrays is to consider the rows $(R_1,R_2,\ldots)$ as an exchangeable sequence such that $(R_1,R_2,\ldots) \stackrel{\mathcal{D}}{=} (h_\pi(R_1),h_\pi(R_2),\ldots)$ for each π, where $h_\pi((x_i)) = (x_{\pi(i)})$. Then Lemma 9.6(i) says that the possible directing random measures α for (R_i) from a RCE array are precisely those with certain invariance properties--see Lynch (1982a). But this approach does not lead to any explicit construction of the general RCE array.

Recall that one version of de Finetti's theorem is: conditional on the tail σ-field of an exchangeable (Z_i), the variables $(Z_1,Z_2,\ldots)$ are i.i.d. So it is natural to study what happens to RCE arrays when we condition on some suitable "remote" σ-field. Define the <u>tail</u> σ-field $\mathcal{T}$ and the <u>shell</u> σ-field $\mathcal{S}$ as follows.

$$\mathcal{T} = \bigcap_n \mathcal{T}_n, \text{ where}$$
$$\mathcal{T}_n = \sigma(X_{i,j}: \min(i,j) > n).$$
$$\mathcal{S} = \bigcap_n \mathcal{S}_n, \text{ where}$$
$$\mathcal{S}_n = \sigma(X_{i,j}: \max(i,j) > n).$$

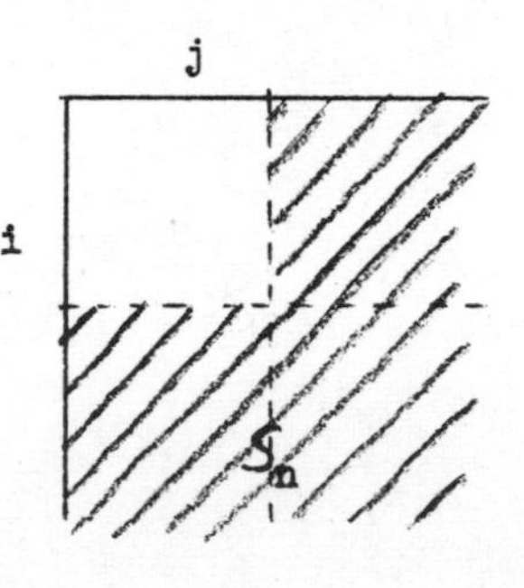

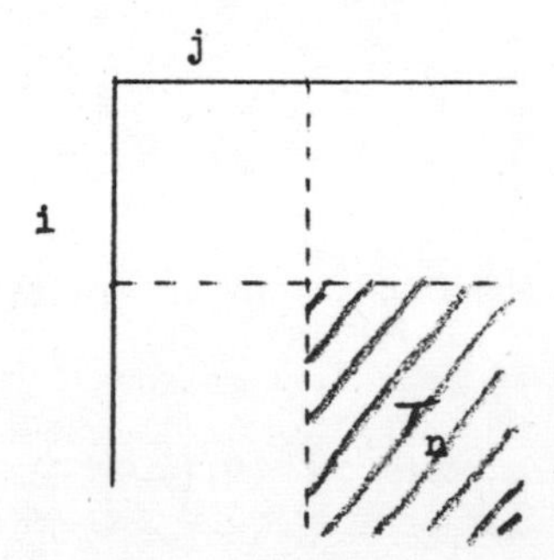

We need another definition. Call an array X <u>dissociated</u> if

(14.5) $(X_{i,j}: \max(i,j) \leq n)$ independent of $(X_{i,j}: \min(i,j) > n)$ for each n.

The next results describe what happens when we condition on these σ-fields.

(14.6) <u>Proposition. Let</u> X <u>be a RCE array. Conditional on</u> T, <u>the array</u> X <u>is RCE and dissociated.</u>

(14.7) <u>Proposition. Let</u> X <u>be a RCE array. Conditional on</u> S, <u>the variables</u> $X_{i,j}$ <u>are independent (but in general not identically distributed).</u>

<u>Proof of Proposition 14.6.</u> Let π_1, π_2 be permutations, and take n so large that π_1 and π_2 do not alter $n+1, n+2, \ldots$. Since $X \stackrel{\mathcal{D}}{=} (X_{\pi_1(i),\pi_2(j)})$, we have

conditional on T_n, the distributions of $(X_{i,j}: i,j \leq n)$ and $(X_{\pi_1(i),\pi_2(j)}: i,j \leq n)$ are identical.

Letting $n \to \infty$, we see that conditional on T, X is RCE.

Now fix M, and consider the diagonal squares $S_k = (X_{i,j}: (k-1)M+1 \leq i,j \leq kM)$. Given k and a permutation π of $\{1,\ldots,k\}$, there exists a permutation ρ of $\{1,\ldots,kM\}$ such that, setting $\hat{X}_{i,j} = X_{\rho(i),\rho(j)}$, we have $(\hat{S}_1,\ldots,\hat{S}_k) = (S_{\pi(1)},\ldots,S_{\pi(k)})$. It follows that $(S_1,\ldots,S_k)$ is exchangeable over T_n $(n > kM)$. Letting $n \to \infty$,

$$(S_1, S_2, \ldots) \text{ is exchangeable over } T.$$

Moreover the tail of (S_i) is contained in T. Using Proposition 3.8 we see that (S_k) is conditionally i.i.d. given T, so in particular

$(X_{i,j}: i,j \leq M)$ and $(X_{i,j}: M < i,j \leq 2M)$ are conditionally independent given T.

This holds for all M; it is easy to deduce X is conditionally dissociated.

Proof of Proposition 14.7. Let $Y_1, Y_2, \ldots$ be an arbitrary enumeration of $(X_{i,j}: i,j \geq 1)$. It is not hard to verify (see Aldous (1982b)) that (Y_i) satisfies the hypotheses of Proposition 6.4. Hence $Y_1, Y_2, \ldots$ are conditionally independent given their tail σ-field T_Y. But $T_Y = S$.

Finally, in example (14.3b) we have $S = \sigma(\xi_1, \xi_2, \ldots) = \sigma(X)$, so that conditional on S the array X is deterministic, and so the entries $X_{i,j}$ are not conditionally identically distributed (except when ξ is degenerate).

Although Propositions 14.6 and 14.7 give useful information, they do not provide a complete description of RCE arrays. Another approach is to use the general theory outlined in Section 12. The general theory says that each RCE array is a mixture of ergodic RCE arrays; the next result identifies the ergodic arrays as the dissociated arrays.

(14.8) Proposition. For a RCE array X the following are equivalent:

(a) X is extreme (= ergodic) in the class of RCE arrays.

(b) T is trivial.

(c) X is dissociated.

Proof. (a)$\Rightarrow$(b). If X is ergodic then the ergodic σ-field E is trivial. But $E \supset T$, just as in the 1-parameter case.

(b)$\Rightarrow$(c). Proposition 14.6.

(c)$\Rightarrow$(a). Suppose $L(X) = \frac{1}{2}L(Y) + \frac{1}{2}L(Z)$ for RCE arrays Y, Z. We must show $L(X) = L(Y)$. Fix M, and consider the diagonal squares

$S_k^X = (X_{i,j}: (k-1)M+1 \leq i,j \leq kM)$. Then $L(S_1^X, S_2^X, \ldots) = \frac{1}{2}L(S_1^Y, S_2^Y, \ldots)$ $+ \frac{1}{2}L(S_1^Z, S_2^Z, \ldots)$. Each of these sequences is exchangeable, and (S_k^X) is i.i.d. by hypothesis; but de Finetti's theorem implies that an i.i.d. sequence is extreme in the class of exchangeable sequences, and hence we must have $S_1^X = S_1^Y$. But this says $(X_{i,j}: i,j \leq M) \stackrel{\mathcal{D}}{=} (Y_{i,j}: i,j \leq M)$, and since M is arbitrary $X \stackrel{\mathcal{D}}{=} Y$.

However, the net effect of Proposition 14.8 and the general theory is to show that each RCE array is a mixture of dissociated RCE arrays--and this was already given by Proposition 14.6.

The results so far have been fairly direct consequences of the circle of ideas around de Finetti's theorem. We now come to the fundamental "characterization theorem," which seems somewhat deeper.

(14.9) <u>Convention</u>. Let $\{\alpha;\ \xi_i,\ i \geq 1;\ \eta_j,\ j \geq 1;\ \lambda_{i,j},\ i,j \geq 1\}$ denote independent $U(0,1)$ random variables.

Now given any $f: (0,1)^4 \rightarrow S$ we can define

(14.10)
$$X_{i,j}^* = f(\alpha, \xi_i, \eta_j, \lambda_{i,j})$$

and this yields an RCE array X^*. Say f <u>represents</u> X^*. This class of arrays is the class obtained from the examples in (14.3), together with the trivial array $X_{i,j} = \alpha$, by the general methods of Section 12. Note that a ϕ-process is represented by $f(a,b,c,d) = 1_{(d \leq \phi(b,c))}$.

(14.11) <u>Theorem. Let</u> X <u>be a RCE array. Then there exists</u> $f: [0,1]^4 \rightarrow S$ <u>such that</u> $X \stackrel{\mathcal{D}}{=} X^*$, <u>where</u> X^* <u>is represented by</u> f.

This result was obtained independently by Aldous (1979, 1981a) and Hoover (1979, 1982); the more general results of Hoover will be described later.

The proof which appears in Aldous (1981a) is a formalized version of an argument due to Kingman (personal communication), which we present essentially verbatim below.

By a <u>coding</u> ξ of a random variable Y we mean a presentation $Y \overset{\mathcal{D}}{=} g(\xi)$ for ξ with $U(0,1)$ distribution.

<u>Kingman's proof of Theorem 14.11.</u> There is no loss of generality in supposing that X is a quadrant of an array $(X_{i,j}: i,j \in \mathbb{Z})$. Write

$$A = (X_{i,j}: i,j \leq 0)$$

$$B_i = (X_{i,j}: j \leq 0) \qquad B = (B_i: i \geq 1)$$

$$C_j = (X_{i,j}: i \leq 0) \qquad C = (C_j: j \geq 1)\ .$$

de Finetti's theorem implies that for an exchangeable sequence $(Y_i: i \in \mathbb{Z})$ the variables $(Y_i: i \geq 1)$ are conditionally i.i.d. given $(Y_i: i \leq 0)$. Applying this to the three sequences

$$Y_i = (X_{i,j}: j \in \mathbb{Z}) = (B_i: X_{i,j},\ j \geq 1)$$

$$Y_i = (B_i, X_{i,j}) \quad (j \text{ fixed})$$

and $\qquad Y_i = B_i\ ,$

we see that

(i) the variables $(B_i, X_{i,1}, X_{i,2}, \ldots)$ for $i \geq 1$ are conditionally i.i.d. given (A,C);

(ii) for fixed $j \geq 1$, the variables $(B_i, X_{i,j})$ for $i \geq 1$ are conditionally i.i.d. given (A, C_j);

(iii) the variables B_i for $i \geq 1$ are conditionally i.i.d. given A. Moreover, the conditional distribution of B_i given (A,C) is expressible as a function of A, so

(iv) B_i is conditionally independent of C given A, for each i.

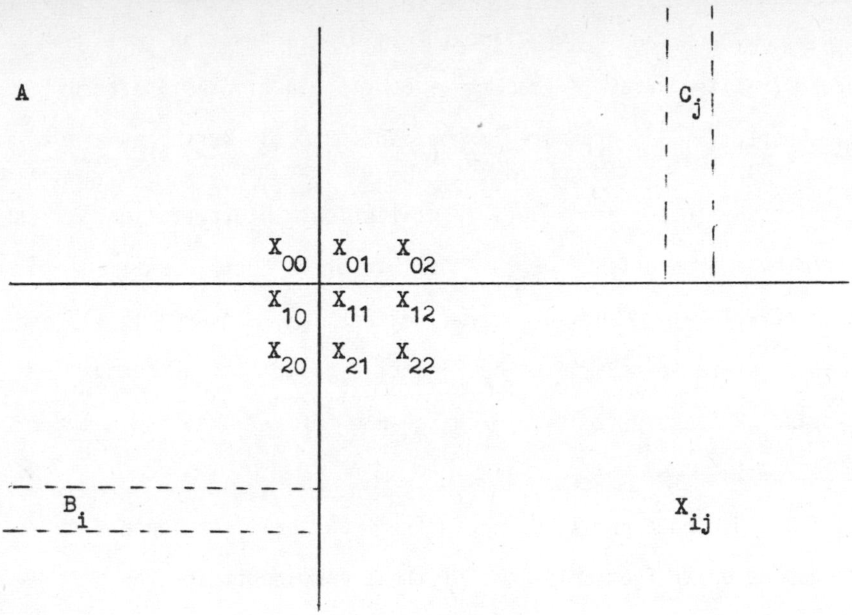

A $\quad C_j$

X_{00} X_{01} X_{02}

X_{10} X_{11} X_{12}

X_{20} X_{21} X_{22}

B_i $\quad X_{ij}$

Corresponding results hold when the roles of i and j are reversed, and standard conditional probability manipulations then show that

(v) $B_1, B_2, \ldots, C_1, C_2, \ldots$ are conditionally independent given A, the conditional distributions of B_i and C_j not varying with i and j respectively;

(vi) the variables $(X_{i,j}: i,j \geq 1)$ are conditionally independent given (A,B,C), the conditional distribution of $X_{i,j}$ depending only on (Λ, B_i, C_j).

Now take a coding α of A, and condition everything on α. Choose a coding ξ_1 for B_1, and let ξ_i be the corresponding coding for B_i; similarly choose codings η_j for C_j (all this conditional on α). Then $\alpha, \xi_1, \xi_2, \ldots, \eta_1, \eta_2, \ldots$ are independent $U(0,1)$, and there is a function $g: (0,1)^3 \rightarrow P(S)$ such that $g(\alpha, \xi_i, \xi_j)$ is the conditional distribution of

$X_{i,j}$ given (A,B,C). The usual construction then yields a function $f: (0,1)^4 \to S$ such that the array X has the same distribution as the array $f(\alpha,\xi_i,\eta_j,\lambda_{i,j})$, where $(\lambda_{i,j})$ are more independent $U(0,1)$ variables.

Remarks. Representation (14.10) has a natural statistical interpretation: $X^*_{i,j}$ is determined by a "row effect" ξ_i, a "column effect" η_j, an "individual random effect" λ_{ij} and an "overall effect" α. Theorem 14.11 may be regarded as a natural extension of de Finetti's theorem if we formulate the latter using (2.5) as: (Z_i) is exchangeable iff $(Z_i) \stackrel{\mathcal{D}}{=} (f(\alpha,\xi_i), i \geq 1)$ for some $f: (0,1)^2 \to S$.

Theorem 14.11 builds the general RCE array from four basic components. The arrays which can be built from only some of these components are the arrays with certain extra properties, as the next few results show.

For $f: [0,1]^3 \to S$ define

$$X^*_{i,j} = f(\xi_i,\eta_j,\lambda_{i,j}) \,. \tag{14.12}$$

Then X^* is a dissociated RCE array. Conversely, we have

(14.13) Corollary. Let X be a dissociated RCE array. Then there exists a function $f: [0,1]^3 \to S$ such that $X = X^*$, for X^* defined at (14.12).

Proof. Theorem 14.11 says X can be represented by some $f: [0,1]^4 \to S$. For each $a \in [0,1]$, let $f_a(b,c,d) = f(a,b,c,d)$. By conditioning on α in the representation (14.10), we see that X is a mixture (over a) of arrays X^a, where $X^a_{i,j} = f_a(\xi_i,\eta_j,\lambda_{i,j})$. But X is dissociated, so by Proposition 14.8 it is extreme in the class of RCE arrays, so $X \stackrel{\mathcal{D}}{=} X^a$ for almost all a.

(14.14) Corollary. *Let* X *be a dissociated* $\{0,1\}$*-valued RCE array. Then* X *is distributed as a* ϕ*-process, for some* $\phi: [0,1]^2 \to [0,1]$.

Proof. Let f be as in Corollary 14.13, and set $\phi(x,y) = P(f(x,y,\lambda_{i,j}) = 1)$.

It is natural to ask which arrays are of the form $f(\xi_i,\eta_j)$ --note the general ϕ-process is not of this form. This result is somewhat deeper. Different proofs appear in Aldous (1981a) and Hoover (1979), and will not be repeated here. See also Lynch (1982b).

(14.15) Corollary. *For a dissociated RCE array* X, *the following are equivalent:*

(a) $X_{1,1} \in S$ a.s.,

(b) $X \underset{\mathcal{D}}{=} X^*$, *where* $X^*_{i,j} = f(\xi_i,\eta_j)$ a.s. *for some* $f: (0,1)^2 \to S$.

An alternative characterization of such arrays, based upon entropy ideas, will be given in (15.28). We remark that although it is intuitively obvious that a non-trivial array of the form $f(\xi_i,\eta_j)$ cannot have i.i.d. entries, there seems no simple proof of this fact. But it is a consequence of Corollary 14.15, since for an i.i.d. array S is trivial.

The next result completes the list of characterizations of arrays representable by functions of fewer than four components.

(14.16) Corollary. *For a dissociated RCE array* X, *the following are equivalent:*

(a) $X \overset{\mathcal{D}}{=} (X_{i,\pi_i(j)}: i,j \geq 1)$ *for all finite permutations* $\pi_1,\pi_2,\dots$.

(b) $X \overset{\mathcal{D}}{=} X^*$, *where* $X^*_{i,j} = f(\xi_i,\lambda_{i,j})$ *for some* $f: (0,1)^2 \to S$.

Proof. Let α_i be the directing random measure for $(X_{i,j}: j \geq 1)$. Corollary 3.9 implies that for each (i,j),

(14.17a) $\quad \alpha_i$ is a r.c.d. for $X_{i,j}$ given $\sigma\{X_{i',j'}: (i',j') \neq (i,j)\}$.

Let N_i be disjoint infinite subsets of $\{1,2,\ldots\}$. Dissociation implies $\sigma(X_{i,j}: j \in N_i)$, $i \geq 1$, are independent, and since $\alpha_i \in \sigma(X_{i,j}: j \in N_i)$ we get

(14.17b) $\quad (\alpha_i: i \geq 1)$ are independent.

Set $X^*_{i,j} = F^{-1}(\alpha_i, \lambda_{i,j})$, where $F^{-1}(\theta,\cdot)$ is the inverse distribution function of θ. Then (14.17a) and (14.17b) imply $X^* \stackrel{\mathcal{D}}{=} X$. Finally, code α_i as $g(\xi_i)$.

Another question suggested by Theorem 14.11 concerns uniqueness of the representing function. Suppose T_i $(1 \leq i \leq 4)$ are measure-preserving functions $[0,1] \to [0,1]$. Then f and $f^*(a,b,c,d) = f(T_1(a),T_2(b),T_3(c),T_4(d))$ represent arrays X and X^* which have the same distribution. It is natural to conjecture that if X and X^* have the same distribution then any representing functions f, f^* must be "equivalent" in the sense above. Hoover (1979) gives a precise statement and proof of this fact.

Finally, let us mention a different type of exchangeability property for arrays. This is motivated by the concept of <u>U-statistics</u>, that is to say sequences (U_n) of the form

$$U_n = \frac{1}{\binom{n}{2}} \sum_{1<i<j<n} g(V_i, V_j) \tag{14.18}$$

where (V_i) is i.i.d. and $g(\cdot,\cdot)$ is symmetric. (Many natural statistical estimators are of this form--see Serfling (1980), Chapter 5.) Now we can regard (U_n) as the partial averages

$$U_n = \frac{1}{\binom{n}{2}} \sum_{\{i,j\} \subset \{1,\ldots,n\}} X_{\{i,j\}}$$

of an array $X = (X_{\{i,j\}})$ indexed by unordered pairs: $X_{\{i,j\}} = g(V_i, V_j)$. Since (V_i) is exchangeable, X has the property

(14.19) $$X \stackrel{\mathcal{D}}{=} (X_{\{\pi(i),\pi(j)\}}), \text{ each finite permutation } \pi.$$

This property has been called _weak exchangeability_. In the spirit of (14.10) we can construct a more general weakly exchangeable array by

(14.20) $$X^*_{\{i,j\}} = g(\alpha, \xi_i, \xi_j, \lambda_{\{i,j\}}) ,$$

where $g(a,\cdot,\cdot,d)$ is symmetric for each (a,d). It is possible to modify the proof of Theorem 14.11 to prove

(14.21) _Theorem. Let_ X _be a weakly exchangeable array. Then_ $X \stackrel{\mathcal{D}}{=} X^*$ _for some array_ X^* _of the form_ (14.20).

And all the other results for RCE arrays have natural analogs for weakly exchangeable arrays.

So far we have considered 2-dimensional arrays. The definitions of RCE and weak exchangeability have natural extensions for k-dimensional arrays, and it is not hard to guess what the analogs of Theorems 14.11 and 14.21 should be. _Proving_ these along the lines of the proof of Theorem 14.11 seems hard. Hoover (1979) uses quite different techniques to establish a general result encompassing Theorems 14.11, 14.21 and their k-dimensional versions. The statement and proof of these results involve ideas from logic which we shall not attempt to present here--we refer the reader to the expository account in Hoover (1982).

15. Partial exchangeability for arrays: complements

Most of the results in Parts 1 and 2 about exchangeable sequences suggest conjectures for similar results for arrays. Rather than attempt any systematic program of extension, we shall present merely a selection of results and open problems which look interesting. Studying the open problems would perhaps make a good Ph.D. thesis.

(15.1) Correlation structure. For a RCE array X with real square-integrable entries, the correlation structure is determined by the three numbers

$$\rho = \rho(X_{1,1},X_{2,2}); \quad \rho_c = \rho(X_{1,1},X_{1,2}); \quad \rho_r = \rho(X_{1,1},X_{2,1}) .$$

(15.2) Proposition. *The possible correlations (ρ,ρ_c,ρ_r) for a RCE array are precisely those satisfying*

$$0 \leq \rho \leq \min(\rho_r,\rho_c); \quad \rho_r + \rho_c \leq 1 + \rho \quad .$$

Proof. Let $\hat{\alpha}$, $(\hat{\xi}_i)$, $(\hat{\eta}_j)$, $(\hat{\lambda}_{i,j})$ be independent $N(0,1)$ and define

$$X_{i,j} = a\hat{\alpha} + b\hat{\xi}_i + c\hat{\eta}_j + d\hat{\lambda}_{i,j} + e . \tag{15.3}$$

Then, setting $\sigma^2 = a^2 + b^2 + c^2 + d^2$, we have

$$\rho = a^2/\sigma^2; \quad \rho_c = (a^2 + c^2)/\sigma^2; \quad \rho_r = (a^2 + b^2)/\sigma^2 .$$

And we can choose (a,b,c,d) to attain any correlations in the range specified.

Conversely, let X be a RCE array. Suppose first that X is dissociated. Then $\rho = 0$. Of course $\min(\rho_r,\rho_c) \geq 0$ by (1.8). We shall prove

$$\rho_r + \rho_c \leq 1 . \tag{15.4}$$

We may suppose $EX_{1,1} = 0$, $EX^2_{1,1} = 1$ and, by Corollary 14.13, that $X_{i,j} = f(\xi_i,\eta_j,\lambda_{i,j})$. Define

$$g(x,y) = Ef(x,y,\lambda_{1,1})$$

$$g_1(x) = Eg(x,\eta_1); \quad g_2(y) = Eg(\xi_1,y) .$$

Then
$$\begin{aligned} \rho_c &= EX_{1,1}X_{1,2} \\ &= Eg(\xi_1,\eta_1)g(\xi_1,\eta_2) \\ &= Eg_1^2(\xi_1) \end{aligned}$$

and similarly $\rho_r = Eg_2^2(\eta_1)$. So

$$\begin{aligned} 1 = EX_{1,1}^2 &\geq Eg^2(\xi_1,\eta_1) \\ &= E(g(\xi_1,\eta_1) - g_1(\xi_1))^2 + Eg_1^2(\xi_1) \\ &\quad \text{by conditioning on } \xi_1 \\ &\geq Eg_2^2(\eta_1) + Eg_1^2(\xi_1) \quad \text{by conditioning on } \eta_1 \\ &= \rho_c + \rho_r, \quad \text{establishing (15.4).} \end{aligned}$$

For general X with $EX_{1,1} = 0$ and $EX_{1,1}^2 = 1$, condition on α in the representation (14.10) and let ρ_r^*, ρ_c^*, ρ^* be the conditional correlations. Then

$$\begin{aligned} \rho^* &= E^2(X_{1,1}|\alpha) \\ \rho_r^* &= \{E(X_{1,1}X_{1,2}|\alpha) - E^2(X_{1,1}|\alpha)\}/var(X_{1,1}|\alpha) \\ &= \{E(X_{1,1}X_{1,2}|\alpha) - \rho^*\}/var(X_{1,1}|\alpha) \end{aligned}$$

and similarly for ρ_c^*, replacing $X_{1,2}$ with $X_{2,1}$. And

$$\begin{aligned} \rho = EX_{1,1}X_{2,2} &= E\,E(X_{1,1}X_{2,2}|\alpha) \\ &= E\{E(X_{1,1}|\alpha)E(X_{2,2}|\alpha)\} \\ &= E\rho^* . \end{aligned}$$

So
$$\begin{aligned} \rho_c = EX_{1,1}X_{1,2} &= E\,E(X_{1,1}X_{1,2}|\alpha) \\ &\geq E\{E(X_{1,1}|\alpha)E(X_{1,2}|\alpha)\} \\ &\quad \text{by (1.8) for the first row of the} \\ &\quad \text{conditioned array} \end{aligned}$$

$$= E\rho^* = \rho,$$

and similarly for ρ_r. Finally, $\rho_r^* + \rho_c^* \leq 1$ by (15.4), so

$$\begin{aligned} E(X_{1,1}X_{1,2}|\alpha) + E(X_{1,1}X_{2,1}|\alpha) &\leq E\ \mathrm{var}(X_{1,1}|\alpha) + 2\rho^* \\ &= \mathrm{var}(X_{1,1}) - \mathrm{var}\ E(X_{1,1}|\alpha) + 2\rho^* \\ &= 1 + \rho^* , \end{aligned}$$

and taking expectations gives $\rho_c + \rho_r \leq 1 + \rho$.

Remarks. (a) Proposition 15.1 could alternatively be derived by analytic methods, without using Theorem 14.11.

(b) Proposition 15.5 implies that the general Gaussian RCE array is of the form (15.3).

(c) One could consider "second order" RCE arrays, in which only the correlation structure is assumed invariant. Such arrays are discussed in Bailey et al. (1984) as part of an abstract treatment of analysis of variance. Invariance of higher moments is discussed by Speed (1982).

(15.5) Estimating the representing function. Corollary 14.14 says that a dissociated $\{0,1\}$-valued RCE array is a ϕ-process. In other words, the family of dissociated $\{0,1\}$-valued RCE arrays can be regarded as a parametric family, parametrized by the set Φ of measurable functions $\phi: [0,1]^2 \to [0,1]$. Can one consistently estimate the parameter? More precisely,

(15.6) Problem. Do there exist functionals $\Lambda_N: \{0,1\}^{\{1,\ldots,N\}\times\{1,\ldots,N\}} \to \Phi$ such that for any ϕ-process X,

$$\Lambda_N(X_{i,j}: i,j \leq N) \to \phi^* \text{ a.s. in } L^1([0,1]^2),$$

for some ϕ^* such that X is a ϕ^*-process?

In view of the non-uniqueness of ϕ representing X, one should really expect a somewhat weaker conclusion: but no results are known.

(15.7) Spherical matrices. Schoenberg's theorem (3.6) asserts that every spherically symmetric infinite sequence is a mixture of i.i.d. $N(0,v)$ sequences. Dawid (1977) introduced the analogous concept for matrices. Call an infinite array $Y = (Y_{i,j}: i,j \geq 1)$ spherical if for each $n \geq 1$,

$$U_1 Y_n U_2 \stackrel{\mathcal{D}}{=} Y_n \quad \text{for all orthogonal } n \times n \text{ matrices } U_1, U_2 ,$$

where Y_n denotes the $n \times n$ matrix $(Y_{i,j}: 1 \leq i,j \leq n)$. Here are two examples of spherical arrays:

(i) a normal array Y, where $(Y_{i,j}: i,j \geq 1)$ are i.i.d. $N(0,1)$;

(ii) a product-normal array Y, where $Y_{i,j} = V_i W_j$ for $(V_1, V_2, \ldots, W_1, W_2, \ldots)$ i.i.d. $N(0,1)$.

Now a spherical array is RCE, by considering permutation matrices. Then Theorem 14.11 can be applied to obtain (with some effort) the following result, conjectured in Dawid (1978) and proved in Aldous (1981a).

(15.8) Corollary. *For an array* $Y = (Y_{i,j}: i,j \geq 1)$ *the following are equivalent:*

(a) Y *is spherical and dissociated, and* $EY_{1,1}^2 < \infty$.

(b) $Y = a_0 Y^0 + \sum a_m Y^m$; *where* $\sum a_m^2 < \infty$, Y^0 *is Normal,* Y^m $(m \geq 1)$ *are product-Normal, and* $(Y^m: m \geq 0)$ *are independent.*

It is clear that (b) implies (a): let us say a few words about the implication (a)$\Rightarrow$(b). Writing $Y = E(Y|\mathcal{S}) + \hat{Y}$, where $\mathcal{S}$ is the shell σ-field of Y, it can be shown that $E(Y|\mathcal{S})$ and $\hat{Y}$ are independent; that $E(Y|\mathcal{S})$ is of the form $a_0 Y^0$; and that $\hat{Y}$ is spherical, dissociated and $\mathcal{S}$-measurable. Corollary 14.15 gives a representation $\hat{Y}_{i,j} = f(\xi_i, \eta_j)$, and the constants

(a_i) appear as the eigenvalues of the integral operator $h(\cdot) \to \int f(\cdot,y)h(y)dy$ associated with f.

(15.9) Finite arrays. Proposition 5.6 gave bounds on how far the initial portion of a finite exchangeable sequence could differ from the initial part of an infinite exchangeable sequence. Analogously, for $m \leq n$ let $c_{m,n}$ be the smallest number such that for any $n \times n$ RCE array X taking values $\{0,1\}$ only there exists an infinite RCE array Y such that

$$\|L(X_{i,j}: 1 \leq i,j \leq m) - L(Y_{i,j}: 1 \leq i,j \leq m)\| \leq c_{m,n} .$$

Weak convergence arguments show that $\lim_{n\to\infty} c_{m,n} = 0$ for each m, but still open is

(15.10) Problem. Give explicit upper bounds for $c_{m,n}$.

(15.11) Continuous-parameter processes. For 2-parameter processes $X_{s,t}$, $0 \leq s,t < \infty$, $X_{0,0} = 0$, we can define analogs of the 1-parameter "processes with interchangeable increments" discussed in Section 10. For a rectangle $B = (s_1,s_2] \times (t_1,t_2]$ and a function $f(s,t)$ let $f(B)$ be the increment of f over B:

$$f(B) = f(s_2,t_2) + f(s_1,t_1) - f(s_1,t_2) - f(s_2,t_1) .$$

For fixed $\delta > 0$ let $B_{i,j} = ((i-1)\delta, i\delta] \times ((j-1), j\delta]$. Say X has separately interchangeable increments if

(15.12) the array $X(B_{i,j})$, $i,j \geq 1$ is RCE (for each δ).

Say X has simultaneously interchangeable increments if

(15.13a) $$X_{s,t} = X_{t,s}$$

(15.13b) $(X(B_{i,j}),\ i,j \geq 1) = (X(B_{\pi(i),\pi(j)}),\ i,j \geq 1)$ for each permutation π (for each δ).

(Condition (15.13b) is slightly more than weak exchangeability.)

Using Theorems 14.11 and 14.21 to describe all processes with these invariance properties is perhaps a feasible project, but looks much harder than the 1-parameter result, Proposition 10.5. Rather than tackle the general case, let us look at two special cases, corresponding to the special case in Corollary 10.6.

(15.14) <u>2-parameter counting processes with separately interchangeable increments</u>
Here are four methods of constructing such processes:

(i) Let $\lambda > 0$. Take a Poisson process on $R^+ \times R^+$ of rate λ.

(ii) Let $\lambda_1, \lambda_2, \ldots > 0$; let $(\xi_i^k : k \geq 1)$ be i.i.d. $U(0,1)$; and let $f: \mathbb{N} \times [0,1] \to [0,\infty)$ be such that $\sum_k \lambda_k f(k,\xi_1^1) < \infty$ a.s. For each k take a Poisson process of horizontal lines of rate λ_k and attach labels $(\xi_i^k : i \geq 1)$ to these lines; independently for different k. On the line labelled ξ_i^k place a Poisson process of points of rate $f(k,\xi_i^k)$, independently for different lines.

(iii) The analog of (ii) with vertical lines, using constants $(\hat{\lambda}_m)$, say, and labels (η_j^m), say.

(iv) Construct both a process of horizontal lines as in (ii), and a process of vertical lines as in (iii). Let $g: \mathbb{N} \times [0,1] \times \mathbb{N} \times [0,1] \to [0,1]$ be such that $\sum_{k,m} \lambda_k \hat{\lambda}_m g(k,\xi_1^1,m,\eta_1^1) < \infty$ a.s. At the intersection of the lines labelled ξ_i^k and η_j^m put a point with probability $g(k,\xi_i^k,m,\eta_j^m)$.

Since the superposition (i.e. sum) of independent invariant processes is invariant, it is natural to make

(15.15) Conjecture. The general ergodic process of type (15.14) is a sum of independent processes of types (i)-(iv) above.

Presumably this can be proved by applying Theorem 14.11 to the arrays of point counts in small squares, and letting the sizes of the squares decrease to zero; but the details look messy.

(15.16) 2-parameter continuous-path processes with simultaneously interchangeable increments. Here are two examples of such processes:

(15.17) $Y_{s,t} = B_s B_t$; where $(B_t: t \geq 0)$ is Brownian motion;

(15.18) On $\nabla = \{0 \leq s \leq t < \infty\}$ let X be 2-parameter Brownian motion; that is, for rectangles B with Lebesgue measure $|B|$, the variable $X(B)$ has Normal $N(0,|B|)$ distribution, and variables $X(B_i)$ are independent for disjoint B_i. For $t < s$ let $X_{s,t} = X_{t,s}$.

Also, the deterministic processes $X_{s,t} = st$ and $X_{s,t} = \min(s,t)$ have the required properties. From these examples we can construct more processes of the form

(15.19) $$Z_{s,t} = \alpha X_{s,t} + \sum_{j=1}^{\infty} \beta_j Y^{(j)}_{s,t} + \gamma st + \delta \min(s,t)$$

where X has distribution (15.18)

$Y^{(j)}$ has distribution (15.17), for each $j \geq 1$

$(X, Y^{(1)}, Y^{(2)}, \ldots)$ are independent

$(\alpha, \beta_1, \beta_2, \ldots, \gamma, \delta)$ are random variables, and $\sum \beta_j^2 < \infty$ a.s.

It is natural to make the

(15.20) Conjecture. Any 2-parameter continuous-path process $Z_{s,t}$ with simultaneously interchangeable increments has a representation of the form (15.19).

These types of processes arise naturally in the study of the asymptotic distributions of U-statistics. Let us describe the simplest cases. Let $(V_1, V_2, \ldots)$ be i.i.d. and let $g(x,y)$ be a symmetric function such that

$$Eg(V_1,V_2) = 0; \quad Eg^2(V_1,V_2) < \infty .$$

As at (14.18), define the U-statistics

$$U_n = \frac{1}{\binom{n}{2}} \sum\sum_{1 \le i < j \le n} g(V_i, V_j)$$

Also define

$$\sigma^2 = E\big(E(g(V_1,V_2)|V_1)\big)^2 \ge 0 .$$

Then we get the following fundamental theorem for the asymptotic behavior of U-statistics. See Serfling (1980), Chapter 5, for proof and a detailed discussion of U-statistics.

(15.21) Theorem. (i) *Suppose* $\sigma^2 > 0$. *Then* $n^{1/2}U_n \xrightarrow{\mathcal{D}} N(0,4\sigma^2)$.

(ii) *Suppose* $\sigma^2 = 0$. *Then* $nU_n \xrightarrow{\mathcal{D}} \sum\lambda_j(W_j^2 - 1)$, *where* $(W_1, W_2, \ldots)$ *are independent* $N(0,1)$ *and* $(\lambda_1, \lambda_2, \ldots)$ *are the eigenvalues of the operator* A *defined by* $(Af)(x) = Eg(x,V_2)f(V_2)$.

These results are analogous to the ordinary central limit theorem; what is the corresponding "process" result? Define processes

$$Z^{(n)}_{s,t} = \frac{1}{\binom{n}{2}} \sum\sum_{\substack{1 \le i \le [ns] \\ 1 \le j \le [nt] \\ i<j}} g(V_i,V_j) \; ; \quad s \le t$$

$$= Z^{(n)}_{t,s} \; ; \quad s \ge t.$$

Then $Z^{(n)}$ has simultaneously interchangeable increments, and $U_n = Z^{(n)}_{1,1}$. Thus we would anticipate that any process arising as a limit of the normalized

$Z^{(n)}$ should be of the form (15.19). This is indeed true; the arguments of Mandelbaum and Taqqu (1983) yield

(15.22) _Theorem_. (i) _Suppose_ $\sigma^2 > 0$. _Then_ $n^{1/2}Z^{(n)} \xrightarrow{\mathcal{D}} 4\sigma^2 X$, _where_ X _has distribution_ (15.18).

(ii) _Suppose_ $\sigma^2 = 0$. _Then_ $nZ^{(n)} \xrightarrow{\mathcal{D}} \sum \lambda_j (Y^{(j)}_{s,t} - \min(s,t))$, _where_ $(Y^{(1)}, Y^{(2)}, \ldots)$ _are independent with distribution_ (15.17), _and_ (λ_j) _are as in Theorem 15.21_.

(15.23) _Exchangeable random partitions_. Consider a weakly exchangeable family $(R_{i,j})$ of events. By the analog of Proposition 14.6 for weak exchangeability, the family is a mixture of dissociated families. As in (14.14), a dissociated family of events can be described as a ϕ-process:

$$P(R_{i,j} | \xi_1, \xi_2, \ldots) = \phi(\xi_i, \xi_j)$$

where $\phi: [0,1]^2 \to [0,1]$ is now symmetric. This leads to an alternative argument for Proposition 11.9. An exchangeable partition (11.4) is a weakly exchangeable array $(R_{i,j})$ with the special property

$$P(R_{i,j} \cap R_{j,k} \cap R^c_{i,k}) = 0; \quad i,j,k \text{ distinct.}$$

This translates into the following property for ϕ:

$$E\, \phi(\xi_1, \xi_2)\phi(\xi_2, \xi_3)(1 - \phi(\xi_1, \xi_3)) = 0 .$$

It is not hard to deduce from this that

$$\phi(x,y) = \sum_n 1_{B_n \times B_n}(x,y) \text{ a.e. for some disjoint } (B_n) .$$

This in turn implies that $(R_{i,j})$ has the "paintbox" distribution ψ_p of Section 11, where $(p_1, p_2, \ldots)$ is the decreasing rearrangement of $(|B_n|)$.

Thus any <u>dissociated</u> exchangeable partition has a "paintbox" distribution, and the general case is a mixture.

(15.24) <u>Ergodic theory techniques</u>. One characterization of RCE arrays of the form $f(\xi_i,\eta_j)$ was given in Corollary 14.15. Here we show how ergodic theory concepts lead to another characterization.

The $\bar{d}$ metric (12.28) on the set of distributions of RCE arrays is

$$d(\mu,\nu) = \inf\{E \min(1,|X_{1,1}-Y_{1,1}|)\}$$

where the infimum is taken over bivariate RCE arrays $Z_{i,j} = (X_{i,j},Y_{i,j})$ such that $L(X) = \mu$, $L(Y) = \nu$. It turns out (Aldous (1982a)) that the $\bar{d}$ topology is strictly stronger than the weak topology; and the $\bar{d}$ topology fits in nicely with the characterization theorem. Consider for simplicity only dissociated arrays. Corollary 14.13 says that any RCE distribution μ has a <u>representation</u>

$$\mu = L(X); \quad X_{i,j} = f(\xi_i,\eta_j,\lambda_{i,j}) ,$$

for some f in the space L^0 of functions $f\colon (0,1)^3 \rightarrow \mathbb{R}$. Give L^0 the topology of convergence in measure.

(15.25) <u>Lemma</u>. $\mu_k \xrightarrow{\bar{d}} \mu_\infty$ <u>iff there exist functions</u> f_k <u>representing</u> μ_k <u>with</u> $f_k \rightarrow f_\infty$.

<u>Proof</u>. The "if" is straightforward, considering $(X_{i,j},Y_{i,j}) = (f_k(\xi_i,\eta_j,\lambda_{i,j}),f_\infty(\xi_i,\eta_j,\lambda_{i,j}))$. Conversely, suppose $\mu_k \xrightarrow{\bar{d}} \mu_\infty$. For each k there exists a bivariate RCE array $(X^k_{i,j},X^\infty_{i,j})$ such that $L(X^k) = \mu_k$, $L(X^\infty) = \mu_\infty$ and

$$E \min(1,|X^k_{1,1}-X^\infty_{1,1}|) \rightarrow 0 . \qquad (15.26)$$

By taking $(X^k: 1 \leq k < \infty)$ conditionally independent given X^∞, we can form a process $(X^\infty; X^k, 1 \leq k < \infty)$ with the bivariate distributions (X^k, X^∞) above and such that $X_{i,j} = (X^\infty_{i,j}; X^k_{i,j}, k \geq 1)$ is a RCE $\mathbb{R}^\infty$-valued array. By Corollary 14.13 we can take $X_{i,j} = g(\xi_i, \eta_j, \lambda_{i,j})$ for some $g: (0,1)^3 \to \mathbb{R}^\infty$. Now set $g_k = g \circ \pi_k$, where $\pi_k((x_r)) = x_k$. Then g_k represents μ_k, and (15.26) implies

$$E \min\left(1, |g_k(\xi_1, \eta_1, \lambda_{1,1}) - g_\infty(\xi_1, \eta_1, \lambda_{1,1})|\right) \to 0 ,$$

so $g_k \to g_\infty$ in L^0.

Next, recall the definition and elementary properties of entropy. A random variable Y with finite range (y_i) has entropy $E(Y) = -\sum P(Y = y_i) \log P(Y = y_i)$. And

(15.27)(a) $E(h(Y)) \leq E(Y)$; any function h.

(b) $E(X,Y) = E(X) + E(Y)$ for independent X, Y.

(c) $E(Y) \geq E\, E(Y|F)$ for any σ-field F, where $E(Y|F)(\omega)$ is the entropy of the conditional distribution $\alpha(\omega, \cdot)$ of Y given F.

For a dissociated RCE array Y such that $Y_{i,j}$ takes values in a finite set, let $E^Y_n = E(Y_{i,j}: 1 \leq i,j \leq n)$. Say Y has _linear entropy_ if $\limsup n^{-1} E^Y_n < \infty$.

(15.28) _Proposition. A dissociated RCE array has a representation as $(f(\xi_i, \eta_j): i,j \geq 1)$ for some f iff it is in the $\bar{d}$-closure of the set of linear entropy arrays._

Proof. Suppose μ is the distribution of an array $(f(\xi_i, \eta_j))$. Let F_k be the set of functions $g: (0,1)^2 \to R$ which are constant on each square of the form $(r2^{-k}, (r+1)2^{-k}) \times (s2^{-k}, (s+1)2^{-k})$. Martingale convergence says there exist $f_k \in F_k$ such that $f_k \to f$ in measure. Let Y^k be the array $(f_k(\xi_i,\eta_j))$. Then $L(Y^k) \xrightarrow{\bar{d}} \mu$ by Lemma 15.25. Now fix k, and set $\hat{\xi}_i = r2^{-k}$ on $\{r2^{-k} < \xi_i < (r+1)2^{-k}\}$, and similarly for $\hat{\eta}_j$. Then $(Y_{i,j}: i,j \leq n)$ is a function of $(\hat{\xi}_i,\hat{\eta}_j: i,j \leq n)$. So

$$\begin{aligned} E_n^Y &\leq E(\hat{\xi}_i,\hat{\eta}_j: i,j \leq n) && \text{by (15.27)(a)} \\ &= 2n \log(2^k) && \text{by (15.27)(b).} \end{aligned}$$

So Y^k has linear entropy.

For the converse we need

(15.29) <u>Lemma</u>. <u>For a finite-valued dissociated RCE array</u> Y, <u>either</u>

(a) <u>there exists</u> $b > 0$ <u>such that</u> $E_n^Y \geq bn^2$, $n \geq 1$; <u>or</u>

(b) <u>each representation</u> f <u>for</u> Y <u>has</u> $f(\xi_1,\eta_1,\lambda_{1,1}) = \bar{f}(\xi_1,\eta_1)$ a.s. <u>for some</u> $\bar{f}$.

<u>Proof</u>. If (b) fails for some representation f, then there exists a subset $B \subset (0,1)^2$ with measure $|B| > 0$ and there exists $\delta > 0$ such that

$$E(f(x,y,\lambda_{1,1})) > \delta, \quad (x,y) \in B.$$

Define
$$\begin{aligned} F_n &= \sigma(\xi_i,\eta_j: i,j \leq n) \\ C_n &= \#\{(i,j): i,j \leq n, (\xi_i,\eta_j) \in B\}. \end{aligned}$$

Then $E(Y_{i,j}: i,j \leq n | F_n) \geq \delta C_n$ by (15.27)(b), and then using (15.27)(c) $E_n^Y \geq \delta E C_n = \delta |B| n^2$.

For the converse part of Proposition 15.28, let $X = (f(\xi_i,\eta_j,\lambda_{i,j}))$ be in the $\bar{d}$-closure of the set of linear entropy arrays. By Lemma 15.26 there exist $f_k: (0,1)^3 \to \mathbb{R}$ such that $f_k \to f$ in measure and f_k represents a linear entropy array. But by Lemma 15.29 $f_k(\xi_1,\eta_1,\lambda_{1,1}) = \bar{f}_k(\xi_1,\eta_1)$ a.s. for some $\bar{f}_k$, and this implies $f(\xi_1,\eta_1,\lambda_{1,1}) = \bar{f}(\xi_1,\eta_1)$ a.s. for some $\bar{f}$.

Remarks. With somewhat more work, one can show that for any dissociated finite-valued RCE array Y represented by f

$$n^{-2}E_n^Y \to \int_0^1\int_0^1 E(f(x,y,\lambda_{1,1}))dxdy .$$

This leads to an alternative characterization in Proposition 15.28. In particular, consider Y of the form $g(\xi_i,\eta_j)$ for some finite-valued g. The assertion above implies E_n^Y is $o(n^2)$:

(15.30) Problem. What is the exact growth rate of E_n^Y in terms of g?

16. The infinite-dimensional cube

Here we present a final example of partial exchangeability where the characterization problem has not been solved--perhaps the examples given here will encourage the reader to tackle the problem.

Let I be the set of infinite sequences $i = (i_1,i_2,\dots)$ of 0's and 1's such that $\#\{n: i_n = 1\} < \infty$; let I_d be the subset of sequences i such that $i_n = 0$ for all $n > d$. Think of I_d as the set of vertices of the d-dimensional unit cube; think of I as the set of vertices of the infinite-dimensional cube. For a permutation π of $\mathbb{N}$ leaving $\{d+1,d+2,\dots\}$ fixed, define $\tilde{\pi}: I \to I$ by

(16.1) $$(\tilde{\pi}i)_n = i_{\pi(n)} .$$

Geometrically, $\tilde{\pi}$ acts on the cube I_d as a rotation about the origin $\underset{\sim}{0}$. For $1 \le s \le d$ define $r_s: I \to I$ by

(16.2) $$\begin{aligned}(r_s i)_n &= i_n \quad , \quad n \ne s\\ &= 1 - i_n , \quad n = s .\end{aligned}$$

Geometrically, r_s acts on the cube I_d as a reflection in the hyperplane

$\{\underset{\sim}{x}: x_s = \frac{1}{2}\}$. The group Γ_d of isometries of the cube I_d is generated by $\{r_s, 1 \leq s \leq d; \tilde{\pi}, \pi$ acting on $\{1,2,\ldots,d\}\}$. And we can regard $\Gamma = \cup \Gamma_d$ as the group of isometries of the infinite-dimensional cube I. Note that I and Γ are both countable.

The pair (I,Γ) fits into the general partial exchangeability setting of Section 12. We are concerned with processes $X = (X_i: i \in I)$, where X_i takes values in some space S, which are invariant in the usual sense

$$X \overset{\mathcal{D}}{=} (X_{\gamma(i)}: i \in I); \quad \text{each } \gamma \in \Gamma$$

For such a process, the processes $X^d = (X_i^d: i \in I_d)$ are invariant processes on the finite-dimensional cubes I_d with the natural consistency property; conversely, any consistent family of invariant processes on the finite-dimensional cubes yields a process on the infinite-dimensional cube.

Here is some more notation. For $i \in I$ let $C_i = \{n: i_n = 1\}$. For $i,j \in I$ let $d(i,j) = \#(C_i \Delta C_j)$, so $d(i,j)$ is the number of edges on the minimal path of edges from i to j. A path in I is a sequence $i^1, i^2, i^3, \ldots$ of vertices such that the sets $C_{i^k} \Delta C_{i^{k+1}}$ are distinct singletons.

As well as the obvious example of i.i.d. processes, there is a related class of invariant processes which involve the "period 2" character of the cube. Given two distributions μ, ν on S let $\theta^0_{\mu,\nu}$ be the distribution of the process (X_i) consisting of independent random variables such that $L(X_i) = \mu$ when $\#C_i$ is even, $L(X_i) = \nu$ when $\#C_i$ is odd. Then the mixture $\theta_{\mu,\nu} = \frac{1}{2}\theta^0_{\mu,\nu} + \frac{1}{2}\theta^0_{\nu,\mu}$ is invariant.

Before proceding further, the reader may like to attempt to construct other examples of invariant processes.

It is interesting to note that an invariant process on the infinite-dimensional cube contains, as subprocesses, examples of other partially exchangeable structures we have described. Let X be invariant.

(16.3) The variables at distance 1 from $\underset{\sim}{0}$, that is $\{X_i: \#C_i = 1\}$, are exchangeable (in fact, exchangeable over $X_{\underset{\sim}{0}}$).

(16.4) The variables at distance 2, that is $\{X_i: \#C_i = 2\}$, form a weakly exchangeable array.

The next result is less obvious. Regard I as a graph. Let $\bar{\Gamma} \supset \Gamma$ be the set of graph-automorphisms of I, that is the set of bijections $\gamma: I \to I$ such that (i,j) is an edge iff $(\gamma(i),\gamma(j))$ is an edge. It is not hard to see that any Γ-invariant process is $\bar{\Gamma}$-invariant.

(16.5) <u>Lemma</u>. <u>There exists a subset</u> $T \subset I$ <u>which is an infinitary tree, in the sense of Section 13, and such that every tree-automorphism</u> γ <u>of</u> T <u>extends to a graph-automorphism</u> θ <u>of</u> I. <u>Hence if</u> $(X_i: i \in I)$ <u>is an invariant process on the cube</u> I <u>then the restriction</u> $(X_i: i \in T)$ <u>is an invariant process on the infinitary tree</u> T.

<u>Proof</u>. As in Section 13 let D be the set of finite sequences $d = (d_1,\ldots,d_m)$ of strictly positive integers. Let $f: D \to \mathbb{N}$ be the prime factorization map $f(d_1,\ldots,d_m) = 2^{d_1}\cdot 3^{d_2}\cdots\cdot$. Now define $\psi: D \to I$ by $C_{\psi(d)} = \{f(d_1), f(d_1,d_2),\ldots,f(d_1,\ldots,d_m)\}$. Then $T = \psi(D)$ is an infinitary tree and $\{\psi(d): d \in D\}$ is a labelling scheme for T.

Now fix a tree-automorphism $\gamma: T \to T$. The map γ induces a map $\tilde{\gamma}: f(D) \to f(D)$ in the following way: if γ maps the edge $(\psi(d),\psi(dq))$ to the edge $(\psi(\hat{d}),\psi(\hat{d}\hat{q}))$ then let $\tilde{\gamma}$ map $f(dq)$ to $f(\hat{d}\hat{q})$. Now define $\theta: I \to I$ as follows. Let $\theta(\underset{\sim}{0}) = \gamma(\underset{\sim}{0}) = \gamma(\psi(\phi))$. For $i \neq \underset{\sim}{0} \in I$ write $C_i = A_i \cup B_i$, where $A_i = C_i \cap f(D)$ and $B_i = C_i \setminus f(D)$. Define $\theta(i)$ by

$$C_{\theta(i)} \,\Delta\, C_{\theta(\underset{\sim}{0})} = \tilde{\gamma}(A_i) \cup B_i \ .$$

By construction θ is an extension of γ. And θ is a graph-automorphism because (i,j) is an edge iff $\#(C_i \Delta C_j) = 1$ iff $\#(C_{\theta(i)} \Delta C_{\theta(j)}) = 1$.

Lemma 16.5 has one noteworthy consequence. For an invariant process on the infinite-dimensional cube with square-integrable real entries, the correlations $\rho(X_i, X_j)$ equal $\rho(d(i,j))$ for some correlation function $\rho(n)$. By (16.5), $\rho(n)$ must be of the form described in Proposition 13.22. Example 16.9 later shows that for each $\lambda \in [-1,1]$ there exists an invariant process with $\rho(n) = \lambda^n$, so by taking mixtures we get

(16.6) _Corollary_. _A sequence_ $(\rho(n): n \geq 0)$ _is the correlation function of some invariant process on the infinite-dimensional cube iff_ $\rho(n) = \int x^n \lambda(dx)$ _for some probability measure_ λ _on_ $[-1,1]$.

This result can be proved by harmonic analysis--see Mansour (1981), who also describes the correlation functions of invariant processes on finite-dimensional cubes. Kingman (personal communication) also has a direct proof of Corollary 16.6.

We now describe a sequence of examples of invariant processes, which we shall loosely refer to as "symmetric random walk models." Here is the basic example, suggested by Kingman.

(16.7) _Example_. _The basic random walk_. Let $(S,+)$ be a compact Abelian group. Let ξ be a random element of S whose distribution is symmetric, in the sense $\xi \overset{\mathcal{D}}{=} -\xi$. Let U be a random element of S whose distribution is Haar measure (i.e. uniform), independent of ξ. Then

$$(\xi+U, U) \overset{\mathcal{D}}{=} (U, \xi+U) . \tag{16.8}$$

For $(U,\xi+U) = (-\xi + (\xi+U),\xi+U)$

$\stackrel{\mathcal{D}}{=} (-\xi+U,U)$ because $\xi+U$ is uniform and independent of ξ

$\stackrel{\mathcal{D}}{=} (\xi+U,U)$ by symmetry.

Now let $\xi_1,\xi_2,\xi_3,\ldots$ be independent copies of ξ, independent of U. For $i \in I$ define

$$X_i = \xi_{n_1} + \xi_{n_2} + \cdots + \xi_{n_m} + U, \quad \text{where } \{n_1,\ldots,n_m\} = C_i = \{n: i_n = 1\} .$$

Then X is invariant: for invariance under the maps $\tilde{\pi}$ of (16.1) is immediate, and invariance under the maps r_s of (16.2) follows from (16.8).

As a particular case of Example 16.7, suppose

$$(16.9) \qquad S = \{1,0\}; \quad P(\xi = 1) = \tfrac{1}{2}(1-\lambda), \quad P(\xi = 0) = \tfrac{1}{2}(1+\lambda);$$
$$P(U = 1) = P(U = 0) = \tfrac{1}{2} .$$

This process has correlation function $\rho(n) = \lambda^n$; indeed, the tree-process which is embedded in X by (16.5) is precisely the tree-process exhibited in the proof of Proposition 13.22.

(16.10) Example. A generalized random walk. Let $(G,\circ)$ be an Abelian group acting on a space S; that is, G consists of functions $g: S \to S$ which form a group under convolution. Let ξ and U be independent random elements of G and S respectively, and suppose

$$(16.11) \qquad (\xi(U),U) \stackrel{\mathcal{D}}{=} (U,\xi(U)) .$$

Now let $\xi_1,\xi_2,\ldots$ be independent copies of ξ, independent of U, and let

$$X_i = \xi_{n_1} \circ \xi_{n_2} \circ \cdots \circ \xi_{n_m}(U), \quad \text{where } \{n_1,\ldots,n_m\} = C_i .$$

Then X is invariant, by the same argument as in the previous example.

This construction can yield processes rather more general than is suggested by the phrase "random walk," as the next example shows.

Remark. We call this a "random walk" model because the values $X_{i^1}, X_{i^2}, \ldots$ along a path $i^1, i^2, \ldots$ in I are a random walk on S, in the usual sense.

(16.12) Example. Randomly-oriented stationary process. Let $U = (\ldots, U_{-1}, U_0, U_1, \ldots)$ be an arbitrary stationary sequence. On the d-dimensional cube I_d choose a diagonal Δ at random (uniformly); each vertex lies in one of $d+1$ hyperplanes $H_0, H_1, \ldots, H_d$ orthogonal to Δ; set $X_i = U_m$ for $i \in H_m$. This describes an invariant process indexed by I_d. As d varies, these are consistent, and so determine an invariant process on I. The process on I has the following alternative description. Regard U as a random element of $S = \mathbb{R}^{\mathbb{Z}}$. Let θ^n be the shift on S; $\theta^n((x_m)) = (x_{m+n})$. Let $G = \{\theta^n : n \in \mathbb{Z}\}$ and let ξ be the random element of G such that $P(\xi = \theta^1) = P(\xi = \theta^{-1}) = \frac{1}{2}$. Then (16.11) holds because

$$\begin{aligned} \mathcal{L}(U, \xi(U)) &= \tfrac{1}{2}\mathcal{L}(U, \theta^1(U)) + \tfrac{1}{2}\mathcal{L}(U, \theta^{-1}(U)) \\ &= \tfrac{1}{2}\mathcal{L}(\theta^{-1}(U), U) + \tfrac{1}{2}\mathcal{L}(\theta^1(U), U) \quad \text{by stationarity} \\ &= \mathcal{L}(\xi(U), U) . \end{aligned}$$

So as in Example 16.10 we can construct an invariant process $\hat{X}$ from U and ξ. Let $g((x_m)) = x_0$. Then the process $X_i = g(\hat{X}_i)$ is the randomly-oriented stationary process described originally.

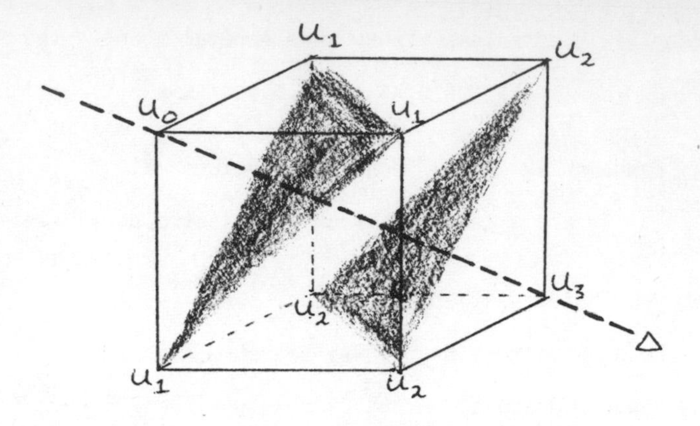

Here is a different generalization of the basic random walk model.

(16.13) Transient random walk. Let $(G,+)$ be a countable Abelian group. Let ξ be a random element of G and let π be a σ-finite measure on G. Suppose

$$(16.14) \qquad \pi(g_1)P(g_1+\xi = g_2) = \pi(g_2)P(g_2+\xi = g_1); \quad \text{all } g_1,\ g_2 \in G\ .$$

This is analogous to (16.8) and (16.11); π is a σ-finite invariant measure for the random walk generated by ξ. This random walk may be transient; consider the particular case

$$(16.15) \qquad G = \mathbb{Z}; \quad P(\xi = -1) = \alpha, \quad P(\xi = 1) = 1-\alpha; \quad \pi(n) = c\left(\frac{1-\alpha}{\alpha}\right)^n .$$

Though the random walk has no stationary distribution in the usual sense, there is a different interpretation. Suppose that at time 0 we place a random number Y_g^0 of particles at each g, where (Y_g^0) are independent and Y_g^0 has distribution Poisson$(\pi(g))$. Then let each particle move independently as a random walk with step distribution ξ. Let Y_g^n be the number of particles at position g at time n, and let $Y^n = (Y_g^n)$, a random element of $S = (\mathbb{Z}^+)^G$. Then it is easy to see, using (16.14),

(16.16) $\quad Y^0, Y^1, Y^2, \ldots$ is a stationary reversible Markov chain.

By adding more detail to the description above, we shall produce a process indexed by the infinite-dimensional cube. Suppose that particle u is initially placed at point $g_0(u)$ and has written on it an i.i.d. sequence $(\xi_1^u, \xi_2^u, \ldots)$ of copies of ξ, representing the successive steps to be made by the particle. So $Y_g^n = \#\{u\colon g_0(u) + \xi_1^u + \cdots + \xi_n^u = g\}$. Now for $i \in I$ define

(16.17) $\quad X_g^i = \#\{u\colon g_0(u) + \xi_{j_1}^u + \cdots + \xi_{j_m}^u = g\}$, where $\{j_1, \ldots, j_m\} = C_i$.

So $X^i = (X_g^i)$ describes the configuration of particles when only the jumps at times in C_i are allowed. It is easy to check that $(X^i\colon i \in I)$ is invariant.

Here is a more concrete example which turns out to be a special case of the construction above.

(16.18) Example. On the d-dimensional cube I_d pick $k = k(d)$ vertices $V_1, \ldots, V_k$ uniformly at random and define

$$X_i^n = \#\{m\colon d(i, V_m) = n + c\}; \quad i \in I_d, \quad n \in \mathbb{Z},$$

for some given $c = c(d)$. Let $X_i = (X_i^n\colon n \in \mathbb{Z})$, taking values in $S = (\mathbb{Z}^+)^{\mathbb{Z}}$. Plainly $X^{(d)} = (X_i\colon i \in I_d)$ is invariant. It can be shown that it is possible to pick $k(d)$ and $c(d)$ such that the processes $X^{(d)}$ converge weakly to some process $\hat{X}$ on the infinite-dimensional cube, and such that

$$d^{-1}\log(k) \to \log(2) + (1-\alpha)\log(1-\alpha) + \alpha\log(\alpha)$$

for any prescribed $0 < \alpha < \frac{1}{2}$. And the limit process $\hat{X}$ is just the particular case (16.15) of the general construction (16.13).

(16.19) Remarks. These "random walk" constructions for invariant processes on the infinite-dimensional cube seem analogous to the constructions $X_{i,j} = f(\xi_i, \eta_j)$ for RCE arrays. Perhaps there is an analog of Corollary 14.15 (resp. Proposition 15.28) which says that an ergodic invariant process on the cube can be represented as a function of some random walk model iff a certain "remote" σ-field contains all the information about the process (resp. iff some "linear entropy" condition holds). On the other hand, it looks plausible that the characterization problem on the cube is rather harder than for RCE arrays, in that the next examples suggest that the general process cannot be obtained from random walk models and independent models.

(16.20) Example. For $1 < k \leq d$ a k-face of I_d is a set of vertices isometric (in I_d) to I_k. Let $X = (X_i : i \in I_d)$ be i.i.d. with $P(X_i = 1) = P(X_i = 0) = \frac{1}{2}$. Let Y $(= Y^{k,d})$ be the process X conditioned on the event

$$\sum_{i \in F} X_i = 0 \bmod 2 \quad \text{for each k-face } F.$$

For fixed k, the processes Y are consistent as d increases, and hence determine a process Y^k on the infinite cube. For $k = 2$ this process Y^k is just Example 16.9 with $\lambda = -1$; for $k \geq 3$ the processes Y^k do not seem to have "random walk" descriptions.

Finally, we can construct invariant processes by borrowing an idea from statistical mechanics (see e.g. Kindermann and Snell (1980)).

(16.21) Example. Ising models. Fix $\alpha \in \mathbb{R}$, $d \geq 1$. For a configuration $\underset{\sim}{x} = (x_i : i \in I_d)$ of 0's and 1's on the d-dimensional cube, define

$$V(\underset{\sim}{x}) = \sum_{\text{edges }(i,j)} 1_{(x_i = x_j)} \, .$$

The function V is invariant under the isometries of the cube, so we can define an invariant distribution by

$$P(X = \underset{\sim}{x}) = c_\alpha \exp(\alpha V(\underset{\sim}{x}))$$

where c_α is a normalization constant. By symmetry, $P(X_i = 0) = P(X_i = 1) = \frac{1}{2}$. Let $\rho_{d,\alpha}$ be the correlation $\rho(X_i, X_j)$ for neighbors i, j. For fixed d, $\rho_{d,\alpha}$ increases continuously from -1 to $+1$ as α increases from $-\infty$ to $+\infty$. There are heuristic arguments which suggest

$$\text{(16.22)} \qquad \rho_{d,\alpha} \to (e^\alpha - 1)/(e^\alpha + 1) \quad \text{as} \quad d \to \infty; \quad \alpha \text{ fixed.}$$

It this is true, then by fixing α, letting $d \to \infty$ and taking (subsequential, if necessary) weak limits we can construct invariant processes on the infinite-dimensional cube with correlation $(e^\alpha - 1)/(e^\alpha + 1)$ between neighbors (and even without (16.22), this holds for some $\alpha(d)$). It would be interesting to get more information about these limit processes; heuristic arguments suggest they are not of the "random walk" types described earlier.

PART IV

17. Exchangeable random sets

In this section we discuss exchangeability concepts for certain types of random subsets M of $[0,1)$ or $[0,\infty)$. Let us start by giving some examples of random subsets M.

(17.1) The zeros of Brownian motion: $M = \{t: W_t = 0\}$.

(17.2) The range of a subordinator: $M = \{X_t(\omega): 0 \le t < \infty\}$, where X_t is a subordinator, that is a Lévy process with $X_0 = 0$ and increasing sample paths.

(17.3) The zeros of Brownian bridge: $M = \{t: W_t^0 = 0\} \subset [0,1]$.

(17.4) An exchangeable interval partition. Take an infinite sequence of constants $c_1 \ge c_2 \ge \cdots > 0$ with $\sum c_i = 1$; take (ξ_i) i.i.d. $U(0,1)$; set

$$L_i = \sum_j c_j 1_{(\xi_j < \xi_i)}, \quad R_i = L_i + c_i .$$

So the intervals (L_i, R_i) have lengths c_i and occur in random order. Let M be the complement of $\bigcup_i (L_i, R_i)$.

These examples all have an exchangeability property we shall specify below. The first three examples are probabilistically natural; the fourth arose in game theory, and attracted interest because certain "intuitively obvious" properties are hard to prove, e.g. the fact (Berbee (1981))

(17.5) $$P(x \in M) = 0 \quad \text{for each } 0 < x < 1.$$

The characterization results for exchangeable sets are roughly similar to those in Section 10 for interchangeable increments processes, but are interesting in that stopping time methods seem the natural tool. Our account closely follows Kallenberg (1982a,b), which the reader should consult for proofs and further results.

Formally, we consider random subsets M of $[0,1]$ or $[0,\infty)$ satisfying

(17.6) $\quad M$ is closed; M has Lebesgue measure zero.

So the complement M^c is a union of disjoint open intervals (L_α, R_α). For each $\varepsilon > 0$ let N_ε be the number of intervals (L_α, R_α) of length at least ε; call these intervals $(L_1^\varepsilon, R_1^\varepsilon), (L_2^\varepsilon, R_2^\varepsilon), \ldots$. Call M <u>exchangeable</u> if for each ε and each $1 \le n \le \infty$ the lengths $(R_i - L_i)$ are, conditional on $\{N_\varepsilon = n\}$, an n-exchangeable sequence.

Consider now the case where M is the closed range of a subordinator (i.e. the closure of M in (17.2)). Set $M^t = \{x-t \colon x \in M,\ x \ge t\}$. The strong Markov property of the subordinator X_t implies that for any stopping time T taking values in $M' = M \setminus (\{L_\alpha\} \setminus \{R_\alpha\})$ we have

(17.7) $\quad M^T$ is independent of $M \cap [0,T]$; $M^T \overset{\mathcal{D}}{=} M$.

Call random subsets satisfying (17.7) <u>regenerative sets</u>. Horowitz (1972) shows a converse: all regenerative sets arise as the closed range of some subordinator. By analogy with (6.1B) and (10.7) consider the condition

(17.8) $\quad M^T \overset{\mathcal{D}}{=} M$; each stopping time $T \in M'$.

Kallenberg calls this <u>strong homogeneity</u>. Kallenberg (1982a), Theorem 4.1, proves

(17.9) <u>Theorem. For unbounded random subsets</u> $M \subset [0,\infty)$ <u>satisfying</u> (17.6), <u>the following are equivalent</u>:

(a) M <u>is exchangeable</u>.

(b) M <u>is strongly homogeneous</u>.

(c) M <u>is a mixture of regenerative sets</u>.

For finite intervals we get a weaker result: Kallenberg (1982a), Theorem 4.2 implies

(17.10) <u>Proposition</u>. <u>For random subsets</u> $M \subset [0,1]$ <u>satisfying</u> (17.6) <u>and with a.s. infinitely many points, the following are equivalent</u>:

(a) M <u>is exchangeable</u>.

(b) M <u>is a mixture of exchangeable interval partitions</u>.

Finally, we remark that the classical theory of local time at the zeros of Brownian motion extends to a theory of local time for regenerative sets, and hence for exchangeable subsets of $[0,\infty)$. For exchangeable interval partitions there is an elementary definition of "local time": in (17.4) set

$$Q_t = \xi_i \quad \text{on} \quad (L_i,R_i) \ .$$

This concept appears useful for tackling problems like (17.5)--see Kallenberg (1983).

18. <u>Sufficient statistics and mixtures</u>

Recall the classical notion of sufficiency. Let $(P_\theta: \theta \in \Theta)$ be a family of distributions on a space S. For notational convenience, let $X: S \to S$ denote the identity map. Then a map $T: S \to \hat{S}$ is a <u>sufficient statistic</u> for the family (P_θ) if the P_θ-conditional distribution of X given T(X) does not depend on θ. More precisely, T is sufficient if there exists a kernel $Q(t,A)$, $t \in \hat{S}$, $A \subset S$, such that for each θ

(18.1) $\qquad Q(T(X),\cdot)$ is a P_θ-r.c.d. for X given T(X).

For instance, if (P_θ) is the family of distributions of i.i.d. Normal sequences $X = (X_1,\dots,X_n)$ on $S = R^n$, then

(18.2)(a) $T_n(x) = (T_{n,1}(x),T_{n,2}(x)) = (\sum x_i,(\sum x_i^2)^{1/2})$ is sufficient, with

(b) $Q_n((t_1,t_2),\cdot)$ the uniform distribution on the surface of the sphere $\{x: T_n(x) = (t_1,t_2)\}$.

The classical interest in sufficiency has been in the context of inference: if $X_1,\dots,X_n$ are assumed to be observations from a known parametric family, then for inference about the unknown parameter one need consider only statistics which are functions of sufficient statistics.

Our interests are rather different. Consider the following general program. Let T_n, Q_n, $n \geq 1$, be a given sequence of maps and kernels. Then study the set M of distributions of sequences $(X_1,X_2,\dots)$ such that for each n

(18.3) $Q_n(T_n(X_1,\dots,X_n),\cdot)$ is a r.c.d. for $(X_1,\dots,X_n)$ given $T_n(X_1,\dots,X_n)$.

For instance, if T_n, Q_n, are the natural sufficient statistics and kernels associated with an exponential family of distributions (P_θ), then by definition M contains the distributions of i.i.d. P_θ sequences. But M is closed under taking mixtures, so M contains the class M_0 of mixtures of i.i.d. P_θ sequences. It generally turns out that $M = M_0$, and so this program leads to a systematic method for characterizing those exchangeable sequences which are mixtures of i.i.d. sequences with distributions from a specified family.

The general program has a much wider scope than the preceding discussion might suggest. First, observe that the class of exchangeable sequences can be defined in this way. For as at (5.2) let $\Lambda_n: R^n \to P(R)$ be the empirical distribution map, and $\Phi_n^{-1}(\mu,\cdot) = L(x_{\pi^*(1)},\ldots,x_{\pi^*(n)})$, where $\mu = \Lambda_n(x)$ and π^* is the uniform random permutation. Then Lemma 5.4 says $(X_1,\ldots,X_n)$ is exchangeable iff $\Phi_n^{-1}(\Lambda_n(X_1,\ldots,X_n),\cdot)$ is a r.c.d. for $(X_1,\ldots,X_n)$ given $\Lambda_n(X_1,\ldots,X_n)$. Thus the class M associated with the sufficient statistics Λ_n and kernels Φ_n^{-1} is precisely the class of infinite exchangeable sequences. Similarly, the other partially exchangeable models in Part III can be fitted into this setting.

Further afield, the study of Markov random fields (as a probabilistic formulation of statistical mechanics problems--Kindermann and Snell (1980)) involves the same ideas: one studies the class of processes $(X_i: i \in \Gamma)$ on a graph Γ such that the conditional distribution of X_i given the distribution at neighboring vertices $(X_j: j \in N_i)$ has a specified form. Yet another subject which can be fitted into the general program is the study of entrance and exit laws for Markov processes.

This general program has been developed recently by several authors, from somewhat different viewpoints: Dynkin (1978), Lauritzen (1982), Diaconis and Freedman (1982), Accardi and Pistone (1982), Dawid (1982). A main theoretical result is a generalization of Theorem 12.10, describing the general distribution in M as a mixture of "extreme" distributions. Our account closely follows that of Diaconis and Freedman (1982): we now state their hypotheses and their version of this main theoretical result.

Let S_i, W_i, $i \geq 1$, be Polish spaces. Let $X_i: \prod_j S_j \to S_i$ be the coordinate map. Let $T_n: \prod_{i=1}^{n} S_i \to W_n$, and let Q_n be a kernel $Q_n(w,A)$, $w \in W_n$, $A \subset \prod_{i=1}^{n} S_i$. Suppose

(18.4)(i) $Q_n(w,\{T_n = w\}) = 1$; $w \in W_n$.

(ii) if $T_n(x) = T_n(x')$ then $T_{n+1}(x,y) = T_{n+1}(x',y)$; $y \in S_{n+1}$.

(iii) for each $w \in W_{n+1}$, $Q_n(T_n(X_1,\ldots,X_n),\cdot)$ is a $Q_{n+1}(w,\cdot)$ r.c.d. for $(X_1,\ldots,X_n)$ given $\sigma(T_n(X_1,\ldots,X_n),X_{n+1})$.

Then let M be the set of distributions P on $\prod_{i\geq 1} S_i$ such that for each n

(18.5) $Q_n(T_n(X_1,\ldots,X_n),\cdot)$ is a P-r.c.d. for $(X_1,\ldots,X_n)$ given $T_n(X_1,\ldots,X_n)$.

Conditions (i) and (ii) are natural: here is an interpretation for (iii). Take the Bayesian viewpoint that (X_i) is an i.i.d. (θ) sequence, where θ has been picked at random from some family. Saying T_n is sufficient is saying that $(X_1,\ldots,X_n)$ and X_{n+1} are conditionally independent given $T_n = T_n(X_1,\ldots,X_n)$. Consider now the conditional distribution of $(X_1,\ldots,X_n)$ given (T_n,X_{n+1},T_{n+1}). By (ii), T_{n+1} is a function of (T_n,X_{n+1}). This and the conditional independence shows that the conditional distribution of $(X_1,\ldots,X_n)$ given (T_n,X_{n+1},T_{n+1}) is the same as the conditional distribution given T_n, which is the kernel distribution $Q_n(T_n,\cdot)$; this is the assertion of (iii). Lauritzen (1982), II.2,3 gives a more detailed discussion.

Next set $S = \bigcap_n \sigma(T_n(X_1,\ldots,X_n),X_{n+1},X_{n+2},\ldots)$, so S is a σ-field on $\prod_{i\geq 1} S_i$. In the context of exchangeable sequences described earlier, S is the exchangeable σ-field. Diaconis and Freedman (1982) prove

(18.6) Theorem. There is a set $S_0 \subset \prod S_i$, $S_0 \in S$, with the following properties:

(i) $P(S_0) = 1$; each $P \in M$.

(ii) $Q(s,\cdot) = \text{weak-limit}_{n\to\infty}\, Q_n(T_n(s),\cdot)$ exists as a distribution on $\prod S_i$; each $s \in S_0$.

(iii) <u>The set of distributions</u> $\{Q(s,\cdot): s \in S_0\}$ <u>is precisely the set of extreme points of the convex set</u> M.

(iv) <u>For each</u> $P \in M$ <u>we have</u> $P(\cdot) = \int_{S_0} Q(s,\cdot)\hat{P}(ds)$, <u>where</u> $\hat{P}$ <u>denotes the restriction of</u> P <u>to</u> S. <u>Thus</u> $Q((X_1,X_2,\ldots),\cdot)$ <u>is a P-r.c.d. for</u> $(X_1,X_2,\ldots)$ <u>given</u> S.

(v) $P \in M$ <u>is extreme iff</u> S <u>is P-trivial</u>.

In the context of exchangeable sequences, S_0 is the set of sequences s for which the limiting empirical distribution $\Lambda(s) = \text{weak-limit } \Lambda_n(s_1,\ldots,s_n)$ exists, and $Q(s,\cdot)$ is the i.i.d. $(\Lambda(s))$ distribution. Thus (iv) recovers a standard form of de Finetti's theorem.

The idea in the proof of Theorem 18.6 is that, if $Q(s,\cdot) = \text{weak-limit } Q_n(T_n(s),\cdot)$ exists, then $Q(s,\cdot)$ defines a distribution in M. Reversed martingale convergence arguments in the spirit of the first proof of de Finetti's theorem show that $Q(s,\cdot)$ exists P-a.s., each $P \in M$. The family of all limiting distributions $Q(s,\cdot)$ is sometimes called the family of <u>Boltzmann laws</u>; this family may contain non-extreme elements of M.

One nice example, outside the context of exchangeability, is the study of mixtures of Markov chains by Diaconis and Freedman (1980b). Let S be a countable set of states. For a sequence $\sigma = (\sigma_1,\ldots,\sigma_n)$ of states and a pair s,t of states let $T_{s,t}(\sigma) = \#\{i: (\sigma_i,\sigma_{i+1}) = (s,t)\}$ be the number of transitions from s to t in the sequence σ. Let $T_n(\sigma) = (\sigma_1; T_{s,t}(\sigma), s,t \in S)$. So $T_n(\sigma) = T_n(\sigma')$ iff σ and σ' have the same initial state and the same transition counts. Now consider a homogenous Markov chain (X_i) on S. Plainly

$$(18.7) \quad P((X_1,\ldots,X_n) = \sigma) = P((X_1,\ldots,X_n) = \sigma') \quad \text{whenever} \quad T_n(\sigma) = T_n(\sigma').$$

Diaconis and Freedman (1980b) prove

(18.8) <u>Proposition</u>. <u>Suppose</u> $X = (X_0, X_1, X_2, \ldots)$ <u>is a process taking values in</u> S <u>which is recurrent, i.e.</u> $P(X_n = X_0$ <u>for infinitely many</u> $n) = 1$. <u>Then</u> X <u>is a mixture of homogenous Markov chains iff</u> X <u>satisfies</u> (18.7).

This fits into the general set-up by making $Q_n(t,\cdot)$ the distribution uniform on the set of sequences σ such that $T_n(\sigma) = t$. Then the set of processes satisfying (18.7) is the set M defined by (18.5); and Proposition 18.8 says that the extreme points of $M \cap \{\text{recurrent processes}\}$ are precisely the recurrent homogenous Markov chains. (A different characterization of such mixtures is in Kallenberg (1982a).)

Another interesting example is the conditional Rasch model discussed by Lauritzen (1982), II.9.7.

We now turn to characterizations of mixtures of i.i.d. sequences. We have already seen one such result, Schoenberg's Theorem 3.6. To fit this into the present context, take $T_n(x_1,\ldots,x_n) = (\sum x_i^2)^{1/2}$, and let $Q_n(t,\cdot)$ be uniform on the surface of the sphere with center 0 and radius t in R^n. Then the set M defined by (18.5) is the set of spherically symmetric sequences. Schoenberg's theorem asserts that each element of M is a mixture of i.i.d. $N(0,\sigma^2)$ sequences; thus the extreme points of M are the i.i.d. $N(0,\sigma^2)$ sequences. There is a related result for general mixtures of i.i.d. Normal sequences. Take T_n, Q_n as at (18.2); then M can be described as the set of sequences (X_i) such that for each n the random vector $(X_1,\ldots,X_n)$ is invariant under the action of all orthogonal $n \times n$ matrices U which preserve the vector $(1,\ldots,1)$. It can be shown (Dawid (1977a); Smith (1981)) that each process in M is a mixture (over μ, σ) of i.i.d. $N(\mu,\sigma^2)$ sequences. These results can in fact be deduced fairly directly from Theorem 18.6; see Diaconis and Freedman (1982); Dawid (1982) for outlines of the argument.

Consider now discrete distributions. For the family of i.i.d. Poisson (λ) sequences, the sufficient statistics are $T_n(x_1,\ldots,x_n) = \sum x_i$ and the kernels are $Q_n(t,(i_1,\ldots,i_n)) = n^{-t}t!/(i_1!\cdots i_n!)$, $\sum i_j = t$. It is natural to hope that M, defined by (18.5), is the class of mixtures of i.i.d. Poisson sequences. This result, and the corresponding results for Binomial and Negative Binomial sequences, are proved in Freedman (1962b). Lauritzen (1982), Section III, gives an abstract treatment of general exponential families.

There are several variations on this theme. One is to consider mixtures of independent non-identically distributed sequences with distributions in a specified family. For example, fix constants $c_i > 0$. For each $\lambda > 0$ let P_λ be the distribution of the independent sequence (X_i), where X_i has Poisson (λ^{c_i}) distribution. Then $T_n(x_1,\ldots,x_n) = \sum c_i x_i$ is sufficient for (P_λ), with kernel $Q_n(t,\cdot)$ being the multinomial distribution of t balls into n equiprobable boxes. Alternatively, for $\mu > 0$ let P_μ be the distribution of the independent sequence (X_i), where X_i has Poisson (μc_i) distribution. Then $T_n(x_1,\ldots,x_n) = \sum x_i$ is sufficient, and the kernel $Q_n(t,\cdot)$ is the multinomial distribution of t balls into n boxes where box i has chance $c_i/\sum c_j$ of being chosen. The structure of M and its extreme points in these examples is discussed in Lauritzen (1982), II.9.20 and in Diaconis and Freedman (1982), Examples 2.5 and 2.6.

So far, we have assumed that both T_n and Q_n are prescribed. Another variant is to prescribe only T_n, and ask what processes are in M for some sequence of kernels Q_n. For instance, it is natural to ask for what classes of exchangeable sequences (X_i) do the partial sums $T_n(x_1,\ldots,x_n) = \sum x_i$ form sufficient statistics; this problem, in the integer-valued case, is discussed in detail in Diaconis and Freedman (1982).

A very recent preprint of Ressel (1983) uses techniques from harmonic analysis on semigroups to obtain characterizations of mixtures of i.i.d. sequences from specific families of distributions. For an infinite sequence $X = (X_j)$ let $\phi_n(\underset{\sim}{t}) = E \exp(\sum_{j=1}^{n} t_j X_j)$. Schoenberg's theorem 3.6 can be stated as

(18.9) If $\phi_n(\underset{\sim}{t}) = f(\sum t_j^2)$ for some function f, then X is a mixture of i.i.d. $N(0,\sigma^2)$ sequences.

Similarly, one can prove the following.

(18.10) If $\phi_n(\underset{\sim}{t}) = f(\sum |t_j|^{\alpha})$ then X is a scale mixture of i.i.d. symmetric stable (α) sequences.

(18.11) If $\phi_n(\underset{\sim}{t}) = f(\Pi(1 + t_j))$ then X is a mixture of i.i.d. Gamma(λ,1) sequences.

Ressel (1983) gives an abstract result which yields these and other characterizations.

19. Exchangeability in population genetics

Perhaps the most remarkable applications of exchangeability are those to mathematical population genetics developed recently by Kingman and others. Our brief account is abstracted from the monograph of Kingman (1980), which the reader should consult for more complete discussion and references.

Consider the distribution of relative frequencies of alleles (i.e. types of gene) at a single locus in a population which is diploid (i.e. with chromosome-pairs, as for humans). Here is the basic Wright-Fisher model for mutation which is neutral (i.e. the genetic differences do not affect fitnesses of individuals).

(19.1) Model. (a) The population contains a fixed number N of individuals (and hence $2N$ genes at the locus under consideration) in each generation.

(b) Each gene is one of a finite number s of allelic types $(A_1,\ldots,A_s)$.

(c) Each gene in the $(n+1)^{st}$ generation can be considered as a copy of a uniformly randomly chosen gene from the nth generation, different choices being independent; except

(d) there is a (small) chance $u_{i,j}$ that a gene of type A_i is mistakenly copied as type A_j (mutation).

Let $X_i^N(n)$ be the proportion of type A_i alleles in the nth generation. Then the vector $(X_1^N(n),\ldots,X_s^N(n))$ evolves as a Markov chain on a finite state space, and converges in distribution as $n\to\infty$ to some stationary distribution

$$(X_1^N,\ldots,X_s^N) . \tag{19.2}$$

We shall consider only the special case where all mutations are equally likely:

(19.3) $\qquad u_{i,j} = v/s \quad (i \neq j), \quad \text{for some} \quad 0 < v$.

Then by symmetry $(X_1^N,\ldots,X_s^N)$ is exchangeable, so $EX_i^N = s^{-1}$. Consider how this distribution varies with the mutation rate v. In the absence of mutation the frequencies $X_i(n)$ evolve as martingales and so eventually get absorbed at 0 or 1; thus $(X_1^N,\ldots,X_s^N) \simeq (1_{(U=1)},\ldots,1_{(U=s)})$, U uniform on $\{1,\ldots,s\}$, as $v \to 0$. On the other hand for large v the mutation effect dominates the random sampling effect, so the allele distribution becomes like the multinomial distribution of $2N$ objects into s classes, so for large v we have $(X_1^N,\ldots,X_s^N) \simeq (1/s,\ldots,1/s) + \text{order } N^{-1/2}$. To obtain more quantitative information, observe that the proportion $X_1^N(n)$ of type 1 alleles evolves as a Markov chain. It is not difficult to get an expression for the variance of the stationary distribution which simplifies to

(19.4) $$\operatorname{var}(X_1^N) \simeq \frac{s^{-1} - s^{-2}}{1 + 4Nv/(s-1)}, \quad N \text{ large}, \quad v \text{ small}.$$

Of course the biologically interesting case is N large, v small, and we can approximate this by taking the limit as

(19.5) $$N \to \infty, \quad v \to 0, \quad 4Nv \to \theta, \quad \text{say}.$$

Then (19.4) suggests we should get some non-trivial limit

(19.6a) $$(X_1^N,\ldots,X_s^N) \xrightarrow{\mathcal{D}} (X_1,\ldots,X_s)$$

where X_i represents the relative frequency of allele A_i in a large population with small mutation rate, when the population is in (time-) equilibrium. This is indeed true, and (Watterson (1976))

(19.6b) $(X_1,\ldots,X_s)$ has the exchangeable Dirichlet distribution (10.22), for $(a,k) = (\theta,s)$.

<u>The infinite-allele model</u>. The s-allele model above describes <u>recurrent</u> mutation, where the effects of one mutation can be undone by subsequent mutation. An opposite assumption, perhaps biologically more accurate, is to suppose that each mutation produces a new allele, different from all other alleles. So consider model (19.1) with this modification, and let v be the probability of mutation. Fix the population size N. It is clear that any given allele will eventually become extinct. So instead of looking at proportions of alleles in prespecified order, look at them in order of frequency; let $Y_1^N(n)$ be the proportion of genes in generation n which are of the most numerous allelic type: $Y_2^N(n)$ the proportion of the second most numerous type, and so on. Again $(Y_1^N(n), Y_2^N(n), \ldots)$ evolves as a finite Markov chain and so converges to a stationary distribution $(Y_1^N, Y_2^N, \ldots)$ with $\sum_i Y_i^N = 1$. Again it is easy to see how this distribution depends on the mutation probability v: as $v \to 0$ we have $Y_1^N \xrightarrow[p]{} 1$; as $v \to 1$ we have each Y_i^N of order (N^{-1}).

What happens as $N \to \infty$? At first sight one might argue that the number of different allelic types in existence simultaneously would increase to infinity, and so the proportions of each type would decrease to zero. But this reasoning is false. In fact, under the assumptions $N \to \infty$, $v \to 0$, $4Nv \to \theta$ used before, we have (see Kingman (1980), p. 40)

(19.7) $(Y_1^N, Y_2^N, \ldots) \xrightarrow{\mathcal{D}} (D_1, D_2, \ldots)$ where (D_i) has the Poisson-Dirichlet(θ) distribution.

Thus for a large population subject to slow, non-recurrent neutral mutation, the proportions of the different alleles present at a particular time, arranged in decreasing order, should follow a Poisson-Dirichlet distribution.

Now consider sampling K genes from such a population. Let a_r be the number of allelic types for which there are exactly r genes of that type in the sample. Then Theorem 11.14 shows that the chance of obtaining a specified $(a_1,a_2,\ldots)$ is given by formula (11.16), the Ewens sampling formula. Indeed, if we consider the partition $\mathcal{R}^K$ into allelic types of a sample of size K from a hypothetical limiting infinite population, these random partitions satisfy the consistency conditions of Theorem 11.14.

Let us outline a method for deriving the infinite-allele result (19.7) from the s-allele result (19.6). Fix the population size N. Imagine that each new allele created by mutation is named by a random variable ξ distributed uniformly on $(0,1)$. So each gene g has a label ξ_g which indicates its allelic type. Thus the genetic composition of generation n can be described by a process $(W_n^N(u)\colon 0\leq u\leq 1)$, where $W_n^N(u)$ is the proportion of genes g for which $\xi_g\leq u$. As $n\to\infty$ this converges to a process $(W^N(u)\colon 0\leq u\leq 1)$, where the jump sizes $(W^N(u)-W^N(u-))$, rearranged in decreasing order, are the variables $(Y_1^N,Y_2^N,\ldots)$ above, and the jump positions are independent uniform. Now fix s, and call an allele "type j", $1\leq j\leq s$, if its name ξ is in the interval $((j-1)/s,j/s)$. If we only take notice of the "type" of alleles, then the infinite-allele model evolves in precisely the same way as the s-allele model. The convergence result (19.6) translates to

(19.8) $$(W^N(0),W^N(1/s),\ldots,W^N(1))\xrightarrow{\mathcal{D}}(Z(0),Z(1/s),\ldots,Z(1)),$$

where Z is the Dirichlet(θ) process. But then

(19.9) $$(W^N(u)\colon 0\leq u\leq 1)\xrightarrow{\mathcal{D}}(Z(u)\colon 0\leq u\leq 1)\ \text{ in } D(0,1),$$

since (19.8) gives convergence of finite-dimensional distributions, and establishing tightness is an exercise in technicalities. But convergence in $D(0,1)$ implies convergence of jump sizes, and this gives (19.7).

Other applications. There are other, quite different, applications of exchangeability to genetics. Suppose the "fitness" of an individual does depend on his genetic type, an individual with gene-pair (A_i, A_j) having fitness $w_{i,j}$. Imagine alleles labelled $A_1, A_2, \ldots$ in order of their creation by mutation. Mutation is a random process, so the $w_{i,j}$ should be regarded as random variables. It is not a priori apparent how to model the distribution $(w_{i,j})$, but it is natural to argue that $(w_{i,j})$ should be weakly exchangeable in the sense of (14.19), and then Theorem 14.21 can be brought to bear. See Kingman (1980), Section 2.5.

Another application is to the gene genealogy of haploid (i.e. single sex) populations. Suppose we sample K individuals from the current generation. For each $n \geq 0$ we can define an exchangeable random partition $\mathcal{R}^K(n)$ of $\{1,\ldots,K\}$, where the components are the families of individuals with a common ancestor in the nth previous generation. Letting the population size increase, K increase, and rescaling time, the process $(\mathcal{R}^K(n): n \geq 0)$ approximates a certain continuous-time partition-valued process $(\mathcal{R}(t): t \geq 0)$, the coalescent. See Kingman (1982a,b).

Finally, Dawson and Hochberg (1982) involve exchangeability ideas in a diffusion analysis of infinite-allele models more complicated than that described here.

20. Sampling processes and weak convergence

Given a finite sequence $x_1, \ldots, x_M$ of real constants, recall that the urn process is the sequence of random draws without replacement:

$X_i = x_{\pi^*(i)}$ where π^* is the uniform random permutation on $\{1,\ldots,M\}$.

By the <u>sampling process</u> we mean the process of partial sums:

$$S_n = \sum_{i=1}^{n} X_i .$$

We shall often consider sampling processes drawn from <u>normalized</u> urns, where

$$\sum x_i = 0 , \quad \sum x_i^2 = 1 .$$

There is of course a vast literature on sampling: we shall merely mention a few results which relate to other ideas in exchangeability. We can distinguish two types of results: "universal" results true for all (normalized) urns, and "asymptotic" results as the individual elements of the urn become negligible. The main asymptotic result, Theorem 20.7, leads naturally to questions about weak convergence of general finite exchangeable sequences.

The most basic universal results are the elementary formulas for moments.

$$(20.1) \qquad \begin{aligned} ES_n &= n\mu/M \\ \mathrm{var}(S_n) &= \frac{n(M-n)(\sigma^2 - \mu^2/M)}{M(M-1)} \quad \text{where } \mu = \sum x_i, \ \sigma^2 = \sum x_i^2 . \end{aligned}$$

Restricting to normalized urns, we have also

$$(20.2) \qquad ES_n^4 = \frac{n(M-n)}{M(M-1)} \sum x_i^4 + \frac{3n(n-1)(M-n)(M-n-1)}{M(M-1)(M-2)(M-3)}(1 - 2\sum x_i^4) .$$

A more abstract universal result involves rescaling the sampling process to make it a continuous-parameter process

$$S_t = S_{[Mt]} , \quad 0 \le t \le 1 .$$

Then we can think of S as a random element of the function space $D(0,1)$ with its usual topology (Billingsley (1968)). In this setting, we have

(20.3) Proposition. <u>The family of processes</u> S_t <u>obtained from all normalized urns is a tight family in</u> $D(0,1)$.

This is implicit in Billingsley (1968), (24.11) and Theorem 15.6. An alternative proof can be obtained from the tightness criteria in Aldous (1978).

In particular, Proposition 20.3 implies that there are bounds on the maxima of sampling processes which are uniform over the family of normalized urns. In other words, there exists a function ϕ with $\phi(\lambda) \to 0$ as $\lambda \to \infty$ and

(20.4) $$P\left(\max_n |S_n| > \lambda\right) \le \phi(\lambda) \; ; \quad \text{all normalized urns.}$$

I do not know what the best possible function ϕ is; here is a crude bound.

(20.5) Lemma. $\phi(\lambda) = 8/\lambda^2$ <u>satisfies</u> (20.4).

Proof. Let $F_k = \sigma(X_1,\dots,X_k)$, let $T = \min\{i: S_i > \lambda\}$. For $k \le m = [M/2]$,

$$E(S_m | F_k) = \frac{(M-m)}{(M-k)} S_k$$

and so

$$E(S_m | F_{T\wedge m}) = \frac{M-m}{M-T\wedge m} S_{T\wedge m} \ge \tfrac{1}{2}\lambda \quad \text{on} \quad \{T \le m\} ,$$

and so $E(S_m^2 | F_{T\wedge m}) \ge \frac{1}{4}\lambda^2$ on $\{T \le m\}$. So

$$P(\max_{i \le m} S_i > \lambda) = P(T \le m) \le 4\lambda^{-2} ES_m^2 \le 2\lambda^{-2} \quad \text{using (20.2).}$$

Using the same inequality on $(-S_i)$, and using the symmetry $(S_i) \overset{\mathcal{D}}{=} (S_{M-i})$, we obtain the desired result.

Another type of universal result relates the sampling process S_n arising from sampling without replacement to the process $S_n^* = \sum_{i=1}^{n} X_i^*$ arising from random draws (X_i^*) from the same urn made with replacement, that is with (X_i^*) an i.i.d. sequence. Proposition 5.6 shows that when n is small compared to $M^{1/2}$ then the total variation distance between S_n and S_n^* is small, and this total variation result cannot be improved. However, there are results which compare the distributions of S_n and S_n^*. Given random variables U, V on R, say U is a <u>dilation</u> of V if there exist random variables $\hat{U}, \hat{V}$ such that

$$\hat{U} \stackrel{\mathcal{D}}{=} U, \quad \hat{V} \stackrel{\mathcal{D}}{=} V, \quad E(\hat{U}|\hat{V}) = \hat{V} .$$

This implies (and in fact is equivalent to--see e.g. Strassen (1965))

$$E\phi(U) \geq E\phi(V), \quad \text{all continuous convex } \phi: R \to R .$$

Informally, the distribution of U is "more spread out" than that of V. The next result extends the familiar result that $\text{var}(S_n) \leq \text{var}(S_n^*)$. See Kemperman (1973) for further related results.

(20.6) <u>Proposition</u>. <u>S_n^* is a dilation of S_n, for each urn and each</u> $n \geq 1$.

<u>Proof</u>. Without essential loss of generality, suppose the urn $\{x_1,\ldots,x_M\}$ contains distinct elements. Fix n and let $(X_1,\ldots,X_n)$ be draws without replacement. For distinct $(y_1,\ldots,y_n) \subset \{x_i\}$ and not necessarily distinct $\{z_1,\ldots,z_n\} \subset \{y_i\}$ define the conditional distribution

$$P\left(X_1^* = z_1,\ldots,X_n^* = z_n | X_1 = y_1,\ldots,X_n = y_n\right) = \frac{\binom{M}{n}}{n^M n! \binom{M-L}{n-L}} \quad \text{where } L = \#\{z_i\} .$$

Then it can be verified that the unconditional distribution of (X_i^*) is the distribution of sampling with replacement. And by symmetry, for $N(y) = \#\{i \leq n: X_i^* = y\}$ we have

$$E(N(y)|X_1 = y_1,\dots,X_n = y_n) = \begin{cases} 1, & y \in \{y_i\} \\ 0 & \text{otherwise} \end{cases}$$

So $E(S_n^*|X_1 = y_1,\dots,X_n = y_n) = \sum y_i$, which gives $E(S_n^*|S_n) = S_n$ as required.

We turn now to asymptotic results. For each $q \geq 1$ let $(x_1^q,\dots,x_{M_q}^q)$ be a normalized urn. Let S^q be the scaled sampling process

$$S_t^q = S_{[tM_q]}^q = \sum_{i=1}^{[tM_q]} X_i \; ; \quad 0 \leq t \leq 1$$

and think of S^q as a random element of $D(0,1)$.

(20.7) Theorem. *If*

(20.8) $$\max_i |x_i^q| \to 0 \quad \text{as} \quad q \to \infty,$$

then $S^q \xrightarrow{\mathcal{D}} W^0$ *in* $D(0,1)$ *as* $q \to \infty$, *where* W^0 *is Brownian bridge.*

This is Theorem 24.1 of Billingsley (1968). Let us sketch a slightly different proof. By Proposition 20.3 we need only show that any subsequential weak limit process Z_t is Brownian bridge. By (20.8) Z has continuous paths, and clearly Z must have interchangeable increments, so by Theorem 10.11 we can take $Z_t = \alpha W_t^0$ for some $\alpha \geq 0$ independent of W^0. But using the moment formulas (20.1, 20.2) we can verify

$$E(S_t^q)^2 \to E(W_t^0)^2; \quad E(S_t^q)^4 \to E(W_t^0)^4 \quad \text{as} \quad q \to \infty .$$

So $E\alpha^2 = 1$ and $E\alpha^4 = 1$, which implies $\alpha = 1$, so Z is indeed Brownian bridge.

Theorem 20.7 leads to results for general finite exchangeable sequences. Consider a triangular array $(Z_{q,i}: 1 \leq i \leq M_q, 1 \leq q)$ such that

(20.9)(a) for each q the sequence $(Z_{q,1},\dots,Z_{q,M_q})$ is exchangeable

(b) $\max_i |Z_{q,i}| \xrightarrow[p]{} 0$ as $q \to \infty$.

As before let $S_t^q = \sum_{i=1}^{tM_q} Z_{n,i}$, let W^0 be Brownian bridge and let W be Brownian motion.

(20.10) Corollary. Suppose $(Z_{q,i})$ satisfies (20.9). Define random variables $\mu_q = \sum_i Z_{q,i}$, $\sigma_q^2 = \sum_i (Z_{q,i} - \mu_q/M_q)^2$. Let $X_t = \sigma W_t^0 + \mu t$, where (σ,μ) is independent of W^0. Then

(a) $S^q \xrightarrow{\mathcal{D}} X$ in $D(0,1)$ iff $(\mu_q,\sigma_q) \xrightarrow{\mathcal{D}} (\mu,\sigma)$.

In particular

(b) $S^q \xrightarrow{\mathcal{D}} W^0$ iff $\sum_i Z_{q,i} \xrightarrow[p]{} 0$ and $\sum_i Z_{q,i}^2 \xrightarrow[p]{} 1$,

(c) $S^q \xrightarrow{\mathcal{D}} W$ iff $\sum_i Z_{q,i} \xrightarrow{\mathcal{D}} N(0,1)$ and $\sum_i Z_{q,i}^2 \xrightarrow[p]{} 1$.

Sketch of proof. Suppose $(\hat{Z}_{q,i})$ satisfies (20.9) and

$$\sum_i \hat{Z}_{q,i} = 0; \quad \sum_i \hat{Z}_{q,i}^2 = 1 . \tag{20.11}$$

Then for each q the process $(\hat{Z}_{q,i})$ is a mixture of normalized urn rpocesses, and Theorem 20.7 implies $\hat{S}^q \xrightarrow[\mathcal{D}]{} W^0$. Now given $(Z_{q,i})$ define $\hat{Z}_{q,i} = (Z_{q,i} - \mu_q/M_q)/\sigma_q$. Then, in the case

$$\sigma > 0 \text{ a.s.} \tag{20.12}$$

the array $(\hat{Z}_{q,i})$ satisfies (20.9) and (20.11), and so $\hat{S}^q \xrightarrow{\mathcal{D}} W^0$. Furthermore, for events of the form $A_q = \{(\mu_q,\sigma_q) \in B_q\}$ with $\lim P(A_q) > 0$, the conditional distributions $(L(\hat{Z}_{q,i} | A_q))$ also satisfy (20.9) and (20.11), so these conditional distributions also converge weakly to W^0. When $(\mu_q,\sigma_q) \xrightarrow{\mathcal{D}} (\mu,\sigma)$ this implies $(S^q,\mu_q,\sigma_q) \xrightarrow{\mathcal{D}} (W^0,\mu,\sigma)$. Since

(20.13) $$S_t^q = \sigma_q \hat{S}_t^q + \mu_q t$$

we finally obtain $S^q \xrightarrow{\mathcal{D}} X$.

In the case $\sigma = 0$ a.s., the maximal inequality (20.5) implies $\max_t |\sigma_q \hat{S}_t^q| \xrightarrow[p]{} 0$ as $q \to \infty$, and then (20.13) implies $S^q \xrightarrow{\mathcal{D}} X$. The general case can be reduced to these two cases, and this gives the "if" assertion of (a). The "only if" assertion follows from the "if" assertion and the facts

(i) tightness of (S^q) implies tightness of (μ_q, σ_q)

(ii) the distribution of $X_t = \sigma W_t^0 + \mu t$ determines the distribution of (σ, μ).

Assertions (b) and (c) are special cases of (a).

There is of course another result on weak convergence which is more celebrated than Theorem 20.7; it is interesting to note that this is essentially just a special case of Theorem 20.7.

(20.14) <u>Corollary</u>. *Let* (ξ_i) *be i.i.d. uniform on* $[0,1]$. *Let* $F^q(\omega,t)$, $0 \le t \le 1$, *be the empirical distribution function of* $(\xi_1(\omega), \ldots, \xi_q(\omega))$. *Let* $Y_t^q(\omega) = q^{1/2}(F_t^q(\omega) - t)$. *Then* $Y^q \xrightarrow{\mathcal{D}} W^0$ *as* $q \to \infty$.

<u>Proof</u>. Let $M_q = q^3$,

$$X_{q,i} = \#\{j: (i-1)/M_q < \xi_j \le i/M_q\} - q/M_q ,$$

so that

(20.15) $$\max_i X_{q,i} \xrightarrow[p]{} 1 \text{ as } q \to \infty .$$

Then the array $Z_{q,i} = q^{-1/2} X_{q,i}$ satisfies the conditions of Corollary 20.10(b), so $S^q \xrightarrow{\mathcal{D}} W^0$. And (20.15) also gives $\sup_t |S_t^q - Y_t^q| \xrightarrow[p]{} 0$, so that $Y^q \xrightarrow{\mathcal{D}} W^0$.

Remark. The weak convergence form of Corollary 20.14 is nowadays rather obsolete, in that much more precise "strong approximation" results are known --see Csörgö and Revesz (1981). Similarly, there must be more precise strong approximation forms of Theorem 20.7, but I do not know any references in the literature.

The previous results do not really tackle the problem of when the sum of a finite exchangeable sequence can be approximated by a Normal distribution. To see that there can be no easy answer to this problem, recall that from an arbitrary sequence $X_1,\dots,X_n$ we can produce an exchangeable sequence $Z_1,\dots,Z_n$ by random permutation, and $\sum Z_i = \sum X_i$. Thus exchangeability imposes no restriction on the possible distribution of $\sum Z_i$.

One way to obtain weak convergence results for exchangeable sequences with little effort is to appeal to the general weak convergence results for martingales. A typical martingale result (Hall and Heyde (1980) Theorem 4.4; Helland (1982) Theorem 3.2) is the following.

(20.16) *Theorem. For each* $q \geq 1$ *let* $(X_{q,i}, F_{q,i}, 1 \leq i \leq M_q)$ *be a martingale difference sequence. If*

$$\sum_{i=1}^{[tM_q]} E(X_{q,i}^2 | F_{q,i-1}) \xrightarrow[p]{} t \; ; \quad \text{each } 0 \leq t \leq 1$$

$$\sum_{i=1}^{M_q} E(X_{q,i}^2 1_{(|X_{q,i}|<\varepsilon)} | F_{q,i-1}) \xrightarrow[p]{} 0 \; ; \quad \text{each } \varepsilon > 0$$

then $S^q \xrightarrow{\mathcal{D}} W$, *where* $S_t^q = \sum_{i=1}^{[tM_q]} X_{q,i}$.

It is not difficult to specialize this to the exchangeable case and obtain the following sufficient conditions for convergence.

(20.17) Corollary. *Let* $(Z_{q,i})$ *satisfy* (20.9). *Let* $F_{q,j} = \sigma(Z_{q,i}: i \le j)$, *and let* $Y_q = Z_{q,M_q} 1(|Z_{q,M_q}| \le 1)$. *Suppose that whenever* $L_q/M_q \to t < 1$ *we have*

$$M_q E(Y_q | F_{q,L_q}) \xrightarrow[p]{} 0$$

$$M_q E(Y_q^2 | F_{q,L_q}) \xrightarrow[p]{} 1 .$$

Then $S^q \xrightarrow{\mathcal{D}} W$.

Of course, by using different forms of the martingale theorem we can obtain different sufficient conditions in Corollary 20.17. We remark also that, under moment conditions, conditions like those of Theorem 20.16 are necessary for convergence of martingales; but there can be no such necessity in Corollary 20.17 for exchangeable arrays, because for each q we could take $Z_{q,1}, \ldots, Z_{q,M_q}$ as a finite mixture of urn processes where the urns contained distinct values, and then $E(Z_{q,M_q} | F_{q,1}) = (S_1^q - Z_{q,1})/(M_q - 1)$.

The situation simplifies somewhat if we assume that the finite exchangeable sequences can be embedded into longer exchangeable sequences. The next result is due to Weber (1980).

(20.18) Proposition. *Let* $(Z_{q,i})$ *satisfy* (20.9). *Suppose for each* q *that* $(Z_{q,i}: 1 \le i \le M_q)$ *extends to an exchangeable sequence* $(Z_{q,i}: 1 \le q \le N_q)$, *where* $M_q/N_q \to 0$ *as* $q \to \infty$. *Suppose also*

$$EZ_{q,1}Z_{q,2} \to 0 \quad \text{as} \quad q \to \infty ;$$

$$\sum_{i=1}^{M_q} Z_{q,i}^2 \xrightarrow[p]{} 1 \quad \text{as} \quad q \to \infty .$$

Then $S^q \xrightarrow{\mathcal{D}} W$.

However, this extendibility condition is very restrictive. The overall picture of central limit theorems for finite exchangeable sequences remains unsatisfactory. It turns out that Poisson limit theorems are somewhat more tractable: Eagleson (1982) gives a survey of such theorems for exchangeable sequences and partially exchangeable arrays. Of course there are many results which deal with special cases of finite exchangeable sequences, e.g. in the theory of random allocations--see Chow and Teicher (1978) Sections 3.2 and 9.2; Kolchin and Sevast'yanov (1978); Quine (1979); Johnson and Kotz (1977).

21. Other results and open problems

In this final section we give references to topics not previously mentioned and speculate on future lines of research.

(21.1) Non-standard versions of probability. Most mathematical probabilists (including the author) work within the "standard" model: (Ω,F,P), Radon measures and all that. There are however numerous alternative formalizations of probability theory, and exchangeability is such a basic concept that it makes sense within any version. Even if the reader is not interested in these alternative versions for their own sake, analysis within an alternative version can sometimes lead to results within the standard version.

Cylinder measures. On infinite-dimensional Hilbert space there are no non-trivial rotationally invariant σ-additive probability measures. Instead, one can consider cylinder measures, and obtain the analog of Schoenberg's Theorem 3.6: every rotationally invariant cylinder measure is a mixture of Gaussian cylinder measures (see e.g. Choquet (1969) Vol. 3). These measures play a fundamental role in certain analytical treatments of Brownian motion (Hida (1980)).

Non-standard analysis. The treatment of partially exchangeable arrays by Hoover (1979, 1982) uses non-standard analysis; perhaps this approach would be useful for other exchangeability problems such as those of Section 16.

Finitely additive measures. At (13.27) we argued informally that the ability to pick an integer uniformly at random, in the finitely additive setting, led to plausibility arguments for several of our characterization results. Can finitely additive measures be employed to give rigorous results in the standard version more simply than the known proofs? See Dubins (1982) for one approach.

Quantum probability. Quantum theorists regard random variables as operators on Hilbert space. There is a version of de Finetti's theorem in this setting: see Accardi and Pistone (1982), Section 7, for an outline.

Function measures. Physicists define Brownian motion as the process on continuous functions $f \in C[0,1]$ with "density" $\psi(f)$ satisfying

$$\log \psi(f) = -\frac{1}{2}\int_0^1 (f'(x))^2 dx .$$

In a similar spirit, the Dirichlet(a) process (Section 10) on increasing functions f with $f(0) = 0$ and $f(1) = 1$ has "density"

$$\log \psi(f) = a\int_0^1 \log(f'(x))dx .$$

The form of these densities makes the interchangeable increments property plain. Does this approach yield other interesting processes with interchangeable increments?

(21.2) Other forms of invariance. There are several naturally-occurring classes of processes defined by invariance properties which we have not yet mentioned.

Self-similar processes. Fix $0 < H < \infty$. A process $(X_t\colon t \geq 0)$ satisfying

$$(X_t\colon t \geq 0) \overset{\mathcal{D}}{=} (X_{t_0+t} - X_{t_0}\colon t \geq 0); \quad \text{each } t_0 > 0,$$

$$(X_t\colon t \geq 0) \overset{\mathcal{D}}{=} (c^{-H} X_{ct}\colon t \geq 0); \quad \text{each } c > 0,$$

is self-similar with stationary increments. Taqqu (1982) surveys this area. See also O'Brien and Vewvaat (1983). As with stationary processes, this class seems too large for any explicit characterization of the ergodic processes to be possible; but perhaps some subclasses are more tractable?

Invariant point processes. Kallenberg (1976, 1976-81) and others have investigated point processes with invariance properties. Here is one natural problem, discussed in Kallenberg (1982c). Consider a group T of transformations of a space S, and suppose there is a unique (up to constant multiples) σ-finite T-invariant measure λ on S. Then a Poisson point process of intensity $c\lambda(\cdot)$ is T-invariant, and hence mixtures (over c) of such processes are also σ-invariant.

Problem. Under what circumstances are all T-invariant point processes mixtures of Poisson processes?

The most famous example, essentially due to Davidson (1974), concerns processes of random lines in $\mathbb{R}^2$ invariant under Euclidean motions of $\mathbb{R}^2$: if such a process has a.s. no parallel lines, then it is a mixture of Poisson line processes.

Further results on invariance of point processes are in Kallenberg (1982a), Section 13.

Sign-invariant processes. Berman (1965) studies processes with the property

$$(X_1,\dots,X_n) \stackrel{\mathcal{D}}{=} (\varepsilon_1 X_1,\dots,\varepsilon_n X_n); \quad \text{all} \quad \varepsilon_i \in \{-1,+1\},$$

and their continuous-time analogs.

(21.3) Special exchangeable sequences. One way to characterize exchangeable sequences which are mixtures of i.i.d. sequences with specific distributions is via sufficient statistics, as in Section 18. But several other types of characterizations exist. One type, natural from the Bayesian viewpoint, involves assumptions of regularity of posterior distributions: see e.g. Diaconis and Ylvisaker (1979); Zabell (1982). Another idea is to consider exchangeable renewal processes and impose regularity conditions on mean residual lifetimes; see Sigalotti (1982).

List of open problems posed in previous sections

1.11	1.12	2.29	3.14	5.13	7.21	9.12	11.26	12.20	12.29
13.10	13.14	13.17	13.28	15.6	15.10	15.15	15.20	15.30	Section 16

APPENDIX

Properties of Conditional Independence

Listed below are the properties of conditional independence we have used. Verification of these properties is a straightforward exercise in the use of properties of conditional expectations (with which we suppose the reader is familiar). Understanding these properties requires good intuition. An excellent elementary account is given by Pfeiffer (1979); a measure-theoretic account is given in Chow and Teicher (1978); more complete discussions and lists of properties are in Dawid (1979), Lauritzen (1982), Dohler (1980).

In what follows, sets are measurable, and functions ϕ are bounded measurable real-valued.

Say X, Y are conditionally independent given F if

(A1) $$P(X \in A, Y \in B | F) = P(X \in A | F) P(Y \in B | F); \quad \text{all } A, B .$$

Each of the following is equivalent to (A1):

(A2) $$E(\phi_1(X)\phi_2(Y) | F) = E(\phi_1(X) | F) E(\phi_2(Y) | F); \quad \text{all } \phi_1, \phi_2.$$

(A3) $$P(X \in A | F, Y) = P(X \in A | F); \quad \text{all } A.$$

(A4) $$E(\phi(X) | F, Y) = E(\phi(X) | F); \quad \text{all } \phi.$$

In these definitions we can replace a random variable X by a σ-field G, by replacing events $\{X \in A\}$ with events G $(G \in G)$, and replacing functions $\phi(X)$ with bounded random variables $V \in G$.

A family $\{X_i : i \in I\}$ is conditionally independent given F if the product formula in (A1) holds for each finite subset of (X_i).

Here are some other properties.

(A5) Suppose that for each $j \geq 1$, X_j and $\sigma(X_i : i > j)$ are conditionally independent given F. Then $(X_i : i \geq 1)$ are conditionally independent given F.

(A6) If X and X are conditionally independent given F then $X \in F$ a.s.

(A7) Suppose X and F are conditionally independent given G, and suppose X and G are conditionally independent given $H \subset G$. Then X and F are conditionally independent given H.

NOTATION

"Positive", "increasing" are used in the weak sense.

$\mathbb{R}$	set of real numbers
$\mathbb{Z}$; $\mathbb{N}$	set of all integers; set of natural numbers
$\#A$	cardinality of set A
1_A	indicator function/random variable: $1_A(x) = 1$ for $x \in A$ $= 0$ else.
$\delta_a(\cdot)$	probability measure degenerate at a: $\delta_a(A) = 1_A(a)$

$\mathcal{F} \subset \mathcal{G}$ a.s. means: for each $G \in \mathcal{G}$ there exists $F \in \mathcal{F}$ such that $P(F \Delta G) = 0$.

$\mathcal{F} = \mathcal{G}$ a.s. means $\mathcal{F} \subset \mathcal{G}$ a.s. and $\mathcal{G} \subset \mathcal{F}$ a.s.

$\mathcal{F}$ is trivial means $\mathcal{F} = \{\phi, \Omega\}$ a.s.

$\mathcal{L}(X)$	distribution of random variable X
$\sigma(X)$	σ-field generated by X
$\xrightarrow[p]{}$	convergence in probability
$\xrightarrow[\mathcal{D}]{}$	convergence in distribution
$N(\mu,\sigma^2)$	Normal distribution
$U(0,1)$	Uniform distribution on (0,1)

REFERENCES

Accardi, L. and Pistone, G. (1982). de Finetti's theorem, sufficiency, and Dobrushin's theory. In Exchangeability in Probability and Statistics (G. Koch and F. Spizzichino, eds.), North-Holland, Amsterdam, 125-156.

Ahmad, R. (1982). On the structure and application of restricted exchangeability. In Exchangeability in Probability and Statistics (G. Koch and F. Spizzichino, eds.), North-Holland, Amsterdam, 157-164.

Aldous, D. J. (1977). Limit theorems for subsequences of arbitrarily-dependent sequences of random variables. Zeitschrift fur Wahrscheinlichkeitstheorie verw. Gebiete **40**, 59-82.

Aldous, D. J. (1978). Stopping times and tightness. Ann. Probability **6**, 335-340.

Aldous, D. J. (1979). Symmetry and independence for arrays of random variables. Preprint.

Aldous, D. J. (1981a). Representations for partially exchangeable arrays of random variables. J. Multivariate Anal. **11**, 581-598.

Aldous, D. J. (1981b). Subspaces of L^1, via random measures. Trans. Amer. Math. Soc. **267**, 445-463.

Aldous, D. J. (1982a). Partial exchangeability and $\bar{d}$-topologies. In Exchangeability in Probability and Statistics (G. Koch and F. Spizzichino, eds.). North-Holland, Amsterdam, 23-38.

Aldous, D. J. (1982b). On exchangeability and conditional independence. In Exchangeability in Probability and Statistics (G. Koch and F. Spizzichino, eds.). North-Holland, Amsterdam, 165-170.

Aldous, D. J. and Eagleson, G. K. (1978). On mixing and stability of limit theorems. Ann. Probability 6, 325-331.

Aldous, D. J. and Pitman, J. (1979). On the zero-one law for exchangeable events. Ann. Probability **7**, 704-723.

Arnaud, J.-P. (1980). Fonctions sphériques et fonctions définies positives sur l'arbre homogène. C. R. Acad. Sci. A 290, 99-101.

Bailey, R., Praeger, C., Speed, T. P. and Taylor, D. (1984). Analysis of Variance. Springer-Verlag, to appear.

Barbour, A. D. and Eagleson, G. K. (1983). Poisson approximations for some statistics based on exchangeable trials. Adv. Appl. Prob. 15, 585-600.

Berbee, H. (1981). On covering single points by randomly ordered intervals. Ann. Probability 9, 520-528.

Berkes, I. and Péter, E. (1983). Exchangeable r.v.'s and the subsequence principle. Preprint, Math. Inst. Hungarian Acad. Sci.

Berkes, I. and Rosenthal, H. P. (1983). Almost exchangeable sequences of random variables. Zeitschrift fur Wahrscheinlichkeitstheorie verw. Gebiete (to appear).

Berman, S. (1965). Sign-invariant random variables and stochastic processes with sign-invariant increments. Trans. Amer. Math. Soc. 119, 216-243.

Billingsley, P. (1968). Convergence of Probability Measures. Wiley, New York.

Blackwell, D. and Freedman, D. (1964). The tail σ-field of a Markov chain and a theorem of Orey. Ann. Math. Statist. 35, 1921-1925.

Blum, J. R. (1982). Exchangeability and quasi-exchangeability. In Exchangeability in Probability and Statistics (G. Koch and F. Spizzichino, eds.). North-Holland, Amsterdam, 171-176.

Blum, J., Chernoff, J., Rosenblatt, M., Teicher, H. (1958). Central limit theorems for interchangeable processes. Canad. J. Math. 10, 222-229.

Breiman, L. (1968). Probability. Addison-Wesley, Reading, Mass.

Brown, G. and Dooley, A. H. (1983). Ergodic measures are of weak product type. Preprint, School of Mathematics, University of New South Wales.

Brown, T. C. (1982). Poisson approximations and exchangeable random variables. In Exchangeability in Probability and Statistics (G. Koch and F. Spizzichino, eds.). North-Holland, Amsterdam, 177-184.

Bru, B., Heinich, H., Lootgitier, J.-C. (1981). Lois de grands nombres pour les variables échangeables. Comptes Rendus Acad. Sci. 293, 485-488.

Bühlmann, H. (1958). Le problème "limite centrale" pour les variables aléatoires échangeables. C. R. Acad. Sci. Paris 246, 534-536.

Bühlmann, H. (1960). Austauschbare stochastische Variabeln und ihre Grenzwertsätze. Univ. Calif. Publ. Statist. 3, 1-35.

Carnal, E. (1980). Indépendence conditionelle permutable. Ann. Inst. Henri Poincaré B 16, 39-47.

Cartier, P. (1973). Harmonic analysis on trees. In Harmonic Analysis on Homogenous Spaces. Amer. Math. Soc. (Proc. Sympos. Pure Math. 26). Providence.

Chatterji, S. D. (1974). A subsequence principle in probability theory. Bull. Amer. Math. Soc. 80, 495-497.

Choquet, G. (1969). Lectures on Analysis. Benjamin, Reading, Mass.

Chow, Y. S. and Teicher, H. (1978). Probability Theory. Springer-Verlag, New York.

Crisma, L. (1982). Quantitative analysis of exchangeability in alternative processes. In Exchangeability in Probability and Statistics (G. Koch and F. Spizzichino, eds.). North-Holland, Amsterdam, 207-216.

Csörgö, M. and Revesz, P. (1981). Strong Approximations in Probability and Statistics. Academic Press, New York.

Daboni, L. (1982). Exchangeability and completely monotone functions. In Exchangeability in Probability and Statistics (G. Koch and F. Spizzichino, eds.). North-Holland, Amsterdam, 39-45.

Dacunha-Castelle, D. (1974). Indiscernability and exchangeability in L^p spaces. Seminar on Random Series, Convex Sets and Geometry of Banach Spaces. Aarhus.

Dacunha-Castelle, D. (1982). A survey of exchangeable random variables in normed spaces. In Exchangeability in Probability and Statistics (G. Koch and F. Spizzichino, eds.). North-Holland, Amsterdam, 47-60.

Dacunha-Castelle, D. and Schreiber, M. (1974). Techniques probabilistes pour l'étude de problèmes d'isomorphismes entre espaces de Banach. Annales Inst. Henri Poincaré **10**, 229-277.

Davidson, R. (1974). Construction of line processes. In Stochastic Geometry (E. F. Harding and D. G. Kendall, eds.). Wiley, New York.

Dawid, A. P. (1977a). Invariant distributions and analysis of variance models. Biometrika **64**, 291-297.

Dawid, A. P. (1977b). Spherical matrix distributions and a multivariate model. J. Roy. Statist. Soc. B **39**, 254-261.

Dawid, A. P. (1978). Extendibility of spherical matrix distributions. J. Multivariate Anal. **8**, 567-572.

Dawid, A. P. (1979). Conditional independence in statistical theory. J. Roy. Statist. Soc. B **41**, 1-31.

Dawid, A. P. (1982). Intersubjective statistical models. In Exchangeability in Probability and Statistics (G. Koch and F. Spizzichino, eds.). North-Holland, Amsterdam, 217-232.

Dawson, D. A. and Hochberg, K. J. (1982). Wandering random measures in the Fleming-Viot model. Ann. Probability **10**, 554-580.

Dellacherie, C. and Meyer, P.-A. (1975, 1980). Probabilités et Potentiel, Ch. I-IV; Ch. V-VIII. Hermann, Paris.

Diaconis, P. (1977). Finite forms of de Finetti's theorem on exchangeability. Synthese **36**, 271-281.

Diaconis, P. and Freedman, D. (1980a). Finite exchangeable sequences. Ann. Probability **8**, 745-764.

Diaconis, P. and Freedman, D. (1980b). de Finetti's theorem for Markov chains. Ann. Probability **8**, 115-130.

Diaconis, P. and Freedman, D. (1980c). de Finetti's generalizations of exchangeability. In Studies in Inductive Logic and Probability II (R. C. Jeffrey, ed.). University of California Press, Berkeley.

Diaconis, P. and Freedman, D. (1981). On the statistics of vision: the Julesz conjecture. J. Math. Psychol. 24, 112-138.

Diaconis, P. and Freedman, D. (1982). Partial exchangeability and sufficiency. Preprint, Department of Statistics, Stanford University (to appear in Sankhyā).

Diaconis, P. and Ylvisaker, Y. (1979). Conjugate priors for exponential families. Ann. Statist. 7, 269-281.

Dohler, R. (1980). On the conditional independence of random events. Theory Probability Appl. 25, 628-634.

Doob, J. L. (1953). Stochastic Processes. Wiley, New York.

Dubins, L. E. (1982). Towards characterizing the set of ergodic probabilities. In Exchangeability in Probability and Statistics (G. Koch and F. Spizzichino, eds.). North-Holland, Amsterdam, 61-74.

Dubins, L. E. (1983). Some exchangeable probabilities which are singular with respect to all presentable probabilities. Zeitschrift fur Wahrscheinlichkeitstheorie verw. Gebiete 64, 1-6.

Dubins, L. E. and Freedman, D. (1979). Exchangeable processes need not be mixtures of independent identically distributed random variables. Zeitschrift fur Wahrscheinlichkeitstheorie verw. Gebiete 48, 115-132.

Dynkin, E. B. (1978). Sufficient statistics and extreme points. Ann. Probability 6, 705-730.

Eagleson, G. K. (1979). A Poisson limit theorem for weakly exchangeable events, J. Appl. Prob. 16, 794-802.

Eagleson, G. K. (1982). Weak limit theorems for exchangeable random variables. In Exchangeability in Probability and Statistics (G. Koch and F. Spizzichino, eds.). North-Holland, Amsterdam, 251-268.

Eagleson, G. K. and Weber, N. C. (1978). Limit theorems for weakly exchangeable arrays. Math. Proc. Cambridge Phil. Soc. **84**, 123-130.

Eaton, M. (1981). On the projections of isotropic distributions. Ann. Statist. **9**, 391-400.

Feller, W. (1971). An Introduction to Probability Theory and its Applications, vol. 2. Wiley, New York.

Ferguson, T. S. (1973). A Bayesian analysis of some nonparametric problems. Ann. Statist. **1**, 209-230.

Ferguson, T. S. (1974). Prior distributions on spaces of probability measures. Ann. Statist. **2**, 615-629.

Figiel, T. and Sucheston, L. (1976). An application of Ramsey sets in analysis. Advances in Math. **20**, 103-105.

de Finetti, B. (1937). La prevision: ses lois logiques, ses sources subjectives. Ann. Inst. H. Poincaré **7**, 1-68.

de Finetti, B. (1938). Sur la condition d'equivalence partielle. Trans. in Studies in Inductive Logic and Probability II (R. C. Jeffrey, ed.). University of California Press, Berkeley.

de Finetti, B. (1972). Probability, Induction and Statistics. Wiley, New York.

de Finetti, B. (1974). Theory of Probability. Wiley, New York.

Freedman, D. (1962a). Mixtures of Markov processes. Ann. Math. Statist. **33**, 114-118.

Freedman, D. (1962b). Invariants under mixing which generalize de Finetti's theorem. Ann. Math. Statist. **33**, 916-923.

Freedman, D. (1963). Invariants under mixing which generalize de Finetti's theorem: continuous time parameter. Ann. Math. Statist. **34**, 1194-1216.

Freedman, D. (1980). A mixture of independent identically distributed random variables need not admit a regular conditional probability given the exchangeable σ-field. Zeitschrift fur Wahrscheinlichkeitstheorie verw. Gebiete **51**, 239-248.

Freedman, D. and Diaconis, P. (1982). de Finetti's theorem for symmetric location families. Ann. Statist. 10, 184-189.

Galambos, J. (1982). The role of exchangeability in the theory of order statistics. In Exchangeability in Probability and Statistics (G. Koch and F. Spizzichino, eds.). North-Holland, Amsterdam, 75-86.

Grigorenko, L. A. (1979). On the σ-algebra of symmetric events for a countable Markov chain. Theor. Probability Appl. 24, 199-204.

Hall, P. and Heyde, C. C. (1980). Martingale Limit Theory and its Application, Academic Press, New York.

Heath, D. and Sudderth, W. (1976). de Finetti's theorem for exchangeable random variables. Amer. Statistician 30, 188-189.

Helland, I. S. (1982). Central limit theorems for martingales with discrete or continuous time. Scand. J. Statist. 9, 79-94.

Heller, A. (1965). On stochastic processes derived from Markov chains. Ann. Math. Statist. 36, 1286-1291.

Hewitt, E. and Savage, L. J. (1955). Symmetric measures on Cartesian products. Trans. Amer. Math. Soc. 80, 470-501.

Hida, T. (1980). Brownian Motion. Springer-Verlag, New York.

Hoover, D. N. (1979). Relations on probability spaces and arrays of random variables. Preprint.

Hoover, D. N. (1982). Row-column exchangeability and a generalized model for exchangeability. In Exchangeability in Probability and Statistics (G. Koch and F. Spizzichino, eds.). North-Holland, Amsterdam, 281-291.

Horowitz, J. (1972). Semilinear Markov processes, subordinators and renewal theory. Zeitschrift fur Wahrscheinlichkeitstheorie verw. Gebiete 24, 167-193.

Hu, Y.-S. (1979). A note on exchangeable events. J. Appl. Prob. 16, 662-664.

Ignatov, Ts. (1982). On a constant arising in the theory of symmetric groups and on Poisson-Dirichlet measures. Theory Prob. Appl. 27, 136-147.

Isaac, R. (1977). Generalized Hewitt-Savage theorems for strictly stationary processes. Proc. Amer. Math. Soc. **63**, 313-316.

Jaynes, E. (1983). Some applications and extensions of the de Finetti representation theorem. In Bayesian Inference and Decision Techniques (P. K. Goel and A. Zellnor, eds.). North-Holland, Amsterdam.

Johnson, N. L. and Kotz, S. (1977). Urn Models and their Applications. Wiley, New York.

Kallenberg, O. (1973). Canonical representations and convergence criteria for processes with interchangeable increments. Zeitschrift fur Wahrscheinlichkeitstheorie verw. Gebiete **27**, 23-36.

Kallenberg, O. (1974). Path properties of processes with independent and interchangeable increments. Zeitschrift fur Wahrscheinlichkeitstheorie verw. Gebiete **28**, 257-271.

Kallenberg, O. (1975). Infinitely divisible processes with interchangeable increments and random measures under convolution. Zeitschrift fur Wahrscheinlichkeitstheorie verw. Gebiete **32**, 309-321.

Kallenberg, O. (1976). Random Measures. Akademie-Verlag, Berlin.

Kallenberg, O. (1976-81). On the structure of stationary flat processes, I-III. Zeitschrift fur Wahrscheinlichkeitstheorie verw. Gebiete **37**, 157-174; **52**, 127-147; **56**, 239-253.

Kallenberg, O. (1982a). Characterizations and embedding properties in exchangeability. Zeitschrift fur Wahrscheinlichkeitstheorie verw. Gebiete **60**, 249-281.

Kallenberg, O. (1982b). A dynamical approach to exchangeability. In Exchangeability in Probability and Statistics (G. Koch and F. Spizzichino, eds.). North-Holland, Amsterdam, 87-96.

Kallenberg, O. (1982c). The stationary-invariance problem. In Exchangeability in Probability and Statistics (G. Koch and F. Spizzichino, eds.). North-Holland, Amsterdam, 293-295.

Kallenberg, O. (1982d). Conditioning in point processes. Preprint, University of Göteborg.

Kallenberg, O. (1983). The local time intensity of an exchangeable interval partition. Preprint, University of Göteborg.

Kelly, F. P. (1979). Reversibility and Stochastic Networks. Wiley, New York.

Kemperman, J. H. B. (1973). Moment problems for sampling without replacement I. Indagationes Math. **35**, 149-164.

Kendall, D. G. (1967). On finite and infinite sequences of exchangeable events. Studia Scient. Math. Hung. **2**, 319-327.

Kerov, S. V. and Vershik, A. M. (1982). Characters of infinite symmetric groups and probability properties of Robinson-Shenstead-Knuth's algorithm. Preprint.

Kimberling, C. H. (1973). Exchangeable events and completely monotonic sequences. Rocky Mtn. J. **3**, 565-574.

Kindermann, R. and Snell, J. L. (1980). Markov Random Fields and their Applications. American Math. Soc., Providence.

Kingman, J. F. C. (1978a). Uses of exchangeability. Ann. Probability **6**, 183-197.

Kingman, J. F. C. (1978b). The representation of partition structures. J. Lond. Math. Soc. **18**, 374-380.

Kingman, J. F. C. (1978c). Random partitions in population genetics. Proc. Roy. Soc. **361**, 1-20.

Kingman, J. F. C. (1980). Mathematics of Genetic Diversity. SIAM, Philadelphia.

Kingman, J. F. C. (1982a). Exchangeability and the evolution of large populations. In Exchangeability in Probability and Statistics (G. Koch and F. Spizzichino, eds.). North-Holland, Amsterdam, 97-112.

Kingman, J. F. C. (1982b). The coalescent. Stochastic Processes Appl. 13, 235-248.

Knuth, D. E. (1981). The Art of Computer Programming, Vol. 2, 2nd Ed. Addison-Wesley, Reading, Mass.

Kolchin, V. F. and Sevast'yanov, B. A. (1978). Random Allocations. Winston, New York.

Komlós, J. (1967). A generalisation of a problem of Steinhaus. Acta Math. Acad. Sci. Hungar. **18**, 217-229.

Lauritzen, S. L. (1982). Statistical Models as Extremal Families. Aalborg University Press.

Letac, G. (1981a). Isotropy and sphericity: some characterizations of the normal distribution. Ann. Statist. **9**, 408-417.

Letac, G. (1981b). Problèmes classiques de Probabilité sur un couple de Gelfand. In Analytic Methods in Probability Theory (D. Dugué et al., eds.). Springer Lecture Notes in Mathematics 861, New York.

Loève, M. (1960). Probability Theory. Van Nostrand, Princeton.

Lynch, J. (1982a). On a representation for row-column-exchangeable arrays. Technical Report 82, Department of Mathematics, University of South Carolina.

Lynch, J. (1982b). Canonical row-column-exchangeable arrays. Technical Report 84, Department of Mathematics, University of South Carolina.

Maitra, A. (1977). Integral representations of invariant measures. Trans. Amer. Math. Soc. **229**, 209-225.

Mandelbaum, A. and Taqqu, M. S. (1983). Invariance principle for symmetric statistics. Preprint, Operations Research, Cornell University.

Mansour, B. (1981). Le cube comme couple de Gelfand. Thesis, University of Toulouse.

Marshall, A. W. and Olkin, I. (1979). Inequalities: Theory of Majorization and its Application. Academic Press, New York.

Milgrom, P. R. and Weber, R. J. (1983). Exchangeable affiliated random variables. Preprint.

Moran, P. A. P. (1973). A central limit theorem for exchangeable variates with geometrical applications. J. Appl. Prob. 10, 837-846.

O'Brien, G. L. and Vervaat, W. (1983). Self-similar processes with stationary increments generated by point processes. Preprint.

Olshen, R. A. (1971). The coincidence of measure algebras under an exchangeable probability. Zeitschrift fur Wahrscheinlichkeitstheorie verw. Gebiete 18, 153-158.

Olshen, R. A. (1973). A note on exchangeable sequences. Zeitschrift fur Wahrscheinlichkeitstheorie verw. Gebiete 28, 317-321.

Ornstein, D. S. (1973). An application of ergodic theory to probability theory. Ann. Probability 1, 43-65.

Palacios, J. (1982). The exchangeable σ-field of Markov chains. Ph.D. thesis, University of California, Berkeley.

Parthasarathy, K. R. (1967). Probability Measures on Metric Spaces. Academic Press, New York.

Pavlov, Y. L. (1981). Limit theorem for a characteristic of a random mapping. Theory Prob. Appl. 26, 829-834.

Pfeiffer, P. E. (1979). Conditional Independence in Applied Probability. UMAP, Educational Development Center, Newton, Mass.

Phelps, R. R. (1966). Lectures on Choquet's Theorem. Van Nostrand, Princeton.

Pitman, J. (1978). An extension of de Finetti's theorem. Adv. Appl. Prob. 10, 268-270.

Pittel, B. (1983). On the distributions related to transitive classes of random finite mappings. Ann. Probability 11, 428-441.

Quine, M. P. (1978). A functional central limit theorem for a class of exchangeable sequences. Preprint, University of Sydney.

Quine, M. P. (1979). A functional central limit theorem for a generalized occupancy problem. Stochastic Proc. Appl. 9, 109-115.

Rényi, A. (1963). On stable sequences of events. Sankhyā Ser. A 25, 293-302.

Rényi, A. and Révész, P. (1963). A study of sequences of equivalent events as special stable sequences. Publ. Math. Debrecen. 10, 319-325.

Ressel, P. (1983). de Finetti type theorems: an analytical approach. Preprint, Department of Statistics, Stanford University.

Ridler-Rowe, C. J. (1967). On two problems on exchangeable events. Studia Sci. Math. Hungar. 2, 415-418.

Ryll-Nardzewski, C. (1957). On stationary sequences of random variables and the de Finetti's equivalence. Colloquium Math. 4, 149-156.

Saunders, R. (1976). On joint exchangeability and conservative processes with stochastic rates. J. Appl. Prob. 13, 584-590.

Schoenberg, I. J. (1938). Metric spaces and positive definite functions. Trans. Amer. Math. Soc. 44, 522-536.

Serfling, R. J. (1980). Approximation Theorems of Mathematical Statistics. Wiley, New York.

Shaked, M. and Tong, Y. L. (1983). Some partial orderings of exchangeable random variables by positive dependence. Preprint, Mathematics Department, University of Arizona.

Shields, P. C. (1979). Stationary coding of processes. IEEE Trans. Information Theory 25, 283-291.

Sigalotti, L. (1982). On particular renewal processes interesting reliability theory. In Exchangeability in Probability and Statistics (G. Koch and F. Spizzichino, eds.). North-Holland, Amsterdam, 303-311.

Slud, E. V. (1978). A note on exchangeable sequences of events. Rocky Mtn. J. 8, 439-442.

Smith, A. M. F. (1981). On random sequences with centered spherical symmetry. J. Roy. Statist. Soc. Ser. B 43, 208-209.

Speed, T. P. (1982). Cumulants, k-statistics and their generalisations. Preprint, University of Western Australia.

Spitzer, F. (1975). Markov random fields on an infinite tree. Ann. Probability 3, 387-398.

Spizzichino, F. (1982). Extendibility of symmetric probability distributions and related bounds. In Exchangeability in Probability and Statistics (G. Koch and F. Spizzichino, eds.). North-Holland, Amsterdam, 313-320.

Stein, C. (1983). Approximate computation of expectations. Unpublished lecture notes.

Strassen, V. (1965). The existence of probability measures with given marginals. Ann. Math. Statist. **36**, 423-439.

Surabian, C. G. (1979). A survey of exchangeability. M.Sc. thesis, Tufts University.

Taqqu, M. S. (1982). Self-similar processes and related ultraviolet and infrared catastrophes. In Random Fields (J. Fritz, ed.). North-Holland, Amsterdam.

Teicher, H. (1982). Renewal theory for interchangeable random variables. In Exchangeability in Probability and Statistics (G. Koch and F. Spizzichino, eds.). North-Holland, Amsterdam, 113-121.

Troutman, B. M. (1983). Weak convergence of the adjusted range of cumulative sums of exchangeable random variables. J. Appl. Prob. **20**, 297-304.

Vershik, A. M. and Schmidt, A. A. (1977). Limit measures arising in the theory of symmetric groups I. Theory Prob. Appl. **22**, 70-85.

Watterson, G. A. (1976). The stationary distribution of the infinite-many neutral alleles diffusion model. J. Appl. Prob. **13**, 639-651.

Weber, N. C. (1980). A martingale approach to central limit theorems for exchangeable random variables. J. Appl. Prob. **17**, 662-673.

Zabell, S. L. (1982). W. E. Johnson's "sufficientness" postulate. Ann. Statist. **10**, 1091-1099.

Zachary, S. (1983). Countable state space Markov random fields and Markov chains on trees. Ann. Probability. to appear.

THEOREMES LIMITES POUR LES MARCHES ALEATOIRES

PAR I.A. IBRAGIMOV

INTRODUCTION

On peut dire en paraphrasant un peu J. Jacod ([28], Introduction) que les théorèmes limites pour les marches aléatoires sont innombrables. Dans ce cours j'ai choisi pour objectif de présenter une théorie qui a été initiée par A.V. Skorodod [42] et développée par A.V. Skorohod et Slobodényuk [43], [44], [45] . Cette théorie étudie les théorèmes limites pour les sommes :

$$\eta_n = \sum_1^n f_n(\zeta_k) \qquad (1)$$

où $\zeta_k = \sum_1^n \xi_j$ et où $\xi_1, \xi_2, \ldots$ sont des variables aléatoires indépendantes de même loi. Les sommes de ce type ont été étudiées par Kallianpur et Robbins [31], Dynkin [15], Dobrushin [12]. Si f est la fonction indicatrice d'un ensemble A, (1) représente le nombre de visites de l'ensemble A par ζ_k (si $A = [0,\infty]$, $\xi_j = \pm 1$ avec probabilité 1/2 nous avons la fameuse loi de l'arc sinus de P. Lévy [35]). A.V. Skorohod considére dans [42] les sommes plus générales que (1) :

$$\eta_n = \sum_1^{n-r} f_n(\zeta_k, \ldots \zeta_{k-r}) \qquad (2)$$

G.N. Sytaya considère le cas "polyadditif" [48], [49] :

$$\eta_n = \sum_{k_1,\ldots k_r=1}^{n} f_n(\xi_{k_1}, \ldots \xi_{k_r}) \qquad (3)$$

La théorie de Skorohod et Slobodenyuk a été exposée par les auteurs dans les chapitres 5, 7 du livre [45]. La méthode des auteurs consiste en l'utilisation des théorèmes limites locaux avec estimation du terme restant. C'est pourquoi ces auteurs demandent que les variables ξ_j aient une densité de probabilité et des moments finis jusqu'à l'ordre 5. Dans ce cours je voudrais présenter la théorie de Skorohod et Slobodenyuk mais en utilisant quelques méthodes différentes. Ces méthodes permettent de démontrer les théorèmes limites pour les sommes (1) sous les hypothèses naturelles : les variables ξ_j appartiennent au domaine d'attraction d'une loi stable. Dans le chapitre 2 nous démontrerons des théorèmes limites pour les sommes (1), dans le chapitre 3 pour les sommes :

$$\sum_i^n f(\zeta_k).$$

Ces chapitres correspondent aux chapitres 5 et 6 du livre [45]. Le chapitre 1 est consacré à des rappels.

Comme J. Jacod (voir [28], Introduction), "faute de temps (et de courage)" je me suis borné au cas (1) et j'ai omis presque entièrement les résultats sur les sommes (2) et (3).

Chaque chapitre est muni d'un commentaire historique mais ce n'est pas une histoire complète. Ces commentaires ont pour but de donner au lecteur une orientation historique très approximative. Je m'excuse par avance de toute omission.

Je voudrais remercier Madame MAY pour la frappe de ce texte. Je remercie tout particulièrement P.L. Hennequin de m'avoir invité à participer à l'Ecole d'Eté de Saint-Flour et de m'avoir soutenu et aidé tout le temps de mon séjour à Saint-Flour.

CHAPITRE I

Ce chapitre est presque entièrement consacré à des rappels. La plupart des résultats sont donnés sans démonstration. Pour les théorèmes limites classiques, nous renvoyons aux livres de Gnedenko et Kolmogorov [23], Feller [19], Loeve [37]et, Ibraguimov et Linnik [27], pour tout ce qui concerne la convergence des processus, nous renvoyons à Billengsley [1], Guihman et Skorohod [21] et Jacod [28] .

On considère ici une suite :

$$\xi_1, \xi_2, \ldots \xi_n, \ldots$$

de variables aléatoires indépendantes de même loi (iid). L'objectif du chapitre est de rappeler des théorèmes limites fondamentals sur les sommes normalisées :

$$S_n = B_n^{-1} \sum_1^n \xi_i - A_n$$

quand $n \to \infty$.

1 - LOIS STABLES. CONDITIONS DE CONVERGENCE VERS UNE LOI STABLE

On dit qu'une loi de probabilité $\mathcal{L}$ est une loi stable de paramètres $\alpha, \beta,\ 0 < \alpha \leqq 2,\ -1 \leqq \beta \leqq 1$, si la fonction caractéristique de cette loi est :

$$\varphi_{\alpha\beta}(\lambda) = \exp\{ia\lambda - b|\lambda|^\alpha (1 + i\beta \operatorname{sign}\lambda\ \omega(\lambda,\alpha))\} ,$$

où :

$$\omega(\lambda,\alpha) = \begin{cases} \operatorname{tg} \frac{\pi\alpha}{2} , & \alpha \neq 1 , \\ \frac{2}{\pi} \ln|\lambda| , & \alpha = 1 . \end{cases} \tag{1.1}$$

On dit que α est l'ordre de la loi stable (1.1). Si, en particulier, $\alpha = 2$ on a la loi de Gauss, si $\alpha = 1$, $\beta = 0$ on a la loi de Cauchy. Nous supposerons toujours le paramètre du déplacement $a = 0$ et le paramètre de l'échelle $b = \alpha^{-1}$. Le théorème suivant explique le rôle des lois stables pour la théorie de l'addition des variables aléatoires indépendantes.

THEOREME 1.1 : *Soit :*

$$\xi_1, \xi_2, \ldots \tag{1.2}$$

une suite de variables aléatoires de même loi. Pour qu'une loi de probabilité $\mathcal{L}$ soit une loi limite des sommes normalisées :

$$B_n^{-1} \sum_1^n \xi_j - A_n \tag{1.3}$$

il faut et il suffit que cette loi $\mathcal{L}$ soit une loi stable. Si la loi limite a l'ordre α , alors les facteurs B_n doivent avoir la forme suivante :

$$B_n = n^{1/\alpha} h(n) \tag{1.4}$$

où $h(n)$ est une fonction à croissance lente au sens de Karamata.

Soit F la loi de probabilité des variables (1.2). On dit que F appartient au domaine d'attraction de la loi stable $\mathcal{L}$ si on peut choisir A_n, B_n de telle manière que les sommes (1.3) convergent en loi vers $\mathcal{L}$. Si on peut prendre dans (1.4) $h(n) \equiv$ const, on dit que F appartient au domaine d'attraction normale de $\mathcal{L}$.

THEOREME 1.2 : *Pour qu'une loi* F *appartienne au domaine d'attraction normale d'une loi stable d'ordre* $\alpha \neq 2$ *il faut et il suffit que :*

$$F(x) = (c_1 a^\alpha + o(1)) |x|^{-\alpha}, \qquad x \to -\infty,$$

$$1-F(x) = (c_2 a^\alpha + o(1)) |x|^{-\alpha}, \qquad x \to \infty$$

où c_1, c_2, a *sont des constantes. Les constantes* c_1, c_2, *ne dépendent que de* α *et* β *et les facteurs normalisés sont* $B_n = an^{1/\alpha}$.

THEOREME 1.3 : *Pour qu'une loi* F *appartienne au domaine d'attraction normale d'une loi de Gauss* ($\alpha = 2$) *il faut et il suffit que :*

$$\int_{-\infty}^{\infty} x^2 \, dF(x) < \infty .$$

On peut réunir ces deux cas sous une forme commune.

THEOREME 1.4 : *Pour qu'une loi* F *appartienne au domaine d'attraction normale de la loi stable (1.1), il faut et il suffit que la fonction caractéristique* φ *de* F *ait la forme suivante :*

$$\varphi(t) = \exp \{ia_1 t - b_1 |t|^\alpha (1+i\beta\omega(t,\alpha)) (1+o(1))\}, \ t \to 0 \qquad (1.5)$$

où a_1, $b_1 > 0$ *sont des constantes.*

On déduit facilement de (1.5) que si F appartient au domaine d'attraction normale d'une loi stable d'ordre α, alors :

$$\int_{-\infty}^{\infty} |x|^\delta \, dF < \infty$$

pour tout $\delta < \alpha$. En particulier, si $\alpha > 1$, alors $\int_{-\infty}^{\infty} x \, dF$ existe, et si dans (1.5) $a_1 = 0$:

$$\int_{-\infty}^{\infty} x \, dF = 0 .$$

On peut trouver les démonstrations des théorèmes 1.1 - 1.3 dans [23], du théorème 1.4 dans [27].

2 - PROCESSUS STABLES. CONDITIONS DE CONVERGENCE VERS UN PROCESSUS STABLE

Soit $\xi(t)$ un processus stochastique à accroissements indépendants. On dit que $\xi(t)$ est un processus stable si tout accroissement $\xi(t) - \xi(s)$ est de loi stable. Nous supposons aussi que la loi de $\xi(t) - \xi(s)$ ne dépend que t-s et que :

$$E \exp \{i\lambda(\xi(t) - \xi(s))\} = \exp\{- \frac{t-s}{\alpha}|\lambda|^{\alpha} (1+i\beta \mathrm{sign}\lambda\,\omega(\lambda,\alpha))\} \quad (2.1)$$

Nous désignons le processus (2.1) par $\xi_{\alpha\beta}(t)$ ou $\xi_{\alpha}(t)$. Notons que $\xi_2(t)$ est le processus de Wiener.

Soit $\{\xi_n\}$ une suite de variables aléatoires indépendantes de même loi F dont la fonction caractéristique a la forme suivante :

$$\varphi(t) = \exp \{-\alpha^{-1}|\lambda|^{\alpha}(1+i\beta\, \mathrm{sign}\lambda\; \omega(\lambda,\alpha))\; (1+o(1))\} \;,\lambda \to 0 \quad (2.2)$$

Posons :

$$\zeta_k = \sum_1^k \xi_j, \quad S_{nk} = n^{-1/\alpha} \sum_i^k \xi_j \;, \qquad \alpha \neq 1 \;,$$

$$S_{nk} = n^{-1} \sum_1^k \xi_j - \frac{2}{\pi} \beta \ln n, \qquad \alpha = 1 \;.$$

On peut définir les processus :

$$S_n(t) = S_{nk} \;, \; \frac{k-1}{n} \quad t < \frac{k}{n} \quad (2.3)$$

Evidemment, sous l'hypothèse (2.2) toutes les lois fini-dimensionnelle de $S_n(t)$ convergent vers celles de $\xi_\alpha(t)$. En fait, des résultats plus forts sont vrais. On peut supposer que toute trajectoire de $S_n(t)$ et $\xi_\alpha(t)$ appartient à l'espace D(0,1) de Skorohod,donc on peut considérer S_n, ξ_α comme des variables aléatoires à valeurs dans l'espace de Skorohod D(0,1) et on peut parler de la convergence en loi de S_n vers ξ_α dans D(0,1) (voir [1] , [21]).

THEOREME 2.1 : *Si la condition (2.2) est vérifiée, alors* $S_n(t)$ *converge en loi dans* D(0,1) *vers le processus* $\xi_\alpha(t)$ *défini par (2.1).*

Ayant des variables ξ_j on peut aussi construire le processus $\tilde{S}_n(t)$ engendré par $\sum_1^k \xi_j$ d'une manière différente. Notamment, soit maintenant $\tilde{S}_n(t)$ une ligne brisée de sommets $(\frac{k}{n}, n^{-1/\alpha} \sum_1^k \xi_j)$. Dans ce cas $\tilde{S}_n \in C(0,1)$. Le processus $\xi_2(t) = W(t)$ est continu avec probabilité 1, et on peut parler de la convergence en loi dans C(0,1) de $\tilde{S}_n$ vers W.

THEOREME 2.2 : *Si* $E\xi_j = 0$, $D\xi_j = 1$, *c'est à dire si :*

$$\varphi(t) = \exp \{- \frac{\lambda^2}{2} (1+o(1))\}, \quad \lambda \to 0,$$

alors $\tilde{S}_n$ *converge en loi dans l'espace* C(0,1) *vers le processus de Wiener* $W(=\xi_2)$.

On peut trouver la démonstration de ces théorèmes très connus dans les livres [1], [21] .

Je voudrais rappeler aussi que, pour démontrer la convergence des processus ξ_n vers un processus ξ dans quelque espace B on procéde souvent en deux étapes :

1° On montre que les lois fini-dimensionnelles de ξ_n convergent vers celles de ξ ;

2° On montre que la suite $\mathcal{L}(\xi_n)$ des lois de ξ_n dans B est tendue dans B.

En rapport avec cela nous utiliserons plus bas un critère de compacité dans C(0,1) de Prohorov.

THEOREME 2.3 : *Supposons qu'une suite* $[\zeta_n]$ *de processus séparables définis sur* $[0,1]$ *satisfait la condition suivante :*

$$E|\xi_n(t) - \xi_n(s)|^p \leqq K|t-s|^q$$

où $p > 1$, $q > 1$, $K > 0$ *sont des constantes positives. Alors si la suite* $\mathcal{L}(\xi_n(t_0))$ *est tendue dans* R^1, *la suite des lois* $\{\mathcal{L}(\xi_n)\}$ *est tendue dans* C(0,1).

Ce critère est basé sur le théorème suivant :

THEOREME 2.4 : *Soit* $\xi(t)$ *un processus séparable défini sur* $[0,1]$ *et tel que :*

$$E\ |\xi(t) - \xi(s)|^p \leqq K(t-s)^q\ , \quad t,s \in [0,1]\ ,$$

où $p > 1$, $q > 1$, $K > 0$. *Alors* $\xi(t)$ *satisfait presque sûrement la condition de Hölder pour tout ordre* $\gamma < \frac{q-1}{p}$. *Qui plus est :*

$$E\left\{\sup \frac{|\xi(t) - \xi(s)|}{|t-s|^\gamma}\right\} \leqq B$$

où B *ne dépend plus de* p, q, K, γ.

Pour la démonstration voir [1] , paragraphe 12.

3 - PROBLEME SUR LA LOI LIMITE DE FONCTIONNELLES DEFINIES SUR UNE MARCHE ALEATOIRE

Soit $\xi_1, \xi_2, \dots \quad \xi_n, \dots$

une suite de variables aléatoires de même loi et à valeurs dans R^k. Elles engendrent une marche aléatoire $\{\zeta_n\}$:

$$\zeta_n = \sum_1^n \xi_j\ .$$

Soient $F_n(x_1,\dots x_n)$ des fonctions définies sur R^{kn} . Elles engendrent des fonctionnelles définies sur la marche aléatoire $\{\zeta_n\}$ par la formule suivante :

$$\eta_n = F_n(\zeta_1, \dots \zeta_n)\ .$$

Il faut trouver des conditions sous lesquelles il existe des constantes normalisées A_n, B_n telles que la variable $B_n^{-1}(\eta_n - A_n)$ ait une loi limite et caractériser cette loi. Bien sûr ce problème est trop général. Il faut poser quelques restrictions sur F_n pour avoir des théorèmes intéressants.

Dans ce cours nous nous restreignons exclusivement au cas de fonctionnelles additives, c'est à dire au cas :

$$\eta_n = \sum_{k=1}^{n-r} f_n(\zeta_k,\ldots\zeta_{k+r}) \tag{3.1}$$

où $r \geqq 0$ est un nombre fixé. De plus nous considérons en détail seulement le cas $r=0$. Voici quelques exemples :

1° Soit $f : R^k \longrightarrow R^1$. On peut considérer :

$$\eta_n = \sum_{j=1}^{n} f(\zeta_j) \; .$$

Si en particulier $f(x) = \mathbf{1}_A(x)$ est la fonction indicatrice d'un ensemble $A \subset R^k$, la fonctionnelle

$$\eta_n = \sum_1^n f(\zeta_k)$$

représente le nombre de visites de l'ensemble A par la marche aléatoire ζ_k. Si $f(x) = \mathbf{1}_A(x) - \mathbf{1}_B(x)$ la fonctionnelle η_n représente la surabondance du nombre de visites de l'ensemble A par rapport à celui de B.

2° Supposons que ξ_j a ses valeurs dans R^1.

Soit :

$$f(x,y) = \begin{cases} 1, & xy < 0 \\ 0, & xy > 0 \; , \end{cases}$$

alors :

$$\eta_n = \sum_1^{n-1} f(\zeta_k,\, \zeta_{k+1})$$

est le nombre d'intersections du niveau zéro par la suite $\{\zeta_k\}$.

3° Soit :

$$f(x,y,z) = \begin{cases} 1, & x < y,\ z < y \\ 0, & y \leqq \max\,(x,z) \end{cases}$$

alors :

$$\eta_n = \sum_1^{n-2} f(\zeta_k,\, \zeta_{k+1},\, \zeta_{k+2})$$

est le nombre de maxima locaux dans la suite $\{\zeta_k\}$.

On peut établir quelques résultats sur les lois limites de (3.1), en utilisant les théorèmes du paragraphe précédent.

THEOREME 3.1 : *Soit* f *une fonction définie et continue sur* R^1. *Sous les conditions du théorème 2.1, la variable :*

$$\eta_n = \frac{1}{n} \sum_1^n f(S_{nk}) = \sum_{k=1}^n f_n(\zeta_k)$$

où $f_n(x) = n^{-1} f(x\, n^{1/\alpha})$ *converge en loi vers* $\int_0^1 f(\xi_\alpha(t))dt$.

En effet, la fonctionnelle :

$$F(x(.)) = \int_0^1 f(x(t))dt$$

est définie et continue dans l'espace de Skorohod D(0,1).

En vertu du théorème 2.1 :

$$\frac{1}{n} \sum_{k=1}^n f(S_{nk}) = \int_0^1 f(S_n(t))dt = F(S_n(.))$$

où $S_n(t) = S_{nk}$, $\frac{k-1}{n} \leqq t < \frac{k}{n}$, converge en loi vers :

$$F(\xi_\alpha) = \int_0^1 f(\xi_\alpha(t))dt.$$

THEOREME 3.2 : *Soit* $f(x_0, \dots x_r)$ *une fonction mesurable à valeurs dans* R^1 *localement bornée et continue sur* $x_0 = x_1 = \dots = x_r$. *Sous les conditions du théorème 2.1 la variable :*

$$\eta_n = \frac{1}{n} \sum_{k=1}^{n-r} f(S_{nk}, \dots S_{n,k+r}) = \sum_1^{n-r} f_n(\zeta_k, \dots \zeta_{k+r})$$

converge en loi vers $\int_0^1 f(\xi_\alpha(t), \dots \xi_\alpha(t))dt$.

Démonstration : On a :

$$\eta_n = \frac{1}{n} \sum_1^{n-r} f(S_{nk}, \dots S_{nk}) + \frac{1}{n} \sum_1^{n-r} \left[f(S_{nk}, \dots S_{n,k+r}) - f(S_{nk}, \dots S_{nk}) \right] =$$

$$= \eta_{n1} + \eta_{n2}$$

On a comme plus haut que η_{n1} converge en loi vers $\int_0^1 f(\xi_\alpha(t), \dots f_\alpha(t))dt$. Montrons que $\eta_{n2} = o(1)$ en probabilité. Soit B une constante positive. Pour tout $\varepsilon > 0$ il existe $\delta > 0$ tel que :

$$|f(x_0, \dots x_r) - f(x_0, \dots x_0)| \leqq \varepsilon$$

si $|x_0| \leqq B$ et $|x_i - x_0| \leqq \delta$. Alors :

$$P\left\{ \frac{1}{n} \sum_1^{n-r} |f(S_{nk}, \dots S_{n,k+r}) - f(S_{nk}, \dots S_{nk})| > 2\,\varepsilon \right\} \leqq$$

$$\leqq P\left\{ \frac{1}{n} \sum_1^{n-r} |f(S_{nk}, \dots S_{n,k+r}) - f(S_{nk}, \dots S_{nk})| > 2\varepsilon,\ \max_k |S_{nk}| \leqq B \right\}$$

$$+ P\left\{ \max_k |S_{nk}| > B \right\} .$$

En vertu du théorème 2.1 :

$$\lim_{B\to\infty} \lim_{n\to\infty} P\{\max_k |S_{nk}| > B\} = \lim_B P\{\sup_{0\leq t\leq 1} |\xi_\alpha(t)| > B\} = 0$$

Puisque $f(x_0,\ldots x_r)$ est bornée dans le domaine $|x_j|\leq B$ il existe un nombre $x > 0$ tel que, pour tout n assez grand,

$$P\{\frac{1}{n}\sum_1^{n-r} |f(S_{nk},\ldots S_{n,k+r}) - f(S_{nk},\ldots S_{nk})| > 2\varepsilon , \max_k |S_{nk}| \leq B\} \leq$$

$$\leq P\{|\xi_{i_1}| \geq \frac{\delta}{r} n^{1/\alpha},\ldots |\xi_{i_p}| \geq \frac{\delta}{r} n^{1/\alpha}, 1 \leq i_1 <\ldots < i_p, p = [nx]\} \leq$$

$$\leq C_n^p(P\{|\xi_1| > \frac{\delta}{r} n^{1/\alpha}\})^p.$$

On déduit du théorème 1.3 que :

$$P\{|\xi_1| > \frac{\delta}{r}\} \leq B(\frac{r}{\delta})^\alpha n^{-1} .$$

Donc :

$$C_n^p(P\{|\xi_1| > \frac{\delta}{r} n^{1/\alpha}\})^p \leq B \frac{n!}{P!(n-p)!} (\frac{r}{\delta})^\alpha n^{-p} \leq B^n \frac{n^n}{p^p(n-p)^p} n^{-p}$$

$$\leq B^n n^{-xn} \xrightarrow[n\to\infty]{} 0 .$$

Le théorème est démontré.

Le théorème 3.1 jouera un rôle important ci-dessous et pour ne pas dépendre du théorème 2.1 qui n'était pas démontré ici nous donnons une autre démonstration de ce théorème pour le cas $\alpha > 1$. Cette démonstration est plus élémentaire ; elle est basée sur le lemme suivant :

LEMME 3.1 : *Soient* $\xi_n(t)$, $\xi(t)$ *des processus mesurables définis sur* $[0,1]$. *Si toute loi fini-dimensionnelle de* $\xi_n(t)$ *converge vers celle de* $\xi(t)$,

$$\sup_{t,n} E|\xi_n(t)| < \infty$$

et :

$$\lim_{h\to 0} \overline{\lim_{n\to\infty}} \sup_{|t_1-t_2|\leq h} E|\xi_n(t_2) - \xi_n(t_1)| = 0 \tag{3.2}$$

alors $\int_0^1 \xi_n(t)dt$ *converge en loi vers* $\int_0^1 \xi(t)dt$.

Démonstration : Soit N un nombre entier. On a :

$$|E\{\exp\{i\lambda \int_0^1 \xi_n(t)dt\}\} - E\{\exp\{i\lambda \int_0^1 \xi(t)dt\}\}| \leq$$

$$\leq E|\exp\{i\lambda \int_0^1 \xi_n(t)dt\} - \exp\{\frac{i\lambda}{N} \sum_1^N \xi_n(\frac{k}{N})\}| +$$

$$+ E|\exp\{i\lambda\int_0^1 \xi(t)dt\} - \exp\{\frac{i\lambda}{N} \sum_1^N \xi(\frac{k}{N})\}|+$$

$$+ \left| E\{\exp \frac{i\lambda}{N} \sum_1^N \xi_n(\tfrac{k}{N})\}\} - E\{\exp\{\frac{i\lambda}{N} \sum_1^N \xi(\tfrac{k}{N})\}\}\right| .$$

Si N est fixé alors :

$$E \exp\{ \frac{i\lambda}{N} \sum_1^N \xi_n(\tfrac{k}{N})\} \underset{n}{\longrightarrow} E \exp\{ \frac{i\lambda}{N} \sum_1^N \xi(\tfrac{k}{N})\} .$$

En vertu de (3.2) :

$$\lim_N \overline{\lim_n} E|\exp \{i\lambda \int_0^1 \xi_n(t)dt - \exp \{\frac{i\lambda}{N} \sum_i^N \xi_n(\tfrac{k}{N})\}| \leq$$

$$\leq \lim_N \overline{\lim_n} |\lambda| E \left|\int_0^1 \xi_n(t)dt - \frac{1}{N} \sum_1^N \xi_n(\tfrac{k}{N})\right| \leq$$

$$\leq \lim_N \overline{\lim_n} \sup_{|t-s|\leq N^{-1}} E|\xi_n(t) - \xi_n(s)| = 0 .$$

D'après le lemme de Fatou :

$$\lim_N E|\exp \{i\lambda \int_0^1 \xi(t)dt\} - \exp\{ \frac{i\lambda}{N} \sum_1^N \xi(\tfrac{K}{N})\}| \leq$$

$$\leq \lim_N \sup_{|t-s|\leq N^{-1}} E|\xi(t) - \xi(s)| \leq \lim_N \overline{\lim_n} \sup_{|t-s|\leq N^{-1}} E|\xi_n(t)-\xi_n(s)| = 0$$

Le lemme est démontré.

Démontrons le théorème 3.1. Soit $\varphi(t)$ la fonction caractéristique de ξ_j. Puisque :

$$\varphi(\lambda) = \exp \{- \frac{|\lambda|^\alpha}{\alpha} (1+i\beta \operatorname{tg} \frac{\pi\alpha}{2} - \operatorname{sign} \lambda) (1+o(1)\} , \quad \lambda \to 0$$

on obtient que pour tout $0 < t_1 < \dots < t_r \leq 1$,

$$E\{\exp \sum_1^r i\lambda_j S_n(t_j)\} \longrightarrow E\{\exp \sum_1^r i\lambda_j \xi_\alpha(t_j)\} .$$

Il reste à montrer pour $S_n(t)$ la propriété (3.2).

<u>LEMME 3.2</u> : *Soit ξ une variable aléatoire d'espérance finie et de fonction caractéristique $\Psi(\lambda)$. Alors :*

$$E|\xi| = \frac{2}{\pi} \int_0^\infty \frac{\mathrm{Re}(1-\psi(\lambda))}{\lambda^2} d\lambda .$$

<u>Démonstration</u> : On a :

$$\frac{2}{\pi} \int_0^\infty \frac{\sin \xi x}{x} dx = \operatorname{sign} \xi .$$

Donc :

$$E |\xi| = \frac{2}{\pi} \int_0^\infty (\mathrm{Im}\, E\, \xi e^{i\xi x}) \frac{dx}{x} = \frac{2}{\pi} \int_0^\infty \mathrm{Im}(\frac{\psi'(x)}{i}) \frac{dx}{x} =$$

$$= - \frac{2}{\pi} \mathrm{Re} \int_0^\infty \frac{\psi'(x)}{x} dx = \frac{2}{\pi} \mathrm{Re} \frac{1-\psi(x)}{x^2} dx .$$

Le lemme est démontré. En vertu de ce lemme on a, pour un $c > 0$:

$$n^{-1/\alpha} E\left|\sum_{l+1}^{m} \xi_j\right| = \frac{2}{\pi} \int_0^\infty \frac{1-(\varphi(\lambda n^{-1/\alpha}))^{m-1}}{\lambda^2} d\lambda \leq$$

$$\leq B\left(\frac{m-l}{n} + \int_{cn^{1/\alpha}}^\infty \lambda^{-2} d\lambda\right) \leq B\,\frac{m-l}{n} + Bn^{-1/\alpha}.$$

Donc :

$$E|S_n(t_2) - S_n(t_1)| \leq B(|t_2 - t_1| + n^{-1/\alpha}).$$

La démonstration est achevée.

On peut déduire du théorème 3.1 quelques résultats sur les sommes :

$$\eta_n = \sum_1^n f(\zeta_k).$$

THEOREME 3.3 : *Soit* $f(x)$ *une fonction homogène d'ordre* γ , *c'est-à-dire telle que pour tout* $u > 0$

$$f(ux) = u^\gamma f(x).$$

Alors sous les conditions du théorème 3.1 les sommes :

$$n^{-1-\gamma/\alpha} \sum_1^n f(\zeta_k)$$

convergent en loi vers $\int_0^1 f(\xi_\alpha(t))dt$.

Démonstration : On a :

$$n^{-1-\gamma/\alpha} \sum_1^n f(\zeta_k) = n^{-1-\gamma/\alpha} \sum_1^n f(S_{nk} n^{1/\alpha}) = \frac{1}{n} \sum_1^n f(S_{nk})$$

et la somme de gauche converge vers $\int_0^1 f(\xi_\alpha(t))dt$ en vertu du théorème 3.1. Bien sûr l'ensemble des fonctions homogènes est très pauvre :

$$f(x) = \begin{cases} A_1\, x^\gamma, & x > 0 \\ A_2\, x^\gamma, & x < 0 \end{cases}$$

Nous montrerons des théorèmes plus généraux dans le chapitre 3.

Nous revenons pour conclure à l'exemple 3.

THEOREME 3.4 : *Soit* $\xi_1, \xi_2, \ldots$ *des variables aléatoires indépendantes de même loi. Soit :*

$$P\{\xi_1 = 0\} = 0 \quad , \; P\{\xi_1 > 0\} = a .$$

Désignons par N_n *le nombre de maxima locaux dans la suite* $\zeta_1, \zeta_2, \ldots \zeta_n$. *Alors la variable :*

$$\frac{N_n - na(1-a)}{\sqrt{n}}$$

converge en loi vers une variable aléatoire normale de moyenne zéro et de variance $a(1-a)\ (1-3a(1-a))$.

Démonstration : Soit :

$$f(x,y) = \begin{cases} 1\ ,\ x > 0\ ,\ y < 0 \\ 0\ ,\ x \leq 0 \ \text{ ou } y \leq 0\ , \end{cases}$$

alors :

$$N_n = \sum_{k=1}^{n-2} f(\zeta_{k+1} - \zeta_k, \zeta_{k+2} - \zeta_{k+1}) = \sum_{k=1}^{n-2} f(\xi_{k+1}, \xi_{k+2}).$$

Soit $X_k = f(\xi_{k+1}, \xi_{k+2})$. Les variables X_k sont 2-dépendantes, c'est-à-dire que les suites $(X_1,\dots X_\ell)$, $(X_{\ell+2},\dots)$ sont indépendantes. La suite des X_k est stationnaire. Le théorème limite central est applicable à de telles suites (voir [1], théorème 20.1). Donc $\frac{1}{\sqrt{n}}(N_n - EN_n)$ converge en loi vers une variable normale ξ telle que $E\xi = 0$ et :

$$\operatorname{Var}\xi = \operatorname{Var} X_1 + 2\sum_{k=1}^{\infty} E(X_{k+1} - EX_{k+1})\ (X_1 - EX_1).$$

On a :

$$EN_n = (n-2)\ Ef(\xi_1,\xi_2) = (n-2)\ P\{\xi_1 > 0,\ \xi_2 < 0\} = (n-2)\ a(1-a)$$

$$= na(1-a) + O(1).$$

En outre :

$$\operatorname{Var} X_1 = EX_1^2 - (EX_1)^2 = a(1-a) - a^2(1-a)^2,\ E(X_1 - EX_1)(X_2 - EX_2) =$$

$$= Ef(\xi_1,\xi_2)\ f(\xi_2,\xi_3) - a^2(1-a)^2 = P\{\xi_1 > 0,\ \xi_2 < 0,\ \xi_2 > 0,\ \xi_3 < 0\} - a^2(1-a)^2 =$$

$$= -a^2(1-a)^2.$$

Donc :

$$\operatorname{Var}\xi = \operatorname{Var} X_1 + 2E(X_1 - EX_1)\ (X_2 - EX_2) = a(1-a)\ (1-3a(1-a)).$$

Le théorème est démontré.

4 - TEMPS LOCAL DE PROCESSUS STABLES

Soit $\zeta(t)$, $t > 0$, un processus stochastique à valeurs dans R^k. Soit Γ un ensemble mesurable sur R^k. Le temps de séjour du processus $\xi(t)$ dans l'ensemble Γ jusqu'au moment T est, par définition :

$$\mu(\Gamma;T;\xi) = \mu(\Gamma;T) \stackrel{\text{def}}{=} \text{mes}\ \{t : \xi(t) \in \Gamma,\ t \in [0,T]\}\ .$$

Pour chaque T fixé, $\mu(\Gamma;T)$ est une mesure sur R^k et $T^{-1}\mu(\Gamma;T)$ est une mesure de probabilité. Si la mesure $\mu(\Gamma;T)$ est absolument continue par rapport à la mesure de Lebesgue λ , on appelle temps local du processus $\xi(t)$ la dérivée :

$$\frac{d\mu}{d\lambda}(x;T) = \mathfrak{L}(x;T) = \mathfrak{L}(x;T;\Gamma).$$

On interprète f comme le temps que le processus $\xi(t)$ a passé dans le point x quand t a changé dans l'intervalle 0,T .

Il est clair que :

$$\int_0^T f(\xi(t))dt = \int_{\mathbb{R}^k} f(x)\, \mu(dx;T;\xi).$$

En particulier, si le temps local $\pounds(x;T;\xi)$ existe, on a :

$$\int_0^T f(\xi(t))dt = \int_{\mathbb{R}^k} f(x)\, \pounds(x;T;\xi)dx \tag{4.1}$$

et les temps locaux apparaitront en général dans ce cours dans des formules du type (4.1).

THEOREME 4.1 : *Soit* $\xi_\alpha(t)$ *un processus stable d'ordre* $\alpha > 1$. *Le temps local* $\pounds_\alpha(x;T)$ *de* ξ_α *existe pour tout* $T > 0$. *Par rapport à* x *le temps local vérifie presque partout la condition de Hölder pour tout ordre* $\lambda < \min(\frac{1}{2}, \alpha-1)$.

Démonstration : Pour simplifier soit T=1, posons :

$$\mu(\Gamma;1;\xi_\alpha) = \mu_\alpha(\Gamma),\ \pounds_\alpha(x;1) = \pounds_\alpha(x).$$

On a :

$$\Psi_\alpha(u) = \int_0^1 e^{i\xi_\alpha(t)u}\, dt = \int_{-\infty}^{\infty} e^{ixu}\, \mu_\alpha(dx).$$

Donc Ψ_α est la fonction caractéristique de la loi de probabilité μ.

Puisque :

$$E|\Psi_\alpha(u)|^2 = \int_0^1 \int_0^1 E \exp\{iu(\xi_\alpha(t)-\xi_\alpha(s))\}\, dtds \leq$$

$$\leq \int_0^1 \int_0^1 \exp\{-\alpha^{-1} |u|^\alpha |t-s|\}dtds \leq B(1+ |u|^{-\alpha})$$

on a :

$$E \int_{-\infty}^{\infty} |\Psi_\alpha(u)|^2\, du = E||\Psi_\alpha||_2^2 < \infty .$$

Ceci montre que $\Psi_\alpha \in L_2(-\infty,\infty)$ avec probabilité 1. Donc, avec probabilité 1, il existe $\frac{d\mu_\alpha}{d\lambda} \in L_2$ et :

$$\pounds_\alpha(x) = \frac{d\mu_\alpha}{d\lambda}(x) = \underset{A}{\mathrm{l.i.m}}\ \frac{1}{2\pi} \int_A^A e^{-iux}\, \Psi_\alpha(u)du .$$

Soit A > 0. Considérons la fonction :

$$\gamma(x;A) = \int_A^{2A} e^{-iux}\, \Psi_\alpha(u)du.$$

Soit k un nombre entier. On a :

$$E|\gamma(x;A)|^{2k} = B \int_A^{2A} \dots \int_A^{2A} \exp\{i(\sum_1^k u_j x - \sum_1^k v_j x)\, du_1 \dots dv_k \,.$$

$$\cdot \int_0^1 \dots \int_0^1 E \exp\{i \sum_j \xi(t_j) u_j + i \sum_j \xi(s_j) v_j\}\, dt_1 \dots ds_k \leqq$$

$$\leqq B \int_A^{2A} \dots \int_{-2A}^{-A} du_1 \dots dv_k \int \dots \int |E \exp\{i \sum \xi(t_j) u_j + i \sum \xi(s_j) v_j\}| dt_s \dots ds_k \leqq$$

$$\leqq B \int_A^{2A} \dots \int_{-2A}^{-A} du_1 \dots dv_k \int_{-\infty}^{\infty} \dots \int \exp\{-t_1 |\sum (u_i + v_i)|^{\alpha} \dots (s_k - s_{k-1}) |v_k|^{\alpha}\}$$

$$\cdot\, dt_1 \dots ds_k \leqq B \int_{-BA}^{BA} \dots \int_A^{\infty} du_1 \dots dv_k \,.$$

$$\cdot \int_{-\infty}^{\infty} \dots \int \exp\{-\sum_i t_j |u_j|^{\alpha} - \sum s_j |v_j|^{\alpha}\} dt_1 \dots ds_k \leqq BA^{-k} \,.$$

De la même manière on a :

$$E|\gamma'(x;A)|^{2k} \leqq B \int_{-BA}^{BA} \int |u_1| \dots |v_k| du_1 \dots dv_k \int_{-\infty}^{\infty} \dots \int \exp\{-\sum_1^k |t_j||u_j|^{\alpha} -$$

$$- \sum_1^k |s_j||v_j|^{\alpha}\}\, dt_1 \dots ds_k \leqq B.A^{2k(2-\alpha)} \,.$$

Soit $x,y \in [-R,R]$, $R < \infty$. Alors :

$$E\{(\sup_{|x-y| \leqq h} |\gamma(x;A) - \gamma(y;A)|)^p\} \leqq \sum_{|j| \leqq B/h} E(\int_{jh}^{(j+1)h} |\gamma'(u;A)| du)^p \leqq$$

$$\leqq B\, A^{(2-\alpha)p}\, h^{p-1} \,.$$

Subdivison l'intervalle $[-R,R]$ en un nombre fini d'intervalles partiels Δ_j. Prenons dans chaque intervalle Δ_j un point $x_j \in \Delta_j$. Alors :

$$E(\sup \gamma(x;A))^p \leqq \sum_j E(\sup_{\Delta j} \gamma(x;A))^p \leqq$$

$$\leqq B \sum_j E|\gamma(x_j;A)|^p + B \sum E \sup_{\Delta_j} |\gamma(x;A) - \gamma(y;A)|^p \leqq$$

$$\leqq Bh^{-1} A^{-p/2} + Bh^{p-2} A^{(2-\alpha)p}$$

On en déduit, en posant $h=2^{-n}$, $A=2^n$ que :

$$E(\sup_x |\int_{2^n}^{2^{n+1}} e^{iux} \Psi_\alpha(u) du|)^p \leqq B(2^{-\frac{np}{2}+n} + 2^{np+2-\alpha np})$$

Soit $\varepsilon > 0$. On peut choisir dans l'inégalité précédente $p=p(\varepsilon)$ assez grand pour que :

$$E\{\sup_x |\int_{2^n}^{2^{n+1}} e^{iux} \Psi_\alpha(u) du|\} \leqq B(2^{-\frac{n}{2}(1-\varepsilon)} + 2^{-(\alpha-1-\varepsilon)n}).$$

On en déduit que $\pounds_\alpha(x)$ vérifie la condition de Hölder d'ordre $\gamma = \min(\frac{1}{2} - \varepsilon, \alpha - 1 - \varepsilon)$. La démonstration est achevée.

Notons que, en fait, $\pounds_\alpha(x;t)$ est continue par rapport aux deux variables (x,t) ; nous le montrerons plus tard (voir § 4 du chapitre II).

Puisque $\pounds_\alpha(x)$ est continue, on a :

$$\frac{1}{2\pi}\int_{-\infty}^{\infty} d\lambda \int_0^1 e^{i\lambda\xi_\alpha(t)}\, dt = \lim_{A\to\infty} \frac{1}{\pi}\int_{-\infty}^{\infty} \pounds_\alpha(x)\, \frac{\sin Ax}{x}\, dx = \pounds_\alpha(0).$$

Puisque $\pounds_\alpha(0)$ vérifie la condition de Hölder, la fonction conjuguée à $\pounds_\alpha$, c'est à dire la transformation de Hilbert de $\pounds_\alpha$

$$\tilde{\pounds}_\alpha(x) = \frac{1}{2\pi}\int_{-\infty}^{\infty} e^{-i\lambda x}\ i \operatorname{sign} \lambda \Psi_\alpha(\lambda) d\lambda = \int_{-\infty}^{\infty} \frac{\pounds_\alpha(x-y)}{y}\, dy$$

est bien définie et vérifie la condition de Hölder (conséquence d'un théorème de Titchmarch ; d'ailleurs, ce résultat figure aussi dans la démonstration du théorème 7.1).

V - CONVERGENCE EN PROBABILITE ET CONVERGENCE EN LOI

Il existe une méthode utile pour établir les théorèmes limites pour les processus, on l'appelle parfois la méthode de l'espace probabilisé commun (**метод одного вероятностного пространства** en russe). Cette méthode consiste à remplacer la convergence en loi par celle en probabilité. Le théorème suivant constitue la base de cette méthode.

THEOREME 5.1 : *Soit* $\{\xi_n(t), n=1,2,..\}$ *une suite de processus à valeurs dans* R^k *définis sur un sous-ensemble* A *de* R^1 *et continus en probabilité. Si :*
1° toute loi fini-dimensionnelle de $\xi_n(t)$ *converge vers celle d'un processus* $\xi_0(t)$;
2° pour chaque $t \in A$ *et chaque* $\varepsilon > 0$

$$\lim_{h\downarrow 0}\ \overline{\lim_{n\to\infty}}\ \sup_{|t-s|\leq h} P\{|\xi_n(t) - \xi_n(s)| > \varepsilon\} = 0$$

alors on peut construire des processus $\{\tilde{\xi}_n(t), n=1,2,...\}$, $\tilde{\xi}_0(t)$ *de telle manière que :*
1° toute loi fini-dimensionnelle de $\tilde{\xi}_n$, $n=0,1,...$ *coïncide avec celle de* $\tilde{\xi}_n$:
2° $\tilde{\xi}_n(t)$ *converge vers* $\tilde{\xi}_0(t)$ *en probabilité pour chaque* $t \in A$.

Ce résultat est presque évident si A contient seulement un point. En effet, soit $F_n(x)$ la fonction de répartition de $\xi_n(t) = \xi_n$. Puisque $F_n(x) \longrightarrow F_0(x)$, on a :

$$\tilde{\xi}_n = F_n^{-1}(\xi) \longrightarrow F_0^{-1}(\xi) = \tilde{\xi}_0 .$$

ξ est une variable équidistribuée sur $[0,1]$.

La démonstration générale n'est pas si simple même si card A=2. On peut trouver la démonstration dans le livre [41] de Skorohod.

On utilise ce théorème de la manière suivante. Soit, par exemple, B un espace de fonctions b(t) définies sur $[0,1]$. Supposons que chaque fonction $b \in B$ peut être définie par ses valeurs sur un sous-ensemble dénombrable R de $[0,1]$ ($0,1 \in R$). Soit $\{\xi_n(t),\ t \in [0,1]\}$ une suite de processus à valeurs dans B. Supposons que toute loi fini-dimensionnelle de ξ_n converge vers celle d'un processus $\xi_0 \in B$. D'après le théorème 5.1, on peut construire des processus $\tilde{\xi}_n$, $\tilde{\xi}_0$ de mêmes lois fini-dimensionnelles que ξ_n, ξ_0 et tels que $\tilde{\xi}_n(t) \longrightarrow \tilde{\xi}_0(t)$ en probabilité. Soit J une topologie dans B. On peut déduire très souvent de la convergence $\tilde{\xi}_n(t) \longrightarrow \tilde{\xi}_0(t)$ en probabilité la convergence de $\tilde{\xi}_n$ vers $\tilde{\xi}_0$ dans la topologie J. Donc si f est une fonctionnelle continue pour la topologie J, on peut montrer sous quelques hypothèses additionnelles,que $f(\tilde{\xi}_n) \longrightarrow f(\tilde{\xi}_0)$ en probabilité et donc que $f(\xi_n) \longrightarrow f(\xi_0)$ en loi.

Considérons l'exemple suivant :

Soit $\{\xi_n(t)\}$ une suite de processus définis sur $[0,1]$ et tels que :

$$E\,|\xi_n(t) - \xi_n(s)|^p \leq K\,|t-s|^q, \quad K > 0\ ,\ p,\ q > 1\ .$$

Si les lois fini-dimensionnelles de ξ_n convergent vers celles d'un processus $\xi_0(t)$, alors ξ_n converge vers ξ_0 en loi dans $C[0,1]$.

On peut construire $\tilde{\xi}_n$, $\tilde{\xi}_0$ de même loi et $\tilde{\xi}_n(t) \longrightarrow \tilde{\xi}_0(t)$ en probabilité. Puisqu'aussi :

$$E|\tilde{\xi}_n(t) - \tilde{\xi}_n(s)|^p \leq K|t-s|^q$$

on peut trouver pour chaque $\varepsilon > 0$ un compact $K_\varepsilon \in C[0,1]$ pour lequel :

$$P\,\{\tilde{\xi}_n \in K_\varepsilon,\ \tilde{\xi}_0 \in K_\varepsilon\} > 1-\varepsilon\ .$$

Puisque K_ε est compact on peut choisir pour chaque $\varepsilon_1, \varepsilon_2 > 0$ des points $t_1, \ldots t_N \in [0,1]$ tels que les inégalités :

$$|b_1(t_i) - b_2(t_i)| \leq \varepsilon_1\ , \quad b_1, b_2 \in K_\varepsilon$$

entraînent l'inégalité :

$$\sup_t |b_1(t) - b_2(t)| \leq \varepsilon_2\ .$$

Mais $\tilde{\xi}_n(t) \longrightarrow \tilde{\xi}_0(t)$ en probabilité, donc :

$$P\,\{\sup_t |\tilde{\xi}_n(t) - \tilde{\xi}_0(t)| > \varepsilon\} \xrightarrow[n]{} 0\ .$$

Donc si f est une fonctionnelle continue dans $C\,[0,1]$, on a :

$$P\,\{|f(\tilde{\xi}_n) - f(\tilde{\xi}_0)| > \varepsilon\} \xrightarrow[n]{} 0$$

et donc :

$$P\,\{f(\xi_n) < x\} \xrightarrow[n]{} P\,\{f(\xi_0) < x\}\ .$$

Revenons aux processus $S_n(t)$ définis par la formule (2.3).

THEOREME 5.2 : *Soient* $S_n(t)$ *les processus définis dans le théorème 2.1. Alors on peut construire des processus* $\tilde{S}_n(t)$ *de même lois fini-dimensionnelles que* $S_n(t)$ *et un processus stable* $\tilde{\xi}_\alpha(t)$ *telles que* $\tilde{S}_n(t) \longrightarrow \tilde{\xi}_\alpha(t)$ *en probabilité.*

Démonstration : C'est une conséquence immédiate du théorème 5.1. En effet, il suffit de prouver que la condition 2 de ce théorème est vérifiée. Mais puisque :

$$\lim_{N\to\infty} P\{N^{-1/\alpha}|\sum_1^N \xi_j| > x\} = P\{|\xi_\alpha(1)| > x\},$$

on a :

$$\sup_{|t-s|\leq h} P\{|S_n(t) - S_n(s)| > x\} \leq \sup_{N\leq nh} P\{N^{-1/\alpha}|\sum_1^N \xi_j| > x(\tfrac{n}{N})^{1/\alpha}\}$$

$$\leq \sup_N P\{N^{-1/\alpha}|\sum_1^N \xi_j| > xh^{-1/\alpha} \underset{h\to 0}{\longrightarrow} 0$$

La démonstration est achevée.

THEOREME 5.3 : *Soit* $\xi_n(t) = (\xi_{n1}(t),\ldots \xi_{nk}(t))$, $n=1,2,\ldots$ *une suite de processus à valeurs dans* R^k *définis sur* $[0,1]$. *Si :*

1° Tout processus $\xi_{nj}(t)$ *est continu stochastiquement à gauche (à droite) ;*

2° Pour chaque $\varepsilon > 0$

$$\lim_{h\ 0} \overline{\lim_{n\to\infty}} \sup_{|t-s|\leq h} P\{|\xi_{nj}(t) - \xi_{nj}(s)| > \varepsilon\} = 0 \quad ;$$

3°

$$\lim_{A\to\infty} \overline{\lim_{n\to\infty}} \sup_t P\{|\xi_{nj}(t)| > A\} = 0 .$$

Alors on peut construire des processus $\tilde{\xi}_n(t) = (\tilde{\xi}_{n1}(t),\ldots \tilde{\xi}_{nk}(t))$, $n=1,2,\ldots,$ $\tilde{\xi}_0(t) = (\tilde{\xi}_{01}(t),\ldots \tilde{\xi}_{0k}(t))$ *tels que :*

1° Toute loi fini-dimensionnelle de $\tilde{\xi}_{nj}(t)$ *coïncide avec celle de*

$$\xi_{nj}(t),\ n=1,2,\ldots,\ j=1,\ldots k \ ;$$

2° Pour une sous-suite $\{n_r\}$ $\tilde{\xi}_{n_{rj}}(t) \longrightarrow \tilde{\xi}_{0j}(t)$ *en probabilité.*

Démonstration : (D'après [45]). Soit N un ensemble dénombrable dense dans $[0,1]$ $N = \{t_1,\ldots t_\ell, \ldots\}$. D'après le principe de sélection de Helly et la condition 3, on peut choisir des sous-suite $\{n_j(p)\}$, $\{n_j(p+1)\} \subset \{n_j(p)\}$ telles que $(\xi_{n_j(p)}(t_1),\ldots. \xi_{n_j(p)}(t_p))$ converge en loi. Soit $n_r = n_r(r)$. Alors toute loi fini-dimensionnelle de $\{\xi_{n_r}(t),\ t\in N\}$ converge vers celle d'un processus $\{\xi_0(t),\ t \in N\}$. D'après la condition 2 du théorème et l'inégalité suivante :

$$P\{|\xi_{0j}(t) - \xi_{0j}(t)| > \varepsilon\} \leq \overline{\lim_r} P\{|\xi_{n_{rj}}(t) - \xi_{n_{rj}}(s)| > \varepsilon\} \tag{5.1}$$

le processus $\xi_0(t)$ est continu uniformément sur N. Donc on peut prolonger $\xi_0(t)$ sur $[0,1]$ en posant :

$$\xi_0(t) = p \lim_{s\to t} \xi_0(s) .$$

Montrons que toute loi finidimensionnelle de $\xi_{n_r}(t)$ converge vers celle de $\xi_0(t)$. Soit $t_1,\dots t_p\in[0,1]$, $s_1,\dots s_p\in N$, $U_{11},\dots U_{1K},\dots U_{p1},\dots U_{pk}$ des nombres réels. On a :

$$\overline{\lim_{r\to\infty}} |E \exp \{i \sum_{l,j} U_{lj}\, \xi_{n_{rj}}(t_l) - E \exp \{i \sum_{l,j} U_{lj}\, \xi_{0j}(t_l)\}| \leq$$

$$\leq \overline{\lim_{r}} |E \exp \{i \sum_{l,j} U_{lj}\, \xi_{n_{rj}}(t_l) - E \exp \{i \sum_{l,j} U_{lj}\, \xi_{n_{rj}}(s_l)| +$$

$$+ \overline{\lim_{r}} |E \exp \{i \sum U_{lj}\, \xi_{n_{rj}}(s_l)\} - E \exp \{i \sum U_{lj}\, \xi_{0j}(s_l)\}| +$$

$$+ |E \exp \{i \sum U_{lj}\, \xi_{0j}(s_l)\} - E \exp \{i \sum U_{lj}\, \xi_{0j}(t_l)\}| = I_1 + I_2 + I_3.$$

Si U_{lj}, t_l sont fixés, alors $I_1 \to 0$ quand $s_l \to t_l$ d'après la condition 2 du théorème, $I_3 \to 0$ d'après (5.1). Puisque $(\xi_{n_r}(s_1),\dots \xi_{n_r}(s_p)$ converge en loi vers $(\xi_0(s_1),\dots \xi_0(s_p)$, on a $I_2 = 0$.

Donc $(\xi_{n_r}(t_1),\dots \xi_{n_r}(t_p))$ converge en loi vers $(\xi_0(t_1),\dots\xi_0(t_p))$. Maintenant pour achever la démonstration il suffit d'utiliser le théorème 5.1.

<u>Remarque</u> : Si, sous les conditions du théorème, les lois limites de $(\xi_{n_r}(t_1),\dots\xi_{n_r}(t_p)$ ne dépendent pas de la sous-suite $\{n_r\}$ alors $(\xi_n(t_1),\dots(\xi_n(t_p))$ converge en loi vers $(\xi_0(t_1),\dots \xi_0(t_p))$ et donc $\tilde{\xi}_n(t) \to \tilde{\xi}_0(t)$ en probabilité.

VI - UNE PROPRIETE CARACTERISTIQUE DU PROCESSUS DE WIENER (UN THEOREME DE P. LEVY)

<u>THEOREME 6.1</u> : *(P. Lévy [36]). Soit $\{\mathcal{F}_t,\ t \in [0,1]\}$ une famille croissante de tribus d'évènements aléatoires. Soit $\xi(t)$, $t\in[0,1]$ un processus adapté à la famille $\mathcal{F}_t$ si :*

1° $\xi(t)$ est continu presque sûrement ;

2° Pour tout $t\in[0,1]$ et pour tout $h > 0$, $0 \leq t < t+h \leq 1$,

$$E \{\xi(t+h) - \xi(t)|\mathcal{F}_t \} = 0 ,$$

$$E \{(\xi(t+h) - \xi(t))^2| \mathcal{F}_t\} = h \;;$$

alors $\xi(t)$ est un processus de Wiener ; de plus ce processus ne dépend pas de $\mathcal{F}_0$.

Pour une démonstration de la première partie voir [14] , chapitre VII § 11. Quant à la démonstration de la seconde partie, on a : si $0 \leq t_0 < \ldots < t_n$, $z_1 \ldots z_n$ sont des nombres réels, alors :

$$E \{\exp \{i \sum_{k=1}^{n} z_k(\xi(t_k) - \xi(t_{k-1}))|\mathcal{F}_0\} =$$

$$= E\{E \ \exp \{i\, z_n(\xi(t_n) - \xi(t_{n-1}))\}|\mathcal{F}_{t_{n-1}}\}$$

$$\exp \{i \sum_{1}^{n-1} z_k(\xi(t_k) - \xi(t_{k-1}))|\mathcal{F}_0\}\} =$$

$$= \exp \{-\frac{1}{2} z_n^2 (t_n - t_{n-1})^2\}.E\{\exp \{i \sum_{k=1}^{n-1} z_k(\xi(t_k) - \xi(t_{k-1}))|\mathcal{F}_{t_0}\} =$$

$$= \exp \{-\frac{1}{2} \sum_{k=1}^{n} (t_k - t_{k-1})\, z_k^2\} .$$

D'où le résultat.

VII - COMMENTAIRES

Paragraphe 1 : Les bases de la théorie générale des lois stables ont été jetées par P. Lévy [34]. La formule (1.1) appartient à A. Ya. Khinchin et P. Lévy [33] , le théorème 1.1 appartient à P. Lévy [34] mais les fonctions à croissance lente ont été utilisées dans cette théorie pour la première fois par W. Doeblin [13] et B.V. Gnedenko [24]. La notion d'attraction normale et les théorèmes 1.3, 1.4 appartiennent à B.V. Gnedenko [24], le théorème 1.5 à I. Ibragimov [27] .

Paragraphe 2 : Le théorème 2.2, premier théorème du principe d'invariance, a été montré par M. Donsker [11], le théorème 2.1 par Yu. V. Prohorov [39] ; le théorème 2.3 appartient à Prohorov [39] .

Paragraphe 3 : Le théorème 3.2 pour le cas $\alpha = 2$ se trouve dans [45], le lemme 3.1 est un cas particulier d'un théorème de I.I. Gihman et A.V. Skorohod (voir [21], chapitre 9, paragraphe 7).Le nombre de maxima locaux d'une marche aléatoire a été étudié par I. Gihman [22] qui a démontré un théorème plus général que le théorème 3.4 ; la démonstration donnée ici est peut être nouvelle.

Paragraphe 4 : Je crois que la première notion de temps local appartient à P. Lévy [36] ; beaucoup de résultats sur ce sujet se trouvent dans le travail [20] de Geman et Horovitz. Le théorème 4.1 est très connu mais je ne sais pas à qui l'attribuer ; la démonstration est peut être nouvelle.

Paragraphe 5 : Tous les résultats de ce paragraphe appartiennent à Skorohod [41] .

CHAPITRE II

I - INTRODUCTION

Soient $\xi_1, \xi_2, \dots. \xi_n, \dots.$ des variables aléatoires indépendantes de même loi. Supposons que les sommes :

$$B_n^{-1} \sum_1^n \xi_j - A_n \tag{1.1}$$

convergent en loi quand $n \to \infty$. Nous allons ici étudier les théorèmes limites pour les sommes :

$$\eta_n = \sum_1^n f_n(S_{nk}) \tag{1.2}$$

quand $n \to \infty$. Nous allons utiliser la méthode suivante :

Soit :

$$\widehat{f}_n(\lambda) = \int_{-\infty}^{\infty} e^{i\lambda x} f_n(x)dx .$$

Alors :

$$\eta_n = \frac{1}{2\pi} \sum_1^n \int_{-\infty}^{\infty} e^{-i\lambda S_{nk}} \widehat{f}_n(\lambda)\, d\lambda =$$

$$= \frac{1}{2\pi} \int_{-\infty}^{\infty} \Big(\frac{1}{n} \sum_1^n e^{-i\lambda S_{nk}}\Big) \Psi_n(\lambda)d\lambda$$

où $\Psi_n(\lambda) = n\,\widehat{f}_n(\lambda)$. Puisque le processus engendré par (1.1) converge en loi vers un processus stable $\xi(t)$, les sommes :

$$\frac{1}{n} \sum_1^n e^{-i\lambda S_{nk}}$$

convergent en loi vers $\int_0^1 e^{-i\lambda\xi} dt$. Donc si $\Psi_n(\lambda)$ converge vers une fonction $\Psi(\lambda)$ alors on peut s'attendre à la convergence de (1.2) vers :

$$\frac{1}{2\pi} \int_{-\infty}^{\infty} \Psi(\lambda)d\lambda \int_0^1 e^{-i\lambda\xi(t)} dt .$$

Nous montrons des théorèmes de ce type (nous les appelons les théorèmes limites du premier type) dans les paragraphes 2 à 4 sous l'hypothèse :

$$\lim_n \int_{-\infty}^{\infty} \frac{|\Psi(\lambda) - \Psi_n(\lambda)|^2}{1 + |\lambda|^\alpha} d\lambda = 0 \tag{1.3}$$

Si la condition (1.3) n'est pas vérifiée, alors les théorèmes limites sont différents. Nous les appelons théorèmes limites du deuxième type et les démontrons dans le paragraphe 5.

Nos méthodes permettent aussi d'étudier les sommes du type :

$$\sum_{1}^{n-r} f_n(S_{nk}, \dots S_{n,k+r}) , \tag{1.4}$$

$$\sum_{1}^{n} f_n(S_{nk_1}, \dots S_{nk_r}) \tag{1.5}$$

mais nous omettons ce sujet. Montrons seulement une méthode de Skorohod qui permet de déduire des résultats sur (1.4) de ceux sur (1.2).

Posons :

$$\Phi_n(x) = Ef(x, x + S_{n1}, \dots, x + S_{nr}) \tag{1.6}$$

$$\bar{\eta}_n = \sum_{1}^{n} \Phi(S_{nk}) \tag{1.7}$$

THEOREME 1.1 : *Si* $f_n \geqq 0$ *et* $\lim_n \sup_{x_i} f_n(x_0, \dots x_r) = 0$ *alors la loi limite de* η_n *défini par (1.4) existe si et seulement si la loi limite de* $\bar{\eta}_n$ *existe. De plus les lois limites de* η_n *et* $\bar{\eta}_n$ *coïncident.*

Démonstration : (D'après [45]). Posons :

$$\eta_{nj} = \sum_{1}^{j} f_n(S_{nk}, \dots S_{n,k+r}) ,$$

$$\bar{\eta}_{nj} = \sum_{1}^{j} \Phi_n(S_{nk})$$

et montrons que si $\bar{\eta}_n$ ou η_n sont bornés en probabilité alors :

$$\sup_{1 \leqq j \leqq n-r} |\eta_{nj} - \bar{\eta}_{nj}| \longrightarrow 0 \tag{1.8}$$

en probabilité. Evidemment le théorème résulte de (1.8).

Posons :

$$\Phi_{nl}(x_0, x_1, \dots x_l) = Ef_n(x_0, x_1, \dots x_l + S_{n1}, \dots x_l + S_{n,r-l}).$$

En particulier :

$$\phi_{n0}(x_0) = \Phi_n(x_0), \ \phi_{nr}(x_0, \dots x_r) = f_n(x_0, \dots x_r).$$

On a :

$$\phi_{n,l-1}(S_{ni}, \dots S_{n,i+l-1}) = E\{ \phi_{nl}(S_{ni}, \dots S_{n,i+l}) \mid S_{ni}, \dots S_{n,i+l-1}\} .$$

Soit :

$$\zeta_{il}(n) = \phi_{nl}(S_{ni}, \dots S_{n,i+l}) - \phi_{n,l-1}(S_{ni}, \dots S_{n,i+l-1}).$$

Alors :

$$\sup_k |\eta_{nk} - \bar{\eta}_{nk}| \leqq \sum_{l=1}^{r} \sup_k \left| \sum_{1}^{k} \zeta_{ik}(n) \right| .$$

Donc il suffit de montrer que :

$$\sup_k \left|\sum_{i=1}^{k} \zeta_{ik}(n)\right| \longrightarrow 0 \qquad (1.9)$$

en probabilité.

Posons $\mu_n = \sup_k f_n$. Alors $0 \leq \phi_{nj} \leq \mu_n$. On a :

$$E\{\zeta_{il}(n) | S_{n1} \dots S_{n,i+l-1}\} = 0, \qquad (1.10)$$

$$E\{\zeta_{il}^2(n) | S_{n1}, \dots S_{n,i+l-1}\} \leq \mu_n \phi_{n,l-1}(S_{ni}, \dots S_{n,i+l-1}).$$

Supposons d'abord que les $\overline{\eta}_n$ sont uniformément bornés.

Puisque $\phi_n \geq 0$ les variables $\overline{\eta}_{nk} \leq \overline{\eta}_n$ sont uniformément bornées. Soit $N > 0$.

Posons :

$$\overline{\eta}_k^N = \begin{cases} 1, & \overline{\eta}_{kn} \leq N \\ 0 & \overline{\eta}_{kn} > N \end{cases}$$

La variable $\overline{\eta}_k^N$ est mesurable par rapport à $S_{n1}, \dots S_{nk}$. On a :

$$\overline{\eta}_1^N \geq \overline{\eta}_2^N \geq \dots \geq \overline{\eta}_n^N .$$

Posons $\alpha_k = \sum_1^k \overline{\eta}_i^N \zeta_{il}(n)$ et montrons que $\alpha_1, \alpha_2, \dots$ est une martingale. En effet, il résulte de (1.10) que :

$$E\{\alpha_{k+1} | \alpha_k, \dots \alpha_1\} = \alpha_k + E\{\eta_{k+1}^N \zeta_{k+1,l}(n) | \alpha_1, \dots \alpha_k\} =$$

$$= \alpha_k + E\{\overline{\eta}_{k+1}^N E\{\zeta_{k+1,l}(n) | S_{n1}, \dots S_{n,k+1}\} | \alpha_1, \dots \alpha_k\} = \alpha_k .$$

On a :

$$P\{\sup_k \left|\sum_{i=1}^{k} \zeta_{il}(n)\right| > \varepsilon\} \leq P\{\sup_k |\alpha_k| > \varepsilon\} +$$

$$+ P\{\overline{\eta}_n^N = 0\} \leq \varepsilon^{-2} E(\sup_k \alpha_k)^2 + P\{\overline{\eta}_n > N\} .$$

La propriété de martingale donne (voir [14], théorème 7.3.4) :

$$E\{(\sup \alpha_k)^2\} \leq 4 E \alpha_{n-r}^2 = 4 E(\sum_{i=1}^{n-r} \overline{\eta}_i^N \zeta_{il}(n))^2 =$$

$$= 4 \sum_{i=1}^{n-r} E\{\eta_i^N E\{\zeta_{il}^2(n) | S_{ni}, \dots S_{n,i+l-1}\} \leq$$

$$\leq 4\mu_n \sum_{i=1}^{n-r} E\{\overline{\eta}_i^N \Phi_n(S_{ni})\} \leq \mu_n N.$$

Donc :

$$P\{\sup_k |\sum_1^k \zeta_{i1}(n)| \geqq \varepsilon\} \leqq N\, \varepsilon^{-2}\, \mu_n + P\{\overline{\eta_n} > N\}$$

et le théorème est démontré dans le cas où η_n converge en loi.

Supposons maintenant que les η_n sont bornés uniformément en probabilité. Posons :

$$\eta_i^N = \begin{cases} 1, & i=1,\dots r \\ 1, & \eta_{i-r,n} \leqq N,\ i > r \\ 0, & \eta_{i-r,n} > N,\ i > r \end{cases}$$

On peut montrer comme plus haut que :

$$P\{\sup_k |\sum_1^n \zeta_{i1}(n)| > \varepsilon\} \leqq B\, \varepsilon^{-2}\, \mu_n \sum_1^{n-r} E\{\eta_i^N f_n(S_{ni}, \dots S_{n,i+r})\}$$

$$+ P\{\eta_n^N = 0\} \leqq B\, \mu_n (N + r\, \mu_n)\varepsilon^{-2} + P\{\eta_n > N\}\,.$$

La démonstration est achevée.

II - THEOREMES LIMITES DU PREMIER TYPE

1 - Enoncé de la condition et du résultat

Nous allons considérer une suite $\{\xi_n\}$ de variables aléatoires indépendantes de même loi. Nous supposons que la loi commune de ces variables appartient au domaine d'attraction d'une loi stable avec $\alpha > 1$, et que $E\xi_n = 0$. La fonction caractéristique $\varphi(t)$ des variables ξ_n au voisinage de zéro a la forme :

$$\begin{aligned}\varphi(t) &= \exp\{- c|t|^{\alpha}(1+i\beta \tfrac{t}{|t|} \operatorname{tg} \tfrac{\pi\alpha}{2})\}\,(1 + o(1)) = \\ &= \exp\{- c|t|^{\alpha}\omega(t)\}(1 + o(1)).\end{aligned} \tag{2.1}$$

Nous supposerons toujours $c = \alpha^{-1}$ pour avoir dans le cas $\operatorname{var}\xi_j < \infty$ une forme standardisée :

$$\varphi(t) = \exp\{-\frac{1}{2} t^2\}\ (1 + o(1)).$$

Désignons par Λ l'ensemble $\{\lambda : \varphi(\lambda) = 1\}$.

Si la loi de ξ_n n'est pas arithmétique l'ensemble Λ contient seulement un point $\lambda = 0$; si cette loi est arithmétique, l'ensemble Λ ou bien contient de nouveau seulement le point $\lambda = 0$, ou bien est une progression arithmétique.

Posons :

$$S_{n,k} = n^{-1/\alpha} \sum_{j=1}^k \xi_j \quad , \ k = 1,2 \dots n$$

Nous allons considérer ici et dans les paragraphes suivants le comportement limite des sommes (1.1) pour r=0, de sorte que :

$$\eta_n = \sum_{k=1}^{n} f_n(S_{kn}) \tag{2.2}$$

Nous allons supposer ci-dessous que la fonction $f_n(x)$ coïncide partout avec la transformée de Fourier de $\widehat{f}_n$:

$$f_n(x) = \frac{1}{2\pi} \int_{-\infty}^{\infty} e^{-i\lambda x} \widehat{f}_n(\lambda) d\lambda \qquad , x \in R^1$$

Posons :

$$\Psi_n(\lambda) = n \sum_{\lambda_k \in \Lambda} \widehat{f}_n(\lambda + \lambda_k n^{1/\alpha}),$$

$$u_n(x) = n \int_0^x f_n(z) \sum_{\lambda_k \in \Lambda} e^{-i\lambda_k z n^{1/\alpha}} \tag{2.3}$$

de sorte que :

$$u_n(x) = \frac{1}{2\pi} \int_{-\infty}^{\infty} \Psi_n(\lambda) \frac{e^{-i\lambda x}-1}{-i\lambda} d\lambda \tag{2.4}$$

THEOREME 2.1 : *Supposons les conditions suivantes vérifiées :*

1° La fonction Ψ_n *est nulle en dehors de l'intervalle* $\left[-an^{1/\alpha}, an^{1/\alpha}\right]$ *où la constante a ne dépend pas de n ;*

2° Il existe une fonction $\Psi(\lambda)$ *telle que :*

$$\int_{-\infty}^{\infty} \frac{|\Psi(\lambda)|^2}{1+|\lambda|^\alpha} d\lambda < \infty; \quad \lim_n \int_{-\infty}^{\infty} \frac{|\Psi_n(\lambda) - \Psi(\lambda)|^2}{1 + |\lambda|^\alpha} d\lambda = 0 \tag{2.5}$$

Alors la loi limite des variables η_n *existe et coïncide avec la loi de la variable :*

$$\eta = \frac{1}{2\pi} \int_{-\infty}^{\infty} \Psi(\lambda) \int_0^1 e^{-i\lambda\xi_\alpha(t)} dt\, d\lambda \tag{2.6}$$

où ξ_α *est un processus stable de fonction caractéristique :*

$$E\, e^{i\lambda\xi_\alpha(t)} = \exp\{- \frac{t}{\alpha}|\lambda|^\alpha(1 + i\beta \frac{t}{|t|} \operatorname{tg} \frac{\pi\alpha}{2})$$

Nous verrons dans la section suivante que sous l'hypothèse du théorème, l'intégrale (2.6) existe bien.

Commençons par une série de lemmes.

2 - Deux Lemmes

LEMME 2.1 : *Soient les fonctions* ψ_n *vérifiant les conditions 1, 2 du théorème 2.1. Alors pour tout* $c > 0$:

$$\lim_{T\to\infty} \sup_n \int_{|v|>T} dv \int_{-\infty}^{\infty} |\Psi_n(v)\ \Psi_n(u-v)| \ . \tag{2.7}$$
$$\cdot \left\{ \frac{1-e^{-c|u|^\alpha}}{|v|^\alpha\ |u|^\alpha} + \frac{|e^{-c|u|^\alpha} - e^{-c|v|^\alpha}|}{|v|^\alpha\ ||u|^\alpha - |v|^\alpha|} \right\} du = 0$$

En particulier :

$$\lim_T \int_{|v|>T} dv \int_{-\infty}^{\infty} |\Psi(v)\ \Psi(u-v)| \left\{ \frac{1-e^{-c|u|^\alpha}}{|v|^\alpha |u|^\alpha} + \frac{|e^{-c|u|^\alpha} - e^{-c|v|^\alpha}|}{|v|^\alpha\ ||u|^\alpha - |v|^\alpha|} \right\} du = 0$$

Démonstration : Dans le domaine d'intégration

$$\frac{|1-e^{-c|u|^\alpha}|}{|u|^\alpha\ |v|^\alpha} \leq \frac{B}{|u|^{\alpha/2}\ \ |v|^{\alpha/2}} \qquad \frac{1-e^{-c|u|^\alpha}}{\sqrt{1+|u-v|^\alpha}} \ (|u|^{-\alpha/2} + |v|^{-\alpha/2})$$

d'où :

$$\int_{|v|>T} \int_{-\infty}^{\infty} |\Psi_n(v)\ \Psi_n(u-v)|\ \frac{1-e^{-c|u|^\alpha}}{|u|^\alpha\ |u|^\alpha} \ du\ dv \leq$$

$$\leq B \int_{-\infty}^{\infty} \frac{1-e^{-c|u|^\alpha}}{|u|^\alpha} du \Big(\int_{|v|>T} \frac{|\Psi_n(v)|^2}{1+|v|^\alpha} dv \int_{-\infty}^{\infty} \frac{|\Psi_n\ (u)|^2}{1+|u|^\alpha} du\Big)^{1/2} +$$

$$+ B\Big(\int_{|v|>T} \frac{|\Psi_n(v)|^2}{1+|v|^\alpha} dv \int_{-\infty}^{\infty} \frac{|\Psi_n(u)|^2}{1+ |u|^\alpha} du\Big) \ .$$

Il existe une constante $c_1 > 0$ telle que pour $\frac{1}{2} \leq (\frac{u}{v})^\alpha \leq 2$

$$\frac{|e^{-c|u|^\alpha} - e^{-c|v|^\alpha}|}{|u|^\alpha - |v|^\alpha} \leq e^{-c_1|u|^\alpha} \leq \frac{B}{1+|u-v|^\alpha} \ .$$

Et si $|u|^\alpha < \frac{1}{2}\ |v|^\alpha$ ou bien $|u|^\alpha > 2\ |v|^\alpha$

$$||u|^\alpha - |v|^\alpha|^{-1} \leq B(1 + |u-v|^\alpha)^{-1} \ .$$

Donc :

$$\int_{|v|>T} \int_{-\infty}^{\infty} |\Psi_n(v)\ \Psi_n(u-v)|\ \frac{|e^{-c|u|^\alpha} - e^{-c|v|^\alpha}|}{|v|^\alpha\ ||u|^\alpha - |v|^\alpha|} \ du\ dv \leq$$

$$\leq B \int_{|v|>T} \frac{|\Psi_n(v)|}{|v|^\alpha} dv \ \Big(\int_{-\infty}^{\infty} \frac{|\Psi_n(u)|^2}{1+ |u|^\alpha} du \int_{-\infty}^{\infty} \frac{du}{1+|u|^\alpha}\Big)^{1/2}$$

$$\leq B \Big(\int_{-\infty}^{\infty} \frac{|\Psi_n(u)|^2}{1 + |u|^\alpha} du\Big)^{1/2} \ . \ T^{\frac{1-\alpha}{2}} \Big(\int_{|v|>T} \frac{|\Psi_n(v)|^2}{1 + |v|^\alpha} dv\Big)^{1/2}$$

En définitive, il vient :

$$\int_{|v|>T} \int_{-\infty}^{\infty} |\Psi_n(v)\ \Psi_n(u-v)| \left\{ \frac{1-e^{-c|u|^\alpha}}{|u|^\alpha\ |v|^\alpha} + \frac{|e^{-c|v|^\alpha}-e^{-c|u|^\alpha}|}{|v|^\alpha\ ||u|^\alpha - |v|^\alpha|} \right\} du\ dv \leq$$

$$\leq B \left(\int_{-\infty}^{\infty} \frac{|\Psi_n(u)|^2}{1 + |\ |^\alpha} du \int_{|v|>T} \frac{|\Psi_n(v)|^2}{1 + |v|^\alpha} dv \right).$$

Cette inégalité avec la condition (2.5) prouve le lemme.

LEMME 2.2 : *Sous les hypothèses du théorème 2.1, les intégrales :*

$$\int_{-\infty}^{\infty} \Psi_n(\lambda) d\lambda \int_0^1 e^{-i\lambda\xi_\alpha(t)} dt, \int_{-\infty}^{\infty} \Psi(\lambda) d\lambda \int_0^1 e^{i\lambda\xi_\alpha(t)} dt$$

sont bien définies et :

$$\lim_n E \left| \int_{-\infty}^{\infty} (\Psi(\lambda) - \Psi_n(\lambda)) d\lambda \int_0^1 e^{-i\lambda\xi_\alpha(t)} dt \right|^2 = 0 \tag{2.8}$$

Démonstration : Soit $\psi(\lambda)$ une fonction telle que :

$$\int_{-\infty}^{\infty} \frac{|\psi(\lambda)|^2}{1 + |\lambda|^\alpha} d\lambda < \infty .$$

Soit C une constante positive assez petite. On a pour $-\infty < a < b < \infty$:

$$E \left| \int_a^b \psi(\lambda) d\lambda \int_0^1 e^{-i\lambda\xi_\alpha(t)} dt \right|^2 =$$

$$= \int_a^b \int_a^b \psi(\lambda)\ \overline{\psi(\mu)} \int_0^1 \int_0^1 E\ e^{i\lambda\xi_\alpha(t) - i\mu\xi_\alpha(s)} dt\ ds\ d\mu\ d\lambda \leq$$

$$\leq B \int_a^b \int_a^b |\psi(\lambda)\ \psi(\mu)| \int_0^1 \left\{ \frac{e^{-ct|\mu-\lambda|} - e^{-ct|\lambda|^\alpha}}{|\lambda|^\alpha - |\mu\cdot\lambda|^\alpha} + \right.$$

$$\left. + e^{-ct|\mu-\lambda|^\alpha} \frac{1-e^{-c(1-t)|\mu|^\alpha}}{|\mu|^\alpha} \right\} d\lambda\ d\mu \leq$$

$$\leq B \int_a^b \int_a^b |\psi(\lambda)\ \psi(\mu)| \left\{ |(|\lambda|^\alpha - |\mu-\lambda|)^{-1} . \right.$$

$$. \left(\frac{1-e^{-c|\mu-\lambda|^\alpha}}{|\mu-\lambda|^\alpha} - \frac{1-e^{-c|\lambda|^\alpha}}{|\lambda|^\alpha} \right) +$$

$$\left. + \left(\frac{1-e^{-c|\lambda-\mu|^\alpha}}{|\mu|^\alpha\ |\lambda-\mu|^\alpha} - \frac{e^{-c|\lambda-\mu|^\alpha} - e^{-c|\mu|^\alpha}}{|\mu|^\alpha (|\mu|^\alpha - |\lambda-\mu|^\alpha)} \right) | \right\} d\lambda\ d\mu =$$

$$= B \int_a^b |\psi(\lambda)\, d\lambda \int_{a-b}^{b-a} |\psi(\lambda-\mu)| \,.$$

$$.\left\{ (|\lambda|^{\alpha} - |\mu|^{\alpha})^{-1} \left(\frac{1-e^{-c|\mu|^{\alpha}}}{|\mu|^{\alpha}} - \frac{1-e^{-c|\lambda|^{\alpha}}}{|\lambda|^{\alpha}} \right) + \right.$$

$$\left. + |\lambda|^{-\alpha} \left(\frac{1-e^{-c|\mu|^{\alpha}}}{|\mu|^{\alpha}} - \frac{e^{-c|\mu|^{\alpha}} - e^{-c|\lambda|^{\alpha}}}{|\lambda|^{\alpha} - |\mu|^{\alpha}} \right) | \right\} d\mu \,.$$

Le lemme 2.2 résulte alors du lemme 2.1.

3 - Démonstration du théorème dans le cas $\Lambda = \{0\}$

Supposons que l'ensemble $\Lambda = \{\lambda : \varphi(\lambda) = 1\} = \{0\}$. Dans ce cas, $\Psi_n(\lambda) = n \,\widehat{f}_n(\lambda)$.

En vertu du théorème on peut supposer que les fonctions aléatoires :

$$W_n(t) = n^{-1/\alpha} \sum_{k \leq nt} \xi_k$$

convergent en chaque point $t \in [0,1]$ vers $\xi_\alpha(t)$ en probabilité. Montrons que sous cet hypothèse la différence :

$$\Delta_n = \eta_n - \frac{1}{2\pi} \int_{-\infty}^{\infty} \Psi_n(\lambda) \int_0^1 e^{-i\lambda\xi_\alpha(t)}\, dt \longrightarrow 0 \qquad (2.9)$$

en probabilité quand $n \to \infty$.

On a :

$$\eta_n = \frac{1}{n} \sum_1^n \frac{1}{2\pi} \int_{-\infty}^{\infty} \Psi_n(\lambda)\, e^{-i\lambda W_n(\frac{k}{n})}\, d\lambda =$$

$$= \frac{1}{2\pi} \int_{-\infty}^{\infty} \Psi_n(\lambda) \left[\frac{1}{n} \sum_1^n e^{-i\lambda W_n(\frac{k}{n})} \right] d\lambda \,.$$

cette égalité entraîne que pour $T,\ \varepsilon > 0$

$$E|\Delta_n|^2 \leq 4 \left\{ E\left| \int_{-T}^{T} \Psi_n(\lambda) \left[\int_0^1 e^{-i\lambda\xi_\alpha(t)}\, dt - \frac{1}{n}\, e^{-i\lambda W_n(\frac{k}{n})}\, d\lambda \right|^2 + \right.\right.$$

$$+ E\left| \int_{|\lambda|>T} \Psi_n(\lambda) \int_0^1 e^{-i\lambda\xi_\alpha(t)} \right] dt\, d\lambda \Big|^2 +$$

$$+ E\left| \frac{1}{n} \int_{T \leq |\lambda| \leq \varepsilon n^{1/\alpha}} \Psi_n(\lambda) . \sum_1^n e^{-i\lambda S_{nj}}\, d\lambda \right|^2 +$$

$$+ E\left| \frac{1}{n} \int_{\varepsilon n^{1/\alpha} \leq |\lambda| \leq a n^{1/\alpha}} \Psi_n(\lambda) \sum_1^n e^{i\lambda S_{nj}}\, d\lambda \right|^2 = I_1 + I_2 + I_3 + I_4$$

Montrons qu'on peut rendre tous les I_j arbitrairement petits en choisissant ε et T convenablement et en faisant tendre n vers l'infini.

Puisque :

$$\frac{1}{n}\sum_{1}^{n} e^{i\lambda W_n(\frac{k}{n})} \longrightarrow \int_0^1 e^{i\lambda\xi_\alpha(t)}\, dt$$

en probabilité et que les deux membres de cette relation sont uniformément bornés on a aussi que :

$$\lim_n E\left|\frac{1}{n}\sum_{1}^{n} e^{i\lambda W_n(\frac{k}{n})} - \int_0^1 e^{i\lambda\xi_\alpha(t)}\, dt\right|^2 = 0 .$$

Donc quand $n \to \infty$

$$I_1 \leqq 8(1 + T^\alpha) . T \int_{-T}^{T} \frac{|\Psi_n(\lambda)|^2}{1 + |\lambda|^\alpha}\, d\lambda .$$

$$\frac{1}{2T} \int_{-T}^{T} E\left|\frac{1}{n}\sum_{1}^{n} e^{i\lambda W_n(\frac{k}{n})} - \int_0^1 e^{i\lambda\xi_\alpha(t)}\, dt\right|^2 d\lambda \leqq$$

$$\leqq B(1 + T^\alpha)\; T . o(1).$$

On déduit de l'inégalité du lemme 2.2 que :

$$I_2 \longrightarrow 0 \quad , \text{ quand } \quad T\to\infty$$

uniformément par rapport à n.

Pour étudier les termes I_3, I_4 notons d'abord qu'ils ont la forme suivante :

$$n^{-2} \int_{-\infty}^{\infty}\int_{-\infty}^{\infty} \Psi_n(\lambda)\, \overline{\Psi_n(\mu)} \sum_{k,j=1}^{n} E \exp\{-i, S_{n,j}\lambda + iS_{n,k}\,\mu\} .$$

Posons $\lambda - \mu = u$, $\lambda = v$, alors :

$$\sum_{k<j} E \exp\{i\lambda S_{n,j} - i\mu\, S_{n,k}\, \mu\} = \sum_{j=2}^{n}\sum_{k=1}^{j-1} \varphi^k(n^{-1/\alpha}u)\, \varphi^{j-k}(n^{-1/\alpha}v) =$$

$$= \frac{\varphi(n^{-1/\alpha}u)\,\varphi(n^{-1/\alpha}v)}{\varphi(n^{-1/\alpha}u) - \varphi(n^{-1/\alpha}v)} \left(\varphi(n^{-1/\alpha}v)\, \frac{1-\varphi^n(n^{-1/\alpha}v)}{1-\varphi(n^{-1/\alpha}v)} - \frac{1-\varphi^n(n^{-1/\alpha}u)}{1-\varphi(n^{-1/\alpha}u)}\right) =$$

$$= (\varphi(n^{-1/\alpha}u).\, \varphi(n^{-1/\alpha}v).\, \frac{1-\varphi^{n+1}\varphi(n^{-1/\alpha}u)}{(1-\varphi(n^{-1/\alpha}v))(1-\varphi(n^{-1/\alpha}u))} - \tag{2.10}$$

$$- \varphi(n^{-1/\alpha}u)\,\varphi(n^{1/\alpha}v) . \frac{\varphi^{n+1}(n^{-1/\alpha}v) - \varphi^{n+1}(n^{-1/\alpha}u)}{(1-\varphi(n^{-1/\alpha}u))\,(1-\varphi(n^{-1/\alpha}v)}$$

et :

$$\sum_{k=1}^{n} E \exp\{i\lambda S_{nk} - i\mu S_{nk}\} = \varphi(n^{-1/\alpha}u)\, \frac{1-\varphi^n(n^{-1/\alpha}u)}{1-\varphi(n^{-1/\alpha}u)} \tag{2.11}$$

Nous utiliserons parfois au lieu de (2.10) la majoration suivante :

$$|\sum_{j=2}^{n}\sum_{k=1}^{j-1}\varphi^{k}(n^{-1/\alpha}u)\,\varphi^{j-k}(n^{-1/\alpha}v)| =$$

$$= |\sum_{k=1}^{n-1}\varphi^{k}(n^{-1/\alpha}u)\sum_{j=k+1}^{n}\varphi^{j-k}(n^{-1/\alpha}v)| \leq \qquad (2.12)$$

$$\leq \frac{|\varphi(n^{-1/\alpha}u)\;\varphi(n^{-1/\alpha}v)|\;(1-|\varphi(n^{-1/\alpha}u)|^{n-1})}{|1-\varphi(n^{-1/\alpha}v)|\;(1-|\varphi(n^{-1/\alpha}v)|)}.$$

Nous allons estimer maintenant I_3. Notons d'abord que si ε est assez petit l'inégalité :

$$|\varphi(n^{-1/\alpha}t)| \leq \exp\{-cn^{-1}|t|^{\alpha}\} \quad , \quad c > 0$$

est satisfaite dans la région $|t| \leq 2\,\varepsilon n^{1/\alpha}$.

Donc pour $|u| \leq 2\varepsilon n^{1/\alpha}$, $|v| \leq 2\varepsilon n^{1/\alpha}$

$$|\frac{1-\varphi^{n}(n^{-1/\alpha}u)}{1-\varphi(n^{-1/\alpha}u)}| \leq \frac{1-|\varphi(n^{-1/\alpha}u)|^{n}}{1-|\varphi(n^{-1/\alpha}u)|} \leq \frac{Bn}{|u|^{\alpha}}(1-e^{-e|u|^{\alpha}}), c > 0,$$

$$|1-\varphi(v\,n^{1/\alpha})|^{-\alpha} \leq Bn|v|^{-\alpha}, \qquad (2.13)$$

$$|\frac{\varphi^{n+1}(n^{-1/\alpha}n)-\varphi^{n+1}(n^{-1/\alpha}v)}{\varphi(n^{-1/\alpha}u)-\varphi(n^{-1/\alpha}v)}| \leq Bn\frac{|e^{-c|u|^{\alpha}}-e^{-c|v|^{\alpha}}|}{||u|^{\alpha}-|v|^{\alpha}|}$$

On déduit de ceci, de (2.10) et de (2.11) que :

$$I_3 \leq Bn^{-2}\int_{T<|\lambda|<\varepsilon n^{1/\alpha}}\int_{T<|\lambda|<\varepsilon n^{1/\alpha}}\Psi_n(\lambda)\,\overline{\Psi_n(\mu)}\,\{\sum_{j\geq k}\exp\{-cxn^{-1}\,.$$

$$.\,|\lambda-\mu|^{\alpha}-c(j-k)n^{-1}|\lambda|^{\alpha}\}+\sum_{j<k}\exp\{-cjn^{-1}|\lambda-\mu|^{\alpha}-$$

$$-\,c(k-j)\,|\mu|^{\alpha}n^{-1}\}\}\,d\lambda\,d\mu \leq$$

$$\leq Bn^{-1}\int_{T<|v|<\varepsilon n^{1/\alpha}}|\Psi_n(v)|\,dv\int_{-\infty}^{\infty}|\Psi_n(u-v)|\,\frac{1-e^{-c|u|^{\alpha}}}{|u|^{\alpha}}\,du\,+$$

$$+\,Bn^{-2}\int_{T<|v|<\varepsilon n^{1/\alpha}}|\Psi_n(v)|\int_{-2\varepsilon n^{1/\alpha}}^{2\varepsilon n^{1/\alpha}}|\Psi_n(u-v)|\,\frac{n}{||u|^{\alpha}-|v|^{\alpha}|}\,.$$

$$. \left| \frac{1 - e^{-c|v|^\alpha}}{1 - e^{-c|v|^\alpha n^{-1}}} - \frac{1 - e^{-c|u|^\alpha}}{1 - e^{-c|u|^\alpha n^{-1}}} \right| du\, dv \leq$$

$$\leq B \int_{|v|>T} |\Psi_n(v)| dv \int_{-\infty}^{\infty} |\Psi_n(u-v)| \left\{ \frac{1 - e^{-c|u|^\alpha}}{|v|^\alpha\, |u|^\alpha} + \frac{|e^{-c|u|^\alpha} - e^{-c|v|^\alpha}|}{|v|^\alpha\, ||u|^\alpha - |v|^\alpha|} \right\} du.$$

En vertu du lemme 2.1, le membre de droite de cette inégalité tend vers zéro quand $T \to \infty$ uniformément en n.

Pour traiter I_4, il faut distinguer deux cas : la loi de ξ_i est arithmétique ou elle ne l'est pas. Supposons d'abord que cette loi n'est pas arithmétique. Alors, il existe $\delta > 0$ tel que $|\varphi(tn^{1/\alpha})| \leq e^{-\delta}$ pour $|t| > \varepsilon n^{1/\alpha}$ (bien sûr δ dépend de ε). Donc pour $\varepsilon n^{1/\alpha} \leq |v| \leq an^{1/\alpha}$:

$$\left| \frac{\varphi^{n+1}(n^{-1/\alpha}v) - \varphi^{n+1}(n^{-1/\alpha}u)}{\varphi(n^{-1/\alpha}v) - \varphi(n^{-1/\alpha}u)} \right| \leq \sum_{j=0}^{\infty} e^{-j\delta} \leq (1-e^{-\delta})^{-1}$$

Ceci implique (voir aussi (2.10) et (2.13) que :

$$I_3 \leq n^{-2} \int_{\varepsilon n^{1/\alpha} > |\lambda| > T} \int_{|\mu|>T} \{\Psi_n(\lambda)\overline{\Psi_n(\mu)} \sum_{j \geq k} \exp\{-ckn^{-1}|\lambda-\mu|^\alpha -$$

$$- c(j-k)n^{-1}\, |\lambda|^\alpha\} + \sum_{j<k} \exp\{-cjn^{-1}\, |\lambda-\mu|^\alpha - c(k-j)|\mu|^\alpha\, n^{-1}\}\, d\mu\, d\lambda \leq$$

$$\leq Bn^{-1} \int_{\varepsilon n^{1/\alpha} > |v| > T} |\Psi_n(v)| dv \int_{-\infty}^{\infty} |\Psi_n(u-v)\; \frac{1 - e^{-c|u|^\alpha}}{|u|^\alpha}\, du +$$

$$+ Bn^{-2} \int_{\varepsilon n^{1/\alpha} > |v| > T} |\Psi_n(v)|\, dv \int_{-2\varepsilon n^{1/\alpha}}^{2\varepsilon n^{1/\alpha}} |\Psi_n(u-v)|\, \frac{n}{||u|^\alpha - |v|^\alpha|}\, .$$

$$. \left| \frac{1 - e^{-c|v|^\alpha}}{1 - e^{-c|v|^\alpha n^{-1}}} - \frac{1 - e^{-c|u|^\alpha}}{1 - e^{-c|u|^\alpha n^{-1}}} \right| du \leq$$

$$\leq B \int_{|v|>T} |\Psi_n(v)| dv \int_{-\infty}^{\infty} |\Psi_n(u-v)| \left\{ \frac{1 - e^{-c|u|^\alpha}}{|v|^\alpha\, |u|^\alpha} + \right.$$

$$\left. + \frac{|e^{-c|u|^\alpha} - e^{-c|v|^\alpha}|}{|v|^\alpha\, ||u|^\alpha - |v|^\alpha|} \right\} du\ .$$

Quand $n \to \infty$ la première intégrale de droite est o(1), la seconde intégrale est inférieure à :

$$\left(\int_{|v|>n^{1/\alpha}\varepsilon} \frac{|\Psi_n(v)|^2}{1+|v|^\alpha} dv\right)^{1/2} \left(\int_{-\infty}^{\infty} du \int_{-\infty}^{\infty} \frac{|\Psi_n(u-v)|^2}{1+|u-v|^\alpha} \frac{1-e^{-c|u|^\alpha}}{|u|^\alpha} du\right)^{1/2}.$$

$$\left(\int_{-\infty}^{\infty} \frac{1-e^{-c|u|^\alpha}}{|u|^\alpha} du\right)^{1/2} = o(1).$$

Donc $I_4 = o(1)$ quand $n \to \infty$.

Supposons maintenant que la loi de ξ_1 est arithmétique et que le pas maximal de cette loi est égal à h mais que l'ensemble $\Lambda = \{0\}$. Dans ce cas la fonction caractéristique $\varphi(t)$ a une période égale à $\frac{2\pi}{h}$ et $|\varphi(t)| < 1$ pour $0 < t < \frac{2\pi}{h}$. Puisque $\Lambda = \{0\}$, on a pour chaque ε , $a > 0$:

$$\sup_{\varepsilon n^{1/\alpha} \leq |v| \leq a n^{1/\alpha}} |1 - \varphi(n^{-1/\alpha} v)|^{-1} \leq B = B(a,\varepsilon) < \infty .$$

Posons :

$$A_k = \{u : (\frac{2\pi k}{h} - \varepsilon) n^{1/\alpha} \leq u < (\frac{2\pi k}{h} + \varepsilon) n^{1/\alpha}\} , \quad k = 0,1 \dots$$

où ε est un nombre positif assez petit. Si $u \notin \bigcup_k A_k$ on a alors l'inégalité $|\varphi(n^{-1/\alpha} u)| \leq e^{-\delta}$, $\delta > 0$. Ecrivons I_4 ainsi :

$$I_4 = 4n^{-2} \int_{\varepsilon n^{1/\alpha} |v| \leq a n^{1/\alpha}} \Psi_n(v) dv \int_{\overline{\cup A_k}} \Psi_n(v-u) (\dots) du +$$

$$+ 4n^{-2} \int_{\varepsilon n^{1/\alpha} \leq |v| \leq a n^{1/\alpha}} \Psi_n(v) dv \int_{A_0} \Psi_n(u-v) (\dots) du +$$

$$+ 4n^{-2} \sum_{k \neq 0} \int_{\varepsilon n^{1/\alpha} \leq |v| \leq a n^{1/\alpha}} \Psi_n(v) dv \int_{A_k} \Psi_n(u-v) (\dots) du.$$

On peut majorer les deux premiers termes de la même manière que dans le cas de non arithmétique. En vertu de (2.12).

$$n^{-2} \left| \int_{\varepsilon n^{1/\alpha} \leq |v| \leq a n^{1/\alpha}} \Psi_n \int_{A_k} \Psi_n(u-v) (\dots) du \right| \leq$$

$$\leq B n^{-2} \int_{n^{1/\alpha} \leq |v| \leq a n^{1/\alpha}} \frac{|\Psi_n(v)| dv}{|v|^{\alpha/2}} \int_{(2\pi k h^{-1} - \varepsilon) n^{1/\alpha}}^{(2\pi k h^{-1} + \varepsilon) n^{1/\alpha}} \frac{|\Psi_n(u-v)|}{\sqrt{1+|u-v|^\alpha}} \frac{1-e^{-c|u - \frac{2\pi k}{h}|^\alpha}}{|u - \frac{2\pi k}{h}|^\alpha} du \leq$$

$$\leq B \left(\int_{|v|>\varepsilon n^{1/\alpha}} \Psi_n(v)|^2 |v|^{-\alpha} dv\right)^{1/2}.$$

$$\cdot \int_{-\infty}^{\infty} \frac{|\Psi_n(v)|^2}{1+|v|^\alpha} dv)^{1/2} \int_{-\infty}^{\infty} \frac{1-e^{-c|u|^\alpha}}{|u|^\alpha} du = o(1).$$

Mais le nombre de tous les ensembles A_k qui sont des sous-ensembles de l'intervalle $\left[-an^{1/\alpha}, an^{1/\alpha}\right]$ est inférieur à $\frac{h\,a}{\pi}$. Donc de nouveau $I_4 = o(1)$.

Ainsi la relation (2.9) est prouvée. D'après le lemme 2.2 :

$$E\left|\int_{-\infty}^{\infty}\Psi_n(\lambda)d\lambda \int_0^1 e^{-i\lambda\xi_\alpha(t)}dt - \int_{-\infty}^{\infty}\Psi(\lambda)d\lambda \int_0^1 e^{-\lambda\xi_\alpha(t)}dt\right|^2 \to 0 .$$

D'où, d'après (2.9) :

$$\lim_n E\left|\eta_n - \frac{1}{2\pi}\int_{-\infty}^{\infty}\Psi(\lambda)d\lambda \int_0^1 e^{-i\lambda\xi_\alpha(t)}dt\right|^2 = 0.$$

Cette dernière égalité démontre le théorème pour des lois telles que $\Lambda = \{0\}$.

4 - Démonstration du théorème dans le cas $\Lambda \neq \{0\}$.

Considérons maintenant des lois telles que $\Lambda \neq \{0\}$. Ce sont des lois arithmétiques. On peut supposer que l'ensemble des valeurs de la variable aléatoire ξ_j a la forme $\{kh + f\}$ où les k sont entiers, $0 \leq f < h$ et h est le pas maximal de la répartition de ξ_j. La fonction caractéristique :

$$\varphi(t) = e^{itf}\sum_k p_k\, e^{ikht}$$

est égale à 1 aux points t pour lesquelles $(kh-f)t = 2\pi\ell_k$, avec ℓ_k entier, pour tout k tel que $p_k \neq 0$. Puisque le pas h est maximal, le plus grand commun diviseur de toutes les différences $k'' - k'$, $p_{k''}, p_{k'} \neq 0$ est égal à 1. Donc on peut trouver des nombres entiers c_j tels que :

$$\sum c_j\,(k''_j - k'_j) = 1 .$$

Mais dans ce cas nécessairement $f = h\,\frac{p}{q}$ où p, q sont deux nombres entiers sans commun diviseur. Donc :

$$\Lambda = \{\, \frac{2\pi}{h}kq,\quad k = 0, \pm 1\ \ldots\,\} .$$

Prenons pour la simplicité $h = 1$, $\Lambda = \{\lambda_k, k=0, \pm 1 \ldots\}$, $\lambda_k = 2\pi kq$.

Notons que :

$$\exp\{i(\lambda + 2\pi\, qkn^{1/\alpha})\, S_{nj}\} = e^{i\lambda S_{nj}} .$$

Donc :

$$\eta_n = \sum_{j=1}^{n}\frac{1}{2\pi}\int_{-an^{1/\alpha}}^{an^{1/\alpha}}\hat{f}_n(\lambda)\, e^{-\lambda S_j}\, d\lambda =$$

$$= \sum_{j=1}^{n}\frac{1}{2\pi}\int_{-\pi qn^{1/\alpha}}^{\pi qn^{1/\alpha}}\sum_k \hat{f}_n(\lambda+n^{1/\alpha}\lambda_k)e^{iS_{nj}\lambda}\, d\lambda =$$

$$= \frac{1}{2\pi n}\int_{-\pi qn^{1/\alpha}}^{\pi qn^{1/\alpha}}\Psi_n(\lambda)\sum_1^n e^{iS_{nj}\lambda}\, d\lambda .$$

Le reste de la preuve coïncide absolument avec le cas analysé $\Lambda = \{0\}$. Notamment, on montre comme ci-dessus que :

$$n^{-1} \frac{1}{2\pi} \int_{-\pi q n^{1/\alpha}}^{\pi q n^{1/\alpha}} \Psi_n(\lambda) \sum_1^n e^{-i\lambda S_{nj}} d\lambda - \frac{1}{2\pi} \int_{-\infty}^{\infty} \Psi_n(\lambda) d\lambda \int_0^1 e^{-i\lambda \xi_\alpha(t)} dt \to 0$$

en probabilité.

Le théorème 2.1 est démontré.

Remarque : On peut montrer sous les hypothèses du théorème 2.1 l'inégalité suivante :

$$E \Big| \sum_{ns<k\leq nt} f_n(S_{nk}) \Big|^r \leq B \, |t-s|^{1+p}$$

où p,r sont deux nombres positifs. On déduit de cette inégalité que les processus engendrés par les lignes brisées aléatoires $(\frac{\ell}{n}, \sum_i^{\ell} f_n(S_{nk}))$ convergent en loi vers la fonction aléatoire $t \frac{1}{2\pi} \int_{-\infty}^{\infty} \Psi_n(\lambda) d\lambda \int_0^1 e^{-i\lambda \xi_\alpha(u)} du$. Nous allons considérer cette question un peu plus tard.

5 - Quelques variantes du théorème 2.1

On peut formuler ce théorème de manière différente. Notamment :

$$\frac{1}{2\pi} \int_{-\infty}^{\infty} \Psi_n(\lambda) \int_0^1 e^{-i\lambda \xi_\alpha(t)} dt \, d\lambda =$$

$$= \int_0^1 dt \, \frac{1}{2\pi} \int_{-\infty}^{\infty} \Psi_n(\lambda) \, e^{-i\lambda \xi_\alpha(t)} d\lambda =$$

$$= \int_0^1 U_n' (\xi_\alpha(t)) dt.$$

En introduisant le temps local $\pounds_\alpha(x)$ du processus $\xi_\alpha(t)$ sur l'intervalle $[0,1]$ on peut écrire que :

$$\int_0^1 U_n' (\xi_\alpha(t)) dt = \int_{-\infty}^{\infty} U_n'(x) \, \pounds_\alpha(x) \, dx = \int_{-\infty}^{\infty} \pounds_\alpha(x) \, dU_n(x).$$

D'où le théorème suivant peut être déduit.

THEOREME 2.2 : *Supposons que les conditions 1,2 du théorème 2.1 et en outre les conditions 3_1 , 3_2 ci-dessous sont satisfaites :*

3_1)
$$\lim_{T\to\infty} \sup_n \int_{|\lambda|>T} \frac{|\Psi_n(\lambda)|^2}{1 + |\lambda|^\alpha} d\lambda < \infty \; ;$$

3_2) *Les fonctions $U_n(x)$ sont à variation bornée sur tout intervalle borné et convergent quand $n\to\infty$ vers une fonction $u(x)$ de variation localement bornée dans tous les points de continuité de $u(x)$.*

Alors la loi limite de η_n existe et coïncide avec la loi de la variable aléatoire :

$$\eta = \int_{-\infty}^{\infty} f_\alpha(x)\ du(x).$$

<u>Démonstration</u> : En démontrant la relation (2.9) nous avons utilisé seulement la condition 3_1. Donc (2.9) est vraie sous les hypothèses du théorèmes 2.2. Puis, avec probabilité 1, $f_\alpha(x)$ est une fonction continue à support compact. Donc :

$$\int_{-\infty}^{\infty} f_\alpha(x)\ du_n(x) \to \int_{-\infty}^{\infty} f_\alpha(x)\ du(x)$$

et le théorème est démontré.

Si $\alpha = 2$, le processus limite $\xi_2(t) = w(t)$ est le processus de Wiener. Posons :

$$F_n(y) = \int_0^y u_n(x)\ dx\ .$$

D'après la formume d'Ito :

$$dF_n(w) = \frac{1}{2} F_n''(w)dt + F_n'(w)dw = \frac{1}{2}\ u_n'(w(t))dt + u_n(w)dw\ ;$$

d'où :

$$F_n(w(1)) = \int_0^{w(1)} u_n(t)dt = \frac{1}{2} \int_0^1\ u_n'(w(t))dt + \int_0^1 u_n(w(dt))dt\ .$$

Donc :

$$\frac{1}{2\pi} \int_{-\infty}^{\infty}\Psi_n(\lambda) \int_0^1 e^{-i\lambda w(t)}dt = \int_0^1\ u_n'(w(t))dt =$$

$$= 2 \int_0^{w(1)}\ u_n(t)dt - \int_0^1\ u_n(w(t))dw(t).$$

Aussi peut-on pour $\alpha = 2$ remplacer la condition 3_2 du théorème 2.2 par la condition suivante : il existe une fonction $u(x)$ telle que :

$$\int_{-c}^{c}\ |u_n(x) - u(x)|^2\ dx \to 0$$

pour tout $c > 0$.

Dans ce cas la loi limite de η_n existe et coïncide avec la loi de la variable :

$$\eta = 2 \int_0^{w(1)}\ u(t)dt - 2 \int_0^1\ u(w(t))dw(t).$$

6 - <u>Exemple</u>

Soient ξ_1, ξ_2 ... des variables aléatoires avec $E\xi_j = 0$, $\text{Var}\xi_j = 1$. Nous allons trouver la loi limite des sommes :

$$\eta_n = \frac{1}{\sqrt{n}} \sum_1^n \frac{\text{Sin}\zeta_k}{\zeta_k}\ ; \quad \zeta_k = \sum_1^k \xi_j$$

en supposant en outre que $\Lambda = \{0\}$.

Posons :

$$f_n(x) = \frac{1}{\sqrt{n}} \frac{\sin x\sqrt{n}}{x\sqrt{n}} ,$$

on a :

$$\eta_n = \sum_1^n f_n(x).$$

Puisque la fonction :

$$u_n(x) = \int_0^{x\sqrt{n}} \frac{\sin y}{y} dy \to u(x) = \frac{\pi}{2} \operatorname{sgn} x$$

on a d'après le théorème 2.2 que la loi limite de η_n coïncide avec la loi de :

$$\eta = \int_{-\infty}^{\infty} \pounds_2(x) \, du(x) = \pi \pounds_2(0).$$

En vertu de la formule (6.1) du chapitre 2 :

$$P\{\eta < x\} = P\{\pounds_2(0) < \frac{x}{\pi}\} = \begin{cases} 0 , & x < 0 \\ \frac{2}{\sqrt{2\pi}} \int_0^{x/\pi} e^{-\frac{z^2}{2}} dz, & x \geq 0 . \end{cases}$$

III - THEOREMES LIMITES DU PREMIER TYPE (Suite)

1 - En continuant à étudier les lois limites des sommes $\eta_n = \sum_1^n f_n(S_{nk})$,

nous ne supposons plus que $\Psi_n(\lambda)$ est une fonction a support compact. En échange de cela il faudra poser quelques restrictions supplémentaires sur la loi de ξ_j. Toutes les notations du paragraphe précédent sont reprises ici. Evidemment,nous pouvons répéter tous les arguments du paragraphe précédent, mais maintenant la formule (2.10) aura un terme complémentaire :

$$I_5 = E\Big| \int_{|\lambda| > an^{1/\alpha}} \hat{f}_n(\lambda) \sum_1^n e^{iS_{nj}\lambda} d\lambda \Big|^2$$

L'analyse de ce terme est l'objet principal de ce paragraphe.

Supposons d'abord $\Lambda = \{0\}$. Dans ce cas :

$$I_5 = n^{-2} E\Big| \int_{|\lambda| > an^{1/\alpha}} \hat{f}_n(\lambda) \sum_1^n e^{iS_{nj}\lambda} d\lambda \Big|^2 =$$

$$\leq Bn^{-1} \int_{|v| > an^{1/\alpha}} |\Psi_n(v)| dv \int_{|u| \leq \varepsilon n^{1/\alpha}} |\Psi_n(u-v)| \frac{1 - e^{-c|u|^\alpha}}{|1 - \varphi(vn^{-1/\alpha})| \; |u|^\alpha} du +$$

$$+ Bn^{-1} \int_{|v| > an^{1/\alpha}} |\Psi_n(v)| \int_{|u| > \varepsilon n^{1/\alpha}} |\Psi_n(u-v)| \frac{du \, dv}{|1 - \varphi(vn^{-1/\alpha})| \, |1 - \varphi(n^{-1/\alpha} u)|}$$

Donc si :

$$\inf_{|t|>\varepsilon} |1-\varphi(t)| > 0 ,$$

alors :

$$I_5 \leq B^{n-1} \int_{|v|>an^{1/\alpha}} |\Psi_n(v)| dv \int_{-\infty}^{\infty} |\Psi_n(u)| \, du \tag{3.1}$$

Si $|\varphi(t)| \leq e^{-\delta}$, $\delta > 0$ pour tout $|t| \geq \varepsilon$, alors, en vertu de (2.12),

$$\begin{aligned} I_5 &\leq Bn^{-1} \int_{|v|>an^{1/\alpha}} |\Psi_n(v)| \, dv . \operatorname{Sup} |\Psi_n(v)| + \\ &+ n^{-2} \int_{|v|>an^{1/\alpha}} |\Psi_n(v)| dv \, . \int_{-\infty}^{\infty} |\Psi_n(u)| du \, . \end{aligned} \tag{3.2}$$

Si $\varphi \in L_1$, alors, en vertu de (2.12),

$$\begin{aligned} I_5 &\leq Bn^{-1} \sup_{|v|>an^{1/\alpha}} |\Psi_n(v)|^2 \int_{|v|>an^{1/\alpha}} |\varphi(vn^{-1/\alpha})| dv + \\ &+ n^{-2} \sup_{|v|>an^{1/\alpha}} |\Psi_n(v)| \sup_u |\Psi_n(u)| \left(\int_{|v|> n^{1/\alpha}} |\varphi(vn^{-1/\alpha})| dv\right)^2 \end{aligned}$$

On peut aussi obtenir des majorations du même type que (3.3) sous les hypothèses $\varphi \in L_p$, $p > 1$. En utilisant (3.1) - (3.3) on peut donner différentes variantes du théorème 2.1 sans supposer que $\Psi_n(\lambda)$ est à support compact. Par exemple,

Supposons que les conditions suivantes sont vérifiées :

1° $\inf_{|t|>\varepsilon} |1-\varphi(t)| > 0$ *pour chaque* $\varepsilon > 0$

2° $\lim_n n^{-1} \int_{|v|>an^{1/}} |\Psi_n(v)| \, dv \int_{-\infty}^{\infty} |\Psi_n(u)| du = 0$ *pour n'importe quel* $a > 0$

3° Il existe une fonction $\Psi(\lambda)$ *telle que :*

$$\int_{-\infty}^{\infty} \frac{|\Psi(\lambda)|^2}{1+|\lambda|^{\alpha}} d\lambda < \infty ,$$

$$\lim_n \int_{-\infty}^{\infty} \frac{|\Psi(\lambda) - \Psi_n(\lambda)|^2}{1 + |\lambda|^2} d\lambda = 0,$$

alors la loi de η_n *existe et coïncide avec la loi de :*

$$\frac{1}{2\pi} \int_{-\infty}^{\infty} \Psi(\lambda) d\lambda \int_0^1 e^{-i\lambda\xi_\alpha(t)} dt .$$

Le cas $\Lambda \neq \{0\}$ se traite même plus simplement. En fait, on peut supposer que : $\Lambda = \{2\pi k\}$. On a :

$$\int_{-\infty}^{\infty} \widehat{f}_n(\lambda) \sum_{j=1}^{n} e^{i\lambda S_{nj}} d\lambda = \int_{-\pi n^{1/\alpha}}^{\pi n^{1/\alpha}} \sum_{k=-\infty}^{\infty} \widehat{f}_n(\lambda + \lambda_k n^{1/\alpha}) \sum_{j=1}^{n} e^{i\lambda S_{nj}} d\lambda =$$

$$= \int_{-\pi n^{1/\alpha}}^{\pi n^{1/\alpha}} \Psi_n(\lambda) \left[\frac{1}{n} \sum_{1}^{n} e^{i\lambda S_{nj}}\right] d\lambda$$

et il suffit de supposer outre la condition 2 du théorème 2.1 que :

$$\int_{-\infty}^{\infty} \widehat{f}_n(\lambda)\, d\lambda < \infty$$

Dans ce cas la série :

$$\Psi_n(\lambda) = n \sum_k \widehat{f}_n(\lambda + \lambda_k n^{1/\alpha})$$

converge absolument.

Soit $\Lambda = \{2\pi k,\ k=0, \pm 1, \ldots\}$. Dans ce cas toutes les valeurs possibles de S_{nj} appartiennent à l'ensemble $\{kn^{-1/\alpha}, k=0, \pm 1,\ldots\}$ et donc seules les valeurs $f_n(kn^{-1/\alpha})$ de la fonction f_n aux points $kn^{-1/\alpha}$ sont importantes. Mais en vertu de la formule de Poisson :

$$\Psi_n(\lambda) = n \sum_k \widehat{f}_n(\lambda + \lambda_k n^{1/\alpha}) = n \sum_k f_n(\lambda_k n^{-1/\alpha})\, e^{ixkn^{-1/\alpha}}$$

c'est à dire que $\Psi_n(\lambda)$ est la transformée de Fourier de la suite $\{n\, f_n(\lambda_k n^{-1/\alpha})\}$ et on peut formuler des résultats pour les termes $\left\{ f_n(\lambda_k n^{-1/\alpha})\right\}$ seulement.

2 - Divergence de l'intégrale de $|f_n(\lambda)|$. On peut penser que nos théorèmes sont restrictifs au sens suivant : il faut supposer l'existence de la transformée de Fourier de f_n. En fait, cette restriction n'est pas très sévère. On peut proposer aux moins deux méthodes pour éviter ces restrictions.

Tout d'abord, posons :

$$f_n^c(x) = \begin{cases} f_n(x) , & |x| \leqq \frac{c}{2} \\ 0 , & |x| > c, \end{cases}$$

sur $\left[-c , -\frac{c}{2}\right], \left[\frac{c}{2} , c\right]$ la fonction f_n^c est définie de manière à conserver la régularité de f_n. Alors :

$$P\left(\sum_{k=1}^{n} f_n^c(S_{nk}) \neq \sum_{k=1}^{n} f_n(S_{nk})\right) \leqq P\left(\sup_k |S_{nk}| > \frac{c}{2}\right) \underset{c\to\infty}{\longrightarrow} 0 .$$

uniformément par rapport à n.

Cette remarque et les inégalités (3.1) (3.2) permettent de formuler quelques théorèmes sur la convergence de η_n en loi. Voici quelques exemples. Désignons par ψ_n^c, u_n^c les homologues des fonctions η_n, u_n construits à partir de f_n^c.

THEOREME 3.1 : *Supposons les conditions suivantes vérifiées :*

1° $\inf_{|t|>\varepsilon} |1-\varphi(t)| > 0$ *pour tout* $\varepsilon > 0$ *(donc* $\Lambda = \{0\}$*)* ;

2° Pour chaque $c > 0$ *il existe une fonction* ψ_n^c *telle que :*

$$\int_{-\infty}^{\infty} \frac{|\psi_n^c(\lambda)|^2}{1+|\lambda|^\alpha} d\lambda < \infty$$

$$\lim_n \int_{-\infty}^{\infty} \frac{|\psi_n^c - \psi^c|^2}{1+|\lambda|^\alpha} d\lambda = 0$$

3°

$$\lim_{a\to\infty} \overline{\lim_n}\ n^{-1} \int_{|v|>an^{1/\alpha}} |\psi_n^c(v)|dv \int_{-\infty}^{\infty} |\psi_n^c(u)|du = 0$$

4° Les lois des variables aléatoires :

$$\eta^c = \frac{1}{2\pi} \psi^c(\lambda)d\lambda \int_0^1 e^{-i\lambda\xi_\alpha(t)} dt$$

convergent quand $c\to\infty$ *vers une loi* $\mathcal{P}$.

Alors les lois de η_n *convergent aussi vers* $\mathcal{P}$ *quand* $n\to\infty$.

THEOREME 3.2 : *Soit* $E\xi^2 < \infty$. *Supposons les conditions suivantes vérifiées :*

1°

$$\int_{-\infty}^{\infty} |\varphi(t)|dt < \infty \text{ (et donc } \Lambda = \{0\}\text{) ;}$$

2° Pour chaque $c > 0$

$$\lim_{a\to\infty} \overline{\lim_n} \Big(\frac{1}{\sqrt{n}} \sup_{|v|>an^{1/\alpha}} |\psi_n^c(v)|^2 \int_{|v|>a} |\varphi(v)|dv +$$

$$+ n^{-1} \sup_n |\psi_n^c(u)| \sup_{|u|>an^{-1/\alpha}} |\psi_n^c(u)|\Big) = 0$$

3° Il existe une fonction $u(x)$ *telle que pour chaque* $c > 0$

$$\int_{-c}^{c} u^2(x)dx < \infty ,$$

$$\lim_n \int_{-c}^{c} |u_n(x) - u(x)|^2 dx = 0.$$

Alors la limite de η_n *existe et coïncide avec la loi de :*

$$2\int_0^{w(1)} u(x)dx - 2\int_0^1 u(w(t))\, dw(t).$$

On peut donner aussi d'autres variantes de ces théorèmes. La deuxième méthode pour éviter des hypothèses sur l'intégrabilité de f_n repose sur l'utilisation de la théorie des distributions de Schwartz. Supposons que $f_n(x)$ est une fonction à croissance lente, c'est-à-dire qu'il existe un entier ℓ tel que :

$$\int_{-\infty}^{\infty} \frac{|f_n(\lambda)|}{(1+\lambda^2)^{\ell}} \, d\lambda < \infty \quad .$$

Soit $\widehat{f}_n$ la transformée de Fourier de f_n au sens de la théorie des distributions. Soit comme précédemment :

$$\Psi_n(\lambda) = n \sum_{\lambda_k \in \Lambda} \widehat{f}_n(\lambda + \lambda_k n^{1/\alpha})$$

$$u_n(x) = n \int_0^x f_n(z) \sum_{\lambda_k \in \Lambda} e^{-i\lambda_k z n^{1/\alpha}} dz \quad .$$

Pour simplifier les calculs nous supposons que Ψ_n est une distribution à support compact appartenant à l'intervalle $\left[-an^{1/\alpha}, an^{1/\alpha}\right]$. Le théorème suivant est un homologue du théorème 2.1.

THEOREME 3.3 : *Supposons les conditions suivantes vérifiées :*

1° La distribution $\widehat{f}_n$ a un support compact appartenant à $\left[-an^{1/\alpha}, an^{1/\alpha}\right]$ où a ne dépend pas de n :

2° Quand $n \to \infty$ Ψ_n converge vers une distribution Ψ ;

3° Il existe un nombre $T_0 > 0$ tel que les restrictions $\widehat{\Psi}_n^T$, $\widehat{\Psi}^T$ de Ψ_n, Ψ à l'extérieur de l'intervalle $[-T,T]$, $T > T_0$, sont des fonctions et :

$$|(\Psi_n^T, h)|^2 \leqq B_T \int_{-\infty}^{\infty} |h(\lambda)|^2 \, (1 + |\lambda|^\alpha) d\lambda \tag{3.4}$$

$$|(\Psi^T, h)|^2 \leqq B_T \int_{-\infty}^{\infty} |h(\lambda)|^2 \, (1 + |\lambda|^\alpha) d\lambda$$

Pour toute fonction h indéfiniment dérivable à support compact.

Alors, la loi limite de η_n existe et coïncide avec la loi de :

$$\eta = \frac{1}{2\pi} (\widehat{\Psi}, \int_0^1 e^{-i\lambda \xi_\alpha(t)} dt).$$

Démonstration : Elle rappelle la démonstration du théorème 2.1 et nous l'ébauchons seulement. Soit $\mathcal{D}$ l'espace de Schwartz de fonctions indéfiniment dérivables à support compact. Soit $h \in \mathcal{D}$ et $h(\lambda) = 1$ si $\lambda \in \left[-an^{1/\alpha}, an^{1/\alpha}\right]$. Ecrivons h comme $h = h_1 + h_2$ où $h_1, h_2 \in \mathcal{D}$, $h_1(\lambda) = 1$ si $[-T,T]$, $h_1(\lambda) = 0$ si $|\lambda| > A$, $A > T > T_0 > 0$.

On a (voir [26], [40]) :

$$\eta_n = \frac{1}{2\pi} \sum_1^n (\widehat{f}_n, e^{-i\lambda S_{nk}} h) = \frac{1}{2\pi} (\widehat{\Psi}_n, h_1 \int_0^1 e^{-i\lambda w_n(t)} dt) +$$

$$+ \frac{1}{2\pi} (\widehat{\Psi}_n, h_2 \frac{1}{n} \sum_1^n e^{-i\lambda S_{nk}}).$$

où comme plus haut :

$$w_n(t) = n^{-1/\alpha} \sum_{k<nt} \xi_k .$$

Comme plus haut nous pouvons supposer que $w_n(t) \to \xi_\alpha(t)$ en probabilité. On déduit facilement de la convergence de Ψ_n vers Ψ et de w_n vers ξ_α que :

$$\frac{1}{2\pi} (\widehat{\Psi}_n, h_1 \int_0^1 e^{-i\lambda w_n(t)} dt) \to \frac{1}{2\pi} (\Psi, h_1 \int_0^1 e^{-i\lambda \xi_\alpha(t)} dt).$$

Les arguments que nous avons utilisés en démontrant le théorème 2.1 montrent que :

$$E(\widehat{\Psi}_n , \frac{h_2}{n} \sum_1^n e^{-i\lambda S_{nk}})^2 \xrightarrow[A\to\infty]{} 0$$

uniformément par rapport à n. Le théorème est établi.

Remarque 1 : En vertu de (3.4) $\widehat{\Psi}$ admet un prolongement tel que $(\Psi, \int_0^1 e^{-i\lambda\xi_\alpha(t)} dt)$ est bien défini.

Remarque 2 : Si les fonctions u_n convergent faiblement vers une fonction u à variation localement bornée on peut écrire η comme :

$$\eta = \int_{-\infty}^{\infty} \pounds_\alpha(x) \, du(x)$$

et si cette fonction u est absolument continue :

$$\eta = \int_{-\infty}^{\infty} \pounds_\alpha(x) u'(x) dx = \int_0^1 u'(\xi_\alpha(t)) dt .$$

3 - Exemples

a) Considérons la somme :

$$\eta_n = \frac{1}{n} \sum_1^n f(S_{nk})$$

où f est une fonction sommable. Dans ce cas :

$$\Psi_n(\lambda) = \widehat{f}(\lambda) \qquad \text{si } \Lambda = \{0\}$$

et :

$$\Psi_n(\lambda) \to \widehat{f}(\lambda) \qquad \text{si } \Lambda \neq \{0\} .$$

Donc la loi limite coïncide avec la loi de

$$\frac{1}{2\pi} \int_{-\infty}^{\infty} \widehat{f}(\lambda) \, d\lambda \int_0^1 e^{-i\lambda\xi_\alpha(t)} dt = \int_0^1 f(\xi_\alpha(t)) dt.$$

Cet exemple n'est pas très intéressant parce que ce résultat (et même un plus général) est une conséquence immédiate du principe de l'invariance (voir chapitre 1).

b) On a un exemple plus intéressant si on considère la somme :

$$\eta_n = n^{\frac{1-\alpha}{\alpha}} \sum_1^n f(\zeta_k) , \quad \zeta_k = \sum_1^n \xi_j$$

Pour utiliser notre théorie réécrivons cette somme comme

$$\eta_n = \sum_1^n f_n(S_{nk})$$

où :

$$f_n(x) = n^{\frac{1-\alpha}{\alpha}} f(x n^{1/\alpha}).$$

Soit d'abord $\Lambda = \{0\}$. Dans ce cas

$$\Psi_n(\lambda) = n.n^{\frac{1-\alpha}{\alpha}} . n^{-\frac{1}{\alpha}} \hat{f}(\lambda n^{-1/\alpha}) = \hat{f}(\lambda n^{-\frac{1}{\alpha}}) \to \psi(\lambda) = \hat{f}(0).$$

Donc si nous supposons par exemple que $||f||_1 + ||\hat{f}||_\infty < \infty$ et que pour chaque $\varepsilon > 0$ $\sup_{|t|>\varepsilon} |\varphi(t)|>\varepsilon$ nous aurons en vertu de (3.2) que la loi limite de η_n coïncide avec la loi de :

$$\frac{1}{2\pi} \int_{-\infty}^{\infty} \Psi(\lambda)d\lambda \int_0^1 e^{-i\lambda\xi_\alpha(t)} dt = f(0)\ \pounds_\alpha(0).$$

Si :

$$\Lambda = \{\lambda_k\} \neq \{0\} \quad \text{et} \quad ||\hat{f}||_1 < \infty ,$$

$$\Psi_n(\lambda) = n \sum_{\lambda_k \in \Lambda} f_n(\lambda + \lambda_k n^{1/\alpha}) = \sum_{\lambda_k \in \Lambda} f(\lambda n^{-1/\alpha} + \lambda_k) .$$

$$\pounds_\alpha(0) \sum_{\lambda_k \in \Lambda} f(\lambda_k).$$

Nous reviendrons sur cet exemple dans le chapitre suivant.

IV - CONVERGENCE DES PROCESSUS ENGENDRES PAR LES SOMMES : $\sum_{k \leq nt} f_n(S_{nk})$

1 - Le résultat principal

Désignons par $\eta_n(t)$, $0 \leq t \leq 1$ la ligne brisée de sommets

$$(\frac{k}{n} , \sum_1^k f_n(S_{nj})), \quad k = 0,1 \dots n .$$

Alors les η_n sont des processus continus et ils engendrent des lois de probabilité dans l'espace $C[0,1]$. Nous allons montrer ici que ces lois convergent faiblement sous les hypothèses des paragraphes précédents vers la loi engendrée par :

$$\eta(t) = \frac{1}{2\pi} \int_{-\infty}^{\infty} \Psi(\lambda)d\lambda \int_0^t e^{-i\lambda\xi_\alpha(u)} du .$$

La fonction $\Psi(\lambda)$ est définie comme précédemment. Cette convergence a toujours lieu sous les hypothèses des théorèmes des paragraphes 2 et 3. Mais dans un but de simplicité nous allons considérer seulement le cas où $\hat{f}_n$ est à support compact.

<u>THEOREME 4.1</u> : *Sous les hypothèses du théorème 2.1* $\eta_n(t)$ *converge en loi vers* η *dans l'espace* $C[0,1]$

<u>Démonstration</u> : On va montrer d'abord les trois lemmes suivants.

<u>LEMME 4.1</u> : *Soit* $0 \leq t_1 < t_2 < \dots < t_k \leq 1$. *Les vecteurs* $(\eta_n(t_1),\dots,\eta_n(t_k))$ *convergent en loi vers le vecteur* $(\eta(t_1),\dots \eta(t_k))$.

LEMME 4.2 : *Il existe une constante* B_k *telle que :*

$$E|\eta_n(t) - \eta_n(s)|^k \leq B_k |t-s|^{\frac{\alpha-1}{2\alpha} k} , \quad 0 \leq t, s \leq 1 \tag{4.1}$$

LEMME 4.3 : *Il existe une cosntante* B_k *telle que :*

$$E|\eta(t) - \eta(s)|^k \leq B_k |t-s|^{\frac{\alpha-1}{2\alpha} k} \tag{4.2}$$

Démonstration du lemme 4.1 : Elle n'a rien de nouveau par rapport à la démonstration du théorème 2.1. En effet, puisque :

$$|f_n(S_{nj})| = \frac{1}{2\pi} \left| \int_{-an^{1/\alpha}}^{an^{1/\alpha}} \hat{f}_n(\lambda) e^{-i\lambda S_{nj}} d\lambda \right| \leq$$

$$\leq \frac{B}{n} \int_{-\infty}^{\infty} |\Psi_n(\lambda)| d\lambda \leq \frac{B_T}{n} \left[\int_{-T}^{T} \frac{|\Psi_n(\lambda)|^2}{1+|\lambda|^\alpha} d\lambda \right]^{1/2} +$$

$$+ B\left(\int_T^\infty \frac{|\Psi_n(\lambda)|^2}{1+|\lambda|^\alpha} d\lambda \right)^{1/2} ,$$

on a :

$$\eta_n(t) = \sum_{k \leq nt} f_n(S_{nk}) + r_n ,$$

où $r_n \to 0$ en probabilité.

Donc, si par exemple, $\Lambda = \{0\}$,

$$\eta_n(t) = \frac{1}{2\pi} \Psi_n(\lambda) d\lambda \int_0^t e^{-i\, w_n(t)} du + o(1)$$

en probabilité,

$$w_n(u) = \sum_{k \leq nu} S_{nk} .$$

Nous avons prouvé qu'on peut construire un espace probabilisé et définir dans cet espace des processus $\tilde{\eta}_n(t)$ de sorte que $(\tilde{\eta}_n(t_1) \dots \tilde{\eta}_n(t_k)) \overset{\mathcal{L}}{=} (\eta_n(t_1) \dots \eta_n(t_k))$ et que $\tilde{\eta}_n(1) \to \eta(1)$ en probabilité. Mais la même démonstration montre aussi que $\tilde{\eta}_n(t) \to \eta(t)$ en probabilité pour tout $t \in [0,1]$. Donc $(\eta_n(t_1) \dots \eta_n(t_k))$ converge en loi vers $(\eta(t_1) \dots \eta(t_k))$. Le lemme 4.1 est ainsi démontré.

Démonstration du lemme 4.2 : Soit $0 \leq s < t \leq 1$. On peut supposer que $t-s > \frac{1}{n}$.
En effet, si $t-s \leq \frac{1}{n}$:

$$E|\eta_n(t)-\eta_n(s)|^k \leq 2^k (t-s)^k \sup_j E|\, n\, f_n(S_{nj})|^k \leq$$

$$\leq 2^k (t-s)^k \left(\int_{-an^{1/\alpha}}^{an^{1/\alpha}} n |\hat{f}_n(\lambda)| d\lambda \right)^k \leq$$

$$\leq B(t-s) \left(\int_{-an^{1/\alpha}}^{an^{1/\alpha}} |\Psi_n(\lambda)| d\lambda \right)^k \leq$$

$$\leq B(t-s)^k n^{\frac{k(1+\alpha)}{2\alpha}} \left(\int_{-\infty}^{\infty} \frac{|\Psi_n(\lambda)|^2}{1+|\lambda|^\alpha} d\lambda \right)^{\frac{k}{2}} \leq B(t-s)^{\frac{k(\alpha-1)}{2\alpha}} .$$

Les mêmes arguments montrent que si $t-s > \frac{1}{h}$

$$E|\eta_n(t) - \eta_n(s)|^k \leq B\, E\left|\int_{-\infty}^{\infty} \hat{f}_n(\lambda) \sum_{ns\leq j\leq nt} e^{-i\lambda S_{nj}} d\lambda\right|^k + B(t-s)^{\frac{k(\alpha-1)}{2\alpha}} \qquad (4.3)$$

Pour estimer le premier temre à droite supposons que $k = 2p$ où p est un nombre entier et considérons séparément les trois cas :

1° La loi de ξ_j n'est pas arithmétique ;

2° La loi de ξ_j est arithémtique mais $\Lambda = \{0\}$;

3° $\Lambda \neq \{0\}$.

<u>Premier cas</u> : Il existe un nombre $\varepsilon > 0$ tel que $|\varphi(\lambda n^{-1/\alpha})| \leq \exp\{-\frac{c|\lambda|^\alpha}{n}\}$ si $|\lambda| \leq \varepsilon n^{1/\alpha}$; dans le domaine $\varepsilon n^{1/\alpha} \leq |\lambda| \leq a n^{1/\alpha}$ on a $|\varphi(\lambda n^{-1/\alpha})| < e^{-\delta}$, $\delta > 0$ puisque la loi de ξ_j n'est pas arithmétique. Donc il existe $c > 0$ tel que :

$$|\varphi(\lambda n^{-1/\alpha})| \leq \exp\{-c \frac{|\lambda|^\alpha}{n}\},\ |\lambda| \leq a n^{1/\alpha} \qquad (4.4)$$

En vertu de ceci :

$$E\left|\int_{-\infty}^{\infty} \hat{f}_n(\lambda) \sum_{ns\leq j\leq nt} e^{-i\lambda S_{nj}}\right|^{2p} =$$

$$= \int_{-\infty}^{\infty}\dots\int_{-\infty}^{\infty} \hat{f}_n(\lambda_1) \dots \hat{f}_n(\lambda_p)\, \overline{\hat{f}_n(\mu_1)} \dots \overline{\hat{f}_n(\mu_p)} .$$

$$\cdot \sum_{ns\leq n_i, m_i \leq nt} E \exp\{- i\lambda_1 S_{nn_1} - \dots - i\lambda_p S_{nn_p} +$$

$$+ i\mu_1 S_{nm_1} + \dots \quad i\mu_p S_{nm_p}\}\, d\lambda_1 \dots d\mu_p \leq \qquad (4.5)$$

$$\leq Bn^{-k} \int_{-\infty}^{\infty} \dots \int_{-\infty}^{\infty} |\Psi_n(\lambda_1)| \dots |\Psi_n(\lambda_k)| .$$

$$\cdot \sum_{ns\leq n_i \leq \dots \leq n_k \leq n_t} E \exp\{i\lambda_1 S_{nn_1} + \dots + i\lambda_k S_{nn_k}\}|\, d\lambda_1 \dots d\lambda_k \leq$$

$$\leq Bn^{-k} \int_{-\infty}^{\infty}\dots \int_{-\infty}^{\infty} |\Psi_n(\lambda_1) \dots |\Psi_n(\lambda_k) .$$

$$\cdot \sum_{ns\leq n_1 \leq nt} |\varphi^{n_1}((\lambda_1 + \dots + \lambda_k) n^{-1/\alpha})|$$

$$\sum_{ns\leq n_2 \leq nt} |\varphi^{n_2-n_1}((\lambda_2 + \dots + \lambda_k)\, n^{-1/\alpha})| .$$

$$\dots \sum_{ns\leq n_k \leq nt} |\varphi^{n_k-n_{k-1}}(\lambda_k)\, d\lambda_1 \dots d\lambda_k \leq$$

$$\leq B \int_{-\infty}^{\infty}\ldots\int_{-\infty}^{\infty}|\Psi_n(\lambda_1-\lambda_2)\ \Psi_n(\lambda_2-\lambda_3)\ldots\Psi_n(\lambda_{k-1}-\lambda_k)\ \Psi_n(\lambda_k)|\ .$$

$$.\ \prod_1^k \frac{1-e^{-c|\lambda_i|^\alpha}}{|\lambda_i|^\alpha}\ d\lambda_1\ldots d\lambda_k\ .$$

En notant que $\dfrac{1+|\lambda_i-\lambda_j|^\alpha}{|\lambda_i|^\alpha|\lambda_j|^\alpha} \leq 2(|\lambda_i|^{-\alpha}+|\lambda_j|^{-\alpha})$ pour max $(|\lambda_i|,|\lambda_j|) \geq 1$, nous

pouvons réécrire la dernière inégalité de la manière suivante :

$$E|\int_{-\infty}^{\infty}\hat{f}_n(\lambda)\sum_{ns\leq j\leq nt} e^{-i\lambda S_{nj}}d\lambda|^{2p} \leq$$

$$\leq B\ |t-s|^k + \int_{-\infty}^{\infty}\ldots\int_{-\infty}^{\infty}\frac{|\Psi_n(\lambda_1-\lambda_2)\ldots\Psi_n(\lambda_{k-1}-\lambda_k)\ \Psi_n(\lambda_k)|}{\sqrt{1+|\lambda_1-\lambda_2|^\alpha}\ldots\sqrt{1+|\lambda_{k-1}-\lambda_k|^\alpha}\sqrt{1+|\lambda_k|^\alpha}}\ .$$

$$.\ \prod_{i=1}^k(1-e^{-c|\lambda_i|^\alpha})\ |\lambda_1|^{-\alpha/2}\ldots|\lambda_k|^{-\alpha/2}\ . \tag{4.6}$$

$$.\ \prod_{i=1}^{k-1}(|\lambda_i|^{-\alpha/2}+|\lambda_{i+1}|^{-\alpha/2})\ d\lambda_1\ldots d\lambda_k\ .$$

On a :

$$|\lambda_1|^{-\alpha/2}\ldots|\lambda_k|^{-\alpha/2}\prod_1^{k-1}(|\lambda_i|^{-\alpha/2}+|\lambda_{i+1}|^{-\alpha/2}) =$$

$$= \sum_{\gamma_1\ldots\gamma_k}|\lambda_1|^{-\alpha\gamma_1}\ldots|\lambda_k|^{-\alpha\gamma_k}$$

où la sommation est étendue aux vecteurs $\gamma = (\gamma_1\ldots\gamma_k)$, $\gamma_i = 0, 1/2, 1$. Ces vecteurs γ doivent aussi satisfaire quelques conditions complémentaires. En particulier, si $\gamma_i = 0$, $\gamma_{i-1}, \gamma_{i+1} > 0$ si $\gamma_i = 1$, $\gamma_{i-1}, \gamma_{i+1} \leq \frac{1}{2}$; si $\gamma_{i-1} = \gamma_{i+1} = 0$, $\gamma_i = 1$; $\gamma_1, \ldots \gamma_k \geq \frac{1}{2}$.

En vertu de ceci et de l'inégalité (4.6), on a :

$$E\ |\int_{-\infty}^{\infty}\hat{f}_n(\lambda)\sum_{ns\leq j\leq nt} e^{-i\lambda S_{nj}}d\lambda|^p \leq B\ |t-s|^k +$$

$$+\sum_{(\gamma_1\ldots\gamma_k)}\int_{-\infty}^{\infty}\ldots\int_{-\infty}^{\infty}\prod_{\gamma_{i=1}}\frac{1-e^{-c(t-s)|\lambda_{i1}|^\alpha}}{|\lambda_{i1}|^\alpha}\ d\lambda_{i1}\ .$$

$$.\int_{-\infty}^{\infty}\ldots\int_{-\infty}^{\infty}\prod_{i:\gamma_{i-1},\gamma_i>0}\frac{\Psi_n(\lambda_i-\lambda_{i-1})}{\sqrt{1+|\lambda_i-\lambda_{i-1}|^\alpha}}\left[\frac{1-e^{-c(t-s)|\lambda_i|^\alpha}}{|\lambda_i|^{\alpha\gamma_i}}\ d\lambda_i\right]^{2(1-\gamma_i)}$$

$$\cdot \prod_{\substack{j:\ \gamma_j=\frac{1}{2} \\ \gamma_{j+1}>0}} \frac{\Psi_n(\lambda_{j+1}-\lambda_j)}{\sqrt{1+|\lambda_{j+1}-\lambda_j|^{\alpha}}} \ \frac{1-e^{-c(t-s)|\lambda_j|^{\alpha}}}{|\lambda_j|^{\alpha\gamma_j}} \, d\lambda_j$$

$$\cdot \int_{-\infty}^{\infty} \cdots \int_{-\infty}^{\infty} \prod_{i:\gamma_i=0} \frac{\Psi_n(\lambda_{i+1}-\lambda_i)\ \Psi_n(\lambda_i-\lambda_{i-1})}{\sqrt{(1+|\lambda_{i+1}-\lambda_i|^{\alpha})\ (1+|\lambda_i-\lambda_{i-1}|^{\alpha}}} \ \frac{d\lambda_i}{\lambda_i^{\alpha\gamma_i}} \ .$$

Puisque $\gamma_{i+1}, \gamma_{i-1} > 0$ si $\gamma_i = 0$ l'intégrale :

$$\int_{-\infty}^{\infty} \cdots \int_{-\infty}^{\infty} \prod_{i:\gamma_i=0} \frac{\Psi_n(\lambda_{i+1}-\lambda_i)\ \Psi_n(\lambda_i-\lambda_{i-1})}{\sqrt{(1+|\lambda_{i+1}-\lambda_i|^{\alpha})\ (1+|\lambda_i-\lambda_{i-1}|^{\alpha})}} \, d\lambda_i \ \leq$$

$$\leq \max_{1\leq j\leq k} \left(\int_{-\infty}^{\infty} \frac{|\Psi_n(\lambda)|^2}{1+|\lambda|^{\alpha}} \, d\lambda\right)^j = B \ .$$

De la même manière l'intégrale :

$$\int_{-\infty}^{\infty} \cdots \int_{-\infty}^{\infty} \prod_{i:\gamma_i,\gamma_{i-1}>0} \frac{\Psi_n(\lambda_i-\lambda_{i-1})}{\sqrt{1+|\lambda_i-\lambda_{i-1}|^{\alpha}}} \ \frac{\left[(1-e^{-c(t-s)|\lambda_i|^{\alpha}})d\lambda_i\right]^{2(1-\gamma_i)}}{|\lambda_i|^{\alpha\gamma_i}}$$

$$\cdot \frac{\left[(1-e^{-c(t-s)|\lambda_{i-1}|^{\alpha}})\ d\lambda_{i-1}\right]^{2(1-\gamma_{i-1})}}{|\lambda_{i-1}|^{\gamma_{i-1}\alpha}} \leq$$

$$\leq B\left(\int_{-\infty}^{\infty} \frac{(1-e^{-c|t-s||\lambda|^{\alpha}})^2}{|\lambda|^{\alpha}} \, d\lambda\right)^{r/2} \leq B(t-S)^{r\frac{\alpha-1}{2\alpha}}$$

où r est le nombre des γ_i qui sont égaux à 1/2. Enfin, l'intégrale :

$$\int_{-\infty}^{\infty} \cdots \int_{-\infty}^{\infty} \prod_{\gamma_i=1} \frac{1-e^{-c(t-s)|\lambda_i|^{\alpha}}}{|\lambda_i|^{\alpha}} \, d\lambda_i \leq B(t-S)^{\ell\frac{\alpha-1}{\alpha}}$$

où ℓ est le nombre des γ_i qui sont égaux à 1.

Ainsi dans le premier cas :

$$E\,|\eta_n(t)-\eta_n(s)|^k \leq B(|t-s|^k + |t-s|^{k\frac{\alpha-1}{2\alpha}} \tag{4.7}$$

$$+ |t-s|^{r\frac{\alpha-1}{2\alpha}}\,|t-s|^{\ell\frac{\alpha-1}{\alpha}} \leq B(t-s)^{\frac{k(\alpha-1)}{2\alpha}}$$

<u>Deuxième cas</u> : Dans ce cas l'ensemble $\Lambda = \{0\}$ mais la loi de ξ_j est arithmétique. Donc la fonction caractéristique $\varphi(\lambda)$ est périodique de période $2\pi/h$. On peut choisir un nombre $\varepsilon > 0$ tel que :

$$|\varphi(\lambda n^{-1/\alpha})| \leq \exp\ \{-\frac{c}{n}\ |\lambda - \frac{2\pi k}{h}|^{\alpha}\}\ ,$$

Si : $|\lambda - \frac{2\pi k}{h}| < \varepsilon n^{1/\alpha}$, $k = 0, \pm 1, \ldots$

Si $\lambda \notin \bigcup_k \{\lambda : |\lambda - \frac{2\pi k}{h}| < \varepsilon\, n^{1/\alpha}\}$, on peut trouver $\delta > 0$ tel que :

$$|\varphi(\lambda n^{-1/2})| \leq e^{-\delta} \leq \exp\ \{-c\ \frac{|\lambda|^{\alpha}}{n}\},\quad c > 0\ .$$

Donc dans ce cas (voir (4.6) si $k = 2p$

$$E\ |\int_{-\infty}^{\infty} \hat{f}_n(\lambda) \sum_{ns \leq j \leq nt} e^{-i\lambda S_{nj}}\ d\lambda|^k \leq \tag{4.8}$$

$$\leq B \sum_{\ell_1,\ldots \ell_k} \int_{-\infty}^{\infty} \ldots \int_{-\infty}^{\infty} \prod_i |\Psi_n\ (\lambda_i + \frac{2\pi\ell_i}{h} - \lambda_{i-1} - \frac{2\pi\ell_{i-1}}{h}|\ .$$

$$.\ \prod_i\ (1 - \exp\ \{-c\ |\lambda_i|^{\alpha}\})\ \ |\lambda_i|^{-\alpha}\ d\lambda_1\ \ldots\ d\lambda_k\ .$$

La sommation étant étendue à tous les entiers $\ell_1,\ldots \ell_k$ tels que $2\pi\ell_j\ h^{-1} < a$ (rappelons que $\Psi_n(\lambda) = 0$ si $|\lambda| > an^{1/\alpha}$).

Le même raisonnement que plus haut montre que, dans ce cas aussi,

$$E|\eta_n(t) - \eta_n(S)|^k \leq B\ |t-s|^{\frac{k(\alpha-1)}{2\alpha}}$$

<u>Troisième cas</u> : Le cas $\Lambda = \{\lambda_i\} \neq \{0\}$. Soit $\ldots < \lambda_{-1} < 0 < \lambda_1 < \ldots$ Dans ce cas (voir paragraphe 2)

$$\int_{-\infty}^{\infty} \hat{f}_n(\lambda) \sum_{ns \leq j \leq nt} e^{-i\lambda S_{nj}}\ d\lambda =$$

$$= \sum_{\lambda_r \in \Lambda} \int_{-\lambda_i\ n^{1/\alpha}}^{\lambda_i n^{1/\alpha}} f_n(\lambda + \lambda_r\ n^{1/\alpha}) \sum_{ns \leq j \leq nt} e^{-i\lambda S_{nj}}\ d\lambda =$$

$$= n^{-1} \int_{-\lambda_1 n^{1/\alpha}}^{\lambda_1\ n^{1/\alpha}} \Psi_n(\lambda) \sum_{ns \leq j \leq nt} e^{-i\lambda S_{nj}}\ d\lambda$$

et donc

$$E|\int_{-\infty}^{\infty} \hat{f}_n(\lambda) \sum_{ns \leq j \leq nt} e^{-i\lambda S_{nj}}\ d\lambda|^k =$$

$$= n^{-k} |E \int_{-\lambda_1 n^{1/\alpha}}^{\lambda_1\ n^{1/\alpha}} \Psi_n(\lambda) \sum_{ns \leq j \leq nt} e^{-i\lambda S_{nj}}\ d\lambda|^{\ k}\ .$$

Ainsi dans ce cas aussi :

$$E|\eta_n(t) - \eta_n(s)|^k \leq B \ |t-s|^{\frac{k(\alpha-1)}{2\alpha}}$$

Le lemme 4.2 est démontré.

Quant à la démonstration du lemme 4.3 elle est calquée sur celle du lemme 4.2 dans le cas nonarithmétique. On peut aussi déduire ce lemme des lemmes 4.1, 4.2. En effet, en vertu de ces lemmes pour chaque s,t :

$$\lim_n |E \ \eta_n(t) - \eta_n(s)|^k = E|\eta(t) - \eta(s)|^k$$

et donc

$$E|\eta(t) - \eta(s)|^k \leq E \ |t-s|^{\frac{k(\alpha-1)}{2\alpha}}$$

Remarque 4.1 : En fait nous avons montré que :

$$E \ |\eta_n(t) - \eta_n(s)|^k \leq B \ \left(\int_{-\infty}^{\infty} \frac{|\Psi_n(\lambda)|^2}{1+|\lambda|^\alpha} d\lambda\right)^{1/2} \ |t-s|^{\frac{k(\alpha-1)}{2\alpha}}$$

$$E \ |\eta(t) - \eta(s)|^k \leq B \ \left(\int_{-\infty}^{\infty} \frac{|\Psi(\lambda)|^2}{1+|\lambda|^\alpha} d\lambda\right)^{1/2} \ |t-s|^{\frac{\alpha-1}{2\alpha} k} .$$

Remarque 4.2 : On déduit immédiatement de (4.6) que si $|\Psi(\lambda)| \leq B$ l'inégalité du lemme 4.3 (ou 4.2) peut être écrite dans la forme plus forte :

$$E \ |\eta_n(t) - \eta_n(s)|^k \leq B_k \ |t-s|^{\frac{\alpha-1}{\alpha} k}$$

$$E \ |\eta(t) - \eta(s)|^k \leq B_k \ |t-s|^{\frac{\alpha-1}{\alpha} k} .$$

Ayant montré les lemmes, on arrive immédiatement à la démonstration du théorème : il est une conséquence immédiate de ces lemmes et du théorème de Prohorov (voir chapitre I).

3° Exemple : Utilisons le lemme 4.3 pour en déduire le théorème sur la continuité du temps local du processus $\xi_\alpha(u)$. Soit x un nombre réel. Choisissons dans le lemme 4.3 $\Psi(\lambda) = e^{i\lambda x}$. En vertu du lemme 2.2, le processus :

$$\eta(t) = \frac{1}{2\pi} \int_{-\infty}^{\infty} \Psi(\lambda) \int_0^1 e^{-i\lambda\xi_\alpha(u)} du \, d\lambda$$

est bien défini. On a :

$$\eta(t) = \lim_{A\to\infty} \frac{1}{\pi} \int_{-\infty}^{\infty} \pounds_\alpha(y,t) \frac{\sin A(x-y)}{x-y} dy$$

ou $\pounds_\alpha(y,t)$ est le temps local de ξ_α.

Nous avons montré (voir chapitre I, paragraphe 4) que pour chaque t la fonction $\mathcal{L}_\alpha(y,t)$ est continue par rapport à y et a un support compact. Alors en vertu d'un théorème classique sur la transformation de Fourier :

$$\eta(t) = \lim_{A\to\infty} \frac{1}{\pi} \int_{-\infty}^{\infty} \mathcal{L}_\alpha(y,t) \frac{\sin A(x-y)}{x-y}\, dy = \mathcal{L}_\alpha(x,t)$$

L'inégalité du lemme 4.3 (voir remarque 4.2) donne alors :

$$E \mid \mathcal{L}_\alpha(x,t) - \mathcal{L}_\alpha(x,s)|^k \leq B|t-s|^{\frac{(\alpha-1)k}{\alpha}}$$

Soient h_1, h_2 deux nombres réels. D'après le théorème du chapitre I et la dernière inégalité :

$$E\,|\mathcal{L}_\alpha(x+h_1,\ t+h_2) - \mathcal{L}_\alpha(x,t)|^p$$

$$\leq 2^{p-1}(E\,|\mathcal{L}_\alpha(x+h_1,\ t+h_2) - \mathcal{L}_\alpha(x,t+h_2)|^p +$$

$$+ E\,|\mathcal{L}_\alpha(x,t+h_2) - \mathcal{L}_\alpha(x,t)|^p) \leq B(h_1^2 + h_2^2)^{\frac{(\alpha-1)p}{\alpha}}$$

Pour achever la démonstration il suffit d'utiliser le théorème du chapitre I. Nous avons montré même que $\mathcal{L}_\alpha(x,t)$ satisfait la condition de Hölder pour tout ordre $\beta < \frac{\alpha-1}{\alpha}$.

V - THEOREMES LIMITES DU DEUXIEME TYPE

1° Enoncé du résultat et schéma de la preuve

Soit comme toujours :

$$\eta_n = \sum_1^n f_n(S_{nk})$$

et :

$$\Psi_n(\lambda) = n \sum_{\lambda_k\in\Lambda} \hat{f}_n(\lambda + \lambda_k n^{1/\alpha}).$$

Nous allons étudier ci-dessous le cas où $\Psi_n(\lambda)$ n'a pas une fonction limite raisonnable. Posons :

$$g_n(x) = \sum_{k=1}^n E\, f_n(x + S_{nk})$$

$$k_n(x) = f_n^2(x) + 2\, f_n(x)\, g_n(x)$$

$$K_n(x) = n \int_0^x k_n(z) \sum_{\lambda_k\in\Lambda} e^{-i\lambda z n^{1/\alpha}}\, dz .$$

Définissons le processus $\eta_n(t)$ comme la ligne brisée de sommets $(\frac{k}{n}, \sum_1^k f_n(S_{nj}))$, k=0,...n . Le processus $\eta_n(t)$ engendre une loi de probabilité dans l'espace C[0,1]. Evidemment, $\eta_n = \eta_n(1)$.

THEOREME 5.1 : *Soient les variables* ξ_j *appartenant au domaine d'attraction de la loi stable de fonction caractéristique* $\exp\{-\frac{1}{\alpha}|\lambda|^{\alpha}(1+i\beta\frac{t}{|t|}\,\mathrm{tg}\,\frac{\pi\alpha}{2})\}$, $\alpha > 1$. *Supposons les conditions suivantes vérifiées :*

1° *La transformée de Fourier* $\widehat{f}_n$ *de* f_n *s'annule hors de l'intervalle* $\left[-an^{1/\alpha}, an^{1/\alpha}\right]$;

2° $$\sup_n \int_{-\infty}^{\infty} \frac{|\Psi_n(\lambda)|^2}{1+|\lambda|^{\alpha}}\, d\lambda < \infty \; ; \quad \lim_n \int_{-\infty}^{\infty} \frac{|\Psi_n(\lambda)|}{1+|\lambda|^{\alpha}}\, d\lambda = 0 \; ;$$

$$\lim_n \int_{-A}^{A} \frac{|\Psi_n(\lambda)|^2}{1+|\lambda|^{\alpha}}\, d\lambda = 0 \quad \textit{pour tout A fixé.}$$

3° *Les fonctions* $K_n(x)$ *convergent faiblement vers une fonction croissante* $K(x)$.

Alors le processus $\eta_n(t)$ converge en loi dans $C[0,1]$ vers un processus $\eta(t) = W(\int_{-\infty}^{\infty} \mathcal{L}_{\alpha}(x,t)\, dK(x))$ où W est un processus de Wiener, $\mathcal{L}_{\alpha}(x,t)$ est le temps local d'un processus stable $\xi_{\alpha}(t)$,

$$E\, e^{i\lambda\xi_{\alpha}(t)} = \exp\{-\frac{|\lambda|^{\alpha}}{\alpha}(1+i\beta\frac{t}{|t|}\,\mathrm{tg}\,\frac{\pi\alpha}{2}\}$$

et les processus W, $\int_{-\infty}^{\infty} \mathcal{L}_{\alpha}(x,t)\, dK(x)$ sont indépendants.

Comme corollaire presque immédiat, on a le

THEOREME 5.2 : *Sous les hypothèses du théorème 5.1 :*

$$\eta_n = \eta_n(1) \longrightarrow \xi\sqrt{\zeta}$$

où ξ *est une variable normale standardisée,* $\zeta = \int_{-\infty}^{\infty} \mathcal{L}_{\alpha}(x,1)dK(x)$ *et les variables* ξ , ζ *sont indépendantes.*

Démonstration : La démonstration est assez longue. Ebauchons d'abord le schéma de la preuve. Introduisons les processus :

$$\eta_n(t) = \sum_{k\leq nt} f_n(S_{nk}),$$

$$\zeta_n(t) = \sum_{k\leq nt} k_n(S_{nk}),$$

$$W_n(t) = n^{-1/\alpha} \sum_{k\leq nt} \xi_k \; .$$

Nous allons montrer que ces processus convergent respectivement vers les processus limites $\eta(t)$, $\zeta(t)$, $\xi(t)$. Pour les processus limites :

$$E\{\eta(t)-\eta(s)|\mathcal{F}_s\} = 0,$$
$$E\{(\eta(t)-\eta(s))^2|\mathcal{F}_s\} = E\{\zeta(t)-\zeta(s)|\mathcal{F}_s\} \qquad (5.1)$$

où $\mathscr{F}_s$ désigne la tribu engendrée par $\{\xi_\alpha(u),\ 0 \leq u \leq 1\ ;\ \eta(u),\ 0 \leq u \leq s\}$.

Nous montrerons aussi que la martingale $\eta(s)$ est continue avec probabilité 1.

Définissons $\tau(s)$ comme la plus petite racine de l'équation :

$$\zeta(\tau(S)) = S.$$

Pour tout s la variable $\tau(s)$ est un temps d'arrêt pour $\xi_\alpha(t)$. Notons $\mathscr{A}_S$ la tribu $\mathscr{F}_{\tau(S)}$. Nous verrons que en vertu de (5.1) $\{\eta(\tau(s)), \mathscr{A}_s\}$ est une martingale et que :

$$E\{(\eta(\tau(s)) - \eta(\tau(t)))^2 | \mathscr{A}_S\} = \zeta(\tau(t)) - \zeta(\tau(s)) = t-s ,$$

On déduit de ceci et du théorème 6.1 du chapitre 1 que le processus $W(s) = \eta(\tau(s))$ est le processus de Wiener qui ne dépend pas de la tribu $\mathscr{A}_0$. Dont $\eta(t) = W(\zeta(t))$ où les processus $W(.)$, $\zeta(.)$ sont indépendants. Le théorème 2.2 résulte alors de la relation :

$$\eta(1) = W(\zeta(1)) = \frac{W(\zeta(1))}{\sqrt{\zeta(1)}} \qquad \sqrt{\zeta(1)}$$

puisque, en vertu du théorème 2.1 :

$$\zeta(1) = \int_{-\infty}^{\infty} \pounds_\alpha(x,1)\, dK(x).$$

La preuve détaillée va être donnée dans les deux sous-paragraphes suivants : le sous-paragraphe 2 contient des lemmes nécessaires ; la démonstration finale du théorème est contenue dans le sous-paragraphe 3. Pour simplifier on raisonnera seulement sur le cas où la loi de ξ_j est nonarithmétique. On a donc toujours $\Lambda = \{0\}$.

2 - Les premiers lemmes assurent la convergence du processus :

$$\zeta_n(t) = \sum_{j \leq nt} k_n(S_{nj})$$

vers le processus :

$$\zeta(t) = \int_{-\infty}^{\infty} \pounds_\alpha\ (x,t) dK(x) \quad .$$

Soit $\kappa_n(\lambda) = n\widehat{k}_n(\lambda)$. La fonction κ_n joue le même rôle pour ζ_n que la fonction Ψ_n pour η_n .

<u>LEMME 5.1</u> : *Sous les hypothèses du théorème :*

$$\sup_n \int_{-\infty}^{\infty} \frac{|\kappa_n(\lambda)|^2}{1 + |\lambda|^\alpha}\, d\lambda < \infty \quad ,$$

$$\sup_n \int_{|\lambda|>T} \frac{|\kappa_n(\lambda)|^2}{1 + |\lambda|^\alpha}\, d\lambda \xrightarrow[T \to \infty]{} 0 \qquad (5.2)$$

Démonstration : On a par définition de K_n :

$$\kappa_n(\lambda) = n(\widehat{f}_n * \widehat{f}_n(\lambda) + 2\widehat{f}_n * \widehat{g}_n(\lambda)).$$

Mais :

$$\widehat{g}_n(\lambda) = \sum_1^n \widehat{f}_n(\lambda) \, E \, e^{-i\lambda S_{nj}} = \widehat{f}_n(\lambda)\, \varphi(-\lambda n^{1/\alpha}) \, \frac{1-\varphi(-\lambda n^{-1/\alpha})}{1-\varphi(-\lambda n^{-1/\alpha})}$$

et puisque la loi de ξ_j n'est pas arithmétique et que $\widehat{f}_n(\lambda) = 0$ si $|\lambda| > an^{1/\alpha}$, on a l'inégalité suivante :

$$|\widehat{g}_n(\lambda)| \leqq Bn|\widehat{f}_n(\lambda) \, \frac{1-e^{-c|\lambda|^\alpha}}{|\lambda|^\alpha} \, , \quad c > 0 \, .$$

Alors pour $T > 0$

$$\int_T^\infty \frac{|\kappa_n(\lambda)|^2}{1+|\lambda|^\alpha} \, d\lambda \leqq$$

$$\leqq Bn^2 \int_T^\infty \frac{d\lambda}{1+|\lambda|^\alpha} \left\{ \left(\int_{-an^{1/\alpha}}^{an^{1/\alpha}} \widehat{f}_n(\lambda-\mu) \, \widehat{f}_n(\mu) d\mu \right)^2 + \right.$$

$$\left. + \left(\int_{-an^{1/\alpha}}^{an^{1/\alpha}} |\widehat{f}_n(\lambda-\mu)| \, |\widehat{f}_n(\mu)| \, \frac{1-e^{-c|\mu|^\alpha}}{|\mu|^\alpha} \, d\mu \right)^2 \right\}.$$

La première intégrale de droite est inférieure à :

$$Bn^2 \int_T^\infty \frac{d\lambda}{1+|\lambda|^\alpha} \left(n \int_{-\infty}^\infty \frac{|\widehat{f}_n(\lambda)|^2}{1+|\lambda|^\alpha} \, d\lambda \right)^2 =$$

$$= B\left(\int_{-\infty}^\infty \frac{|\Psi_n(\lambda)|^2}{1+|\lambda|^\alpha} \right)^2 \int_T^\infty \frac{d\lambda}{1+|\lambda|^\alpha} \leqq B \int_T^\infty \frac{d\lambda}{1+|\lambda|^\alpha}$$

en vertu de la condition du théorème.

Pour avoir une majoration de la deuxième intégrale on note d'abord que :

$$n^2 \int_T^\infty \frac{d\lambda}{1+|\lambda|^\alpha} \left(n \int_{-T/2}^{T/2} |\widehat{f}_n(\lambda-\mu) \, \widehat{f}_n(\mu)| \, \frac{1-e^{-c|\mu|^\alpha}}{|\mu|^\alpha} \, d\mu \right)^2 \leqq$$

$$\leqq \int_T^\infty \frac{d\lambda}{1+|\lambda|^\alpha} \int_{-T/2}^{T/2} \frac{|\Psi_n(\lambda-\mu)|^2}{1+|\mu|^\alpha} \, d\mu \int_{-T/2}^{T/2} \frac{|\Psi_n(\mu)|^2}{1+|\mu|^\alpha} \, d\mu \leqq$$

$$\leqq B \int_{-T/2}^{T/2} \frac{|\Psi_n(\mu)|^2}{1+|\mu|^\alpha} \, d\mu \; \sup_n \int_{-\infty}^\infty \frac{|\Psi_n(\lambda)|^2}{1+|\lambda|^\alpha} \, d\lambda \xrightarrow[n\to\infty]{} 0$$

si T est fixé. Puis que

$$n^2 \int_T^\infty \frac{d\lambda}{1+|\lambda|^\alpha} \left(n \int_{|\mu|>T/2} |f_n(\lambda-\mu)\, f_n(\mu)|\, \frac{1-e^{-\epsilon|\mu|^\alpha}}{|\mu|^\alpha}\, d\mu\right)^2 \leq$$

$$\leq B \int_T^\infty \frac{d\lambda}{1+|\lambda|^\alpha} \int_{|\mu|> T/2} \frac{|f_n(\lambda-\mu)|^2}{1+|\mu|^\alpha}\, d\mu \leq$$

$$\leq B \int_T^\infty d\lambda \int_{|\mu|>T/2} \frac{|\Psi_n(\lambda-\mu)|^2}{1+|\lambda-\mu|^\alpha} \left(\frac{1}{1+|\mu|^\alpha} + \frac{1}{1+|\lambda|^\alpha}\right) d\mu \leq$$

$$\leq B \sup_n \int_{-\infty}^{\infty} \frac{|\Psi_n(\lambda)|^2}{1+|\lambda|^\alpha}\, d\lambda \,.\, \int_{T/2}^{\infty} \frac{d\mu}{1+|\mu|^\alpha} .$$

Le lemme est démontré.

<u>LEMME 5.2</u> : *Sous les hypothèses du théorème, les processus* $\zeta_n(t)$ *convergent en loi dans l'espace* $D(0,1)$ *de Skorohod vers le processus*

$$\zeta(t) = \int_{-\infty}^{\infty} \mathcal{L}_\alpha(x,t)\, dK(x).$$

<u>Démonstration</u> : On déduit du lemme 5.1, que :

$$\lim_T \overline{\lim_n}\; E\left|\int_{|\lambda|>T} \widehat{K}_n(\lambda) \sum_j e^{-i\lambda S_{nj}}\, d\lambda\right|^2 = 0,$$

et que la loi limite de $(\zeta_n(t_1)\ldots\, \zeta_n(t_k))$ coïncide avec la loi limite de :

$$\left(\frac{1}{2\pi} \int_{-\infty}^{\infty} \kappa_n(\lambda)\, d\lambda \int_0^{t_1} e^{-i\lambda\xi_\alpha(u)}\, du, \ldots, \frac{1}{2\pi} \int_{-\infty}^{\infty} \kappa_n(\lambda)\, d\lambda \int_0^{t_k} e^{-i\lambda\xi_\alpha(u)}\, du\right) =$$

$$= \left(\int_{-\infty}^{\infty} \mathcal{L}_\alpha(x,t_1)\, dK_n(x)\, ,\, \ldots,\, \int_{-\infty}^{\infty} \mathcal{L}_\alpha(x,t_k)\, dK_n(x)\right).$$

Mais puisque toute fonction $\mathcal{L}_\alpha(x,t_j)$ de x a un support compact et puisque $K_n(x)$ converge faiblement, les variables

$$\int_{-\infty}^{\infty} \mathcal{L}_\alpha(x,t_j)\, dK_n(x) \underset{n\to\infty}{\longrightarrow} \int_{-\infty}^{\infty} \mathcal{L}_\alpha(x,t_j)\, dK(x)$$

en probabilité. Donc $(\zeta_n(t_1)\ldots\, \zeta_n(t_x)) \xrightarrow{(\mathcal{L})} (\zeta(t_1)\ldots\zeta(t_k))$.

Puis en vertu des lemmes 5.1 et 4.2 :

$$E|\zeta_n(t) - \zeta_n(S)|^k \leq B_k\, |t-S|^{\frac{k(\alpha-1)}{2}} \qquad (5.3)$$

pour tout K > 0. Ainsi $\{\zeta_n\}$ converge fini-dimensionnellement vers ζ et la suite $\{\zeta_n\}$ est tendue. Donc ζ_n converge vers ζ en loi pour la topologie de Skorokhod (et même pour la topologie uniforme).

LEMME 5.3 : *Pour chaque nombre positif* p *il existe une constante* B_k *telle que pour* $t,s \in [0,1]$

$$E|\eta_n(t) - \eta_n(S)|^p \leq B_k \; |t-S|^{\frac{p(\alpha-1)}{2\alpha}} \qquad (5.4)$$

Démonstration : Ce lemme est une conséquence directe du lemme 4.2 (voir remarque 4.1) et de la condition 2 du théorème.

LEMME 5.4 : *Soit* $0 = t_0 < t_1 < \ldots < t_r = 1$. *Si* $j \leq r$, $G(y_0 \ldots y_{j-1} \; ; \; x_0 \ldots x_r)$ *est une fonction continue à support compact, alors* :

$$\lim_n \; E\,\{(\eta_n(t_j) - \eta_n(t_{j-1}))\; G(\eta_n(t_0)\ldots\eta_n(t_{j-1}), W_n(t_0)\ldots W_n(t_r))\} = 0 \qquad (5.5)$$

Démonstration : En vertu du théorème de Weierstrass sur l'approximation d'une fonction continue par des polynômes trigonométriques on peut trouver un polygône trigonométrique $G_\varepsilon(y_0 \ldots y_{j-1}, x_0 \ldots x_r)$ tel que :

$$\sup_{y_i,\, x_i} |G_\varepsilon - G| < \varepsilon$$

ou ε est un nombre positif donné. D'après le lemme 5.3 :

$$E\{|\eta_n(t_j) - \eta_n(t_{j-1})| \, . \, |G - G_\varepsilon|\} \leq$$

$$\leq \varepsilon \sup_{t,s} \; E\,|\eta_n(t) - \eta_n(s)| \xrightarrow[\varepsilon \to 0]{} 0 .$$

Donc on peut se borner au cas :

$$G(y_0 \ldots y_{j-1}, x_0 \ldots x_r) = \exp\{i\,(\sum_0^{j-1} \lambda_k y_k + \sum_0^{r} \mu_k x_k)\} \; .$$

Supposons que les nombres t_j, t_{j-1} sont tels que :

$$\eta_n(t_j) - \eta_n(t_{j-1}) = \sum_{k \leq nt_j} f_n(S_{nk}) - \sum_{k \leq nt_{j-1}} f_n(S_{nk}) =$$

$$\sum_{\ell_1}^{\ell_2} f_n(S_{nk}) = \sum_{k=0}^{\ell_2-\ell_1} f_n(S_{n\ell_1} + \sum_{\ell_1+1}^{k} \xi_j \, n^{-1/\alpha}) .$$

Alors :

$$E\,\{(\eta(t_j) - \eta(t_{j-1}))G\} = E\,\{\exp\{i \sum_1^{j-1} \lambda_k \eta(t_k) + i \sum_1^{j-1} \mu_k W_n(t_k)\} \; .$$

$$. \; E\{ \sum_0^{\ell_2-\ell_1} f_n(S_{n\ell_1} + \sum_{\ell_1+1}^{k} \xi_j \, n^{-1/\alpha}) . \exp\{i\mu(W_n(t_j) - W_n(t_{j-1})) +$$

$$+ \; i \sum_{k=j} \tilde{\mu}_k \, (W_n(t_{k+1}) - W_n(t_k))\} \,|S_0, \ldots S_{\ell_1}\} \; .$$

Toutes les différences $W_n(t_{k+1}) - W_n(t_k)$, $k \geq j$ et toutes les sommes $\sum_{\ell+1}^{k} \xi_j$ ne dépendent pas de $S_{n1} \ldots S_{n\ell_1}$. Donc il suffit de montrer que pour tout μ fixé :

$$\lim_n \sup_{\substack{x \\ 1\leq\ell\leq n}} E\{\sum_{k=0}^{\ell} f_n(x + \sum_1^k n^{1/\alpha}\xi_j) \exp\{i\mu S_{n\ell}\}\} = 0$$

On a :

$$|E\{\sum_0^{\ell} f_n(x + S_{nk}) e^{i\mu S_{n\ell}}\}| =$$

$$= \frac{1}{2\pi} \left|\int_{-\infty}^{\infty} \widehat{f}_n(\lambda) e^{-i\lambda x} \sum_{k=0}^{\ell} \varphi^k (n^{-1/\alpha}(\mu-\lambda)) \varphi^{\ell-k}(n^{-1/\alpha}\mu) d\lambda\right| \leq$$

$$\leq B_n \int_{-\infty}^{\infty} \frac{|f_n(\lambda)|}{1+|\mu-\lambda|^\alpha} d\lambda \leq B_\mu \int_{-\infty}^{\infty} \frac{|\Psi_n(\lambda)|}{1+|\lambda|^\alpha} d\lambda \to 0$$

d'après la condition 2 du théorème. Le lemme est démontré.

<u>LEMME 5.5</u> : *Soit* $0 = t_0 < t_1 < \ldots < t_r = 1$. *Si* $G(y_0 \ldots y_{j-1}, x_0 \ldots x_r)$, $j \leq r$ *est une fonction continue à support compact, alors*

$$\lim_n E\{[(\eta_n(t_j) - \eta_n(t_{j-1}))^2 - (\zeta_n(t_j) - \zeta_n(t_{j-1}))] \tag{5.6}$$

$$. G(\eta_n(t_0) \ldots \eta_n(t_{j-1}), W_n(t_0) \ldots W_n(t_r))\} = 0.$$

<u>Démonstration</u> : Comme plus haut il suffit de considérer seulement le cas :

$$G(y_0 \ldots y_{j-1}, x_0 \ldots x_r) = \exp\{i \sum_0^{j-1} \lambda_k y_k + i \sum_0^r \mu_k x_k\}$$

et de vérifier que pour tout μ fixé :

$$\sup_{x,\ell} E\{[(\sum_0^{\ell} f(x + S_{nj}))^2 - \sum_0^{\ell} k_n(x + S_{nj})] . e^{i\mu S_{n\ell}} \xrightarrow[n\to\infty]{} 0$$

On peut écrire l'espérance mathématique à gauche comme :

$$I = 2\{E e^{i\mu S_{n\ell}} (\sum_{1\leq k\leq j\leq \ell} f_n(x+S_{nk}) f_n(x+S_{nj}) - f_n(x+S_{nk}) g_n(x+S_{nj})) =$$

$$= \frac{1}{2\pi^2} \iint_{-an^{1/\alpha}}^{an^{1/\alpha}} \widehat{f}_n(u) \widehat{f}_n(v) . e^{-ixu-ixv} (\sum_{k=1}^{\ell-1} \varphi^k((\mu-u-v)n^{-1/\alpha})).$$

$$. \{\sum_{j=k+1}^{\ell} \varphi^{j-k}((\mu-u)n^{-1/\alpha}) \varphi^{\ell-j}(\mu n^{-1/\alpha}) -$$

$$- \varphi(-un^{-1/\alpha}) \varphi^{\ell-k}(\mu n^{-1/\alpha}) \frac{1-\varphi^n(-un^{-1/\alpha})}{1-\varphi(-un^{-1/\alpha})}\} du\, dv .$$

Puis :

$$I \leq B \int \int |\widehat{f}_n(u)\ \widehat{f}_n(v)\ \sum_1^{\ell}\ |\varphi^k((\mu+u+v)n^{-1/\alpha})|\ .$$

$$.\ |\varphi^{\ell-k}(un^{-1/\alpha})|\ |\ \frac{1-\varphi^{n-\ell+k}(un^{-1/\alpha})}{1-\varphi(un^{-1/\alpha})}\ |\ +$$

$$+\ B \int\int|\widehat{f}_n(u)\ \widehat{f}_n(v)\ |\ \sum_1^{\ell}|\varphi^k((\mu+u+v)n^{-1/\alpha})|\ .$$

$$|\sum_1^{\ell-k}\ (\ \frac{\varphi((\mu+u)n^{-1/\alpha})}{\varphi(\mu n^{-1/\alpha})}\ )^s\ -\ (\varphi(un^{-1/\alpha}))^s\ |du\ dv = I_1 + I_2\ .$$

Puisque dans le domaine ou l'intégrale est positive$|\ \varphi(un^{-1/\alpha})|\leq e^{-\frac{c}{n}|u|^\alpha}$, $c > 0$

et puisque dans ce domaine on a $|\mu + u + v| > c\ |v|$ ou $|u| > c\ |v|$, $c > 0$, on a :

$$I_1 \leq B \int_{-\infty}^{\infty}\int_{-\infty}^{\infty}|\widehat{f}_n(u)\ \widehat{f}_n(v)|\ \frac{n(1-e^{-c|u|^\alpha})}{|u|^\alpha}\ .$$

$$.\ n\ \min\ \{\ \frac{1-\exp\ \{-c\ |u+v+\mu|^\alpha\}}{|u+v+\mu|^\alpha}\ ;\ \frac{1-e^{-c|u|^\alpha}}{|u|^\alpha}\ \}\ du\ dv\ \leq$$

$$\leq B\ (\int_{-\infty}^{\infty}\ \frac{|\Psi_n(\lambda)|^2}{1+|\lambda|^\alpha}\ d\lambda \xrightarrow[n\to\infty]{} 0\ .$$

Quant à I_2 on note d'abord que :

$$|(\frac{\varphi((\mu+u)n^{-1/\alpha})}{\varphi(\mu n^{-1/\alpha})})^s\ -\ (\varphi(un^{-1/\alpha}))^s\ |\leq \tag{5.7}$$

$$\leq B_\mu |\ \varphi((\mu+u)n^{-1/\alpha}) - \varphi(\mu n^{-1/\alpha})\ \varphi(un^{-1/\alpha})|\ .$$

$$.\ \sum_{j=0}^{s-1}|\ \varphi((\mu+u)n^{-1/\alpha})|^j|\ \varphi(un^{-1/\alpha})|^{s-1-j} \leq$$

$$\leq B_\mu\ (\ \frac{1}{n} + |\varphi((\mu+u)n^{-1/\alpha}) - \varphi(un^{-1/\alpha})|\ ).$$

$$.\ \sum_{j=0}^{s-1}\ |\ \varphi((\mu+u)n^{-1/\alpha})|^j|\ \varphi(un^{-1/\alpha})|^{s-1-j}\ .$$

Un peu plus tard nous montrerons que :

$$|\varphi((\mu+u)n^{-1/\alpha}) - \varphi(un^{-1/\alpha})|\ \leq B_\mu\ (1+|u|)n^{-1} \tag{5.8}$$

En supposant que cette inégalité est vraie, utilisons la dans (5.7). On a :

$$I_2 \leq \frac{B_\mu}{n} \int_{-an^{1/\alpha}}^{an^{1/\alpha}} (1+|u|)|\hat{f}_n(u)|du \int_{-an^{1/\alpha}}^{an^{1/\alpha}} |\hat{f}_n(v)|dv \, .$$

$$\cdot \sum_1^{\ell} |\varphi^k((u+v+\mu)n^{-1/\alpha})| \sum_{s=1}^{\ell-k} \sum_{j=0}^{s-1} |\varphi((\mu+u)n^{-1/\alpha})|^j |\varphi(un^{-1/\alpha})|^{s-1-j}$$

D'où :

$$I_2 \leq B_\mu \{ \frac{1}{n} \iint (1+|u|)|\hat{f}_n(u)| \, |\hat{f}_n(v)| du\, dv +$$

$$+ \iint_{|u+v|\leq \varepsilon n^{1/\alpha}} (1+|u|)|\hat{f}_n(u)| \frac{|\hat{f}_n(v)|}{1+|u-v|^\alpha} du\, dv +$$

$$+ n^2 \iint_{-\varepsilon n^{1/\alpha}}^{\varepsilon n^{1/\alpha}} (1+|u|^{\alpha/2}) \frac{|\hat{f}_n(u)| \, |\hat{f}_n(v)|}{(1+|u|^{2\alpha})(1+|u-v|^\alpha)} du\, dv \, .$$

Puis :

$$\frac{1}{n} \iint (1+|u|)|\hat{f}_n(u)| |\hat{f}_n(v)| du\, dv \leq B\left(\int \frac{|\Psi_n(u)|}{1+|u|^\alpha} du\right)^2 = o(1),$$

$$\iint_{|u-v|\leq \varepsilon n^{1/\alpha}} (1+|u|) \, |\hat{f}_n(u)| \frac{|\hat{f}_n(v)|}{1+|u-v|^\alpha} du\, dv \leq$$

$$\leq \frac{B}{n} \int (1+|u|^\alpha) \, |\hat{f}_n(u)| \, du \leq B \int \frac{|\Psi_n(u)|}{1+|u|^\alpha} du = o(1),$$

$$n^2 \iint (1+|u|^{\alpha/2}) \frac{|\hat{f}_n(u)| \, |\hat{f}_n(v)|}{(1+|u|^{2\alpha})(1+|u-v|^\alpha)} du\, dv \leq$$

$$\leq B \int \frac{1+|u|^\alpha}{1+|u|^{2\alpha}} \cdot n \cdot |\hat{f}_n(u)| \, du = o(1).$$

La démonstration est finie sauf pour l'inégalité (5.8). Nous allons la démontrer. Soit $F(x)$ la fonction de répartition de ξ_j. En vertu des théorèmes 1.3, 1.4 du chapitre I :

$$1 - F(x) + F(-x) = G(x) = O(|x|^{-\alpha}) \, , \quad x \to \infty \quad .$$

On a :

$$\varphi(S+t) - \varphi(t) = \int_{-\infty}^{\infty} [e^{ix(s+t)} - e^{ixt}] \, dF(x) =$$

$$= \int_0^\infty (1-F(x)) \, (i(s+t) \, e^{ix(s+t)} - it \, e^{ixt}) \, dx -$$

$$- \int_{-\infty}^0 F(x) \, (i(s+t)e^{ix(s+t)} - ite^{ixt}) \, dx.$$

Soit A un nombre positif assez grand que nous choisirons définitivement plus tard. On a en vertu du second théorème de la moyenne :

$$\int_A^\infty (1-F(x))\ (i(s+t)e^{ix(s+t)} - ite^{ixt})dx = O(A^{-\alpha}), \qquad (5.10)$$

$$\int_{-\infty}^{-A} F(x)\ (i(S+t)e^{ix(S+t)} - ite^{ixt})\ dx = O(A^{-\alpha}).$$

Puisque :

$$\int_{-\infty}^\infty xdF(x) = \int_0^\infty (1-F(x))dx - \int_0^\infty F(x)dx = 0 \quad ,$$

on a :

$$\int_0^A (1-F(x))dx - \int_{-A}^0 F(x)dx = O\ (A^{-\alpha+1}) \qquad (5.11)$$

On déduit des relations (5.9) - (5.11) que :

$$|\varphi(t) - \varphi(S)| \leq |t||s| \int_0^A x\ G(dx +$$

$$+ |s|\ |t+s| \int_0^A x\ G(x)\ dx + O(A^{-\alpha}) + O(A^{-\alpha+1})$$

$$\leq B(|t|\ |s|\ A^{-\alpha+2} + s^2 A^{-\alpha+2} + |s|\ A^{-\alpha+1} + A^{-\alpha})$$

En choisissant dans cette inégalité $A = n^{1/\alpha}(1+|u|)^{-1/2}$, $t = un^{-1/\alpha}$, $s = \mu n^{-1/\alpha}$, on trouve que :

$$|\varphi((u+\mu)n^{-1/\alpha}) - \varphi(un^{-1/\alpha})| \leq B_\mu\ n^{-1}\ (1 + |u|^{\alpha/2})$$

c'est à dire l'inégalité (5.8). Le lemme est démontré.

<u>LEMME 5.6</u> : *Sous les hypothèses du lemme 5.5 :*

$$\lim_n E\{[(\eta_n(t_j) - \eta_n(t_{j-1}))^4 - 3(\zeta_n(t_j) - \zeta_n(t_{j-1}))^2].$$

$$.\ G(\eta_n(t_0) \ldots \eta_n(t_{j-1}), W_n(t_0) \ldots W_n(t_r))\} = 0 \qquad (5.12)$$

<u>Démonstration</u> : Elle est calquée sur celle des lemmes 5.4 et 5.5 et ne demande rien de nouveau mais les calculs sont plus fastidieux et nous ne les donnons pas ici.

<u>Démonstration du théorème</u> : D'après le lemme 5.3, l'ensemble des lois engendrées dans l'espace C [0,1] par les lignes brisées η_n est relativement compact. Soit $\{\eta_{n(m)}\}$ une sous-suite qui converge en loi dans C [0,1] vers un processus η. D'après le théorème 5.3 du chapitre I, on peut construire un espace de probabilité et définir sur cet espace des processus $\tilde{\eta}_{n(m)}(t)$, $\tilde{\zeta}_{n(m)}(t)$, $\tilde{W}_{n(m)}(t)$ de sorte que :

1° Toutes les lois fini-dimensionnelles des $\tilde{\eta}_{n(m)}$, $\tilde{\zeta}_{n(m)}$, $\tilde{W}_{n(m)}$ coïncident avec les mêmes lois des processus $\eta_{n(m)}$, $\zeta_{n(m)}$, $W_{n(m)}$;

2° Pour tout nombre $t \in [0,1]$ $\tilde{\eta}_{n(m)}(t)$, $\tilde{\zeta}_{n(m)}(t)$, $\tilde{W}_{n(m)}(t)$ convergent en probabilité vers les valeurs $\tilde{\eta}(t)$, $\tilde{\zeta}(t)$, $\tilde{W}(t)$ des processus $\tilde{\eta}$, $\tilde{\zeta}$, $\tilde{W}$. Bien sur, on peut prendre pour $\tilde{W}$ et $\tilde{\zeta}$ respectivement :

$$\tilde{W}(t) = \xi_\alpha(t)$$
$$\tilde{\zeta}(t) = \int_{-\infty}^{\infty} \pounds_\alpha(x,t)\, dK(x) \tag{5.13}$$

où $\pounds_\alpha$ est le temps local de $\xi_\alpha(t)$ (voir le lemme 5.2).

On déduit des lemmes 5.4, 5.5 que pour toute fonction continue $G(y_0 \ldots y_{j-1}, x_0, \ldots x_r)$ à support compact :

$$E\{(\tilde{\eta}(t_j) - \tilde{\eta}(t_{j-1}))\, G(\eta(t_0)..\eta(t_{j-1}), \tilde{W}(t_0)\ldots \tilde{W}(t_r))\} = 0 ,$$

$$E\{([\tilde{\eta}(t_j) - \tilde{\eta}(t_{j-1})]^2 - (\tilde{\zeta}(t_j) - \tilde{\zeta}(t_{j-1}))) \tag{5.14}$$

$$. G(\tilde{\eta}(t_0) .. \tilde{\eta}(t_{j-1}), \tilde{W}(t_0)\ldots \tilde{W}(t_r))\} = 0 ,$$

$$0 = t_0 < t_1 < \ldots < t_r = 1 .$$

Evidemment ces égalités restent valables pour toute G mesurable bornée.

Puis, on déduit de (5.14) que pour $i < j$

$$E\{(\tilde{\eta}(t_j) - \eta(t_i))\, G(\tilde{\eta}(t_0)\ldots\tilde{\eta}(t_i) , \tilde{W}(s_0)\ldots \tilde{W}(s_r))\} = 0 ,$$

$$E\{[\tilde{\eta}(t_j) - \eta(t_i))^2 - (\zeta(t_j) - \zeta(t_i))] \tag{5.15}$$

$$. G(\tilde{\eta}(t_0)\ldots \tilde{\eta}(t_i), \tilde{W}(s_0)\ldots \tilde{W}(s_r))\} = 0$$

pour toute G mesurable bornée.

Notons $\mathcal{F}_t$ la tribu engendrée par $\{\tilde{W}(s), 0 \leq s \leq 1 ; \tilde{\eta}(s), 0 \leq s \leq t\}$. Les égalités (5.15) signifient que pour $0 \leq s < t \leq 1$

$$E\{\tilde{\eta}(t) - \eta(s) | \mathcal{F}_s\} = 0,$$

$$E\{(\tilde{\eta}(t) - \eta(s))^2 | \mathcal{F}_s\} = E\{\tilde{\zeta}(t) - \tilde{\zeta}(s) | \mathcal{F}_s\} \tag{5.16}$$

Soit :

$$\tau(s) = \inf\{t : \zeta(t) = s\} .$$

Evidemment, $\tau(s)$ est un temps d'arrêt par rapport à la famille croissante des tribus engendrées par $\{\tilde{W}(u), u \leq t\}$. Soit $\mathcal{A}_s = \mathcal{F}_{\tau(s)}$. Puisque $\zeta(t)$ croît, $\mathcal{A}_s$ est une famille croissantes de tribus. On déduit de (5.16) et des propriétés des temps d'arrêt (voir [14]) que si $t > s$, alors :

$$E\{\tilde{\eta}(\tau(t)) - \tilde{\eta}(\tau(s)) | \mathcal{A}_s\} = 0$$

$$E\{(\tilde{\eta}(\tau(s)) - \tilde{\eta}(\tau(t)))^2 | \mathcal{A}_s\} = \zeta(\tau(t)) - \zeta(\tau(s)) = t-s$$

On a aussi en vertu du lemme 5.6 que :

$$E\{[\eta(\tau(t)) - \eta(\tau(S))]^4 | \mathcal{A}_s\} = 3(t-s)^2 .$$

Donc le processus $\lambda(S) = \eta(\tau(S))$ est une martingale continue satisfaisant la relation :

$$E\{(\lambda(t) - \lambda(s))^2 | \mathcal{F}_s\} = t-s .$$

D'après le théorème de P. Lévy (théorème 6.1 du chapitre 1) $\lambda(s) = W(s)$ est un processus de Wiener qui ne dépend pas de la tribu $\mathcal{A}_0 = \mathcal{F}_0$ c'est à dire que les processus W et $\tilde{\zeta}$ sont indépendants.

Nous avons finalement l'égalité :

$$\tilde{\eta}(s) = W(\tilde{\zeta}(s)) = W(\int_{-\infty}^{\infty} £_\alpha(x,t)\, dK(x)$$

avec W, $\tilde{\zeta}$ indépendants, cette égalité signifie que :

$$\eta_{n(m)}(.) \xrightarrow{(\mathcal{L})} W\, (\int_{-\infty}^{\infty} £_\alpha(x,t)\, dK(x))$$

et puisque la loi limite ne dépend pas de la sous-suite $\{\eta_{n(m)}\}$ choisie, on a que le processus :

$$\eta_n \xrightarrow{(\mathcal{L})} W\, (\int_{-\infty}^{\infty} £_\alpha(x,t)\, dK(x).$$

Le théorème 5.1 est démontré.

Quant au théorème 5.2, on a :

$$\eta_n = \eta_n(1) \xrightarrow{(\mathcal{L})} W(\zeta(1)) = \frac{W(\zeta(1))}{\sqrt{\zeta(1)}} \quad \sqrt{\zeta(1)} .$$

Ici la variable aléatoire $\xi = \dfrac{W(\zeta(1))}{\sqrt{\zeta(1)}}$ a une loi normale standardisée et les variables $\xi, \sqrt{\zeta(1)}$ sont indépendantes. En effet, puisque les processus W, ζ sont indépendants, on a :

$$P\{\xi\in A, \sqrt{\zeta(1)}\in B\} = E\{P\{\sqrt{\zeta(1)}\in B \mid \zeta(1)\}\}$$

$$P\{\frac{W(\zeta(1)}{\sqrt{\zeta(1)}}\in A|\zeta(1)\}\} = \frac{1}{2\pi} P\{\sqrt{\zeta(1)}\in B\} \int_A e^{-\frac{u^2}{2}} du .$$

Le théorème 5.2 est démontré.

Remarque 5.1 : On peut affaiblir les restriction introduites dans les hypothèses du théorème 5.1 de la même manière que plus haut : considérer des fonctions f_n^c, utiliser la théorie des distributions etc...

4° Exemple : Considérons les sommes :

$$\eta_n = n^{-1/2}\sum_1^n \sin S_k \quad ; \quad S_k = \sum_1^k \xi_j .$$

En posant $f_n(x) = n^{-1/2} \sin x\sqrt{n}$ on a $\eta_n = \sum_1^n f_n(S_{nk})$.

Ici :

$$f_n(x) = \int_{\sqrt{-n}}^{\sqrt{n}} e^{i\lambda x} \quad \Psi_n(dx)$$

ou la mesure Ψ_n est concentrée en deux points $\pm\sqrt{n}$ et $\Psi_n(\pm\sqrt{n}) = \frac{1}{2\sqrt{n}}$.

La fonction :

$$k_n(x) = \frac{\sin^2 x\sqrt{n}}{n} + 2 \sin x\sqrt{n} \ \operatorname{Im} \ e^{ix\sqrt{n}} \varphi(1) \frac{1-\varphi^n(1)}{1-\varphi(1)} ,$$

et :

$$\lim_n K_n(x) = \lim_n \int_0^\infty nk_n(y)\, dy = (1 + 2 \operatorname{Re} \frac{\varphi(1)}{1-\varphi(1)}) .$$

$$. \lim_n \int_0^x \sin^2 x\sqrt{n}\, dx = \frac{1}{2} \operatorname{Re} \frac{1+\varphi(1)}{1-\varphi(1)} x .$$

On ne peut utiliser le théorème 5.1 parce que la mesure Ψ_n n'est pas absolument continue par rapport à la mesure de Lebesgue. Mais on voit facilement que la méthode de la démonstration marche bien si $|\varphi(1)| < 1$. Donc le processus $\eta_n(t)$ converge vers $W(\int_{-\infty}^\infty \pounds_\alpha(x,t)\, dK(x))$. Mais :

$$\int_{-\infty}^\infty \pounds_\alpha(x,t)\, dx = t$$

et le processus limite en loi est $W(\frac{1}{2} \operatorname{Re} \frac{1+\varphi(1)}{1-\varphi(1)} t)$ où W est un processus de Wiener, en particulier :

$$\lim_n P\{n^{-1/2} \sum_1^n \sin S_k < x\} = \frac{1}{\sigma\sqrt{2\pi}} \int_{-\infty}^x e^{-\frac{\sigma^2 y^2}{2}} dy,$$

$$\sigma = \frac{1}{2} \operatorname{Re} \frac{1+\varphi(1)}{1-\varphi(1)} .$$

Nous verrons plus tard que des résultats de ce type sont vrais presque sans restrictions sur la loi de ξ_j. La raison en est la périodicité de la fonction f (dans notre cas $f(x) = \sin x$).

VI - COMMENTAIRE

Ce chapitre correspond au chapitre 5 du livre [45] .

Paragraphe 1, le théorème 1.1 et la démonstration sont empruntés à Skorohod et Slobodenyuk [45] .

Paragraphes 2,3, la théorie générale des théorèmes limites pour les sommes $\sum f_n(S_{nk},\dots S_{n,k+r})$ a été développée par Skorohod et Slobodenyuk [42] - [47] . Ces auteurs utilisaient comme méthode de démonstration les théorèmes limites locaux avec terme résiduel.C'est pourquoi ils supposaient que les variables ξ_j avaient des moments d'ordre 4 ou 5 et une densité de probabilité appartenant à L_2 . Les résultats des paragraphes 2, 3 sont récents.

G.N. Sytaja a démontré quelques théorèmes limites pour les sommes $\sum_{k_1,k_2=1}^{n} f_n(S_{k_1}, S_{k_2})$ en utilisant les méthodes de Skorohod et Slobodenyuk [48] .

Notons que nos méthodes permettent de démontrer des théorèmes limites pour les sommes :

$$\sum_{k=1}^{n-r} f_n(S_{nk},\dots S_{n,k+r}) , \quad \sum_{k_1,\dots k_r=1}^{n} f_n(S_{nk_1},\dots S_{nk_r}) .$$

Les points limites de ces sommes ont respectivement les formes suivantes :

$$\frac{1}{(2\pi)^{r+1}} \int_{\mathbb{R}^{r+1}} \Psi(\lambda_0,\dots\lambda_r)d\lambda_0\dots d\lambda_r \int_0^1 \exp\{-\sum_0^r \lambda_i\, \xi_\alpha(t)\}\, dt,$$

$$\frac{1}{(2\pi)^r} \int_{\mathbb{R}^r} \Psi(\lambda_1,\dots\lambda_r)\, d\lambda_1 \dots d\lambda_r \int_0^1 \int_0^1 \exp\{-\sum_1^r \lambda_i\, \xi_\alpha(t_i)\}\, dt_1\dots dt_r$$

où Ψ représente $\lim n \widehat{f_n}$ ou $\lim n^r \widehat{f_n}$.

Les théorèmes limites démontrés, il reste encore une chose à faire : calculer les lois limites. Celles-ci sont les lois de fonctionnelles additives de processus stables. Il existe une littérature assez vaste sur ce sujet, spécialement sur les fonctionnelles des processus de Wiener, voir par exemple [5] , [21] , [29] , [30] ,[44] , [45] .

Voici quelques exemples (voir [4] , [45]).

1° Si :

$$\eta(t) = |W(t)| - \int_0^t \text{sign } W(s)\, dW(s) =$$

$$= \int_0^{W(t)} \text{signx}dx - \int_0^t \text{sign } W(x)dW(x) = \frac{1}{2}\int_{-\infty}^{\infty} \pounds_2(x;t)d \text{ sign } x = \pounds_2(0,t)$$

Alors :

$$P\{\eta(t) < x\} = \begin{cases} \frac{2}{\sqrt{2\pi t}} \int_0^x \exp\{-\frac{y^2}{2t}\}\, dy, & x > 0 \\ 0, & x < 0 \end{cases} \qquad (6.1)$$

2° Soit $r(x) = 0,\ x < 0,\ r(x) = 1,\ x > 0$ et

$$\eta(t) = \int_0^t r(W(S))\, dS = \int_0^{\infty} \pounds_2(x;t)\, dx$$

Alors :

$$P\{\eta(t) < x\} = \begin{cases} 0, & x \leqq 0 \\ \frac{2}{\pi} \text{ arc sin}\sqrt{\frac{x}{t}}, & 0 < x \leqq t, \\ 1, & x > t \end{cases}$$

3° Soit $\eta = \int_0^1 W^2(t)\, dt$ alors :

$$E \exp\{i\lambda\, \eta\} = \prod_1^{\infty} (1 - \frac{8i\lambda}{(2k+1)^2\pi^2})^{-1/2} = (\cos\sqrt{2i\lambda})^{-1/2} .$$

Dans [6] on peut trouver la démonstration de la formule :

$$P(x) = \left[\pi(1-\frac{1}{\alpha})\right]^{-1} \sum_1^{\infty} \frac{(-1)^{k-1}}{k!} x^{k-1} \sin(\pi(1-\frac{1}{\alpha})k) \frac{\Gamma(1+k(1-\frac{1}{\alpha}))}{\left[x\,\Gamma(1-\frac{1}{\alpha})\right]^k}$$

pour la densité de probabilité de $\pounds_\alpha$ (0,1. Ici :

$$x = \frac{1}{2\pi} \int_{-\infty}^{\infty} \exp\{-\frac{|\lambda|^\alpha}{\alpha}(1 + i\beta \text{ sign } \lambda \text{ tg } \frac{\pi\alpha}{2})\}\, d\lambda \ .$$

Ces lois s'appellent lois de Mittag-Leffler [38].

Paragraphe 4, résultat nouveau.

Paragraphe 5, le phénomène d'existence des théorèmes limites de deuxième type a été découvert par R.L. Dobruchin [12] qui a démontré un théorème limite pour les sommes.

$$\sum_1^n f(\zeta_k),\ \zeta_k = \sum_1^k \xi_j,\ P\{\xi_j = \pm 1\} = \frac{1}{2},\ \sum f(k) = 0.$$

La théorie générale appartient à Skorohod [42] , [45] mais il ne considérait pas la convergence des processus. Les résultats de ce paragraphe sont récents.

CHAPITRE III

I - INTRODUCTION

Soient $\xi_1, \xi_2, \ldots$ des variables aléatoires indépendantes à valeurs dans R^k. Toutes les ξ_j ont la même loi. Ces variables définissent une marche aléatoire dans R^k :

$$\zeta_0 = 0 \; , \; \zeta_1 = \zeta_0 + \xi_1 , \ldots , \zeta_k = \zeta_{k-1} + \xi_k , \ldots$$

Nous étudions dans ce chapitre des lois limites des sommes du type :

$$\sum_1^n f(\zeta_k), \quad n \longrightarrow \infty .$$

Nous nous intéressons exclusivement aux cas des marches récurrentes. On attend dans ce cas que, comme en règle générale,

$$\sum_1^n f(\zeta_k) \xrightarrow[n\to\infty]{} \infty$$

Donc pour avoir des théorèmes limites raisonnables,il faut centrer et normer ces sommes de la manière suivante :

$$\eta_n = B_n^{-1}\Big(\sum_1^n f(\zeta_k) - A_n\Big).$$

Donc nous avons les problèmes suivants :

1° Trouver B_n, A_n pour lesquels les η_n ont une loi limite propre.

2° Trouver cette loi limite.

Ce chapitre est consacré à l'investigation de ces problèmes. Une méthode générale consiste à utiliser les résultats du chapitre précédent. En effet :

$$\eta_n = \sum_1^n f_n(S_{nk})$$

où :

$$f_n(x) = B_n^{-1}\Big(f(xn^{1/\alpha}) - \frac{A_n}{n}\Big).$$

II - THEOREMES LIMITES DANS LE CAS $A_n = 0$

On suppose ici que les variables ξ_n appartiennent au domaine d'attraction normale d'une loi stable d'indice $\alpha > 1$ de fonction caractéristique :

$$\exp \Big\{- \frac{|\lambda|^\alpha}{\alpha} \Big(1 + i\beta \frac{\lambda}{|\lambda|} \operatorname{tg} \frac{\pi\alpha}{2}\Big)\Big\} .$$

On suppose aussi que $A_n = 0$. Or, soit :

$$\eta_n = B_n^{-1} \sum_1^n f(\zeta_k) = \sum_1^n f_n(S_{nk})$$

$$f_n(x) = B_n^{-1} f(xn^{1/\alpha}).$$

En utilisant les notations du chapitre précédent, on a :

$$\Psi_n(\lambda) = n \sum_{\lambda_k \in \Lambda} \hat{f}_n(\lambda + \lambda_k n^{1/\alpha}) = B_n^{-1} n^{1-1/\alpha} \sum_{\lambda_k \in \Lambda} \hat{f}(\lambda n^{-1/\alpha} + \lambda_k)$$

où comme toujours les λ_k sont les racines de $\varphi(\lambda) = 1$, et où $\varphi(\lambda)$ désigne la fonction caractéristique de ξ_j.

THEOREME 2.1 : *Pour que* $\Psi_n(\lambda)$ *converge quand* $n\to\infty$ *pour un choix convenable de* B_n *vers une fonction mesurable* $\Psi(\lambda)$ *en chaque point* λ *il suffit que dans un voisinage de zéro :*

$$\sum_{\lambda_k \in \Lambda} \hat{f}(\lambda + \lambda_k) = |\lambda|^\gamma (c_1 + c_2 \text{ sign } \lambda) h(\lambda) \tag{2.1}$$

où c_1, c_2 *sont des constantes et où* h *est une fonction à croissance lente au sens de Karamata. On peut prendre :*

$$B_n = n^{1-1/\alpha - \gamma/\alpha} \times (h(n^{-1/\alpha}))^{-1} \quad si \quad \alpha - 1 > \gamma .$$

Si :

$$\lim_{\lambda\to 0} \frac{\sum \hat{f}(\lambda + \lambda_k)}{\sum \hat{f}(\lambda(1+o(1)) + \lambda_k)} = 1 ,$$

la condition (2.1) est aussi nécessaire pour que $\Psi_n(\lambda) \to \Psi(\lambda)$.

Démonstration : (D'après [45]). Il suffit de montrer la deuxième partie du théorème, la première partie est évidente. Pour simplifier on raisonnera sur le cas où l'ensemble $\Lambda = \{0\}$. Soit $\Psi(\lambda) = \lim_n \Psi_n(\lambda)$. Montrons que pour tout $a > 0$ la fonction $\Psi(\lambda)$ satisfait l'équation fonctionnelle :

$$\Psi(\lambda)\, \Psi(a\mu) = \Psi(a\lambda)\, \Psi(\mu) , \quad \lambda,\mu \neq 0 \tag{2.2}$$

Si $\Psi(\lambda) = \Psi(\mu) = 0$. (2.2) est satisfaite. Soit $\Psi(\mu) \neq 0$. On a :

$$\frac{\Psi(\lambda)}{\Psi(\mu)} = \lim_n \frac{\hat{f}(\lambda n^{-1/\alpha})}{\hat{f}(\mu n^{-1/\alpha})} = \lim_n \frac{\hat{f}(\lambda a(na^\alpha)^{-1/\alpha})}{\hat{f}(\mu a(na^\alpha)^{-1/\alpha})} \tag{2.3}$$

D'où si $\Psi(\mu a) \neq 0$, on déduit (2.2). Si $\Psi(\mu a) = 0$, (2.3) donne $\Psi(\lambda a) = 0$, c'est-à-dire (2.2).

Montrons maintenant que toutes les solutions de (2.2) ont la forme suivante :

$$\Psi(\lambda) = |\lambda|^\gamma (\gamma_1 + \gamma_2 (\gamma_1 - \gamma_2) \text{ sign } \lambda)$$

où γ est une constante réelle et où γ_1, γ_2 sont des constantes complexes. En éliminant le cas $\Psi(\lambda) \equiv 0$, on peut supposer $\Psi(\lambda) \neq 0$, $\lambda > 0$. Soit $\lambda > 0$, $\mu > 0$, on déduit de (2.2) :

$$\frac{\Psi(a\lambda)}{\Psi(\lambda)} = \frac{\Psi(a\mu)}{\Psi(\mu)} = Q(a)$$

$$\Psi(a\lambda) = \Psi(\lambda)\ Q(a).$$

En supposant ici $\lambda=1$, on a $Q(a) = \Psi(a)\ (\Psi(1))^{-1}$.

Donc Q est une solution de :

$$Q(\lambda a) = Q(\lambda)\ Q(a)\ ,\quad \lambda\ ,\ a > 0\ .$$

Mais toutes les solutions mesurables de cette équation ont la forme $Q(\lambda) = |\lambda|^{\gamma}$.
Donc :

$$\Psi(\lambda) = \Psi(1)\ Q(\lambda) = \gamma_1\ |\lambda|^{\gamma},\ \lambda > 0\ .$$

On montre de la même manière que :

$$\Psi(\lambda) = \gamma_2\ |\lambda|^{\gamma_1},\quad \lambda < 0$$

et (2.2) donne $\gamma_1 = \gamma$. Donc :

$$\Psi(\lambda) = \frac{1}{2}\ |\lambda|^{\gamma}\ (\gamma_1 + \gamma_2 + (\gamma_1 - \gamma_2)\ \mathrm{sign}\ \lambda).$$

Si maintenant on suppose $\hat{f}(\lambda) = |\lambda|^{\gamma} h(\lambda)$ on a :

$$\lim_{n} \frac{h(\lambda n^{-1/\alpha})}{h(\mu n^{-1/\alpha})} = 1$$

qui signifie que h est une fonction à croissance lente.
Le théorème est démontré.

Pour utiliser les résultats du chapitre précédent, nous supposons pour simplifier que $\hat{f}(\lambda)$ a un support compact. Nous supposerons désormais que $\hat{f}(\lambda)$ satisfait (2.1). Il faut distinguer deux cas : $2\gamma \geq \alpha-1$ et $2\gamma < \alpha-1$. Si $2\gamma \geq \alpha-1$:

$$\int \frac{|\Psi\ (\lambda)|^2}{1+\ |\lambda|^{\alpha}}\ d\lambda\ = \infty$$

et on ne peut pas utiliser directement les théorèmes du chapitre II. Nous considérerons ce cas plus tard.

<u>THEOREME 2.2</u> : *Soit $\hat{f}(\lambda)$ à support compact et telle que pour tout $\varepsilon > 0$ $\hat{f} \in L_2(\varepsilon,\infty)$. Supposons (2.2) satisfaite avec $2\gamma < \alpha - 1$. Alors la loi limite de :*

$$\eta_n = n^{-1+1/\alpha + \gamma/\alpha}\ (h(n^{-1/\alpha}))^{-1}\ \sum_{1}^{n}\ f(\zeta_k) \tag{2.3}$$

existe et coïncide avec la loi de :

$$\eta = \frac{1}{2\pi}\ \int_{-\infty}^{\infty}\theta(\lambda)d\lambda\ \int_0^1 e^{-i\xi_\alpha(t)\lambda}\ dt\ =\ \int_{-\infty}^{\infty} \mathfrak{L}_\alpha(x)\ \hat{\theta}(x)\ dx \tag{2.4}$$

où :

$$\theta(\lambda) = |\lambda|^{\gamma}(c_1 + c_2 \text{ sign } \lambda)$$

et :

$$\widehat{\theta}(x) = \frac{1}{2\pi} \int_{-\infty}^{\infty} e^{-ix\lambda} \theta(\lambda) d\lambda \tag{2.5}$$

Ce théorème est un corollaire immédiat du théorème 2.1 du chapitre 2. Bien sûr, il faut comprendre (2.4) et (2.5) du point de vue de la théorie des distributions. On peut calculer la partie droite de (2.4) plus précisément.

LEMME 2.1 : *On a pour la transformée de Fourier de* $|\lambda|^{\gamma}$ *et de* $|\lambda|^{\gamma}$ *sign* γ *les formules suivantes (du point de vue de la théorie des distributions) :*

$$\frac{1}{2\pi} \int_{-\infty}^{\infty} e^{-ix\lambda} |\lambda|^{\gamma} d\lambda = \frac{1}{2\pi} C(\gamma) |x|^{-\gamma-1}, \quad \gamma \neq -1, -3, \ldots$$

$$\frac{1}{2\pi} \int_{-\infty}^{\infty} e^{-ix\lambda} |\lambda|^{\gamma} \text{ sign}\lambda \, d\lambda = - \frac{i}{2\pi} D(\gamma) |x|^{-\gamma-1} \text{ sign } x, \quad \gamma \neq -2, -4, \ldots$$

$$\frac{1}{2\pi} \int_{-\infty}^{\infty} e^{-ix\lambda} |\lambda|^{\gamma} \text{ sign}\lambda \, d\lambda = \frac{i}{2\pi} \left[-d_0^{(-\gamma)} x^{-\gamma-1} + d_{-1}^{(-\gamma)} x^{-\gamma-1} \ln |x| \right],$$
$$\gamma = -2, -4, \ldots$$

$$\frac{1}{2\pi} \int_{-\infty}^{\infty} e^{-ix\lambda} |\lambda|^{\gamma} d\lambda = \frac{1}{2\pi} \left[c_0^{(-\gamma)} x^{-(\gamma+1)} - c_{-1}^{(-\gamma)} x^{-(\gamma+1)} \ln |x| \right],$$
$$\gamma = -1, -3, \ldots$$

Ici :

$$C(\gamma) = -2 \sin \frac{\gamma\pi}{2} \Gamma(\gamma+1), \quad D(\gamma) = 2 \cos \frac{\gamma\pi}{2} \Gamma(\gamma+1),$$

$$c_0^{(n)} = 2 \text{ Re} \left\{ \frac{i^{n-1}}{(n-1)!} \left[1 + \frac{1}{2} + \ldots + \frac{1}{n-1} + \Gamma'(1) + \frac{i\pi}{2} \right] \right\},$$

$$d_0^{(n)} = 2 \text{ Im} \left\{ \frac{i^{n-1}}{(n-1)!} \left[1 + \frac{1}{2} + \ldots + \frac{1}{n-1} + \Gamma'(1) + \frac{i\pi}{2} \right] \right\},$$

$$c_{-1}^{(n)} = \frac{2(-1)^{n-1}}{(n-1)!} \cos (n-1) \frac{\pi}{2}, \quad d_{-1}^{(n)} = \frac{2(-1)^{n}}{(n-1)!} \sin (n-1) \frac{\pi}{2} .$$

On peut trouver la démonstration de ces formules par exemple dans [26], table de transformées de Fourier. Ces formules donnent la possibilité de réécrire le résultat du théorème d'une manière plus concrète. Par exemple, soit $0 < \gamma < \frac{\alpha-1}{2}$. Les distributions $|x|^{-\gamma-1}$ et $|x|^{-\gamma-1}$ sign x sont définies sur l'espace $\mathcal{D}$ des fonctions φ indéfiniment dérivables et à support compact par les formules :

$$(|x|^{-\gamma-1}, \varphi) = \int_{-\infty}^{\infty} |x|^{-\gamma-1} \varphi(x) \, dx =$$
$$= \int_{-1}^{1} |x|^{-\gamma-1} (\varphi(x) - \varphi(0)) dx + \int_{|x|>1} |x|^{-\gamma-1} \varphi(x) \, dx ;$$

$$(|x|^{-\gamma-1} \text{ sign } x, \varphi) = \frac{1}{2} \int_{-\infty}^{\infty} |x|^{-\gamma-1} \text{ sign } x \, (\varphi(x) - \varphi(-x))$$

Il est évident que ces distributions peuvent être prolongées à toutes les fonctions φ qui satisfont une condition de Hölder d'ordre supérieur à γ . Dans ce cas la loi limite est donc la loi de :

$$\eta = \int_{-\infty}^{\infty} f_\alpha(x)\,\widehat{\theta}(x)dx = -\frac{1}{\pi}\Big\{C_1 \sin\frac{\gamma\pi}{2} \,.\, \Gamma(\gamma+1) \times$$

$$\times \Big(\int_{-1}^{1}|x|^{-\gamma-1}(f_\alpha(x) - f_\alpha(0))dx + \int_{|x|>1}|x|^{-\gamma-1} f_\alpha(x)dx\Big) +$$

$$+ iC_2 \cos\frac{\gamma\pi}{2}\,\Gamma(\gamma+1)\int_{-\infty}^{\infty}|x|^{-\gamma-1} \operatorname{sign} x \,\frac{f_\alpha(x)-f_\alpha(-x)}{2}\, dx\Big\}.$$

Si $\gamma = 0$,

$$\eta = C_1 \int_{-\infty}^{\infty}\delta(x)\, f_\alpha(x)\, dx + \frac{C_2}{i\pi}\int_{-\infty}^{\infty}\frac{f_\alpha(x)}{x}\, dx =$$

$$= C_1\, f_\alpha(0) - iC_2\, \tilde{f}_\alpha(0). \tag{2.6}$$

Ici $\tilde{g}$ désigne la transformée de Hilbert d'une fonction

$$\tilde{g}(x) = \frac{1}{\pi}\int_{-\infty}^{\infty}\frac{g(\lambda)}{\lambda - x}\, d\lambda \quad .$$

Nous allons étudier ce cas plus en détail dans le paragraphe suivant.

Enfin, si $\gamma < 0$ et si γ n'est pas un entier, on a :

$$\eta = -\frac{1}{\pi}\ \Big(C_1 \sin\frac{\gamma\pi}{2}\,\Gamma(\gamma+1)\int_{-\infty}^{\infty}|x|^{-\gamma-1} f_\alpha(x)\, dx +$$

$$+ iC_2 \cos\frac{\gamma\pi}{2}\,\Gamma(\gamma+1)\int_{-\infty}^{\infty}|x|^{-\gamma-1} \operatorname{sign} x\, f_\alpha(x)\, dx\Big) \tag{2.7}$$

et les intégrales sont bien définies parce que f_α est une fonction à support compact.

Soit $\eta_n(t)$ la ligne brisée de sommets :

$$\Big(\frac{k}{n}\,, n^{-1+1/\alpha+\gamma/\alpha}(h(n^{-1/\alpha}))^{-1}\sum_{1}^{k} f(\zeta_j)\Big).$$

Il résulte du paragraphe 4, chapitre II, que sous les hypothèses du théorème 2.2 $\eta_n(t)$ converge en loi dans l'espace $C(0,1)$ vers :

$$\eta(t) = \int_{-\infty}^{\infty}\widehat{\theta}(x)\, f_\alpha(x;t)dx\ .$$

Nous avons supposé que $\widehat{f}$ est une fonction à support compact. Les résultats du paragraphe 3, chapitre II, donnent des possibilité de traiter aussi des cas où $\widehat{f}$ n'est pas à support compact. Bien sûr, les lois limites sont les mêmes mais il faut imposer quelques conditions à la fonction caractéristique.

Considérons d'abord le cas $\Lambda \neq \{0\}$. Dans ce cas toutes les valeurs possibles de ξ_n appartiennent à une progression arithmétique. On peut supposer sans perdre de généralité que ces valeurs possibles sont des nombres entiers et que le pas maximal est égal à 1. Donc $\Lambda = \{2\pi k,\ k=0, \pm 1, \ldots\}$. Les sommes $\sum f(\zeta_k)$ sont définies par les valeurs $f(n)$ de la fonction f aux points $n = 0, \pm 1, \ldots$ Les valeurs $f(z)$, z non entier, peuvent être arbitraires et on peut utiliser convenablement cette liberté.

Posons :

$$f(x) = \frac{1}{\pi} \sum_{n=-\infty}^{\infty} f(n) \frac{\sin(x-n)\pi}{x-n} .$$

Il est évident que f prend aux points entiers les valeurs nécessaires. La transformée de Fourier de f est :

$$\hat{f}(\lambda) = g(\lambda) \sum_{n=-\infty}^{\infty} f(n)e^{in\lambda} ,$$

où :

$$g(\lambda) = \frac{1}{\pi} \int_{-\infty}^{\infty} e^{i\lambda x} \frac{\sin \pi x}{x} dx = \begin{cases} 1 , & |x| < \pi \\ 0 , & |x| > \pi \end{cases}$$

Donc :

$$\sum \hat{f}(\lambda + \lambda_k) = \sum_{n=-\infty}^{\infty} f(n) e^{in\lambda} .$$

Le théorème 2.2 donne alors le résultat suivant :

<u>THEOREME 2.3</u> : *Supposons que* ξ_j *prend ses valeurs dans une progression arithmétique* $\{kd, k=0, \pm 1, \ldots\}$ *et que d est le pas maximal. Si :*

1° $\sum_k |f(kd)| < \infty$;

2° $\sum_k f(kd) e^{ikd\lambda} = |\lambda|^{\gamma}(C_1 + C_2 \text{ sign } \lambda) h(\lambda)$

dans un voisinage de zéro ; alors les processus $\eta_n(t)$ *engendrés par les sommes* $n^{-1+1/\alpha+\gamma/\alpha}(h(n^{-1/\alpha}))^{-1} \sum_1^k f(\zeta_j)$ *convergent en loi vers :*

$$\eta(t) = \int_{-\infty}^{\infty} \hat{\theta}(x) \, \pounds_{\alpha}(x;t) \, dx .$$

Considérons maintenant le cas où $\overline{\lim_{t\to\infty}} |\varphi(t)| < 1$. Supposons qu'on puisse écrire f(x) comme :

$$f(x) = f_1(x) + f_2(x)$$

où la fonction $f_1(x)$ satisfait les hypothèses du théorème 2.2 et la fonction f_2 a comme transformée de Fourier une fonction $\hat{f}_2$ à support en dehors d'un intervalle $[-\varepsilon,\varepsilon]$. Définissons B_n comme plus haut :

$$B_n = n^{1-1/\alpha-\gamma/\alpha} h(n^{-1/\alpha}) .$$

On a :

$$\eta_n = B_n^{-1} \sum_1^n f(\zeta_k) = B_n^{-1} \sum_1^n f_1(\zeta_k) + B_n^{-1} \sum_1^n f_2(\zeta_k) = \eta_{n1} + \eta_{n2}.$$

En vertu du théorème 2.2, η_{n1} converge en loi vers $\eta = \int_{-\infty}^{\infty} \pounds_{\alpha}(x) \hat{\theta}(x) dx$. Tachons de trouver des conditions sous lesquelles $\eta_{n2} \to 0$. On a :

$$E|\eta_n|^2 \leq 2 B_n^{-2} \iint_{\substack{|\lambda|\geq\varepsilon \\ |\mu|\geq\varepsilon}} \hat{f}_2(\lambda) \hat{f}_2(\mu) \sum_{1\leq k\leq \ell\leq n} \varphi^k(\lambda-\mu) \varphi^{\ell-k} (-\mu) d\lambda \, d\mu .$$

Donc si on suppose que $\hat{f}_2 \in L_2 \cap L_\infty$ alors :

$$E|\eta_n|^2 \leq 2B_n^{-2} B\int_{-\infty}^{\infty} \frac{1-e^{-n|\mu|^\alpha}}{|\mu|^\alpha} d \leq B(h(n^{-1/\alpha}))^{-1} n^{-\frac{2\gamma}{\alpha}+\frac{1}{\alpha}-1} = o(1),$$

puisque $2\gamma < \alpha - 1$.

Soit maintenant $\varphi \in L_p$ pour quelque $p > 0$. Soit m un entier plus grand que p. Soit $\hat{f} \in L_2$. On a :

$$E|B_n^{-1} \sum_{k=m}^{n} f_2(\zeta_k)|^2 \leq$$

$$\leq 2 B_n^{-2} \iint_{\substack{|\lambda| > \varepsilon \\ |\mu| > \varepsilon}} \hat{f}(\lambda) \overline{\hat{f}(\mu)} \sum_{m \leq k \leq \ell \leq n} \varphi^k(\lambda-\mu)\, \varphi^{\ell-k}(-\mu) d\lambda d\mu \leq$$

$$\leq B.B_n^{-2} \int_{-\infty}^{\infty}\int_{-\infty}^{\infty} \hat{f}(\lambda-\mu)\, \hat{f}(\mu)\, |\sum_m^n |\varphi^k(\mu) d\mu \leq$$

$$\leq B.B_n^{-2}\, ||\hat{f}||_2^2 \sum_m^n \int_{-\infty}^{\infty} |\varphi(\mu)|^k d\mu .$$

Mais on peut choisir $\delta > 0$, $c > 0$ de sorte que :

$$\int_{-\infty}^{\infty} |\varphi(\mu)|^k d\mu \int_{|\mu| \leq \delta} |\varphi(\mu)|^k + (\sup_{|\mu|>\delta} |\varphi(\mu)\,)^{k-m} \int_{-\infty}^{\infty} |\varphi(\mu)|^m d\mu \leq$$

$$\leq \int_{-\infty}^{\infty} e^{-kc|\mu|^\alpha} d\mu + \sup_{|\mu|>\delta} |\varphi(\mu)|^{k-m} \int_{-\infty}^{\infty} |\varphi(\mu)|^m d\mu.$$

On déduit de ceci que :

$$E|B_n^{-1} \sum_m^n f_2(\zeta_k)|^2 \leq B\, B_n^{-2} ||f_2||^2 (1 + ||\varphi||_m^m)\, n^{1-1/\alpha} = o(1).$$

Il est évident que :

$$B_n^{-1} \sum_1^m f(\xi_k) = o(1)$$

en probabilité. Donc nous avons montré le résultat suivant (comparer au théorème 2.3).

<u>THEOREME 2.4</u> : *Supposons que la fonction caractéristique* φ *de* ξ_n *appartient à* $L_p(R^1)$. *Si* $f \in L_2$ *et* :

$$\int_{-\infty}^{\infty} e^{i\lambda x} f(x) dx = |\lambda|^\gamma (c_1 + c_2 \operatorname{sign} \lambda)\, h(\lambda)$$

dans un voisinage de zéro, alors η_n *converge en loi vers* $\int_{-\infty}^{\infty} \hat{\theta}(x)\, \pounds_\alpha(x) dx$.

On peut montrer aussi que sous les hypothèses du théorème, $\eta_n(t)$ converge vers $\int_{-\infty}^{\infty} \hat{\theta}(x)\, \pounds_\alpha(x;t) dx$.

III - THEOREME LIMITES POUR DES FONCTIONS SOMMABLES

Considérons $\sum_1^n f(\zeta_k)$ où $f \in L_1 \cap L_2$. Dans ce cas la transformée de Fourier

$$\hat{f}(\lambda) = \int_{-\infty}^{\infty} e^{i\lambda x} f(x)$$

est une fonction continue. Nous supposons que :

$$\hat{f}(\lambda) \neq 0 .$$

Dans ce cas, il faut choisir $B_n \sim \sqrt{n}$.

Notons $\eta_n(t)$ le processus dans $C[0,1]$ engendré par $\frac{1}{\sqrt{n}} \sum_1^k f(\zeta_k)$. Il résulte du théorème 2.2 que si f est à support compact alors $\eta_n(t)$ converge en loi vers :

$$\eta(t) = f(0)\ \mathfrak{L}_\alpha(0;t) .$$

Nous allons considérer ici des cas où $\hat{f}$ n'est pas une fonction à support compact.

THEOREME 3.1 : *Soit* $|f(x)| \leq B(1+|x|^{1+c})$, $c > 0$. *Supposons* $\mathfrak{f}$ *intégrable au sens de Riemann, alors* $\eta_n(t)$ *converge en loi vers* $\eta(t)$.

Démonstration : En vertu du théorème 2.2, il suffit de considérer le cas $\Lambda = \{0\}$.

1 - Soit d'abord f(x) une fonction continue. Soit $\varepsilon > 0$. On peut construire une fonction g (x) à support compact telle que :

$$\int_{-\infty}^{\infty} f(x) - g_\varepsilon(x)|dx < \varepsilon , \quad \sup_x |f(x) - g_\varepsilon(x)| < \varepsilon.$$

On peut, par exemple, prendre pour g_ε l'intégrale de Fejer :

$$g_\varepsilon(x) = \frac{2}{\pi T}\int_{-\infty}^{\infty} \frac{\sin^2 \frac{T}{2}(x-y)}{(x-y)^2} f(y)dy$$

où $T = T(\varepsilon)$, $T(\varepsilon) \underset{\varepsilon\to 0}{\longrightarrow} \infty$.

La fonction :

$$g(x) = \sum n^{-1-c/2} \frac{\sin^2\pi(n-x)}{(n-x)^2}$$

est une fonction à support compact et :

$$c_1(1+|x|)^{-1-\frac{c}{2}} \leqq g(x) \leqq c_2\ (1+|x|)^{-1-\frac{c}{2}}$$

où c_1, c_2 sont des constantes positives.

Donc si $\delta > 0$ est donné on peut trouver $\varepsilon > 0$ de la manière suivante :

$$f(x) + \delta > f_\delta^+(x) = g_c(x) + \varepsilon g(x) > f(x),$$

$$f(x) - \delta < f_\delta^-(x) = g_\varepsilon(x) - \varepsilon g(x) < f(x),$$

$$\int_{-\infty}^{\infty} |f(x) - f_\delta^+(x)|\ dx \leq \delta, \quad \int_{-\infty}^{\infty} f(x) - f_\delta^-(x)\ dx \leq \delta.$$

Notons $\eta_n^-(t)$, $\eta_n^+(t)$ les processus dans $C[0,1]$ engendrés par f_δ^-, f_δ^+. Alors :

$$\eta_n^-(t) \leq \eta(t) \leq \eta_n^+(t)$$

et les processus η_n^-, η_n^+ convergent en loi vers :

$$\mathfrak{L}_\alpha(0;t) \int_{-\infty}^{\infty} f_\delta^-(x)dx \quad , \ \mathfrak{L}_\alpha(0;t) \int_{-\infty}^{\infty} f_\delta^+(x)dx .$$

Puisque :

$$\int_{-\infty}^{\infty} [f_\delta^+ (x) - f_\delta^-(x) - f_\delta^-(x)]\ dx < 2\delta$$

et que δ est arbitrairement petit, on en déduit que $\eta_n(t)$ converge en loi vers $\eta(t)$.

2 - Soit f une fonction continue par morceaux. Soit $\delta > 0$. On peut trouver deux fonctions continues f_δ^+, f_δ^- qui satisfont les hypothèses du théorème et pour lesquelles

$$f_\delta^-(x) \leq f(x) \leq f_\delta^+(x) \;;$$

$$\int_{-\infty}^{\infty}\left[f_\delta^+(x) - f_\delta^-(x)\right] dx \leq \delta .$$

En vertu du point 1 de la démonstration les processus η_n^+, η_n^- engendrés par f_δ^+, f_δ^- convergent en loi vers :

$$\pounds_\alpha(0;t) \int_{-\infty}^{\infty} f_\delta^+(x)dx \;,\; \pounds_\alpha(0;t) \int_{-\infty}^{\infty} f_\delta^-(x)dx.$$

Donc η_n converge en loi vers η .

3 - Soit enfin f une fonction intégrable au sens de Riemann. Soit $\delta > 0$. On peut trouver deux fonctions f_δ^- , f_δ^+ qui satisfont les hypothèses du théorème, se composent d'un nombre fini de morceaux continus et pour lesquelles :

$$f_\delta^-(x) \leq f(x) \leq f_\delta^+(x),$$

$$\int_{-\infty}^{\infty}\left[f_\delta^+(x) - f_\delta^-(x)\right] dx \leq \delta.$$

De ceci et du point 2, le théorème résulte.

Exemple : Soit :

$$f(x) = \begin{cases} 1 & , \; a < x < b \\ \frac{1}{2} & , \; x = a,b \\ 0 & , \; x \notin [a,b] \end{cases}$$

Soit d'abord $\Lambda = \{0\}$. Dans ce cas la somme $\sum_1^n f(\zeta_k)$ est le nombre de visites d'une marche aléatoire dans l'intervalle $[a,b]$. La somme normée $\eta_n = \frac{1}{\sqrt{n}} \sum_1^n f(\zeta_k)$ convergz en loi vers $\pounds_\alpha(0).(b-a)$. Si $\Lambda = \{\lambda_k\} \neq \{0\}$, les valeurs des ξ_n appartiennent à une porgression arithmétique $\{kd, k = 0, \pm 1,\ldots\}$ et η_n converge en loi vers :

$$f_\alpha(0) \left\{ \sum_{a<kd<b} f(kd) + \frac{1}{2} \sum_{kd=a,b} f(kd)\right\} .$$

IV - CAS $\gamma > \frac{\alpha-1}{2}$

Considérons à nouveau $\sum_1^n f(\zeta_k)$ et supposons maintenant que $\sum_k \hat{f}(\lambda + \lambda_k)$ a une racine d'ordre supérieur à $\frac{1}{2}(\alpha-1)$ au point $\lambda = 0$. La méthode du paragraphe 2 ne marche plus. On utilise dans ce cas le résultat du paragraphe 5, chapitre II.

THEOREME 4.1 : *Supposons que la fonction* f *a une transformée de Fourier* $\hat{f}$ *à support compact et que :*

$$\int_{-\infty}^{\infty} \Big|\sum_k f(\lambda + \lambda_k)\Big|^2 \, |\lambda|^{-\alpha} d\lambda < \infty \tag{4.1}$$

Alors les processus $\eta_n(t)$ *en* $C(0,1)$ *engendrés par* $n^{-\frac{\alpha-1}{2\alpha}} \sum_1^k f(\zeta_j)$ *convergent en loi vers le processus :*

$$w(\pounds_\alpha(0;t).b)^{1/2} \;,\; b = \int_{-\infty}^{\infty} \sum_k f(\lambda+\lambda_k)\, \overline{f(\lambda)}\, \frac{1+\varphi(\lambda)}{1-\varphi(\lambda)}\, d\lambda,$$

où le processus de Wiener w *et le processus* $\pounds_\alpha(0;t)$ *sont indépendants. En particulier,* $\eta_n = n^{-\frac{\alpha-1}{2\alpha}} \sum_1^n f(\zeta_j)$ *converge en loi vers* $\xi\,(\pounds_\alpha(0)b)^{1/2}$ *où* $\xi \in \mathcal{N}(0,1)$ *et* ξ , $\pounds_\alpha(0)$ *sont indépendants.*

Démonstration : Montrons que toutes les hypothèses du théorème 5.1, chapitre II, sont réalisées. On a, en utilisant les notations de ce théorème :

$$\Psi_n(\lambda) = n^{-1/2(1-1/\alpha)-1/\alpha}\, n \sum_k \hat{f}(\lambda n^{-1/\alpha} + \lambda_k).$$

Donc :

$$\sup_n \int_{-\infty}^{\infty} \frac{|\Psi_n(\lambda)|^2}{1+|\lambda|^\alpha}\, d\lambda = \sup_n\, n \int_{-\infty}^{\infty} \frac{|\sum \hat{f}(\lambda+\lambda_k)|^2}{1+n|\lambda|^\alpha}\, d\lambda \leqq$$

$$\leqq \int_{-\infty}^{\infty} |\sum \hat{f}(\lambda+\lambda_k)|^2\, |\lambda|^{-\alpha} d\lambda < \infty \tag{4.2}$$

De la même manière, pour chaque $\varepsilon > 0$

$$\int_{-\infty}^{\infty} \frac{|\Psi_n(\lambda)|}{1+|\lambda|^\alpha}\, d\lambda = n^{1/2 + 1/2\alpha} \int_{-\infty}^{\infty} \frac{|\sum \hat{f}(\lambda+\lambda_k)|^2}{1+n|\lambda|^\alpha}\, d\lambda \leqq \tag{4.3}$$

$$\leqq B_\varepsilon\, n^{\frac{1}{2\alpha}-\frac{1}{2}} + n^{\frac{1}{2}+\frac{1}{2\alpha}} \left(\int_{-\infty}^{\infty} \frac{d\lambda}{1+|\lambda|^\alpha n}\right)^{1/2} \left(\int_{-\infty}^{\varepsilon} \frac{|\sum \hat{f}(\lambda+\lambda_k)|^2}{n|\lambda|^\alpha}\, d\lambda\right)^{1/2} \leqq$$

$$\leqq B_\varepsilon\, n^{\frac{1}{2\alpha}-\frac{1}{2}} + B \int_{-\varepsilon}^{\varepsilon} \frac{|\sum \hat{f}(\lambda+\lambda_k)|^2}{|\lambda|^\alpha}\, d\lambda .$$

Donc la condition 3 du théorème 5.1 est vérifiée.

Par définition de la fonction k_n, on a :

$$\widehat{nk_n}(\lambda) = n \int_{-\infty}^{\infty} e^{i\lambda x}\, n^{\frac{1}{\alpha}-1} f^2(xn^{1/\alpha})\, dx +$$

$$+ 2n^{1/\alpha} \int_{-\infty}^{\infty} e^{i\lambda x} f(xn^{1/\alpha}) \sum_1^n E\, f(n^{1/\alpha}(x+S_{nk}))\, dx =$$

$$= \int_{-\infty}^{\infty} \hat{f}(\mu+\lambda n^{1/\alpha})\, \overline{\hat{f}(\mu)}\, d\mu + 2\int_{-\infty}^{\infty} \hat{f}(\mu+\lambda n^{1/\alpha})\, \overline{\hat{f}(\mu)}\, \frac{1-\varphi^n(\mu)}{1-\varphi(\mu)}\, d\mu .$$

D'où :

$$b_n(\lambda) = n \sum \hat{k}_n(\lambda + n^{1/\alpha}\lambda_k) =$$

$$= \int_{-\infty}^{\infty} \hat{f}(\mu) \sum_k \hat{f}(\mu + \lambda_k + n^{1/\alpha}\lambda)\, \frac{1+\varphi(-\mu)-2(\varphi(\mu))^n}{1-\varphi(\mu)}\, d\mu .$$

Soit d'abord $\Lambda = \{0\}$. Nous allons montrer que :

$$\int_{-\infty}^{\infty} \frac{|b_n(\lambda) - b|^2}{1+|\lambda|^\alpha}\, d\lambda \longrightarrow 0 \tag{4.4}$$

Il suffit de montrer que :

$$\int_{-\infty}^{\infty} \frac{|b_n(\lambda) - b_n(0)|^2 \, d\lambda}{1 + |\lambda|^\alpha} \quad \lambda \longrightarrow 0 \tag{4.5}$$

Soit $\varepsilon > 0$. On a :

$$|b_n(\lambda) - b_n(0)|^2 \leq 2 \left\{ \left[2 \int_{|\mu|>\varepsilon} |\hat{f}(\mu)| \, |\hat{f}(\mu+n^{-1/\alpha}\lambda) - \hat{f}(\mu)| \frac{d\mu}{|1-\varphi(\mu)|} \right]^2 + \right.$$

$$\left. + B \int_{-\varepsilon}^{\varepsilon} \frac{|\hat{f}(\mu)|^2}{|\mu|^\alpha} d\mu \int_{-\varepsilon}^{\varepsilon} |\hat{f}(\mu+\lambda n^{-1/\alpha}) - \hat{f}(\mu)|^2 \frac{(1-e^{-cn|\mu|^\alpha})^2}{|\mu|^\alpha} d\mu \right\}$$

La première intégrale à droite est plus petite que :

$$B_\varepsilon ||\hat{f}||_2 \int_{-\infty}^{\infty} |\hat{f}(\mu + n^{-1/\alpha}\lambda) - \hat{f}(\mu)|^2 \, d\mu$$

et tend vers zéro uniformément par rapport à λ . Donc :

$$\int_{-\infty}^{\infty} \frac{|b_n(0) - b_n(\lambda)|^2}{1 + |\lambda|^\alpha} d\lambda \leq$$

$$\leq B \int_{-\varepsilon}^{\varepsilon} \frac{|\hat{f}(\mu)|^2}{|\mu|^\alpha} d\mu \int_{-\infty}^{\infty} \frac{d\lambda}{1+|\lambda|^\alpha} \int_{-\infty}^{\infty} \frac{|\hat{f}(\mu+n^{-1/\alpha}\lambda)|^2}{|\mu|^\alpha} (1-e^{-cn|\mu|^\alpha}) d\mu +$$

$$+ B \int_{-\varepsilon}^{\varepsilon} \frac{|\hat{f}(\mu)|^2}{|\mu|^\alpha} + B_\varepsilon . \, O(1).$$

Or, il suffit de montrer que l'intégrale :

$$\int_{-\infty}^{\infty} \frac{d\lambda}{1+|\lambda|^\alpha} \int_{-\infty}^{\infty} \frac{|\hat{f}(\mu+n^{-1/\alpha}\lambda)|^2}{|\mu|^\alpha} (1-e^{-cn|\mu|^\alpha}) \, d\mu$$

est bornée. Mais cette intégrale est plus petite que :

$$n^{1-1/\alpha} \int_{-\infty}^{\infty} \frac{d\lambda}{1+|\lambda|^\alpha} \int_{-\infty}^{\infty} \frac{|\hat{f}(vn^{-1/\alpha})|^2}{|v-\lambda|^\alpha} (1 - e^{-c|v-\lambda|^\alpha}) \, dv =$$

$$= n^{1-1/\alpha} \int_{-\infty}^{\infty} |\hat{f}(vn^{-1/\alpha})|^2 \, dv \int_{-\infty}^{\infty} \frac{(1-e^{-c|v-\lambda|^\alpha})}{|v-\lambda|^\alpha (1+|\lambda|^\alpha)} d\lambda \leq$$

$$\leq Bn^{1-1/\alpha} \int_{-\infty}^{\infty} \frac{|\hat{f}(vn^{-1/\alpha})|^2}{|v|^\alpha} dv \leq B \int_{-\infty}^{\infty} \frac{|\hat{f}(v)|^2}{|v|^\alpha} dv < \infty .$$

La relation (4.5) et donc (4.4) est démontrée. On déduit de (4.4) que le processus engendré par $\sum_1^k g_n(S_{nj})$:

$$g_n(x) = n^{\frac{\alpha-1}{\alpha}} \left[f^2(xn^{1/\alpha}) + 2 \sum_{j=1}^{n} f(xn^{1/\alpha}) \, Ef((x+S_{nj})n^{1/\alpha}) \right]$$

converge vers $\mathcal{L}_\alpha(0;t)$. Donc toutes les hypothèses du théorème 5.1, chapitre II, sont vérifiées et notre théorème est démontré dans le cas $\Lambda = \{0\}$.

Soit maintenant $\Lambda = \{\lambda_k, k=0, \pm 1, \ldots\} \neq \{0\}$. Dans ce cas $\lambda_k = k\lambda$ et ces nombres sont aussi les périodes de φ . En vertu de cela :

$$b_n(\lambda) = \int_{-1/2\,\lambda_1}^{1/2\,\lambda_1} \sum_k \overline{\hat{f}(\mu+\lambda_k)} \sum_j f(\mu+\lambda_j+n^{-1/\alpha}\lambda) \times \frac{1+\varphi(-\mu) - 2^{\,n}(-\mu)}{1-\varphi(-\mu)}\, d\mu$$

et comme plus haut on peut montrer que :

$$\int_{-\infty}^{\infty} \frac{|b_n(\lambda) - b|^2}{1 + |\lambda|^\alpha}\, d\lambda \longrightarrow \infty .$$

La démonstration est terminée.

THEOREME 4.2 : *Supposons que toutes les valeurs possibles de* ξ_j *appartiennent à une progression arithmétique* $\{kd, k=0, \pm 1, \ldots\}$ *et que* d *est le pas maximal. Si :*

1° $$\sum_k |f(kd)|^2 < \infty ;$$

2° $$\int_{-\pi/d}^{\pi/d} \frac{|\Psi_1(\lambda)|^2}{|\lambda|^\alpha}\, d\lambda < \infty$$

$\Psi_1(\lambda) = \sum f(kd)\, e^{ikd\lambda}$, *alors les processus engendrés par* $\{n^{-\frac{\alpha-1}{2\alpha}} \sum_1^k f(\zeta_j)\}$ *convergent en loi vers* $w(\pounds_\alpha(0;t).b)$,

$$b = \int_{-\pi/d}^{\pi/d} |\Psi_1(\lambda)|^2\, \frac{1+\varphi(\lambda)}{1-\varphi(\lambda)}\, d\lambda .$$

Démonstration : Elle est calquée sur celle du théorème 2.3. On réduit le problème au cas où $\hat{f}$ est à support compact.

THEOREME 4.3 : *Supposons que la fonction caractéristique* $\varphi \in L_p(-\infty,\infty)$ *pour un* $p > 0$. *Si :*

$$\int_{-\infty}^{\infty} \left(\frac{|\hat{f}(\lambda)|^2}{|\lambda|^\alpha} + |\hat{f}(\lambda)|^2 \right) d\lambda < \infty,$$

alors $\eta_n = n^{-\frac{\alpha-1}{2\alpha}} \sum_1^n f(\zeta_k)$ *converge en loi vers* $\eta = \xi\sqrt{\pounds_\alpha(0)\, b}$, *où* $\xi \in \mathcal{N}(0,1)$,

$$b = \int_{-\infty}^{\infty} |\hat{f}(\lambda)|^2\, \frac{1+\varphi(\lambda)}{1-\varphi(\lambda)}\, d\lambda .$$

Démonstration : Soit $T > 0$. Posons :

$$f_1(x) = \frac{1}{2\pi} \int_{-T}^{T} e^{-i\lambda x}\, \hat{f}(\lambda)\, d\lambda , \quad f_2(x) = \frac{1}{2\pi} \int_{|\lambda|>T} e^{-i\lambda x}\, \hat{f}(\lambda) d\lambda .$$

Alors :

$$\eta_n = n^{-\frac{\alpha-1}{2\alpha}} \sum_1^n f_1(\zeta_k) + n^{-\frac{1}{2\alpha}} \sum_1^n f_2(\zeta_k) = \eta_{n1} + \eta_{n2} .$$

En vertu du théorème 4.1 η_n converge en loi vers $\xi \sqrt{\pounds_\alpha(0)}\, b_T$ où :

$$b_T = \int_{-T}^{T} |\widehat{f}(\lambda)|^2 \frac{1+\varphi(\lambda)}{1-\varphi(\lambda)} d\lambda .$$

Soit m un entier plus grand que p. Ecrivons η_{n2} comme $n^{\frac{1-\alpha}{2\alpha}} \sum_1^{m-1} f(\zeta_k) + \eta_{n3}$. Il est évident que $n^{\frac{1-\alpha}{2\alpha}} \sum_1^{m-1} |f(\zeta_k) \longrightarrow 0$ en probabilité. Estimons $E|\eta_{n3}|^2$. On a :

$$E|\eta_{n3}|^2 \leq 2n^{-\frac{1-\alpha}{\alpha}} \int_{|\lambda|\geq T}\int_{|\mu|\geq T} |\widehat{f}(\lambda)|\, |\overline{\widehat{f}(\mu)}| \times \sum_{m\leq k\leq \ell\leq n} \varphi^k(\lambda-\mu)\varphi^{\ell-k}(-\mu)\, d\lambda d\mu \leq$$

$$\leq B\, n^{-\frac{1-\alpha}{\alpha}} \int_{-\infty}^{\infty} |\varphi(\mu)|^m \frac{1-e^{-cn|\mu|^\alpha}}{|\mu|^\alpha} d\mu \int_{|\lambda|\geq T} |\widehat{f}(\lambda)|\,|\widehat{f}(\lambda-\mu)|\, d\lambda \leq$$

$$\leq Bn^{-\frac{1-\alpha}{\alpha}} (||\varphi||_m^m\, ||f||^2 + n^{\frac{1-\alpha}{\alpha}} ||\widehat{f}|| \int_{|\lambda|\geq T} |\widehat{f}(\lambda)|^2\, d\lambda).$$

En choisissant T assez grand, on peut rendre b_T très proche de b et $E|\eta_{n3}|^2$ très proche de zéro. Le théorème est démontré.

Exemple : Soit $f(x) = 0$, $|x| > 1$, $f(x) = \text{sign } x$, $|x| \leq 1$. Dans ce cas la somme $\sum_1^n f(\zeta_k)$ est le nombre de visites de $[0,1]$ moins celles de $[-1,0]$. Supposons que $\varphi \in L_p$. En vertu du théorème précédent $n^{-\frac{\alpha-1}{2\alpha}} \sum_1^n f(\zeta_k)$ converge en loi vers $\xi \sqrt{\pounds_\alpha(0) b}$ où :

$$b = \int_{-\infty}^{\infty} \frac{16 \sin^4 \frac{\lambda}{2}}{|\lambda|^{2+\alpha}} \frac{1+\varphi(\lambda)}{1-\varphi(\lambda)} d\lambda .$$

V - MARCHE ALEATOIRE DE CAUCHY

Nous supposons dans ce paragraphe que la loi des pas ξ_j de la marche aléatoire $\{\zeta_k\}$ appartient au domaine d'attraction de la loi de Cauchy. Les théorèmes généraux du chapitre II ne contiennent pas ce cas. Mais puisque la marche continue à être récurrente on peut espérer quelques théorèmes limites raisonnables pour les sommes $B_n^{-1} \sum_1^n f(\zeta_k)$.

THEOREME 5.1 : *Supposons que les variables aléatoires ξ_j appartiennent au domaine d'attraction de la loi de Cauchy de fonction caractéristique $e^{-|t|}$. Soit f une fonction sommable de transformée de Fourier à support compact. Soit :*

$$\sum_{\lambda_j \in \Lambda} \widehat{f}(\lambda_j) = a \neq 0 .$$

Alors la somme normée :

$$\eta_n = \frac{\pi}{a \ \ell n \ n} \sum_1^n f(\zeta_k)$$

converge en loi vers une variable aléatoire de densité e^{-x}, $x \geq 0$.

Démonstration : La seule loi F de moments :

$$\int_{-\infty}^{\infty} x^k \, dF(x) = k!$$

est la loi de densité e^{-x}, $x \geq 0$. Donc il suffit de montrer que :

$$\lim_{n\to\infty} \eta_n^k = k! \quad , \ k = 1,2 \ , \ \ldots$$

Si $\varphi(t)$ est la fonction caractéristique de ξ_j on a :

$$\varphi(t) = \exp \{- |t| \ (1 + o(1))\} \quad , \quad t \to 0 \ .$$

Donc dans un voisinage de zéro, disons $|t| \leq \varepsilon$, on a :

$$|\varphi(t)| \leq \exp \{ - \frac{|t|}{2} \}.$$

Considérons les trois cas suivants :

1° La loi de ξ_j n'est pas arithmétique. Dans ce cas $\Lambda = \{0\}$ et pour $0 < \varepsilon \leq |t| \leq c$

$$|\varphi(t)| \leq \exp \ \{-\delta\} \quad , \quad 0 < \delta = \delta(c,\varepsilon) \ .$$

Soit :

$$Z_n = \sum_1^n f(\zeta_k).$$

Si r est un nombre entier et si $\hat{f}(\lambda) = 0$ pour $|\lambda| > c$ on a :

$$E \, Z_n^{2r} = (2\pi)^{-2r} \int_{-c}^{c} \ldots \int_{-c}^{c} \hat{f}(\lambda_1) \ldots \hat{f}(\lambda_{2r}).$$

$$\cdot \sum_{i_1,\ldots, i_{2r}} E \, e^{-i\lambda_1 \zeta_{i_1}} \ldots e^{-i\lambda_{2r} \zeta_{i_{2r}}} d\lambda_1 \ldots d\lambda_{2r} \leq$$

$$\leq B \sum_{\ell=1}^{2r} \sum_{1 \leq i_1 < \ldots < i_\ell < n} \int_{-\infty}^{\infty} \ldots \int_{-\infty}^{\infty} e^{-i_1|\lambda_1|} \ldots e^{-(i_\ell - i_{\ell-1})|\lambda_\ell|} d\lambda_1 \ldots d\lambda_2 \leq$$

$$\leq B(\ell n \ n)^{2r} \ .$$

De la même manière :

$$E\, Z_n^r = r!\ (2\pi)^{-r} \sum_{1\leq i_1<\ldots<i_r\leq n} \int_{-c}^{c} \ldots \int_{-c}^{c} \hat{f}(\lambda_1)\ldots \hat{f}(\lambda_r).$$

$$.\ \varphi^{i_1}(-(\lambda_1+\ldots+\lambda_r))\,\varphi^{i_2-i_1}(-(\lambda_2+\ldots+\lambda_r))\ldots\varphi^{i_r-i_{r-1}}(-\lambda_r)d\lambda_1\ldots d\lambda_r +$$

$$+ B(\ell n\ n)^{r-1} =$$

$$= r!\ (2\pi)^{-r} \sum_{1\leq i_1<\ldots<i_r\leq n} \int_{-\varepsilon}^{\varepsilon}\ldots\int_{-\varepsilon}^{\varepsilon} \hat{f}(\lambda_1)\ldots\hat{f}(\lambda_r)\varphi^{i_1}\ldots\varphi^{i_r-i_{r-1}}\, d\lambda_1\ldots d\lambda_r +$$

$$+ B(\ell\ n\ n)^{r-1}\ .$$

On peut supposer $a > 0$. On a, en vertu de l'inégalité précédente, que :

$$r!\ (2\pi)^{-r}(a+\delta)^r \int_{-\infty}^{\infty}\ldots\int_{-\infty}^{\infty} \sum_{1\leq i_1<\ldots<i_r\leq n} \exp\{-(1-\delta)\ i_1|\lambda_1 + \ldots + \lambda_r| -$$

$$- \ldots - (1-\delta)\ (i_r - i_{r-1})\ |\lambda_r|\}\ d\lambda_1\ \ldots\ d\lambda_r + B(\ell n\ n)^{r-1} \geq E\, Z_n^r \geq$$

$$\geq r!\ (2\pi)^{-r}\ (a-\delta)^r \int_{-\infty}^{\infty}\ldots\int_{-\infty}^{\infty} \sum_{1\leq i_1<\ldots<i_\ell\leq n} \exp\{-(1+\delta).$$

$$.\ i_1|\lambda_1 + \ldots + \lambda_r| - \ldots - (1+\delta)\ (i_\ell - i_{\ell-1})|\lambda_r|\}\ d\lambda_1\ \ldots\ d\lambda_r +$$

$$+ B(\ell n\ n)^{r-1}\ .$$

Puisque :

$$\frac{1}{2\pi}\int_{-\infty}^{\infty} e^{-|\lambda|}\, d\lambda = \frac{1}{\pi}\ ,$$

et :

$$\sum_{1\leq i_1<\ldots<i_r\leq n} \frac{1}{i_1}\ \ldots\ \frac{1}{i_r - i_{r-1}} = (\ell n\ n)^r\ (1+o(1))$$

On a :

$$\lim_{n\to\infty} E\left(\frac{Z_n \pi}{a\ell n\ n}\right)^r = r! = \int_0^\infty x^r e^{-x}\, dx\ .$$

Ce qui achève la démonstration pour le cas nonarithmétique.

2° Supposons maintenant que la loi de ξ_j est arithmétique mais que $\Lambda = \{0\}$. Soit $\varepsilon_1 > 0$ un nombre positif assez petit. Posons :

$$\eta_n = \eta_{n1} + \eta_{n2}$$

où

$$\eta_{n1} = \frac{\pi}{a\ell n\ n} \sum_1^n \frac{1}{2\pi} \int_{-\varepsilon_1}^{\varepsilon} \hat{f}(\lambda)\ e^{-i\lambda\zeta_k}\, d\lambda\ ,$$

$$\eta_{n_2} = \frac{\pi}{a\ell n\ n} \sum_1^n \frac{1}{2\pi} \int_{|\lambda|>\varepsilon_1} \hat{f}(\lambda)\ e^{-i\lambda\zeta_k}\, d\lambda\ .$$

Choisissons $\varepsilon_1 > \varepsilon$ si petit que $\sup_{\varepsilon_1 \geq |t| \geq \varepsilon} |\varphi(t)| < 1$. En raisonnant comme dans le point 1, nous aurons que η_{n1} converge en loi vers une loi de densité e^{-x}, $x \geq 0$.

Quant à η_{n2}, on a :

$$E\,\eta_{n2}^2 \leq B\ell_n^{-2}\, n\, E\left(\sum_1^n \int_{|\lambda|>\varepsilon_1} \hat{f}(\lambda)\, e^{-i\lambda\zeta_k} d\lambda\right)^2 \leq$$

$$\leq B\,\ell_n^{-2}\, n \int_{|\lambda|>\varepsilon_1} \int_{|\mu|>\varepsilon_1} \hat{f}(\lambda)\, \overline{\hat{f}(\mu)} \sum_{1\leq k\leq \ell\leq n} \varphi^k(\mu-\lambda)\varphi^{\ell-k}(\lambda)\, d\lambda d\mu .$$

Puisque $\Lambda = \{0\}$

$$\sup_{c\geq|\lambda|\geq\varepsilon_1} \sup_k \left|\sum_1^k \varphi^j(\lambda)\right| \leq 2 \sup_{c\geq|\lambda|\geq\varepsilon_1} |1-\varphi(\lambda)|^{-1} < \infty$$

si $|\varphi(u_0)| = 1$

$$|\varphi(u-u_0)| < \exp\{-\tfrac{1}{2}|u-u_0|\}$$

dans un voisinage de u_0. Donc :

$$E\eta_{n2}^2 \leq \frac{B}{\ell n^2 n} \int_{-c}^{c} \frac{1-e^{-\frac{1}{2}n|u|}}{|u|}\, du \leq \frac{B}{\ell n\, n} \underset{n\to\infty}{\longrightarrow} 0$$

et la loi limite de η_n coïncide avec la loi limite de η_{n1} .

3° Soit enfin $\Lambda = \{\ldots, -\lambda_1, 0, \lambda_1, \ldots\} \neq \{0\}$. Parce que :

$$\exp\{-i(\lambda \pm \lambda_j)\zeta_k\} = \exp\{-i\lambda\zeta_k\} ,$$

on a :

$$\eta_n = \frac{\pi}{a\,\ell n\, n} \sum_{k=1}^n \frac{1}{2\pi} \int_{-\infty}^{\infty} e^{-\lambda\zeta_k} \hat{f}(\lambda)\, d\lambda =$$

$$= \frac{\pi}{a\,\ell n\, n} \sum_{k=1}^n \int_{-\lambda 1/2}^{\lambda 1/2} e^{-i\lambda\zeta_k} \sum_{\lambda_j\in\Lambda} \hat{f}(\lambda+\lambda_j)\, d\lambda .$$

Dans la région $\varepsilon \leq |\lambda| \leq \lambda_{1/2}$: $\sup |\varphi(\lambda)| < 1$

et on peut raisonner comme dans le point 1. Le théorème est démontré.

On peut maintenant considérer des fonctions f pour lesquelles $\hat{f}$ n'a pas un support compact. Parce que les démonstrations de ces théorèmes sont claquées sur celle des théorèmes des 2,3 , nous ne les donnons pas ici.

THEOREME 5.2 : *Supposons que* ξ_j *prend ses valeurs dans une progression arithmétique* $\{kd, k=0,\pm 1, \ldots\}$ *et que* d *est un pas maximal. Si :*

1° $\sum_k |f(kd)| < \infty$, $\sum_k f(kd) = a \neq 0$

alors les sommes $\frac{\pi}{a \ln n} \sum_1^n f(\zeta_k)$ *convergent en loi vers une loi de densité* e^{-x}, $x \geq 0$.

Démonstration : Elle coïncide avec la démonstration du théorème 2.3.

THEOREME 5.3 : *Soit* f *une fonction intégrable localement au sens de Riemann et*

$$|f(x)| \leq B(1 + |x|)^{1 + \gamma}$$

Soit $\Lambda = \{0\}$. *Si :*

$$\int_{-\infty}^{\infty} f(x)\, dx = a \neq 0 ,$$

alors les sommes $\frac{\pi}{a \ln n} \sum_1^n f(\zeta_k)$ *convergent en loi vers une loi de densité* e^{-x}, $x \geq 0$.

Démonstration : Elle coïncide avec celle du théorème 3.1. Notons que le cas $\Lambda \neq \{0\}$ est contenu dans le théorème 5.2.

THEOREME 5.4 : *Supposons que* $\int |\varphi(t)|^p dt < \infty$ *pour tout* $p < \infty$. *Si la fonction* $f \in L_2$ *et si sa transformée de Fourier* $\hat{f}$ *est continue au point* $\lambda=0$ *avec* $\hat{f}(0) = a \neq 0$, *alors* $\eta_n = \frac{\pi}{a \ln n} \sum_1^n f(\zeta_k)$ *converge en loi vers une loi de densité* e^{-x}, $x \geq 0$.

La Démonstration est la même que celle du théorème 2.4.

VI - MARCHE ALEATOIRE DANS R^2

Nous supposons ici que les variables aléatoires ξ_j -les pas de la marche aléatoire $\{\zeta_k\}$ -prennent leurs valeurs dans R^2. La marche aléatoire $\{\zeta_k\}$ est encore récurrente si $E\, \xi_j = 0$, $E|\xi_j|^2 < \infty$ et on peut chercher des théorèmes limites pour les sommes normées $B_n^{-1} \sum_1^n f(\zeta_k)$. Soit R, la matrice des covariances des variables aléatoires ξ_j. Bien sûr, il faut supposer que $\det R \neq 0$. Soit $\varphi(\lambda)$ la fonction caractéristique des ξ_j. Comme toujours $\Lambda = \{\lambda : \varphi(\lambda) = 1\}$.

THEOREME 6.1 : *Soit* f *une fonction sommable de transformée de Fourier à support compact. Soit :*

$$\sum_{\lambda_j \in \Lambda} \hat{f}(\lambda_j) = a \neq 0$$

Alors :

$$P\left\{\frac{2\pi}{a \det R \ \ell n n} \sum_{1}^{n} f(\zeta_k) > x\right\} \longrightarrow e^{-x}, \quad x \geq 0 .$$

Démonstration : Elle coïncide avec celle du théorème 5.1. On peut supposer que $R = \begin{pmatrix} 1 & 0 \\ 0 & 1 \end{pmatrix}$. Dans ce cas :

$$\varphi(\lambda) = \exp\{-\frac{1}{2} |\lambda|^2 (1 + o(1))\}$$

dans un voisinage de zéro. Donc il existe un nombre positif $\varepsilon > 0$ tel que :

$$|\varphi(\lambda)| < \exp\{-\frac{1}{4} |\lambda|^2\} \quad , \quad |\lambda| < \varepsilon .$$

Puisque la démonstration est presque calquée sur celle du théorème 5.1, celle-ci ne sera pas ébauchée.

Pour préciser on raisonnera sur le cas où $|\varphi(\lambda)| < 1$, $\lambda \neq 0$ et donc $\Lambda = \{0\}$.

Soit : $$z_n = \sum_{1}^{n} f(\zeta_k).$$

Si r est un nombre entier et si $\hat{f}(\lambda) = 0$ pour $|\lambda'| > c$, $|\lambda''| > c$, $\lambda = (\lambda', \lambda'')$ on a :

$$E\, z_n^{2r} = (4\pi^2)^{-2r} \int_{R^2} \cdots \int_{R^2} \hat{f}(\lambda_1) \ldots \hat{f}(\lambda_{2r}) .$$

$$\cdot \sum_{i_1, \ldots, i_{2r}=1}^{n} E \exp\{- i\lambda_1 \zeta_{i_1} - \ldots - i\lambda_{2r} \zeta_{i_{2r}}\} d\lambda_1 \ldots d\lambda_{2r} \leq$$

$$\leq B \sum_{\ell=1}^{2r} \sum_{1 \leq i_1 < \ldots < i_\ell \leq n} \int_{R^2} \cdots \int_{R^2} e^{-\frac{i_1 |\lambda_1|^2}{2}} \ldots e^{-\frac{(i_\ell - i_{\ell-1}) |\lambda_\ell|^2}{2}} d\lambda_1 \ldots d\lambda_\ell \leq$$

$$\leq B(\ell n\, n)^{2r} .$$

En suivant la méthode de démonstration du théorème 5.1, on a pour n'importe quel $\delta > 0$ $(a > 0)$

$$r!\ (4\pi)^{-r} (a+\delta)^r \int_{R^2} \cdots \int_{R^2} \sum_{1 \leq i_1 < \ldots < i_r \leq n} \exp\{- \frac{(1-\delta) i_1}{2} |\lambda_1 + \ldots + \lambda_r|^2 -$$

$$- \ldots - \frac{1}{2} (1-\delta) (i_r - i_{r-1}) ||\lambda_r|^2\} d\lambda_1 \ldots d\lambda_r +$$

$$+ B(\ell n\, n)^{r-1} \geq E\, z_n^r \geq r!\ (4\pi)^{-r} (a-\delta)^r .$$

$$\cdot \int_{R^2} \cdots \int_{R^2} \exp\{ - \frac{(1+\delta) i_1}{2} | \lambda_1 + \ldots + \lambda_r|^2 - \ldots - \frac{(1+\delta)(i_r - i_{r-1})}{2}$$

$$\cdot |\lambda_r|^2 \} d\lambda_1 \ldots d\lambda_r + B (\ell n\, n)^{r-1} .$$

Puisque :

$$\int_{R^2} \exp\{-\frac{1}{2} |\lambda|^2\} \, d\lambda = 2\pi ,$$

on déduit de cette inégalité que :

$$\lim E\left(\frac{2\pi Z_n}{a\ell n\, n}\right)^r = r!$$

La démonstration est achevée.

THEOREME 6.2 : *Supposons que* ξ_j *prend ces valeurs dans* $\mathbb{Z}^2$ *et que* $d = (1,1)$ *est un pas maximal. Si :*

1° $\sum_{k \in \mathbb{Z}^2} |f(k)| < \infty$ *2°* $\sum_{k \in \mathbb{Z}^2} f(k) = a \neq 0$

alors les sommes $\frac{2\pi}{a\ell n\, n} \sum_1^n f(\zeta_j)$ *convergent en loi vers une loi de densité* $e^{-x}, x \geq 0$.

THEOREME 6.3 : *Soit* f *une fonction intégrable localement au sens de Riemann et* $|f(x)| \leq B(1 + |x|^2)^{-1+\gamma}$, $|x| \to \infty$. *Soit* $\Lambda = \{0\}$. *Si :*

$$\int_{R^2} f(x)\, dx = a \neq 0.$$

alors les sommes $\frac{2\pi}{a\, \ell n\, n} \sum_1^n f(\zeta_k)$ *convergent en loi vers une loi de densité* e^{-x}, $x \geq 0$.

THEOREME 6.4 : *Supposons que* $\int_{R^2} |\varphi(x)|^p\, dx < \infty$ *pour tout* $p < \infty$. *Si la fonction* $f \in L_2(R^2)$ *et si sa transformée de Fourier* $\hat{f}(\lambda)$ *est continue au point* $\lambda=0$ *avec* $\hat{f}(0) = a \neq 0$ *alors* $\eta_n = \frac{2\pi}{a\, \ell n\, n} \sum_1^n f(\zeta_k)$ *converge en loi vers une loi de densité* e^{-x}, $x \geq 0$.

Les démonstrations de ces théorèmes sont calquées sur celles des théorèmes 5.2-5.4.

VII - LE CAS DES FONCTIONS PERIODIQUES

1° Nous allons étudier ici les lois limites des sommes :

$$Z_n = \frac{1}{\sqrt{n}} \sum_1^n f(\zeta_k) \tag{7.1}$$

où comme toujours $\{\zeta_k\}$ est une marche aléatoire et où f est une fonctions périodique. Nous avons vu dans le paragraphe 5 du chapitre II que $\frac{1}{\sqrt{n}} \sum_1^n \sin \zeta_k$ converge en loi vers une variable Gaussienne si les ξ_j appartiennent au domaine d'attraction normale d'une loi stable. Nous verrons ici que la convergence vers une loi Gaussienne des sommes (7.1) est une règle générale.

On suppose ci-dessous que f est une fonction périodique de période $2\pi\tau$ et que cette fonction a la série de Fourier :

$$f(x) = \sum_{j\neq 0} c_j e^{ixaj}, \quad \sum |c_j| < \infty, \quad a = \frac{1}{\tau} \tag{7.2}$$

quand nous parlons de la période de f nous entendons que a est une période minimale c'est à dire que le plus grand commun diviseur de $\{j : c_j \neq 0\}$ est égal à 1. Nous ne supposons aucune restriction sur la loi commune des ξ_j.

THEOREME 6.1 : *Soit* f *une fonction périodique de période* $2\pi\tau$, $a = \tau^{-1}$. *Soit* (7.2) *la série de Fourier de* f. *Supposons que la fonction caractéristique de* ξ_j *que nous notons comme toujours* $\varphi(\lambda)$ *vérifie la condition* $|\varphi(a)| < 1$. *Si la série :*

$$\sum_j \frac{|c_j|}{|1-\varphi(aj)|} \tag{7.3}$$

converge, alors les sommes normées $Z_n = \frac{1}{\sqrt{n}} \sum_1^n f(\zeta_k)$ *convergent en loi vers une variable Gaussienne de moyenne* 0 *et de variance :*

$$\sigma^2 = \sum_{p,q:\,\varphi(a(p-q))=1} c_p \bar{c}_q \frac{1-\varphi(ap)\overline{\varphi(aq)}}{(1-\varphi(ap)(1-\varphi(aq))} \tag{7.4}$$

En particulier, si $\varphi(ap) \neq 1$ *pour* $p \neq 0$, *alors :*

$$\sigma^2 = \sum_p |c_p|^2 \frac{1+\varphi(ap)}{1-\varphi(ap)}$$

Remarque : Bien sûr, quand nous parlons de la convergence de la série (7.3), nous supposons que la sommation est étendue aux nombres k pour lesquels $c_k = 0$. En particulier, il est possible que $\varphi(a_k) = 1$ si $c_k = 0$.

Le théorème va être démontré dans les trois sous-paragraphes suivants.

2° On calcule ici les deux premiers moments.

LEMME 7.1 : $EZ_n = o(1)$ *quand* $n \to \infty$.

Démonstration : En vertu de la convergence de la série (7.3), on a :

$$EZ_n = \frac{1}{\sqrt{n}} \sum_{k=1}^{n} \sum_q c_q E e^{iaq\zeta_k} =$$

$$= \frac{1}{\sqrt{n}} \sum_q \varphi(aq) \frac{1-\varphi^n(aq)}{1-\varphi(aq)} \to 0 .$$

LEMME 7.2 : *Quand* $n \to \infty$.

$$\operatorname{Var} Z_n = \sum_{(p,q):\varphi(a(p-q))=1} c_p c_q \frac{1-\varphi(ap)\overline{\varphi(aq)}}{(1-\varphi(ap))(1-\varphi(aq))} + o(1) =$$

$$= \sigma^2 + o(1).$$

Démonstration : En vertu du lemme 7.1 :

$$\operatorname{Var} Z_n = EZ_n^2 + o(1) .$$

Puis :

$$EZ_n^2 = \frac{1}{n} \{ \sum_{q\neq 0} |c_q|^2 E |\sum_{\nu=1}^{n} e^{iaq\zeta_\nu}|^2 +$$

$$+ \sum_{p\neq q} c_p \overline{c_q} E \sum_{\nu=1}^{n} e^{iap\zeta_\nu} \sum_{\mu=1}^{n} e^{-iaq\zeta_\mu} \} = S_1 + S_2 . \tag{7.5}$$

On écrit la somme S_1 comme :

$$S_1 = \frac{1}{n} \sum_{q\neq 0} |c_q|^2 (n + \sum_{\nu=2}^{n} \sum_{\mu=1}^{\nu-1} (\varphi(aq))^{\nu-\mu} +$$

$$+ \sum_{\mu=2}^{n} \sum_{\nu=1}^{\mu-1} (\varphi(-aq))^{\nu-\mu}) =$$

$$= \sum |c_q|^2 (1 + (1-\frac{1}{n}) (\frac{\varphi(aq)}{1-\varphi(aq)} + \frac{\varphi(-aq)}{1-\varphi(-aq)})) \tag{7.6}$$

$$+ O(\frac{1}{n} \sum |c_q|^2 \frac{|1-\varphi^{n-1}(aq)|}{|1-\varphi(aq)|^2} =$$

$$= \sum_{q\neq 0} |c_q|^2 \frac{1+\varphi(aq)}{1-\varphi(aq)} + o(1).$$

Soit l'ensemble $A = \{p : \varphi(ap) = 1\}$. On peut écrire S_2 comme la somme :

$$S_2 = S_{21} + S_{22} ,$$

où :

$$S_{21} = \frac{1}{n} \sum_{p-q\in A} c_p \overline{c_q} \{n + \sum_{\nu=2}^{n} \sum_{\mu=1}^{\nu-1} \varphi^{\nu-\mu}(ap) +$$

$$+ \sum_{\mu=2}^{n} \sum_{\nu=1}^{\mu-1} \varphi^{\mu-\nu}(-aq)\} = \tag{7.7}$$

$$= \sum_{p-q\in A} c_p \overline{c_q} \frac{1-\varphi(ap)\varphi(-aq)}{(1-\varphi(ap))(1-\varphi(-aq))} + o(1).$$

On va montrer que $S_{22} \to 0$ quand $n \to \infty$. On a :

$$S_{22} = \frac{1}{n} \sum_{p-q \in A} c_p \overline{c_q} \; \{ \sum_{\nu=1}^{n} \varphi^{\nu}(a(p-q)) +$$

$$+ \sum_{\nu=2}^{n} \sum_{\mu=1}^{\nu-1} \varphi^{\nu-\mu}(ap)\varphi^{\mu}(a(p-q)) + \tag{7.8}$$

$$+ \sum_{\mu=2}^{n} \sum_{\nu=1}^{\mu-1} \varphi^{\nu}(a(p-q))\varphi^{\mu-\nu}(-aq) \} = S_{221} + S_{222} + S_{223} .$$

Soit :

$$\alpha_T = \sup\{ |1 - \varphi(a(p-q))|^{-1} : p-q \notin A , |p-q| \leq T\} .$$

Puisque $\alpha_T < \infty$ pour tout T on peut définir une suite $T = T(n)$ de sorte que

$\alpha_T n^{-1} \to 0$. Alors :

$$S_{221} \leq \frac{1}{n} \sum_{p-q \notin A} |c_p c_q| \; | \sum_{\nu=1}^{n} \varphi^{\nu}(a(p-q)) | \leq$$

$$\leq \frac{2\alpha_T}{n} (\sum |c_p|)^2 + \sum_{|p-q|>T} |c_p c_q| \leq \tag{7.9}$$

$$\leq \frac{2\alpha_T}{n} (\sum |c_p|)^2 + \sum_{|p| > \frac{T}{2}} |c_p| \sum |c_p| \xrightarrow[n\to\infty]{} 0.$$

Nous allons majorer S_{222}. Soit :

$$B = \{(p,q) : \varphi(ap) = \varphi(a(p-q))\} .$$

Montrons que si $(p,q) \in B$, ou bien $|\varphi(ap)| < 1$, ou bien $\varphi(aq) = 1$. En effet soit $\varphi(ap) = e^{i\alpha}$, $p > 0$, $0 \leq \alpha < 2\pi$; soit :

$$P_0 = \min \{p : p > 0 , \varphi(ap) = e^{i\alpha}\} .$$

Dans ce cas toutes les valeurs possibles des variables ξ_j appartiennent à un ensemble dénombrable $\{x_k\}$ où les x_k ont la forme :

$$x_k = \alpha + 2\pi k \,(ap_0)^{-1}$$

Donc si $(p,q) \in B$ les égalités suivantes doivent être satisfaites :

$$apx_k = \alpha + 2\pi t_k$$

$$apx_k - aq\, x_k = \alpha + 2\pi S_k$$

où t_k, S_k sont des entiers. Donc pour tout x_k :

$$aq\, x_k = 2\pi (t_k - S_k) = 2\pi r,$$

où r est un entier. Donc $\varphi(aq) = 1$.

Ecrivons S_{222} comme :

$$S_{222} = \sum_1 + \sum_2 \quad ,$$

ici, $\sum_1$ est la partie de S_{222} qui correspond à la sommation étendue aux couples (p,q) tels que $(p-q \notin A)$ et $(p,q) \in B$. On a :

$$\sum_1 \leq \frac{1}{n} \sum_{\substack{p-q \notin A \\ (p-q) \in B}} \{ |c_p \, c_q| \, . \, |1 - \varphi(ap)|^{-1} \, .$$

$$. \frac{|(\varphi(ap) - \varphi(a(p-q))) + \varphi(ap)\,\varphi(a(p-q))\,(\varphi^{n-1}(ap) - \varphi^{n-1}(a(p-q))) - (\varphi^{n}(ap) - \varphi^{n}(a(p-q)))|}{|1 - \varphi(a(p-q))| \, . \, |\varphi(ap) - \varphi(a(p-q))|}$$

Définissons :

$$\beta_T = \sup \frac{|(\varphi(ap) - \varphi(a(p-q))) + \varphi(ap)\varphi(a(p-q))\,(\varphi^{n-1}(ap) - \varphi^{n-1}(a(p-q))) - (\varphi^{n}(ap) - \varphi^{n}(a(p-q)))|}{|1 - \varphi(a(p-q))| \, . \, |\varphi(ap) - \varphi(a(p-q))|}$$

où on prend le sup sur les couples (p,q) qui appartiennent au domaine de sommation de $\sum_1$ et pour lesquels $|p-q| \leq T$. On peut choisir $T = T(n)$ de sorte que $\beta_T \, n^{-1} \to 0$.

Posons :

$$c_p (1 - \varphi(ap))^{-1} = b_p \quad .$$

En vertu de la condition du théorème $\sum |b_p| < \infty$. Soit enfin :

$$\Gamma_n = \sum_{|z_1| \leq 1, |z_2| \leq 1} \frac{1}{n} \; \frac{|(z_1 - z_2) + z_1 z_2 (z_1^{n-1} - z_2^{n-1}) - (z_1^n - z_2^n)|}{|1 - z_1| \; |z_1 - z_2|} \, .$$

Alors :

$$\sum_1 \leq \frac{\beta_T}{n} \sum_p |b_p| \sum_q |c_q| + \Gamma_n \sum_{|p-q| > T} |b_p| \; |c_q|$$

et pour prouver que $\sum_1 = o(1)$, il suffit de montrer que :

$$\sup_n \; \Gamma_n < \infty \quad .$$

La fonction :

$$g(z_1, z_2) = \frac{(z_1 - z_2) + z_1 z_2 (z_1^{n-1} - z_2^{n-1}) - (z_1^n - z_2^n)}{(1 - z_1) \; (z_1 - z_2)}$$

est holomorphe dans le produit des disques $|z_1| \leq 1$, $|z_2| \leq 1$ et donc $|g(z_1, z_2)|$ possède un maximum quand (z_1, z_2) parcourt le produit des cercles $|z_1| = 1$, $|z_2| = 1$.

Donc :

$$\Gamma_n \leqq \sup_{\substack{-\pi\leqq\alpha<\pi \\ -\pi<\beta<\pi}} \frac{1}{n} \frac{|(e^{i\alpha}-e^{i\beta}) + e^{i(\alpha+\beta)}(e^{i\alpha(n-1)}-e^{i\beta(n-1)}) - (e^{in\alpha}- e^{in\beta})|}{|1 - e^{i\alpha}|.\ |e^{i\alpha} - e^{i\beta}|} \leqq$$

$$\leqq \sup_{\alpha} \frac{\alpha}{|1-e^{i\alpha}|} \frac{1}{n} \sup_{\alpha,\beta} \{| \frac{(1-e^{i\alpha})\ (1-e^{i(\beta-\alpha)})}{\alpha(e^{i\alpha} - e^{i\beta})} | +$$

$$+ |\frac{(e^{i\alpha}-1)e^{i\beta}\ (e^{i(n-1)\alpha}-e^{i(n-1)\beta}}{\alpha(e^{i\alpha} - e^{i\beta})}| + |\frac{(1-e^{i(\beta-\alpha)})(1-e^{in\alpha})}{\alpha(e^{i\alpha} - e^{i\beta})}|\}\leqq \Gamma < \infty.$$

Ainsi $\sum_1 \to 0$, $n \to\infty$. Quand à $\sum_2$ puisque $(p,q) \in B$, on a $\varphi(ap) = \varphi(a(p-q))$ et donc :

$$\sum_2 \leqq \frac{1}{n} \sum_{\substack{p-q\notin A \\ (p,q)\in B}} |c_p\ c_q|\ |\sum_{\nu=2}^{n} (\nu-1)\varphi^{\nu}(ap)| \leqq$$

$$\leqq \frac{1}{n} \sum_{\substack{p-q\notin A \\ (p,q)\in B}} |c_p\ c_q|\ |\frac{1 -\varphi^n(ap)}{|(1- \varphi(ap))^2}| + \sum_{\substack{p-q\notin A \\ (p,q)\in B}} |c_p\ c_q| \frac{|\varphi^{n-1}(ap)|}{|1- \varphi(ap)|}.$$

Soit de nouveau $c_p(1- \varphi(ap))^{-1} = b_p$. En raisonnant comme plus haut on trouve :

$$\frac{1}{n} \sum |c_p\ c_q|\ | \frac{1 - \varphi^n(ap)}{(1- \varphi(ap))^2} | \leqq$$

$$\leqq \sup_{|p|\leqq T} \frac{2}{n} |1- \varphi(ap)|^{-2} (\sum |c_p|)^2 + \sum_{|p|> T} |b_p| \sum_q |c_q| \to 0,\ n\to\infty .$$

On a remarqué plus haut que $(p,q)\in B$ et $c_q \neq 0$ alors $|\varphi(ap)| < 1$. C'est pourquoi :

$$\sum_{(p,q)\in B} |c_p\ c_q| \frac{|\varphi^{n-1}(ap)|}{|1- \varphi(ap)|} \leqq \sup_{\substack{0\leqq p\leqq T \\ |\varphi(ap)|\neq 1}} |\varphi(ap)|^{n-1} \sum |b_p| \sum|c_q| +$$

$$+ \sum_{|p|>T} |b_p| \sum |c_q| \to 0,\ n\to\infty .$$

Ainsi $\sum_1$ et $\sum_2$ sont tous deux $o(1)$ quand $n\to\infty$.

Donc :

$$S_{222} = \sum_1 + \sum_2 = o(1).$$

Enfin $S_{223} = \overline{S_{222}} = o(1)$. Ceci donne que :

$$DZ_n = S_1 + S_{21} + o(1),$$

et en vertu de (7.6) et (7.7) la démonstration du lemme 7.2 est achevée.

3° On va chercher ici des majorants pour les moments $E|z_n|^p$, $p > 2$.

LEMME 7.3 : *Soit* $\sup_{1\leq k\leq N} |\varphi(k)| = \alpha < 1$. *Alors pour tout entier* r, $1 \leq r \leq N$

$$E\left|\sum_{\nu=1}^{n} e^{i\zeta_\nu}\right|^{2r} \leq \frac{B_r}{(1-\alpha)^r}\, n^r \tag{7.10}$$

où B_r *ne dépend que de* r .

Démonstration : On a :

$$E\left|\sum_{\nu=1}^{n} e^{i\zeta_\nu}\right|^{2r} = \sum_{\substack{t_1+..+t_n=r \\ \tau_1+..+\tau_n=r}} \frac{r!}{t_1!..t_n!}\,\frac{r!}{\tau_1!\,...\,\tau_n!}\,.$$

$$. \, E\exp\{it_1\zeta_1 + ... + it_n\zeta_n - i\tau_1\zeta_1 - ... - i\tau_n\zeta_n\} \leq$$

$$\leq (r!)^2 \sum_{\substack{t_1+..+t_n=r \\ \tau_1+..+\tau_n=r}} \left|\varphi\Big(\sum_{1}^{n}(t_i-\tau_i)\Big)\varphi\Big(\sum_{2}^{n}(t_i-\tau_i)\Big)..\,\varphi(t_n-\tau_n)\right|$$

$$= (r!)^2 \sum_{\substack{r=\lambda_n\geq\,..\,\geq\lambda_1>0 \\ r=\mu_n\geq\,..\,\geq\mu_1>0}} \left|\varphi(\lambda_1-\mu_1)\varphi(\lambda_2-\mu_2)...\varphi(\lambda_n-\mu_n)\right| \tag{7.11}$$

Si m au moins des différences $\lambda_i-\mu_i$ ne sont pas égales à zéro, alors :

$$|\varphi(\lambda_1-\mu_1)\,...\,\varphi(\lambda_n-\mu_n)| \leq \alpha^m.$$

On va trouver un majorant pour le nombre des couples de vecteurs entiers $\lambda = (\lambda_1,...,\lambda_n)$, $\mu = (\mu_1,..., \mu_n)$ qui satisfont les conditions :

a) $r = \lambda_n \geq \lambda_{n-1} \geq .. \geq \lambda_1 > 0$; $r = \mu_n \geq \mu_{n-1} \geq ... \geq \mu_1 > 0$

b) il y a k des couples (λ_j, μ_j) avec $\lambda_j = \mu_j$.

On peut associer à chaque vecteur λ une courbe en escalier $\lambda(t)$ de sorte que $\lambda(n-k) = \lambda_k$. La fonction $\lambda(t)$ est définie par les points de saut et, si les points de saut sont donnés, par les valeurs des sauts. Si les points de saut sont fixés le nombre des valeurs possibles des sauts est borné supérieurement par un nombre qui ne dépend que de r. Cette interprétation montre que le nombre de couples satisfaisant aux conditions a), b) est majoré par le nombre de possibilités de choisir ℓ points entiers $i_1 < i_2 < ... < i_\ell$, $\ell \leq 2r$ de manière que l'on puisse choisir $p \leq r$ des intervalles $[i_j, i_{j+1}]$ de longueur commune égale à k.

Si les longueurs des intervalles choisis sont fixées et égales à

$t_1, \dots, t_p$, $t_1 + \dots + t_p = k$, alors le nombre de ces intervalles est inférieur à

$(n-k)\ (n-t_1 - (k-t_1))\dots = (n-k)^p \leq (n-k)^r$.

Le nombre des longueurs possibles $t_1, \dots, t_p$ est inférieur à $k^p \leq k^r$.

Donc le nombre des couples (λ,μ) sous les conditions a), b) est inférieur à

$B_r(n-k)^r\, k^r \leq B_r(n-k)^r\, n^r$ où B_r ne dépend que de r. Ainsi :

$$\sum |\varphi(\lambda_1-\mu_1)\dots\varphi(\lambda_n-\mu_n)| \leq c_r^1\, n^r \sum_{k\leq n}^{\infty} (n-k)^r\, \alpha^{n-k} \leq c_r^1\, n^r \sum_{k=0}^{\infty} k^r \alpha^k \leq$$

$$\leq c_r^1\, n^r \sum_{k=0}^{\infty} (k+r)\ (k+r+1)\dots(k+1)\alpha^k = \frac{c_r^1\, n^r}{(1-x)^r}\ (r-1)\ !\ .$$

La démonstration du lemme est achevée.

<u>LEMME 7.4</u> : *Soit :*

$$f(x) = \sum_{-N}^{N} c_j\, e^{ixa_j}\ ,\ c_o = 0,$$

et $\sup_{|n|\leq Nr} |\varphi(ak)| = \alpha<1$. *Alors pour tout* $p \leq r$:

$$E\ |\sum_{j=1}^{n} f(\zeta_j)|^{2p} \leq (2N+1)^{2p-1} \sum_{-N}^{N} |c_j|^{2p}\ \frac{c_p}{(1-\alpha)^p}\ n^p \tag{7.12}$$

<u>Démonstration</u> : On déduit l'inégalité (7.12) immédiatement de l'inégalité de Hölder et du lemme 7.3.

<u>Remarque</u> : Il est évident, voir (7.11),qu'on peut écrire les moments (7.12) de la manière suivante :

$$n^{-r}\ E\ |\sum_{j=1}^{n} f(\zeta_j)|^{2r} = H_{nr}(\varphi(a),\ \varphi(2a),\dots\ \varphi(Nra),\ \varphi(-a),\ \dots,\ \varphi(-Nra))$$

où $H_{nr}(u_1,\dots u_{Nr},\ v_1,\dots,\ v_{Nr})$ est un polynôme des 2Nr variables complexes u_i, v_i.
En démontrant les lemmes 7.3, 7.4 nous avons utilisé seulement que $|\varphi(ja)| < \alpha < 1$ et la construction des polynômes H_{nr} ; aucune propriété spéciale des fonctions caractéristiques n'a été utilisée. Donc on peut reformuler le lemme 7.4 de manière formellement plus générale.

<u>LEMME 7.4'</u> : *Dans le produit des disques* $|u_i| \leq \alpha < 1$, $|v_i| \leq \alpha < 1$ *les polynômes* H_{nr} *satisfont les inégalités suivantes :*

$$|H_{nr}\ (u_1,\dots,\ u_{Nr},\ v_1,\ \dots,\ v_{Nr})| \leq (2N+1)^{2r-1}\ \frac{B_r}{(1-\alpha)^r} \sum_{-N}^{N} |c_j|^{2r} \tag{7.13}$$

4° On achève ici la démonstration du théorème. Distinguons deux cas :

A) Il existe un entier $p > 1$ pour lequel $|\varphi(ap)| = 1$

B) Pour tout entier $k \neq 0$ $|\varphi(ak)| < 1$.

A) Soit $|\varphi(ap)| = 1$. Dans ce cas toutes les valeurs possibles des ξ_j appartiennent à une progression arithmétique :

$$h + k\frac{2\pi}{ap} = h + \frac{2\pi\tau}{p} k, \quad k=0, \pm 1, \ldots$$

On suppose que $\frac{2\pi}{ap}$ est un pas maximal (notez que, en vertu de la condition du théorème, $p > 1$). Notons C l'ensemble des nombres $\exp\{\frac{2\pi ik}{p}\}$, $k=0,1,\ldots, p-1$. Si card C désigne le nombre des éléments de C, alors $1 \leq \text{card } C \leq p$. Soit : $\xi'_j = \xi_j - h$, $\zeta'_\nu = \sum_1^\nu \xi'_j$. Considérons la suite des variables aléatoires $y_\nu = \exp\{ia\, \xi'_\nu\}$, $\nu = 1,2,\ldots$

Les variables $y_1, y_2, \ldots$ consitituent une chaîne de Markoff d'ensemble des états C. Les états de la chaîne constituent une classe positive sans sous-classe. Si τ_k désigne le moment du premier retour dans l'état $\frac{2\pi ik}{p}$, alors $E\tau_k^m < \infty$ pour tout nombre positif m.

LEMME 7.5. : *Soient $\varphi_\nu(y)$ des fonctions uniformément bornées définies sur les états de la chaîne $\{y_\nu\}$. Si :*

$$E\sum_1^n \varphi_\nu(y_\nu) = o(\sqrt{n}),$$

$$\text{Var}\left(\sum_1^n \varphi_\nu(y_\nu)\right) = \sigma^2 n + o(n), \quad \sigma > 0,$$

alors les sommes $\frac{1}{\sqrt{n}}\sum_1^n \varphi_\nu(y_\nu)$ convergent en loi vers une variable Gaussienne de moyenne 0 et variance σ^2.

Pour des φ_ν qui ne dépendent pas de ν ce lemme est un cas particulier du théorème de Doeblin (voir [7], chapitre 16) ; la démonstration du cas général est la même que la démonstration de ce théorème de Doeblin. Nous l'omettons.

Posons maintenant $\varphi(z) = \sum c_j z^j$, les c_j sont les coefficients de Fourier de f. Soit $\varphi_\nu(z) = (z\, e^{i\nu h})$. Alors les sommes :

$$\frac{1}{\sqrt{n}}\sum_{\nu=1}^n f(\zeta_\nu) = \frac{1}{\sqrt{n}}\sum_1^n \varphi_\nu(z_\nu)$$

convergent en loi vers une variable de $N(0,\sigma^2)$ et le théorème est montré pour le cas cas A.

B) Soit maintenant $|\varphi(ak)| < 1$, $k=1,2,\ldots$ Supposons d'abord que $f(x) = \sum_{-N}^{N} c_j e^{ixaj}$ est un polynôme trigonométrique. Nous avons noté que les moments :

$$E\left|\frac{1}{\sqrt{n}} \sum_{1}^{n} f(\zeta_\nu)\right|^{2r} = E|Z_n|^{2r} = H_{nr}(\varphi(a),\ldots,\varphi(-aNr))$$

où $H_{nr}(u_1,\ldots, v_{Nr}) = H_{nr}(u,v)$ est un polynôme des variables complexes u_i, v_i qui ne dépend que de f,r et n. D'après le lemme 7.4', les polynômes $H_{nr}(u,v)$ sont uniformément bornés dans le produit des disques $|u_i| \leq \alpha$, $|v_i| \leq \alpha$. Donc l'ensemble des fonctions holomorphes $\{H_{nr}(u,v),\ n=1,2,\ldots\}$ est compact dans ce produit des disques.

Considérons avec les variables ξ_j les variables aléatoires :

$$\xi_{jp} = \frac{2\pi k\tau}{p} \qquad \text{si } \frac{2\pi k\tau}{p} \leq \xi_j < \frac{2\pi(k+1)\tau}{p} .$$

Ces variables satisfont les conditions du point A : si $\varphi_p(\lambda) = E\exp\{i\lambda\xi_{1p}\}$ alors $|\varphi_p(ap)| = 1$. D'après A les sommes $Z_{np} = \frac{1}{\sqrt{n}} \sum_{j=1}^{n} \xi_{jp}$ sont asymptotiquement Gaussiennes de moyenne 0 et de variance σ_p^2. Puisque pour p assez grand $|\varphi_p(aj)| < 1$, $j=1,\ldots,2N$. (N est fixé) pour de tels p

$$\sigma_p^2 = \sum_{-N}^{N} |c_j|^2 \frac{1+\varphi_p(aj)}{1-\varphi_p(aj)} .$$

Les moments $E(Z_{np})^{2r}$ sont comme auparavant les valeurs des polynômes $H_{nr}(u,v)$ aux points $u_1 = \varphi_p(a),\ldots, v_1 = \varphi_p(-a),\ldots$ En particulier si p est assez grand le point (u,v) appartient à un polydisque $|u_i| < \alpha$, $|v_i| < \alpha$ et tous les moments $E|Z_{np}|^{2r}$ sont bornés uniformément en n. Donc :

$$\lim_{n\to\infty} E|Z_{np}|^{2r} = b_{pr} = \frac{2r!}{2^r . r!} \sigma_p^2 . \tag{7.14}$$

On déduit de la forme de σ_p^2 qu'on peut écrire :

$$b_{pr} = h_r(\varphi_p(a),\ldots,\varphi_p(-Nra))$$

où $h_r(u,v)$ est holomorphe dans le produit de disques $|u_i| < \alpha$, $|v_i| < \alpha$. D'après (7.14) :

$$H_{nr}(u,v) \longrightarrow h_r(u,v)$$

sur l'ensemble O des points (u,v) du polydisque $|u_i| < \alpha$, $|v_i| < \alpha$, qui peuvent être représentés comme $u_i = \varphi_p(ja)$, $v_i = \varphi_p(-ja)$ pour n'importe quelle fonction caractéristique φ. Parce que l'ensemble des fonctions holomorphes $\{H_{nr}\}$est compact, on a que :

$$H_{nr}(u,v) \longrightarrow h_r(u,v)$$

dans la fermeture de l'ensemble O.

Donc sous la condition B :

$$E|z_n|^{2r} \longrightarrow \frac{(2r)!}{2^r.r!} \left(\sum |c_j|^2 \frac{1+\varphi(aj)}{1-\varphi(aj)}\right)^2 .$$

Une démonstration analogue montre que :

$$E\, z_n^{2r+1} \longrightarrow 0 .$$

Ainsi le théorème est démontré pour le cas où f est un polynôme trigonométrique.

Dans le cas général on peut écrire f comme la somme :

$$f(x) = \sum_{-N}^{N} c_j e^{iajx} + \sum_{|j|>N} c_j e^{ia_jx} = f_N(x) + g_N(x)$$

et z_n comme la somme :

$$z_n = \frac{1}{\sqrt{n}} \sum_1^n f_N(\zeta_\nu) + \frac{1}{\sqrt{n}} \sum_1^n g_N(\zeta_\nu) = z_{n1} + z_{n2} .$$

La variable z_{n1} converge en loi vers une variable Gaussienne de moyenne 0 et de variance :

$$\sigma_N^2 = \sum_{-N}^{N} |c_j|^2 \frac{1+\varphi(aj)}{1-\varphi(aj)} .$$

D'après le lemme 7.2 pour tout n assez grand :

$$E|z_{nr}|^2 \leq 2 \sum_{|j|>N} |c_j|^2 \frac{1+\varphi(aj)}{1-\varphi(aj)} \underset{N\to\infty}{\longrightarrow} 0$$

La démonstration est achevée.

En raisonnant de la même manière on peut montrer un résultat analogue pour des fonctions presque périodiques. Soit :

$$f(x) = \sum_j c_j e^{i\lambda_j x} .$$

Pour simplifier on suppose que $\Lambda = \{0\}$.

THEOREME 7.2 : *Si la série :*

$$\sum_j \frac{|c_j|}{|1-\varphi(\lambda_j)|}$$

converge, alors la somme $\frac{1}{\sqrt{n}} \sum_1^n f(\zeta_\nu)$ *converge en loi vers une variable Gaussienne de moyenne* 0 *et de variance :*

$$\sigma^2 = |c_j|^2 \frac{1+\varphi(\lambda_j)}{1-\varphi(\lambda_j)} .$$

Remarque : On peut montrer aussi que sous les conditions des théorèmes 7.1, 7.2 les processus $z_n(t)$ engendrés par les lignes brisées de sommets $(\frac{k}{n}, \frac{1}{\sqrt{n}} \sum_1^n f(\zeta_j))$ convergent en loi dans l'espace $C[0,1]$ vers le processus $w(\sigma t)$, où w est le processus de Wiener.

VIII - COMMENTAIRE

Ce chapitre correspond au chapitre 6 du livre [45], paragraphe 1.4 P. Lévy a démontré le théorème limite pour le nombre des termes positifs de la marche aléatoire simple $\{\zeta_k\}$ [35] c'est à dire pour les sommes $\sum_1^n f(\zeta_k)$ (8.1) où $f(x) = 0$, $x \leq 0$, $f(x) = 1$, $x > 0$. W. Feller a étudié les mêmes sommes mais pour $f(x) = 0$, $x \neq x_0$, $f(x) = 1$, $x = x_0$ [18], R.L. Dobruchin [12] a considéré ces sommes pour :

$$f : \sum |f(k)| = 1$$

et E.B. Dynkin pour $f(k) \sim c|k|^\alpha$, $k \to \infty$, [15]. P. Erdös et M. Kac ont considéré le cas de marches aléatoires dont les pas ξ_j appartiennent au domaine d'attraction d'une loi normale pour les fonctions :

$$f(x) = 0,\ x \leq 0,\quad f(x) = 1,\ x > 0\ [16]\ ;\ f(x) = |x|\ ,\ f(x) = x^2\ [17]$$

Les théorèmes limites pour le cas où f est une fonction homogène résultent des théorèmes de M. Donsker [11] et Yu V. Prohorov [39].

Chung et Kac [8] ont obtenu les résultats limites pour le cas où les ξ_j sont des variables de loi stable symétrique et où $f(x) = \mathbf{1}_{[-a,a]}(x)$. G. Kallianpur et H. Robbins ont étendu ces résultats au cas de variablès ξ_j appartenant au domaine d'attraction normale d'une loi stable symétrique et de f à support compact intégrable au sens de Riemann [31]. La théorie générale a été développée par Skorohod et Slobodenyuk [44], [45]. Ils supposent que $E|\xi_j|^5 < \infty$. Yu. A. Davydov [10] a étendu quelques résultats de Skorohod et Slovodenyuk au cas de convergence vers une loi stable. Les résultats de ces paragraphes appartiennent à l'auteur. Ils sont des généralisations des résultats de Skorohod et Slobodenyuk [45].

Soient ξ_j des variables à valeurs entière et $E\,\xi_j^2 < \infty$ A.N. Borodin [3] a étudié la convergence du champ aléatoire :

$$\hat{t}_n(x,t) = \frac{1}{\sqrt{n}} \sum_1^{nt} \mathbf{1}_{\{[x\sqrt{n}]\}}(\zeta_k)$$

vers le champ aléatoire $\mathfrak{L}_2(x,t)$. Il a donné aussi des applications très intéressantes de ces résultats [3], [4] (voir aussi [2], [6], [32]).

Les résultats des paragraphes 5, 6 sont des généralisations de ceux de Kallianpur et Robbins [31]. On peut montrer que sous les hypothèses des théorèmes les processus engendrés par $\{\sum_1^k f(\zeta_k)\}$ convergent en loi vers une constante aléatoire η dans $C[\varepsilon, 1]$, $\varepsilon > 0$ ou dans $D[0,1]$. Bien sûr $P\{\eta > x\} = e^{-x}$.

Paragraphe 7 : Skorohod et Slobodenyuk pour le cas $E\,|\xi_j|^5 < \infty$ ont déduit le résultat des théorèmes limites du paragraphe 5 chapitre 2. Les théorèmes de ce paragraphe appartiennent à l'auteur. Bien sûr, on peut considérer toujours les sommes $\sum_1^n f(\zeta_k)$ comme les sommes de valeurs de la fonction f définie sur la chaîne de Markov $\{\zeta_k\}$. Mais si f est périodique cette chaîne peut être traitée comme une chaîne de Markov sur un groupe compact. M.I. Gordin et B.A. Lifšic ont montré une variante du théorème limite central pour ce cas [25]. Ils ont montré aussi indépendamment les théorèmes 7.1, 7.2.

NOTATIONS

$P\{.\}$, $E\{.\}$, $\mathrm{Var}\{.\}$ désignent respectivement la probabilité d'un évènement dans { }, l'espérance mathématique ou la variance d'une variable aléatoire dans { }.

$\xi_\alpha(t)$ désigne un processus stable de fonction caractéristique :

$$E\exp\{i\lambda(\xi_\alpha(t)-\xi_\alpha(s))\} = \exp\{-(t-s)\frac{|\lambda|^\alpha}{\alpha}(1+i\beta\,\mathrm{sign}\lambda\ \omega(\lambda,\alpha))\},$$

$$\text{avec} \qquad \omega(\lambda,\alpha) = \begin{cases} \mathrm{tg}\,\frac{\pi\alpha}{2}, & \alpha \neq 1, \\ \frac{2}{\pi}\,\ell n|\lambda|, & \alpha = 1 \end{cases}$$

$w(t) = \xi_2(t)$ désigne le processus de Wiener.

$\pounds_\alpha(x,t)$ désigne le temps local du processus $\xi_\alpha(u)$, $0 \leq u \leq t$.

$\pounds_\alpha(x)$ désigne $\pounds(x,1)$.

$\mathbf{1}_A(x)$ désigne la fonction indicatrice d'un ensemble A.

On note B des "constantes", c'est à dire des nombres qui ne dépendent pas des paramètres sur lesquels on raisonne et dont les valeurs précises ne sont pas importantes. On utilise C pour noter une constante strictement positive.

Chaque nouveau symbole qu'on rencontré pour la première fois dans une formule sans l'expliquer est défini par cette formule.

BIBLIOGRAPHIE (*)

[1] P. BILLINGSLEY, Convergence of Probability Measures, J. Wiley and Sons, 1968.

[2] A.N. BORODIN, A limit theorem for sums of independent random variables defined on a recurrent random walk, Dokl. Akad. Nauk SSSR 246, n°4, 1979, 786-788 ; Soviet Math. Dokl. 20, (4), 1978, 528-530.

[3] A.N. BORODIN, On the asymptotic behavior of local times of recurrent random walks with finite variance, Teor. Verojatnost. i Primenen, 26, n°4, 1981, 769-783 ; Theor. Probability Appl. 26, (4), 1981, 758-772.

[4] A.N. BORODIN, Limit theorems for sums of independent random variables defined on a recurrent random walk, Teor. Verojatnost. i Primenen, 28, n°1, 1983, 98-114 ; Theor. Probabality Appl. 28, (1), 1983, 105-121.

[5] A.N. BORODIN, Distribution of integral functionals of Brownian motion, Zapiski Nauchn Seminarov LOMI, t. 119, 1982, 19-38.

[6] A.N. BORODIN, On the asymptotic behavior of local times of recurrent random walks with infinite variance, Teor. Verojatnost. i Primenen, 29, n°1, 1984.

[7] K.L. CHUNG, Markov Chains with Stationary Probabilities, Springer Verlag, 1967.

[8] K.L. CHUNG, M. KAC, Remarks on fluctuations of sums of independent random variables, Mem. Amer. Math. Soc., v.6, 1951.

[9] K.L. CHUNG, G.A. HUNT, On the zeros of $\sum_1^n \pm 1$, Ann. Math., 50, 1949, 385-400.

[10] Yu.A. DAVYDOV, On limit behavior of additive functionals of semi-stable processes and processes attracted to semi-stable ones, Zapiski Nauchn Seminarov LOMI, t. 55, 1976, 102-112.

[11] M. DONSKER, An invariance principle for certain limit theorems, Mem. Amer. Math. Soc., 6, 1951, 1-12.

[12] R.L. DOBRUSHIN, Two limit theorems for the simplest random walk, Uspekhi Mat. Nauk, t. 10, 3, 1955, 139-146.

[13] W. DOEBLIN, Sur l'ensemble des puissances d'une loi de probabilité, Studia Math., 9, 1940, 71-96.

[14] J.L. DOOB, Stochastic processes, Wiley and sons, 1953.

[15] E.B. DYNKIN, On some limit theorems for Markov chains, Ukrain. Mat. Zh., t.6, I, 1954, 285-307.

[16] P. ERDOS, M. KAS, On the number of positive sums of independent random variables, Bull. Amer. Math. Soc., 53, 10, 1947, 1011-1020.

[17] P. ERDOS, M. KAS, On certain limit theorems of the theory of probability, Bull. Amer. Soc., 52, 4, 1946, 292-302.

[18] W. FELLER, Fluctuations theory of recurrent events, Trans. Amer. Math. Soc., 67, 1949, 98-119.

[19] W. FELLER, An introduction to Probability and its applications, 2nd v., J. Wiley and Sons, 1966.

[20] D. GEMAN, J. HOROWITZ, Occupation densities, Ann. Prob., v.8, N 1, 1980, 1967.

[21] I.I. GIKHMAN, A.V. SKOROHOD, Introduction to the theory of random processes, Moscou, 1965, (Trad. Anglaise : Saunders, Ph. 1969).

[22] I.I. GIKHMAN, A limit theorem for the number of maxima in the sequence of random variables in a Markov chain, Teor. Verojatnost. i Primenen, t.3, 2, 1968, 166-172 ; Theor. Probability Appl. 3, (2), 1968, 154-160.

[23] B.V. GNEDENKO, A.N. KOLMOGOROV, Limit distributions for sums of independent random variables, Moscou, 1949 (Trad. anglaise : Addison-Wesley, 1954).

[24] B.V. GNEDENKO, On the theory of domains of attraction of stable laws, Uchenye Zapiski Moskov. Gos. Univ., t. 30, 1939, 61-72.

[25] M.I. GORDIN, V.A. LIFSIC, Remark on Markov process with normal transition operator, Thés, 3d International Vilnius Conf. on Prob. and Math. Stat., Vilnius, 1982, 147-148.

[26] I.M. GELFAND, G.E. SHILOV, Generalized functions, V I, Moscou, 1958, (trad. Anglaise : Acad. Press, 1964).

[27] I.A. IBRAGIMOV, Yu. V. LINNIK, Independent and stationary sequences of random variables, Moscou, 1965, (trad. Anglaise : Wolters-Noordhoff, 1971).

[28] J. JACOD, Théorèmes limite pour les processus, Ecole d'été de Calcul des Prob. de St.-Flour 1983, Lect. Notes Math., 1984.

[29] M. KAC, On distributions of certain Wiener functionals, Trans. Amer. Math. Soc., v. 65, 1949, 1-13.

[30] M. KAC, On some connections between probability theory and differential and integral equations, Proc. 2nd Berkeley Symp. on Math. Stat. and Prob., 1951, 189-215.

[31] G. KALLIANPUR, H. ROBBINS, The sequence of sums of independent random variables, Duke Math. J., v. 21, N 2, 1954, 285-307.

[32] H. KESTEN, F. SPITZER, A limit theorem related to a new class of self similar processes, Z. Wahrscheinlichkeitstheorie und verw. Gebiete, 50, 1979, 5-85.

[33] A. KHINCHIN, P. LEVY, Sur les lois stables, C.R. Acad. Sci. Paris, t. 202, 1936, 701-702.

[34] P. LEVY, Calcul des probabilités, Paris, 1925.

[35] P. LEVY, Sur certains processus stochastiques homogènes, Compositio Math. 7, 1939, 283-339.

[36] P. LEVY, Processus stochastiques et mouvement brownien, Paris, 1948.

[37] M. LOEVE, Probability Theory, Springer-Verlag, 1978, (4 th. ed.).

[38] H. POLLARD, The completely monotonic character of the Mittag-Leffler function, Bull. Amer. Math. Soc., 54, 1948, 1115-1116.

[39] Yu. V. PROHOROV, Convergence of random processes and limit theorems in Probability Theory, Teor. Verojatnost. i Primenen I, n°2, 1956, 177-238 ; Theor. Probability Appl. I, (2), 1956, 157-214.

[40] L. SCHWARTZ, Théorie des distributions, Paris, 1966.

[41] A.V. SKOROHOD, Studies in the theory of random processes, Kiev University Press, 1961 (Traduction anglaise, Addison-Wesley 1961).

[42] A.V. SKOROHOD, Some limit theorems for additive functionals of a sequence of sums of independent random variables, Ukrain. Mat. Zh., 13, n°4, 1961, 67-78.

[43] A.V. SKOROHOD, N.P. SLOBODENYUK, Limit theorems for additive functionals of a sequence of sums of identically distributed independent lattice random variables, Ukrain. Mat. Zh. 17, n°2, 1965, 97-105.

[44] A.V. SKOROHOD, N.P. SLOBODENYUK, Limit theorems for random walks I, II, Teor. Verojatnost. i Primenen 10, n°4, 1965, 660-672 et 11, n°1, 1966, 56-67 ; Theor. Probability Appl. 10, (4), 1965, 596-606 et 11, (1), 1966, 46-57.

[45] A.V. SKOROHOD, N.P. SLOBODENYUK, Limit theorems for random walks, Kiev, 1970.

[46] A.V. SKOROHOD, N.P. SLOBODENYUK, On the asymptotic behavior of some functionals of Brownian motion, Ukrain. Mat. Zh. 18, n°4, 1966, 60-71.

[47] N.P. SLOBODENYUK, Some limit theorems for additive functionals of a sequence of sums of independent random variables, Ukrain. Mat. Zh. 16, n°1, 1964, 41-60.

[48] G.N. SITAYA, On the limit distribution of a certain class of functionals of a sequence of sums of independent random variables, Ukrain. Mat. Zh. 16, n°6, 1964, 799-810.

[49] G.N. SITAYA, Limit theorems for some functionals of random walks, Teor. Verojatnost. i Primenen, 12, n°3, 1967, 483-492, Theor. Probability Appl. 12, (3), 1967, 432-442.

(*) NDLR .- Bien que le Professeur IBRAGIMOV nous ait communiqué une bibliographie de certaines références, en russe, nous avons préféré donner ici la traduction en anglais des références

THEOREMES LIMITE POUR LES PROCESSUS

PAR J. JACOD

INTRODUCTION

Les théorèmes limite pour les processus stochastiques sont innombrables: au gré de ses besoins, et souvent motivé par des applications (statistiques, biologie, files d'attente, modélisation de phénomènes physiques,...), chaque auteur démontre (ou redémontre) à partir de zéro le théorème limite qui lui est nécessaire.

Pourtant, il n'existe que peu de méthodes différentes permettant d'établir ces théorèmes limite; et parmi celles-ci, les méthodes utilisant les martingales jouent un rôle absolument prépondérant: c'est dire que ce sont essentiellement les mêmes deux ou trois théorèmes de base qui sont redémontrés constamment, sous des hypothèses variées, avec des conditions plus ou moins restrictives sur le processus limite.

Dans ce cours, nous avons pour objectif de présenter ces quelques théorèmes de base, *sous des conditions aussi générales que possible* ((à une exception près, importante pour la théorie mais anodine pour les applications, à savoir que le processus limite sera toujours supposé quasi-continu à gauche), et en nous restreignant aux résultats qu'on peut obtenir *en utilisant les martingales*. Que l'utilisateur éventuel ne se méprenne pas: il s'agit de théorèmes généraux, donc souvent inapplicables tels quels dans la pratique; néanmoins, la plupart du temps on peut se ramener, au prix de transformations plus ou moins astucieuses, à vérifier les hypothèses de ces théorèmes (nous en donnons quelques exemples dans le texte).

Dans le chapitre I, on rappelle l'essentiel sur la topologie de Skorokhod, et surtout on démontre des critères de compacité faciles à vérifier dans les applications.

Le chapitre II concerne les processus à accroissements indépendants: c'est un sujet sans doute peu intéressant pour les applications, mais qui nous semble d'un assez grand intérêt théorique (outre le fait qu'historiquement, les premiers théorèmes limite concernent les sommes de variables indépendantes: le livre fondamental [15] de Gnedenko et Kolmogorov est tout entier consacré à ce sujet). Nous démontrons en particulier une condition nécessaire et suffisante de convergence.

Les deux chapitres suivants constituent le coeur de ce cours. Après des rappels sur les semimartingales et leurs caractéristiques locales (§III-1), nous donnons des théorèmes de convergence de plus en plus généraux: d'abord vers un processus à accroissements indépendants (chapitre III), puis vers une semimartingale (presque!) quelconque (chapitre IV). On pourrait faire l'économie du chapitre III, dont les résultats sont des cas particuliers de ce qui est fait au chapitre IV: nous avons préféré exposer les deux choses; en effet le cas où une suite de semimartingales converge vers un processus à accroissements indépéndants peut se traiter aussi bien par la méthode des "problèmes de martingales" du chapitre IV, que par les méthodes spécifiques aux processus à accroissements indépendants (qui reposent sur la convergence fini-dimensionnelle).

Enfin dans le chapitre V nous donnons quelques indications sur les conditions nécessaires de convergence. Il n'y a que des résultats très partiels, car il s'agit d'un sujet encore largement ouvert. En particulier, faute de place, nous avons laissé complètement de coté une notion de convergence, introduite récemment par D. Aldous et I. Helland, qui est un peu plus forte que la convergence en loi, et qui est précisément faite pour obtenir des conditions nécessaires et suffisantes de convergence.

Faute de temps (et de courage!) nous avons aussi omis un sujet fort important pour les applications: celui de l'évaluation des vitesses de convergence. Là aussi, seuls quelques résultats très partiels sont connus actuellement dans ce domaine.

Un mot sur la bibliographie, pour finir: bien que déjà longue, elle est très loin d'être complète! on peut la considérer comme une mise à jour (dans le domaine de la convergence fonctionnelle) de la bibliographie considérable du livre [20] de Hall et Heyde. Profitons-en pour dire que, sous certains aspects, ce livre est très proche du cours ci-dessous (et il contient bien d'autres sujets!): les méthodes sont les mêmes, nous avons simplement accentué ici les aspects "théorie générale" et "convergence fonctionnelle", notamment vers des processus limite discontinus.

I

TOPOLOGIE DE SKOROKHOD ET CONVERGENCE EN LOI DE PROCESSUS

1 - L'ESPACE DE SKOROKHOD

Nous avons pour but d'étudier la convergence en loi de processus à valeurs dans R^d, dont les trajectoires sont indicées par R_+ et sont continues, ou continues à droite et pourvues de limites à gauche (on dira: càdlàg). Ces processus sont donc des variables aléatoires à valeurs dans l'un des espaces suivants:

$C^d = C(R_+,R^d)$ = ensemble des fonctions continues de R_+ dans R^d,

$D^d = D(R_+,R^d)$ = ensemble des fonctions càdlàg de R_+ dans R^d.

On note α le point générique de ces espaces, et $\alpha(t)$ la valeur de la fonction α au point t; on note aussi $\Delta\alpha(t) = \alpha(t) - \alpha(t-)$ le "saut" en t, avec $\Delta\alpha(0) = 0$.

On munit C^d (resp. D^d) de la tribu $\underline{\underline{C}}^d$ (resp. $\underline{\underline{D}}^d$) engendrée par toutes les applications: $\alpha \rightsquigarrow \alpha(t)$ de C^d (resp. D^d) dans R^d.

La convergence en loi n'est pas vraiment maniable si on n'est pas sur un espace polonais (= métrique complet séparable). Il s'agit donc de munir C^d et D^d de topologies polonaises pour lesquelles $\underline{\underline{C}}^d$ et $\underline{\underline{D}}^d$ sont les tribus boréliennes. Pour C^d, c'est facile, on prend la topologie de la convergence uniforme sur les compacts. Pour D^d une topologie convenable a été introduite par Skorokhod [51] sous le nom de topologie J1. Une référence plus récente est le livre [3] de Billingsley. Ci-dessous nous allons rappeler (sans démonstration) l'essentiel des résultats de ce livre qui sont utiles ici, et donner (avec démonstration) quelques compléments.

Signalons toutefois une différence avec [3]. Skorokhod et Billingsley ont défini une topologie sur l'espace $D([0,N],R^d)$ des fonctions càdlàg sur $[0,N]$ pour tout N fini: le point N y joue un rôle très particulier. L'extension à $D(R_+,R^d)$ a été faite par Stone [52] et Lindvall [34].

§a - LA TOPOLOGIE UNIFORME SUR LES COMPACTS. Cette topologie est associée à la distance définie sur C^d ou D^d par

1.1 $$d_u(\alpha,\beta) = \sum_{N\geq 1} 2^{-N} \{1 \wedge \operatorname{Sup}_{t\leq N} |\alpha(t) - \beta(t)|\}.$$

Il est facile de voir que, pour cette topologie, C^d est polonais, de tribu borélienne $\underline{\underline{C}}^d$. Le théorème d'Ascoli donne une caractérisation simple des compacts de C^d. A cet effet, posons

1.2 $\qquad w(\alpha,I) = \mathrm{Sup}_{r,s\in I} |\alpha(r) - \alpha(s)| \quad$ pour tout intervalle I,

1.3 $\qquad w_N(\alpha,\delta) = \mathrm{Sup}\{w(\alpha,[t,t+\delta]) : t\geqq 0,\ t+\delta\leqq N\} \quad$ pour $N\in\mathbb{N}^*$, $\delta>0$,

qui est le "module d'uniforme continuité" de α sur $[0,N]$. $w(\alpha,I)$ et $w_N(\alpha,\delta)$ ont un sens pour toute fonction α sur R_+, et on a:

1.4 $$\alpha\in C^d \iff \forall N\in N^*,\ \lim_{\delta\downarrow 0} w_N(\alpha,\delta) = 0.$$

On a alors:

THEOREME 1.5: *Une partie* A *de* C^d *est relativement compacte* (pour la topologie associée à d_u) *si et seulement si*

(*i*) $\mathrm{Sup}_{\alpha\in A} |\alpha(0)| < \infty$

(*ii*) $\forall N\in\mathbb{N}^*,\ \lim_{\delta\downarrow 0} \mathrm{Sup}_{\alpha\in A} w_N(\alpha,\delta) = 0.$

Dans ce cas, on a aussi: $\mathrm{Sup}_{\alpha\in A,t\leq N} |\alpha(t)| < \infty$ *pour tout* $N\in\mathbb{N}^*$.

L'espace D^d, muni de la distance d_u, est aussi complet mais il n'est pas séparable: les fonctions $\alpha^s(t) = 1_{[s,\infty[}(t)$, pour $s\in R_+$, sont en nombre non dénombrable et $d_u(\alpha^s,\alpha^{s'}) = 1/2$ si $s = s'$, $s\leqq 1$, $s'\leqq 1$.

§b - LA TOPOLOGIE DE SKOROKHOD. Pour cette topologie, deux fonctions α et β sont proches l'une de l'autre s'il existe un "petit" changement d'échelle des temps, tel que la fonction α changée de temps et la fonction β soient uniformément proches; ainsi, les fonctions α^s et $\alpha^{s'}$ définies plus haut seront proches si $|s-s'|$ est petit, ce qui permettra d'obtenir la séparabilité.

Plus précisément, on note Λ l'ensemble des changements de temps, i.e. des fonctions $\lambda : R_+ \to R_+$ continues, strictement croissantes, nulles en 0 et vérifiant $\lim_{t\uparrow\infty} \lambda(t) = \infty$. On commence par caractériser la convergence des suites:

DEFINITION 1.6: On dit que la suite (α_n) *converge vers* α *pour la topologie de Skorokhod de* D^d s'il existe des $\lambda_n\in\Lambda$ tels que

(i) $\mathrm{Sup}_t |\lambda_n(t) - t| \to 0$,

(ii) $\forall N\in\mathbb{N}^*,\ \mathrm{Sup}_{t\leqq N} |\alpha_n\circ\lambda_n(t) - \alpha(t)| \to 0 \qquad (\iff d_u(\alpha_n\circ\lambda_n,\alpha)\to 0).$

Il découle facilement de cette définition que

1.7 Si $\alpha_n\to\alpha$ pour la topologie de Skorokhod, on a:

a) $\alpha_n(t)\to\alpha(t)$ pout tout t tel que $\Delta\alpha(t) = 0$;

b) $\forall t,\ \exists t_n$ avec $t_n\to t$ et $\alpha_n(t_n)\to\alpha(t)$ et $\Delta\alpha_n(t_n)\to\Delta\alpha(t)$.

La définition de la topologie elle-même (et pas seulement de la convergence des suites) n'offre pas grand intérêt pour les applications. Nous la donnons pour être complet, mais elle ne sera pas utilisée dans la suite. Soit d'abord les fonctions

$$g_N(t) = \begin{cases} 1 & \text{si } t \leq N \\ N+1-t & \text{si } N<t<N+1 \\ 0 & \text{si } N+1 \leq t. \end{cases}$$

Soit ensuite

$$d_s(\alpha,\beta) = \sum_{N\geq 1} 2^{-N} \{1 \wedge \mathrm{Inf}(a: \exists \lambda \in \Lambda \text{ avec } \mathrm{Sup}_{s\neq t} |\mathrm{Log}\frac{\lambda(t)-\lambda(s)}{t-s}| \leq a \text{ et } \mathrm{Sup}_t |g_N(t)(\alpha\circ\lambda(t) - \beta(t))| \leq a)\}.$$

THEOREME 1.8: *La fonction* d_s *est une distance sur* D^d *, pour laquelle cet espace est complet et séparable, et pour laquelle* $\underline{\underline{D}}^d$ *est la tribu borélienne; on a* $d_s(\alpha_n,\alpha) \to 0$ *si et seulement si les conditions de 1.6 sont satisfaites.*

Pour la preuve, nous renvoyons à Billingsley [3, pp. 111-115]; il y a quelques modifications triviales à effectuer, car on travaille sur $D(R_+,R^d)$ au lieu de $D([0,N],R^d)$ (c'est pour cela qu'on introduit, par exemple, les fonctions g_N).

1.9 L'espace D^d n'est pas un espace vectoriel topologique pour la topologie de Skorokhod: par exemple les fonctions $\alpha_n = 1_{[1-1/n,\infty[}$ convergent dans D^1 vers $\alpha = 1_{[1,\infty[}$; de même les fonctions $\alpha'_n = 1_{[1+1/n,\infty[}$ convergent vers α ; par contre $\alpha_n + \alpha'_n$ ne converge pas vers 2α (ni vers aucune autre limite).

1.10 Notons α^i la $i^{\text{ième}}$ composante de $\alpha \in D^d$ $(i = 1,..,d)$. L'application $\alpha \rightsquigarrow \alpha^i$ est continue de D^d dans D^1; par contre si $\alpha^i_n \to \alpha^i$ dans D^1 pour tout $i \leq d$ il se peut que α_n ne converge pas dans D^d (exemple: $d=2$, $\alpha^1_n = 1_{[1-1/n,\infty[}$ et $\alpha^2_n = 1_{[1,\infty[}$).

1.11 La topologie uniforme sur les compacts est plus fine que la topologie de Skorokhod.

1.12 Si $\alpha_n \to \alpha$ pour la topologie de Skorokhod et si α est continue, alors $d_u(\alpha_n,\alpha) \to 0$ (en particulier, en restriction à C^d, les distances d_s et d_u définissent la même topologie): supposons en effet qu'on ait les conditions 1.6; soit $\varepsilon>0$, $N \in \mathbb{N}^*$. Il existe $\delta>0$ tel que $w_N(\alpha,\delta) \leq \varepsilon$; il existe n_o tel que $\sup_t |\lambda_n(t) - t| \leq \varepsilon$ et $\sup_{t\leq N} |\alpha_n \circ \lambda_n(t) - \alpha(t)| \leq \varepsilon$ si $n>n_o$. Comme $|\alpha(t) - \alpha\circ\lambda_n(t)| \leq \varepsilon$ on a $\sup_{t\leq N} |\alpha_n(t) - \alpha(t)| \leq 2\varepsilon$ pour $n>n_o$.

§c - CARACTERISATION DES COMPACTS. Il existe sur D^d un théorème analogue au théorème d'Ascoli. A cet effet il faut définir un module de continuité semblable à w_N, mais adapté à D^d (et notamment "plus petit" que w_N à cause de 1.4). On pose

<u>1.13</u> $w'_N(\alpha,\delta) = \operatorname{Inf}\{\operatorname{Max}_{1\leq i\leq r} w(\alpha,[t_{i-1},t_i[): 0=t_o<..<t_r=N, \inf_{i\leq r-1}(t_i-t_{i-1})\geq\delta\}.$

Ce module est défini pour toute fonction sur R_+, et on vérifie que

<u>1.14</u> $$w'_N(\alpha,\delta) \leq w_N(\alpha,2\delta)$$

<u>1.15</u> $$\alpha \in D^d \iff \forall N\in\mathbb{N}^*,\ \lim_{\delta\downarrow 0} w'_N(\alpha,\delta) = 0.$$

<u>THEOREME 1.16</u>: *Une partie* A *de* D^d *est relativement compacte pour la topologie de Skorokhod si et seulement si:*

(i) $\forall N\in\mathbb{N}^*,\ \operatorname{Sup}_{\alpha\in A, t\leq N} |\alpha(t)| < \infty$

(ii) $\forall N\in\mathbb{N}^*,\ \lim_{\delta\downarrow 0} \operatorname{Sup}_{\alpha\in A} w'_N(\alpha,\delta) = 0$.

La démonstration se trouve dans Billingsley [3, pp.116-118]; là encore il y a des modifications dues à ce qu'on travaille sur $D(R_+,R^d)$ et qui se traduisent par une définition légèrement différente de w'_N: dans [3] on impose $t_i-t_{i-1}\geq\delta$ pour $i=r$ aussi.

<u>§d - QUELQUES COMPLEMENTS UTILES.</u> Dans ce paragraphe nous rassemblons quelques résultats que nous utiliserons de temps en temps; ce sont essentiellement des lemmes techniques.

<u>1.17</u> Soit $\alpha_n \to \alpha$, $\beta_n \to \beta$ pour la topologie de Skorokhod. Si β est continu, alors $\alpha_n+\beta_n \to \alpha+\beta$ pour cette topologie: en effet, soit les $\lambda_n\in\Lambda$ associés aux α_n par 1.6; on a

$$|(\alpha_n+\beta_n)\circ\lambda_n(t) - (\alpha+\beta)(t)| \leq |\alpha_n\circ\lambda_n(t) - \alpha(t)| + |\beta_n\circ\lambda_n(t)-\beta\circ\lambda_n(t)| + |\beta\circ\lambda_n(t) - \beta(t)|.$$

En utilisant 1.12 et l'uniforme continuité de β sur les compacts, on en déduit que 1.6.(ii) est aussi satisfaite par la suite $(\alpha_n+\beta_n)$, relativement à λ_n.

Le résultat suivant précise 1.7, il est très facile à montrer en utilisant 1.6:

<u>1.18</u> Soit $\alpha_n \to \alpha$ pour la topologie de Skorokhod; soit $t>0$.

a) Il existe une suite (t_n) telle que $t_n \to t$, $\alpha_n(t_n) \to \alpha(t)$ et $\Delta\alpha_n(t_n) \to \Delta\alpha(t)$; si $\Delta\alpha(t) \neq 0$, toute autre suite vérifiant les mêmes propriétés coïncide avec t_n pour tout n assez grand.

b) $t'_n \leq t_n$, $t'_n \to t \implies \alpha_n(t'_n-) \to \alpha(t-)$.

c) $t'_n < t_n$, $t'_n \to t \implies \alpha_n(t'_n) \to \alpha(t-)$.

d) $t'_n \geq t_n$, $t'_n \to t \implies \alpha_n(t'_n) \to \alpha(t)$.

e) $t'_n > t_n$, $t'_n \to t \implies \alpha_n(t'_n-) \to \alpha(t)$.

f) Si $\alpha'_n(s) = \alpha_n(s) - \Delta\alpha_n(t_n)1_{\{t_n \leq s\}}$ et si $\alpha'(s) = \alpha(s) - \Delta\alpha(t)1_{\{t \leq s\}}$, alors $\alpha'_n \to \alpha'$ pour la topologie de Skorokhod (montrer que les suites α_n et α'_n vérifient 1.6 pour la même suite λ_n).

g) Si $a<b$ on a: $\limsup_n \mathrm{Sup}_{a \leq s \leq b} |\Delta\alpha_n(s)| \leq \mathrm{Sup}_{a \leq s \leq b} |\Delta\alpha(s)|$.

1.19 Soit $u>0$. Si $\alpha \in D^d$ on pose

$$t^0(\alpha,u) = 0, \dots , \quad t^{p+1}(\alpha,u) = \mathrm{Inf}(t> t^p(\alpha,u): |\Delta\alpha(t)|>u)$$

$$\alpha^u(s) = \alpha(s) - \sum_{p \geq 1 : t^p(\alpha,u) \leq s} \Delta\alpha(t^p(\alpha,u))$$

(α^u est la fonction α, amputée de ses sauts d'amplitude $>u$). Il est clair que $\alpha^u \in D^d$. Supposons que $\alpha_n \to \alpha$ pour la topologie de Skorokhod, et que $|\Delta\alpha(s)| \neq u$ pour tout $s>0$ (cette propriété est satisfaite pour tout $u>0$, sauf au plus une infinité dénombrable). Alors

a) $t^p(\alpha_n,u) \to t^p(\alpha,u)$

b) $\alpha_n(t^p(\alpha_n,u)) \to \alpha(t^p(\alpha,u))$, $\Delta\alpha_n(t^p(\alpha_n,u)) \to \Delta\alpha(t^p(\alpha,u))$, si $t^p(\alpha,u)<\infty$.

c) $\alpha^u_n \to \alpha^u$ pour la topologie de Skorokhod.

Preuve. Soit $t^p = t^p(\alpha,u)$, $t^p_n = t^p(\alpha_n,u)$. Supposons que $t^p_n \to t^p$ pour une valeur de p, et soit $s = \liminf_n t^{p+1}_n$. On a $s \geq t^p$; si $s=t^p$ il existe une sous-suite telle que $t^{p+1}_{n_k} \to t^p$, et comme $t^p_n \to t^p$ et $t^{p+1}_n > t^p_n$ cela contredit 1.18-d,e. On a donc $s>t^p$.

Ensuite, pour tout intervalle fermé $I \subset]t^p,t^{p+1}[$ on a $\mathrm{Sup}_{r \in I} |\Delta\alpha(r)|<u$ d'après l'hypothèse faite sur u. Donc 1.18-g entraine: $\limsup_n \mathrm{Sup}_{r \in I} |\Delta\alpha_n(r)|<u$; comme $s>t^p$ on en déduit que $s \geq t^{p+1}$. Une nouvelle application de 1.18 entraine alors que $t^{p+1}_n \to t^{p+1}$, et que $\alpha_n(t^{p+1}_n) \to \alpha(t^{p+1})$ et $\Delta\alpha_n(t^{p+1}_n) \to \Delta\alpha(t^{p+1})$ si $t^{p+1}<\infty$. Cela donne (a) et (b), par récurrence sur p.

Finalement, notons $\alpha^{u,q}_n$ et $\alpha^{u,q}$ les fonctions définies comme α^u_n et α^u, mais en sommant p de 1 à q seulement. D'après (a), (b) et (1.18-f) on a $\alpha^{u,q}_n \to \alpha^{u,q}$ pour chaque q. Comme pour tout $N \in \mathbb{N}^*$ il existe q tel que $t_q>N$ et $t^q_n>N$ pour n assez grand, on en déduit que $\alpha^u_n \to \alpha^u$. ∎

1.20 Soit $\underline{\underline{D}}^d_t$ la tribu engendrée par les fonctions $\alpha \rightsquigarrow \alpha(s)$ pour tout $s \leq t$. Soit $\underline{\underline{D}}^d_{t-} = \bigvee_{s<t} \underline{\underline{D}}^d_s$. Alors $\underline{\underline{D}}^d_{t-}$ est engendrée par les fonctions réelles bornées sur D^d qui sont mesurables par rapport à $\underline{\underline{D}}^d_{t-}$ et continues pour la topologie de Skorokhod.

Preuve. Il suffit de montrer que si $s<t$ et si f est continue bornée sur R^d,

alors $f(\alpha(s))$ est limite simple de fonctions mesurables par rapport à $\underline{\underline{D}}^d_{t-}$, bornées, continues pour la topologie de Skorokhod. La suite

$$g_n(\alpha) = n\int f(\alpha(r))\, 1_{\{s<r<t\wedge(s+1/n)\}}\, dr$$

remplit ces conditions. ∎

<u>1.21</u> Soit D^+_o l'ensemble des fonctions de D^1 qui sont croissantes et nulles en 0. Si $\alpha_n, \alpha \in D^+_o$ et si α est *continue*, il y a équivalence entre:

(i) $\alpha_n \to \alpha$ pour la topologie de Skorokhod

(ii) $\alpha_n \to \alpha$ uniformément sur les compacts

(iii) $\alpha_n(t) \to \alpha(t)$ pour tout t appartenant à un ensemble dense de R_+.

<u>Preuve</u>. On a vu que (i)$\Longleftrightarrow$(ii) (voir 1.12) et (ii)$\Longrightarrow$(iii) trivialement. Supposons que $\alpha_n(t) \to \alpha(t)$ pour tout $t \in A$ avec A dense. Soit $\varepsilon > 0$ et $N \in \mathbb{N}^*$. Soit $0 = t_o < t_1 < \ldots < t_r$ avec $t_i \in A$, $t_r \geqq N$ et $|\alpha(t_i) - \alpha(t_{i-1})| \leqq \varepsilon$. Il existe n_o tel que $|\alpha_n(t_i) - \alpha(t_i)| \leqq \varepsilon$ pour tout $i \leqq r$, si $n > n_o$. Comme α_n et α sont croissantes on en déduit que $\mathrm{Sup}_{t \leqq N} |\alpha_n(t) - \alpha(t)| \leqq 3\varepsilon$ pour $n > n_o$. ∎

<u>2 - CONVERGENCE EN LOI DE PROCESSUS</u>

<u>§a - GENERALITES SUR LA CONVERGENCE EN LOI</u>. Pour tout ce qui concerne la convergence étroite des probabilités sur un espace polonais, nous renvoyons à Billingsley [3] ou à Parthasarathy [44].

Si Y est une variable aléatoire définie sur un espace $(\Omega, \underline{\underline{F}}, P)$ et à valeurs dans un espace polonais $(E, \underline{\underline{E}})$, on note $\mathcal{L}(Y)$ ou P^Y sa loi, qui est une probabilité sur $(E, \underline{\underline{E}})$. Un processus X indicé par R_+, à trajectoires càdlàg, est considéré comme une variable aléatoire à valeurs dans l'espace polonais $(D^d, \underline{\underline{D}}^d)$ muni de la topologie de Skorokhod.

Si (X^n) est une suite de tels processus, on dit que (X^n) <u>tend en loi vers</u> X et on écrit $X^n \xrightarrow{\mathcal{L}} X$, si (P^{X^n}) converge étroitement vers P^X. Noter que ces processus peuvent être définis sur des espaces $(\Omega^n, \underline{\underline{F}}^n, P^n)$ différents, quoique ce ne soit pas une restriction que de les supposer tous définis sur le même espace (prendre par exemple le produit tensoriel des $(\Omega^n, \underline{\underline{F}}^n, P^n)$). Lorsqu'on considère plusieurs suites X^n, Y^n,... et qu'on effectue des opérations comme $X^n + Y^n$, ..., il est bien entendu que pour chaque n, X^n et Y^n sont définis sur le même espace: tout cela ne sera pas nécessairement répété dans les énoncés.

Une autre "convergence en loi" est utile pour les processus: si $A \subset R_+$ on dit que (X^n) converge fini-dimensionnellement vers X le long de A, et on écrit $X^n \xrightarrow{\mathcal{L}(A)} X$, si

2.1 $$\forall t_1,\dots,t_p \in A, \quad \mathcal{L}(X^n_{t_1},\dots,X^n_{t_p}) \longrightarrow \mathcal{L}(X_{t_1},\dots,X_{t_p}).$$

A tout processus X on associe l'ensemble $J(X)$ de ses temps de discontinuités fixes:

2.2 $$\begin{cases} J(X) = \{t>0: P(\Delta X_t \neq 0) > 0\} \\ D(X) = R_+ \setminus J(X). \end{cases}$$

Il est facile de voir que $J(X)$ est au plus dénombrable; en effet il existe une suite (T_n) de variables aléatoires à valeurs dans $\overline{R}_+$ qui "épuise" les sauts de X, ce qui veut dire que pour tous ω, t tels que $\Delta X_t(\omega) \neq 0$ il existe n avec $t = T_n(\omega)$; alors

$$J(X) = \bigcup_n \{t: P(T_n = t, \Delta X_t \neq 0) > 0\}.$$

PROPOSITION 2.3: *Si* $X^n \xrightarrow{\mathcal{L}} X$, *on a* $X^n \xrightarrow{\mathcal{L}(A)} X$ *pour* $A = D(X)$.

Preuve. Soit $t_1,\dots,t_p \in A = D(X)$. D'après 1.7 l'application: $\alpha \rightsquigarrow (\alpha(t_1),..,\alpha(t_p))$ est continue sur D^d en tout point α tel que $\Delta\alpha(t_i) = 0$ pour $i = 1,..,p$. Cette application est donc P^X-p.s. continue, d'où le résultat. ∎

La question essentielle abordée dans ce cours est la suivante: comment montrer que $X^n \xrightarrow{\mathcal{L}} X$. Pour cela, à part certains cas très particuliers, la méthode constante consiste à montrer:

2.4
(i) que la suite $\{\mathcal{L}(X^n)\}$ est tendue, i.e. relativement compacte pour la convergence étroite sur $(D^d, \underline{\underline{D}}^d)$;

(ii) que $\mathcal{L}(X)$ est le seul point limite de cette suite.

(Noter que 2.4) est nécessaire et suffisant pour que $X^n \xrightarrow{\mathcal{L}} X$). Pour montrer (ii) nous verrons plusieurs méthodes; l'une est basée sur le lemme bien connu suivant:

LEMME 2.5: *Soit* A *une partie dense de* R_+. *Soit* X *et* X' *deux processus càdlàg à valeurs dans* R^d. *Si* $\mathcal{L}(X_{t_1},\dots,X_{t_p}) = \mathcal{L}(X'_{t_1},\dots,X'_{t_p})$ *pour tous* $t_1,\dots,t_p \in A$, *alors* $\mathcal{L}(X) = \mathcal{L}(X')$.

Preuve. A étant dense dans R_+, la tribu $\underline{\underline{D}}^d$ est engendrée par les applications $\alpha \rightsquigarrow \alpha(t)$ pour $t \in A$. Un argument de classe monotone montre alors le résultat. ∎

Ainsi, la convergence $X^n \xrightarrow{\mathcal{L}} X$ équivaut à:

2.6 | (i) la suite $\{\mathcal{L}(X^n)\}$ est tendue,
| (ii) $X^n \xrightarrow{\mathcal{L}(A)} X$ pour une partie dense A de R_+.

§b - RELATIVE COMPACITE: RESULTATS GENERAUX. Le reste du chapitre I est consacré à l'étude du problème 2.4-(i). Commençons par des résultats de base; les modules de continuité w_N et w'_N (voir 1.3 et 1.13) peuvent être calculés pour chaque trajectoire du processus X^n, donnant ainsi des variables aléatoires $w_N(X^n,\delta)$, $w'_N(X^n,\delta)$.

THEOREME 2.7: *Pour que la suite* $\{\mathcal{L}(X^n)\}$ *soit tendue, il fautet il suffit que:*

(i) $\forall N\in\mathbb{N}^*$, $\varepsilon>0$, *il existe* $n_o\in\mathbb{N}^*$, $K\in R_+$ *avec*

2.8 $$n > n_o \Longrightarrow P^n(\mathrm{Sup}_{t\leq N} |X^n_t| > K) \leq \varepsilon;$$

(ii) $\forall N\in\mathbb{N}^*$, $\varepsilon>0$, $\eta>0$, *il existe* $n_o\in\mathbb{N}^*$, $\delta>0$ *avec*

2.9 $$n > n_o \Longrightarrow P^n(w'_N(X^n,\delta) > \eta) \leq \varepsilon$$

(bien que ces conditions soient exprimées en terme des X^n, elles ne dépendent en fait que des lois P^{X^n}). On verra qu'on peut toujours prendre $n_o = 0$ dans 2.8 et 2.9.

Preuve. Condition nécessaire: Soit $\varepsilon>0$. D'après le théorème de Prokhorov, il existe un compact $\tilde{K}$ de D^d tel que $P^n(X^n\notin\tilde{K})\leq\varepsilon$ pour tout n. Appliquons le théorème 1.16 avec $N\in\mathbb{N}^*$ et $\eta>0$ fixés. $K = \mathrm{Sup}_{t\leq N,\alpha\in\tilde{K}} |\alpha(t)|$ est fini, et il existe $\delta>0$ avec $\mathrm{Sup}_{\alpha\in\tilde{K}} w'_N(\alpha,\delta)\leq\eta$. On a donc 2.8 et 2.9 avec $n_o=0$.

Condition suffisante: Supposons (i) et (ii). La famille finie $\{\mathcal{L}(X^n)\}_{n\leq n_o}$ étant tendue, elle vérifie 2.8 et 2.9 avec des constantes K' et δ'. Quitte à remplacer K et δ par $K\vee K'$ et $\delta\wedge\delta'$, on peut donc supposer 2.8 et 2.9 avec $n=n_o$. Soit $\varepsilon>0$, $N\in\mathbb{N}^*$; soit $K_{N\varepsilon}\in R_+$, $\delta_{Nk\varepsilon}>0$ tels que

$$\sup_n P^n(\sup_{t\leq N} |X^n_t|>K_{N\varepsilon}) \leq 2^{-N}\frac{\varepsilon}{2}, \qquad \sup_n P^n(w'_N(X^n,\delta_{Nk\varepsilon})>1/k) \leq 2^{-N-k}\frac{\varepsilon}{2}.$$

Soit $A_{N\varepsilon} = \{\alpha\in D^d: \sup_{t\leq N} |\alpha(t)|\leq K_{N\varepsilon}$ et $w'_N(\alpha,\delta_{Nk\varepsilon})\leq 1/k$ pour tout $k\geq 1\}$ et $A_\varepsilon = \bigcap_{N\geq 1} A_{N\varepsilon}$. Par construction A_ε vérifie les conditions de 1.16, donc est relativement compact. On conclut en utilisant une nouvelle fois le théorème de Prokhorov et les inégalités:

$$P^n(X^n \notin A_\varepsilon) \leq \sum_{N\geq 1} P^n(X^n \notin A_{N\varepsilon}) \leq \sum_{N\geq 1}\{P^n(\sup_{t\leq N}|X^n_t|>K_{N\varepsilon}) + \sum_{k\geq 1} P^n(w'_N(X^n,\delta_{Nk\varepsilon})>1/k)\} \leq \varepsilon. \blacksquare$$

Etant donnés 1.5 et 1.12, lorsque tous les processus X^n sont continus, la même démonstration montre que $\{\mathcal{L}(X^n)\}$ est tendue si et seulement si on a les conditions précédentes, avec w_N au lieu de w'_N. On peut dire un peu mieux:

DEFINITION 2.10: La suite $\{\mathcal{L}(X^n)\}$ est dite *C-tendue* si elle est tendue et si ses points limite sont des probabilités qui ne chargent que le sous-espace C^d de D^d.

PROPOSITION 2.11: *Il y a équivalence entre:*

a) la suite $\{\mathcal{L}(X^n)\}$ *est C-tendue;*

b) on a 2.7-(i), et pour tous $N\in\mathbb{N}^*$, $\varepsilon>0$, $\eta>0$ *il existe* $n_o\in\mathbb{N}^*$, $\delta>0$ *avec*

2.12 $$n > n_o \implies P^n(w_N(X^n,\delta) > \eta) \leq \varepsilon$$

c) la suite $\{\mathcal{L}(X^n)\}$ *est tendue et pour tous* $N\in\mathbb{N}^*$, $\varepsilon>0$ *on a:*

2.13 $$\lim_n P^n(\sup_{t\leq N} |\Delta X^n_t| > \varepsilon) = 0 .$$

Preuve. (a)$\implies$(c): La suite $\{\mathcal{L}(X^n)\}$ étant tendue, il suffit de montrer 2.13 pour toute sous-suite convergeant en loi. Supposons donc que $X^n \xrightarrow{\mathcal{L}} X$, avec X continu par hypothèse. D'après 1.19 la fonction $\alpha \rightsquigarrow \sup_{t\leq N}|\Delta\alpha(s)|$ est continue pour la topologie de Skorokhod en tout point α tel que $\Delta\alpha(N) = 0$, donc P^X-p.s. Par suite $\sup_{t\leq N} |\Delta X^n_t| \xrightarrow{\mathcal{L}} \sup_{t\leq N} |\Delta X_t|$, qui est nul car X est continu: d'où 2.13.

(c)$\implies$(b): cela découle de 2.7 et de l'inégalité suivante:

2.14 $$w_N(\alpha,\delta) \leq 2\, w'_N(\alpha,\delta) + \sup_{t\leq N} |\Delta\alpha(t)| .$$

(b)$\implies$(a): Comme $w'_N(\alpha,\delta)\leq w_N(\alpha,2\delta)$, le théorème 2.7 entraine que la suite $\{\mathcal{L}(X^n)\}$ est tendue. Quitte à prendre une sous-suite, on peut supposer que $X^n \xrightarrow{\mathcal{L}} X$, et il faut démontrer que X est continu. Mais il est évident que: $\sup_{t\leq N}|\Delta\alpha(t)| \leq w_N(\alpha,\delta)$ pour tout $\delta>0$, de sorte que 2.12 implique que $\sup_{t\leq N} |\Delta X^n_t| \to 0$ en probabilité. On a vu ci-dessus que $\sup_{t\leq T} |\Delta X^n_t| \xrightarrow{\mathcal{L}} \sup_{t\leq T} |\Delta X_t|$ pour tout $T\in D(X)$: il s'ensuit que $\sup_{t\leq T}|\Delta X_t| = 0$ p.s., donc X est continu.■

Comme D^d n'est pas un espace vectoriel pour la topologie de Skorokhod, la relative compacité des suites $\{\mathcal{L}(X^n)\}$ et $\{\mathcal{L}(Y^n)\}$ n'entraine pas celle de la suite $\{\mathcal{L}(X^n+Y^n)\}$. Il y a cependant quelques résultats partiels dans cette direction. Le premier est trivial:

2.15 Si $\lim_n P^n(\sup_{t\leq N}|Y^n_t|>\varepsilon) = 0$ pour tous $N>0$, $\varepsilon>0$, alors la suite (Y^n) converge en loi vers le processus nul (et même en probabilité pour la topologie uniforme sur les compacts).

LEMME 2.16: *Si la suite* $\{\mathcal{L}(X^n)\}$ *est tendue (resp. converge vers* $\mathcal{L}(X)$ *) et si la suite* (Y^n) *vérifie 2.15, alors la suite* $\{\mathcal{L}(X^n+Y^n)\}$ *est tendue (resp. converge vers* $\mathcal{L}(X)$ *).*

Preuve. L'assertion concernant la relative compacité se montre facilement (elle découle aussi du lemme suivant, avec $U^{nq} = X^n$, $V^{nq} = 0$, $W^{nq} = Y^n$). Si $X^n \xrightarrow{\mathcal{L}} X$ on a aussi $X^n \xrightarrow{\mathcal{L}(A)} X$ avec $A = D(X)$ et 2.15 implique que $Y_t^n \to 0$ en probabilité pour tout t. Donc $X^n + Y^n \xrightarrow{\mathcal{L}(A)} X$ et comme $\{\mathcal{L}(X^n+Y^n)\}$ est tendue on en déduit que $X^n + Y^n \xrightarrow{\mathcal{L}} X$. ∎

LEMME 2.17: *Supposons que pour chaque* $q \in \mathbb{N}$ *on ait une décomposition*

$$X^n = U^{nq} + V^{nq} + W^{nq}$$

avec: (*i*) $\{\mathcal{L}(U^{nq})\}_{n\geq 1}$ *est tendue;*

(*ii*) $\{\mathcal{L}(V^{nq})\}_{n\geq 1}$ *est tendue, et il existe une suite de réels* a_q *tendant vers* 0, *telle que:* $\lim_n P^n(\sup_{t\leq N} |\Delta V_t^{nq}|>a_q) = 0$;

(*iii*) $\forall N \in \mathbb{N}^*, \forall \varepsilon>0$, $\lim_{q\uparrow\infty} \limsup_n P^n(\sup_{t\leq N} |W_t^{nq}|>\varepsilon) = 0$.

Alors, la suite $\{\mathcal{L}(X^n)\}$ *est tendue.*

Preuve. Il est évident que la suite (X^n) vérifie la condition 2.7-(i). Par ailleurs on vérifie aisément que

$$w'_N(\alpha+\beta,\delta) \leq w'_N(\alpha,\delta) + w_N(\beta,2\delta)$$

$$w_N(\alpha,\delta) \leq 2 \operatorname{Sup}_{t\leq N} |\alpha(t)| .$$

Ces inégalités, jointes à 2.14, entrainent:

$$w'_N(X^n,\delta) \leq w'_N(U^{nq}+V^{nq},\delta) + w_N(W^{nq},2\delta)$$

$$\leq w'_N(U^{nq},\delta) + 2\, w'_N(V^{nq},2\delta) + \operatorname{Sup}_{t\leq N} |\Delta V_t^{nq}| + 2 \operatorname{Sup}_{t\leq N} |W_t^{nq}|.$$

Soit $\varepsilon>0$, $\eta>0$. On choisit q de sorte que: $\limsup_n P^n(\sup_{t\leq N} |W_t^{nq}|>\eta) \leq \varepsilon$ et $a_q \leq \eta$. Appliquant ensuite 2.7-(ii), on choisit n_o et $\delta>0$ tels que si $n>n_o$ on ait:

$$\begin{cases} P^n(w'_N(U^{nq},\delta)>\eta) \leq \varepsilon, \quad P^n(w'_N(V^{nq},2\delta)>\eta) \leq \varepsilon, \quad P^n(\sup_{t\leq N} |\Delta V_t^{nq}|>2\eta) \leq \varepsilon, \\ P^n(\sup_{t\leq N} |W_t^{nq}|>\eta) \leq 2\varepsilon . \end{cases}$$

Par suite $P^n(w'_N(X^n,\delta)> 8\eta) \leq 5\varepsilon$, donc (X^n) vérifie la condition 2.7-(ii). ∎

COROLLAIRE 2.18: *Supposons la suite* $\{\mathcal{L}(Y^n)\}$ *C-tendue, et la suite* $\{\mathcal{L}(Z^n)\}$ *tendue* (*resp. C-tendue*). *Alors la suite* $\{\mathcal{L}(Y^n+Z^n)\}$ *est tendue* (*resp. C-tendue*).

Preuve. Il suffit d'appliquer le lemme précédent avec $U^{nq} = Z^n$, $V^{nq} = Y^n$ et $a_q = 1/q$, $W^{nq} = 0$, et d'utiliser la proposition 2.11. ∎

§c - PROCESSUS CROISSANTS.

DEFINITION 2.19: - On appelle *processus croissant* un processus réel à trajectoires càdlàg, croissantes, nulles en 0 (i.e., à trajectoires dans l'ensemble D_0^+ défini en 1.21).

- Si X et Y sont deux processus croissants, on dit que X *domine fortement* Y, et on écrit $Y \prec X$, si le processus X - Y est lui-même croissant.

PROPOSITION 2.20: *Supposons que pour chaque* n *le processus croissant* X^n *domine fortement le processus croissant* Y^n. *Si la suite* $\{\mathcal{L}(X^n)\}$ *est tendue (resp. C-tendue), il en est de même de la suite* $\{\mathcal{L}(Y^n)\}$.

Preuve. Il suffit de remarquer que $|Y_t^n| \leq |X_t^n|$, $w_N'(Y^n,\delta) \leq w_N'(X^n,\delta)$ et $w_N(Y^n,\delta) \leq w_N(X^n,\delta)$, et d'appliquer 2.7 ou 2.11.∎

La même démonstration donne aussi la

PROPOSITION 2.21: *Soit* $X^n = (X^{n,i})_{i\leq d}$ *des processus à valeurs dans* R^d, *dont chaque composante* $X^{n,i}$ *est à variation finie; soit* $\mathrm{Var}(X^{n,i})$ *le processus variation de* $X^{n,i}$. *Si la suite* $\{\mathcal{L}(\sum_{i\leq d} \mathrm{Var}(X^{n,i}))\}$ *est tendue (resp. C-tendue), il en est de même de la suite* $\{\mathcal{L}(X^n)\}$.

3 - UN CRITERE DE COMPACITE ADAPTE AUX PROCESSUS ASYMPTOTIQUEMENT QUASI-CONTINUS A GAUCHE

Dans ce paragraphe on suppose donnée une suite (X^n) de processus càdlàg à valeurs dans R^d, chaque X^n étant défini sur l'espace $(\Omega^n,\underline{\underline{F}}^n,P^n)$. Il s'agit de formuler des critères de relative compacité pour la suite $\{\mathcal{L}(X^n)\}$, qui soient plus maniables que le théorème général 2.7.

Voici un exemple de tel critère, tiré (à une modification mineure près) de Billingsley [3, p.128].

THEOREME 3.1: *Supposons que:*

(i) la suite $\{\mathcal{L}(X_0^n)\}$ *soit tendue* (ces lois sont des probabilités sur R^d);

(ii) pour tout $\varepsilon>0$, $\lim_{\delta\downarrow 0} \limsup_n P^n(|X_\delta^n - X_0^n|>\varepsilon) = 0$;

(iii) il existe une fonction croissante continue F *sur* R_+ *et des constantes* $\gamma\geq 0$, $\alpha>1$ *telles que*

3.2 $\quad \forall \lambda>0 \,,\ \forall s<r<t \,,\ \forall n \,, \quad P^n(|X^n_r - X^n_s| \geq \lambda,\ |X^n_t - X^n_r| \geq \lambda) \leq \lambda^{-\gamma} \{F(t) - F(s)\}^{\alpha}.$

Alors, la suite $\{\mathcal{L}(X^n)\}$ *est tendue.*

Ce critère est raisonnablement général pour les applications: il s'applique par exemple lorsque les processus X^n sont des processus de diffusion (continus ou non), lorsque les différents paramètres (ou coefficients) caractérisant les X^n sont bornés, uniformément en n. Cependant, il souffre de deux limitations importantes:

1) la majoration 3.2 est uniforme en n;

2) la majoration 3.2 impose un contrôle déterministe des accroissements des X^n.

La limitation 1) pourrait être levée en faisant dépendre la fonction F de l'indice n (et en imposant un certain mode de convergence des F_n). Par contre la limitation 2) est intrinsèque au critère. Or, on sait bien que dans de nombreux cas les accroissements de X^n ne sont pas contrôlables de manière déterministe (par exemple, si X^n est une diffusion à coefficients non-bornés).

C'est pourquoi nous allons ci-dessous démontrer un critère de nature assez différente, dû à Aldous [1]. Ce critère nécessite une structure supplémentaire sur les espaces $(\Omega^n, \underline{\underline{F}}^n, P^n)$:

3.3 L'espace $(\Omega^n, \underline{\underline{F}}^n, P^n)$ est muni d'une filtration $(\underline{\underline{F}}^n_t)_{t\geq 0}$, i.e. une suite croissante continue à droite de sous-tribus de $\underline{\underline{F}}^n$. On suppose aussi que X^n est adapté à cette filtration, i.e. X^n_t est $\underline{\underline{F}}^n_t$-mesurable pour chaque $t\geq 0$.

§a - LE CRITERE DE COMPACITE D'ALDOUS. On suppose qu'on a 3.3. Si $N \in \mathbb{N}^*$, on note $\underline{\underline{T}}^n_N$ l'ensemble des $(\underline{\underline{F}}^n_t)$-temps d'arrêt sur Ω^n qui sont majorés par N.

THEOREME 3.4: *Pour que la suite* $\{\mathcal{L}(X^n)\}$ *soit tendue, il suffit qu'on ait:*

(*i*) $\forall N \in \mathbb{N}^*, \forall \varepsilon>0$, *il existe* $n_o \in \mathbb{N}^*$, $K \in R_+$ *avec*

3.5 $$n > n_o \implies P^n(\sup_{t\leq N} |X^n_t| > K) \leq \varepsilon.$$

(*ii*) $\forall N \in \mathbb{N}^*, \forall \varepsilon>0, \forall \eta>0$, *il existe* $n_o \in \mathbb{N}^*$, $\delta>0$ *avec*

3.6 $$n > n_o \implies \operatorname{Sup}_{S,T\in \underline{\underline{T}}^n_N;\ S\leq T\leq S+\delta} P^n(|X^n_T - X^n_S| > \eta) \leq \varepsilon.$$

REMARQUE 3.7: La compacité n'ayant a-priori rien à voir avec les diverses filtrations dont on peut munir l'espace $(\Omega^n, \underline{\underline{F}}^n)$, cet énoncé peut paraitre étrange. Cependant,

1) plus la filtration $(\underline{\underline{F}}^n_t)$ est "petite", moins il y a de temps d'arrêt, et moins forte est la condition (ii); il est ainsi judicieux de choisir la plus petite filtra-

tion rendant le processus X^n adapté: dans ce cas, un élément de $\underline{\underline{T}}_N^n$ est une fonction de la trajectoire de X^n et la condition porte en fait sur les lois $\mathcal{L}(X^n)$. Cependant dans certains cas le contexte impose une filtration plus grande.

2) Si au contraire on prend la plus grosse filtration possible, $\underline{\underline{F}}_t^n = \underline{\underline{F}}^n$ pour tout t, les conditions (i) et (ii) impliquent que la suite $\{\mathcal{L}(X^n)\}$ est C-tendue.∎

Preuve. Les condition 3.4-(i) et 2.7-(i) étant identiques, il reste à montrer que 3.4-(ii) implique 2.7-(ii). Fixons $N\in\mathbb{N}^*$, $\varepsilon>0$, $\eta>0$. Pour tout $\rho>0$ il existe $\delta(\rho)>0$ et $n(\rho)\in\mathbb{N}^*$ tels que

3.8 $$n > n(\rho), \quad S,T\in \underline{\underline{T}}_N^n, \quad S \leq T \leq S+\delta(\rho) \Longrightarrow P^n(|X_T^n - X_S^n|\geq\eta) \leq \rho.$$

Soit les temps d'arrêt $S_0^n = 0, \ldots, S_{k+1}^n = \mathrm{Inf}(t>S_k^n: |X_t^n - X_{S_k^n}^n|\geq\eta),\ldots$ On applique 3.8 à $\rho = \varepsilon$, $S = S_k^n\wedge N$ et $T = S_{k+1}^n \wedge (S_k^n+\delta(\varepsilon))\wedge N$ en remarquant que $|X^n(S_{k+1}^n) - X^n(S_k^n)| \geq \eta$ si $S_{k+1}^n<\infty$, ce qui donne:

3.9 $$n > n(\varepsilon),\ k \geq 1 \Longrightarrow P^n(S_{k+1}^n\leq N,\ S_{k+1}^n\leq S_k^n+\delta(\varepsilon)) \leq \varepsilon.$$

Choisissons ensuite $q\in\mathbb{N}^*$ tel que $q\delta(\varepsilon) > 2N$. Le même raisonnement que ci-dessus montre que si $\theta = \delta(\varepsilon/q)$ et $n_o = n(\varepsilon)\vee n(\varepsilon/q)$, on a:

3.10 $$n > n_o,\ k \geq 1 \Longrightarrow P^n(S_{k+1}^n\leq N,\ S_{k+1}^n\leq S_k^n+\theta) \leq \varepsilon/q .$$

Comme $S_q^n = \sum_{1\leq k\leq q} (S_k^n - S_{k-1}^n)$, on a pour $n > n_o$:

$$\begin{aligned} N P^n(S_q^n \leq N) &\geq E^n\{\textstyle\sum_{1\leq k\leq q}(S_k^n - S_{k-1}^n)\, 1_{\{S_q^n\leq N\}}\} \\ &\geq \textstyle\sum_{1\leq k\leq q} E^n\{(S_k^n - S_{k-1}^n)\, 1_{\{S_q^n\leq N,\ S_k^n-S_{k-1}^n>\delta(\varepsilon)\}}\} \\ &\geq \textstyle\sum_{1\leq k\leq q} \delta(\varepsilon)\, \{P^n(S_q^n\leq N) - P^n(S_q^n\leq N,\ S_k^n - S_{k-1}^n\leq\delta(\varepsilon))\} \\ &\geq \delta(\varepsilon)\, q P^n(S_q^n \leq N) - \delta(\varepsilon)\, q\varepsilon , \end{aligned}$$

la dernière inégalité venant de 3.9. Comme $\delta(\varepsilon)q > 2N$ il s'ensuit que

3.11 $$n > n_o \Longrightarrow P^n(S_q^n < N) \leq 2\varepsilon .$$

Soit alors $A^n = \{S_q^n \geq N\}\cap\left[\bigcap_{1\leq k\leq q}\{S_k^n - S_{k-1}^n > \theta\}\right]$. Pour $\omega\in A^n$ fixé, on considère la subdivision $0=t_o<\ldots<t_r=N$ avec $t_i = S_i^n(\omega)$ si $i\leq r-1$ et $r = \inf(i: S_i^n(\omega)\geq N)$; on a $w(X^n(\omega),]t_{i-1},t_i]) \leq 2\eta$ par construction des S_i^n, et $t_i-t_{i-1} \geq \theta$ pour $i\leq r-1$ car $\omega\in A^n$: donc $w_N'(X^n(\omega),\theta) \leq 2\eta$. Par suite 3.10 et 3.11 entrainent:

$$n > n_o \Longrightarrow P^n(w_N'(X^n,\theta)>2\eta) \leq P^n((A^n)^c) \leq P^n(S_q^n<N) + \textstyle\sum_{1\leq k\leq q} P^n(S_k^n\leq N, S_k^n-S_{k-1}^n\leq\theta)$$

$$n > n_o \Longrightarrow P^n(w'_N(X^n,\theta)>2\eta) \leq 3\varepsilon$$

et on a 2.7-(ii).∎

Il nous reste à expliquer pourquoi ce critère est adapté aux processus "asymptotiquement quasi-continus à gauche". Nous nous contentons d'ailleurs d'une explication partielle. Rappelons d'abord la

DEFINITION 3.12: Un processus càdlàg X défini sur un espace probabilisé filtré $(\Omega,\underline{\underline{F}},(\underline{\underline{F}}_t),P)$ est dit *quasi-continu à gauche* (relativement à la filtration $(\underline{\underline{F}}_t)$) s'il vérifie l'une des deux conditions équivalentes suivantes:

(i) pour tout temps d'arrêt prévisible fini T, on a $\Delta X_T = 0$ p.s.

(ii) pour toute suite croissante (T_n) de temps d'arrêt telle que $T = \text{Sup}\, T_n$ soit fini, on a $X_{T_n} \to X_T$ p.s.

3.13 Si la suite stationnaire $X^n = X$ pour tout n vérifie 3.4-(ii), le processus X est quasi-continu à gauche. En effet, si ce n'était pas le cas, il existerait un temps d'arrêt prévisible $T\in \underline{\underline{T}}_N$ pour un $N\in \mathbb{N}^*$, et il existerait $\eta>0$, $\varepsilon>0$ tels que $P(|\Delta X_T|>2\eta) \geq 3\varepsilon$; il existe aussi $\delta>0$ tel que $P(\text{Sup}_{T-\delta\leq s<T} |X_s - X_{T-}|>\eta)\leq\varepsilon$. Comme T est prévisible, il existe des temps d'arrêt S_n croissant vers T et vérifiant $S_n < T$; il existe alors n avec $P(S_n<T-\delta) \leq \varepsilon$, et

$$P(|X_{(S_n+\delta)\wedge T} - X_{S_n}|>\eta) \geq P(|\Delta X_T|>2\eta,\ S_n\geq T-\delta,\ \text{Sup}_{T-\delta\leq s<T} |X_s - X_{T-}|\leq\eta)$$
$$\geq \varepsilon ,$$

ce qui contredit 3.4-(i).∎

A l'inverse, on pourrait montrer que si X est quasi-continu à gauche, la suite stationnaire $X^n = X$ vérifie 3.4-(ii) (et aussi 3.4-(i), évidemment). Plus généralement, si la suite (X^n) vérifie les conditions de 3.4, alors les points limite de $\{\mathcal{L}(X^n)\}$ sont des lois de processus quasi-continus à gauche pour leur filtration propre.

Ainsi, le critère 3.4 souffre lui aussi d'une limitation intrinsèque, mais qui n'a que très peu d'importance pour les applications.

§b - APPLICATION AUX MARTINGALES. Le critère 3.4 apparait encore comme très abstráit. Nous allons voir qu'il s'applique cependant très simplement lorsque tous les processus X^n sont des martingales localement de carré intégrable.

Commençons par quelques rappels. D'abord, un processus càdlàg X sur l'espace probabilisé filtré $(\Omega,\underline{\underline{F}},(\underline{\underline{F}}_t),P)$ est une *martingale localement de carré intégrable* (resp. une *martingale locale*) s'il existe une suite (T_n) de temps d'arrêt croissant vers $+\infty$, telle que chaque processus arrêté $X_t^{T_n} = X_{t \wedge T_n}$ soit une martingale de carré intégrable (resp. une martingale).

Si X est une martingale localement de carré intégrable, on lui associe sa *variation quadratique prévisible*, notée $\langle X,X\rangle$: c'est le seul processus croissant prévisible tel que $X^2 - \langle X,X\rangle$ soit une martingale locale (décomposition de Doob-Meyer de la sousmartingale locale X^2). Lorsque X est de carré intégrable, le processus $X^2 - \langle X,X\rangle$ est une martingale uniformément intégrable et d'après le théorème d'arrêt de Doob, on a pour tout temps d'arrêt fini T :

3.14 $$E(X_T^2) = E(\langle X,X\rangle_T) .$$

Lorsque X est seulement localement de carré intégrable, on applique 3.14 aux temps d'arrêt $T \wedge T_n$, et le lemme de Fatou permet d'obtenir:

3.15 $$E(X_T^2) \leqq E(\langle X,X\rangle_T) \quad \text{(éventuellement } = +\infty)$$

En second lieu, nous démontrons deux inégalités qui joueront un rôle très important dans la suite, et qui sont dues à Lenglart [33] et Rebolledo [46].

LEMME 3.16: *Soit* X *un processus càdlàg adapté à valeurs dans* R^d *et* A *un processus croissant adapté. On suppose que pour tout temps d'arrêt borné* T *on a*

3.17 $$E(|X_T|) \leqq E(A_T)$$

(*on dit que* A *domine* X *au sens de Lenglart*).

a) *Si* A *est prévisible, pour tous* $\varepsilon>0$, $\eta>0$ *et tout temps d'arrêt* T *on a*

3.18 $$P(\sup_{s\leqq T} |X_s|\geqq\varepsilon) \leqq \frac{\eta}{\varepsilon} + P(A_T\geqq\eta) .$$

b) *Pour tout* $\varepsilon>0$, $\eta>0$ *et tout temps d'arrêt* T *on a:*

3.19 $$P(\sup_{s\leqq T} |X_s|\geqq\varepsilon) \leqq \frac{1}{\varepsilon}\{\eta + E(\sup_{s\leqq T} \Delta A_s)\} + P(A_T\geqq\eta).$$

Preuve. Si T est un temps d'arrêt quelconque, on peut démontrer 3.18 et 3.19 pour chaque $T\wedge n$ puis faire tendre n vers l'infini. Autrement dit, il suffit de montrer ces inégalités pour T borné. Soit $R = \mathrm{Inf}(s: |X_s|\geqq\varepsilon)$ et $S = \mathrm{Inf}(s: A_s\geqq\eta)$. On a $\{\sup_{s\leqq T} |X_s|\geqq\varepsilon\} \subset \{A_T\geqq\eta\} \cup \{R\leqq T<S\}$, donc

3.20 $$P(\sup_{s\leqq t} |X_s|\geqq\varepsilon) \leqq P(A_T\geqq\eta) + P(R\leqq T<S) .$$

a) Supposons A prévisible. Le temps d'arrêt S est alors prévisible, et il existe une suite (S_n) de temps d'arrêt croissant vers S et vérifiant $S_n<S$. Par suite

$$P(R \leq T < S) \leq \lim_n P(R \leq T < S_n) \leq \lim_n P(|X_{R \wedge T \wedge S_n}| \geq \varepsilon)$$
$$\leq \frac{1}{\varepsilon} \lim_n E(|X_{R \wedge T \wedge S_n}|) \leq \frac{1}{\varepsilon} \lim_n E(A_{R \wedge T \wedge S_n})$$

d'après 3.17. Mais $R \wedge T \wedge S_n < S$, donc $A_{R \wedge T \wedge S_n} \leq \eta$ et $P(R \leq T < S) \leq \eta/\varepsilon$. On déduit alors 3.18 de 3.20.

b) Supposons simplement maintenant que A est optionnel. Il vient

$$P(R \leq S < T) \leq P(|X_{R \wedge T \wedge S}| > \varepsilon) \leq \frac{1}{\varepsilon} E(|X_{R \wedge T \wedge S}|) \leq \frac{1}{\varepsilon} E(A_{R \wedge T \wedge S})$$

et on a $E(A_{R \wedge T \wedge S}) \leq \eta + E(\sup_{s \leq T} \Delta A_s)$ par définition de S. On déduit alors 3.19 de 3.20. ∎

COROLLAIRE 3.21: *Soit* $X = (X^i)_{i \leq d}$ *un processus dont les composantes sont des martingales localement de carré intégrable; soit* $A = \sum_{i \leq d} \langle X^i, X^i \rangle$. *Pour tous* $\varepsilon > 0$, $\eta > 0$ *et tous temps d'arrêt finis* $S \leq T$ *on a*

3.22
$$P(\sup_{S \leq s \leq T} |X_s - X_S| > \varepsilon) \leq \frac{\eta}{\varepsilon^2} + P(A_T - A_S \geq \eta)$$

Preuve. Les composantes du processus $X'_t = X_t - X_{t \wedge S}$ sont encore des martingales localement de carré intégrable, et $\langle X'^i, X'^i \rangle_t = \langle X^i, X^i \rangle_t - \langle X^i, X^i \rangle_{t \wedge S}$. Donc si $A'_t = A_t - A_{t \wedge S}$ et si $Y = |X'|^2$, 3.15 montre que pour tout temps d'arrêt T on a

$$E(Y_T) = \sum_{i \leq d} E\{(X'^i_T)^2\} \leq E(A'_T) .$$

Comme A' est prévisible, 3.22 découle de 3.19 appliqué à Y, A', ε^2 et η. ∎

Revenons au critère de compacité. Pour chaque n, on suppose que les composantes $X^{n,i}$ de X^n sont des martingales localement de carré intégrable. Soit

3.23
$$A^n = \sum_{i \leq d} \langle X^{n,i}, X^{n,i} \rangle .$$

On a alors le résultat suivant, dû à Rebolledo [46]:

THEOREME 3.24: *Avec les hypothèses ci-dessus, pour que la suite* $\{\mathcal{L}(X^n)\}$ *soit tendue il suffit que:*

(i) la suite $\{\mathcal{L}(X^n_0)\}$ *soit tendue,*

(ii) la suite $\{\mathcal{L}(A^n)\}$ *soit C-tendue.*

Preuve. Soit $N \in \mathbb{N}^*$, $\varepsilon > 0$. D'après (i) et (ii) il existe $K \in \mathbb{R}_+$ avec

$$\sup_n P^n(|X^n_0| > K) \leq \varepsilon , \qquad \sup_n P^n(A^n_N > K) \leq \varepsilon$$

pour tout n; 3.22 appliqué avec $S = 0$, $T = N$, $\eta = K$ entraine

$$P^n(\sup_{t\leqq N} |X_t^n| > K + \frac{\sqrt{K}}{\sqrt{\varepsilon}}) \leqq P^n(|X_0^n|>K) + P^n(\sup_{t\leqq N} |X_t^n - X_0^n| > \frac{\sqrt{K}}{\sqrt{\varepsilon}})$$

$$\leqq \varepsilon + K(\frac{\sqrt{\varepsilon}}{\sqrt{K}})^2 + P^n(A_N^n>K) \leqq 3\varepsilon,$$

si bien qu'on a la condition 3.4-(i).

Soit $N\in\mathbb{N}^*$, $\varepsilon>0$, $\eta>0$. D'après (ii) et 2.11 il existe $\delta>0$, $n_o\in\mathbb{N}^*$ avec

$$n > n_o \Longrightarrow P^n(w_N(A^n,\delta) \geqq \varepsilon\eta^2) \leqq \varepsilon .$$

D'autre part si $w_N(A^n,\delta) < \varepsilon\eta^2$ et si $S,T\in \underline{\underline{T}}_N^n$ vérifient $S\leqq T\leqq S+\delta$, on a $A_T^n - A_S^n \leqq \varepsilon\eta^2$. Par suite, en appliquant 3.22 on obtient:

$$n > n_o,\ S,T\in \underline{\underline{T}}_N^n,\ S\leqq T\leqq S+\delta \Longrightarrow P^n(|X_T^n - X_S^n|>\eta) \leqq P^n(\sup_{S\leqq s\leqq T} |X_s^n - X_S^n|>\eta)$$

$$\leqq \frac{1}{\eta^2}(\varepsilon\eta^2) + P^n(A_T^n - A_S^n\geqq\varepsilon\eta^2)$$

$$\leqq \varepsilon + P^n(w_N(A^n,\delta)\geqq\varepsilon\eta^2) \leqq 2\varepsilon,$$

de sorte qu'on a aussi la condition 3.4-(ii). ∎

§c - REMARQUE FINALE. Les premiers critères de compacité faisant intervenir la structure des filtrations sont dus à Grigelionis [17] et Billingsley [4]. La démonstration du théorème 3.4 suit exactement celle de Métivier [41]. On pourra consulter [27] pour des critères de compacité permettant des limites non quasi-continues à gauche: ils sont basés sur un résultat du même type que 3.4, mais bien plus compliqué à énoncer. Voici, à titre d'exemple, comment le théorème 3.24 se généralise:

THEOREME: *On suppose que les* X^n *sont des martingales localement de carré intégrable;* A^n *est défini par 3.23. Pour que la suite* $\{\mathcal{L}(X^n)\}$ *soit tendue, il suffit qu'on ait 3.24-(i) et qu'il existe des processus croissants* G^n *dominant fortement les* A^n *et vérifiant* <u>*l'une*</u> *des conditions suivantes:*

(C1) La suite $\{\mathcal{L}(G^n)\}$ *est tendue et ses points limite sont des masses de Dirac.*

(C2) La suite $\{\mathcal{L}(G^n)\}$ *est tendue et, pour tout point limite* Q *de cette suite, le processus canonique* $X_t(\alpha) = \alpha(t)$ *sur* D^1 *est prévisible relativement à la filtration* $(\overline{\underline{\underline{D}}}_t^1)$, *où* $\overline{\underline{\underline{D}}}_t^1$ *est la tribu engendrée par* $\underline{\underline{D}}_t^1$ *et par les négligeables de la Q-complétion de* $\underline{\underline{D}}^1$.

(C3) Les espaces $(\Omega^n,\underline{\underline{F}}^n,(\underline{\underline{F}}_t^n),P^n)$ *sont tous égaux, et* G^n *converge en probabilité pour la topologie de Skorokhod vers un processus prévisible.*

On a (C1) $\Longrightarrow$ (C2); la condition 3.24-(ii) entraine également (C2). Des résultats un peu différents, mais de la même veine, se trouvent dans l'article [32] de V. Lebedev, et l'article de revue [18] contient un certain nombre de compléments.

II

CONVERGENCE DES PROCESSUS A ACCROISSEMENTS INDEPENDANTS

1 - LES CARACTERISTIQUES D'UN PROCESSUS A ACCROISSEMENTS INDEPENDANTS

L'objectif de ce chapitre est de démontrer une condition nécessaire et suffisante pour qu'une suite de processus à accroissements indépendants converge en loi. Cette condition sera exprimée en terme des "caractéristiques" que nous allons définir ci-dessous. Ces caractéristiques sont plus ou moins bien connues depuis Lévy (au moins pour les processus sans discontinuités fixes), seule la formulation donnée ici est un peu différente de la formulation classique. On utilisera librement le livre [10] de Doob, en ne démontrant que les résultats qui ne figurent pas explicitement dans ce livre.

Soit $(\Omega,\underline{\underline{F}},(\underline{\underline{F}}_t),P)$ un espace filtré. Un *processus à accroissements indépendants* (en abrégé: PAI) est un processus X indicé par R_+, à valeurs dans R^d, adapté à $(\underline{\underline{F}}_t)$ et tel que les accroissements $X_{t+s} - X_t$ soient indépendants de la tribu $\underline{\underline{F}}_t$ pour tous $s,t \geq 0$. Cette notion dépend donc de la filtration: en général, mais pas toujours, cette filtration est celle engendrée par le processus lui-même. Comme en définitive on ne s'intéresse qu'à la convergence en loi des processus, ceux-ci doivent être càdlàg, et on fait donc en outre l'hypothèse:

1.1 $\qquad X$ est à trajectoires càdlàg, nulles en 0.

(la condition $X_0 = 0$ sert à éviter des complications sans intérêt).

Fixons quelques notations. Si $x,y \in R^d$ on note $x.y$ le produit scalaire, $|x|$ la norme euclidienne, et x^j la $j^{\text{ième}}$ composante de x. Si de plus c est une matrice $d\times d$, on note $x.c.y$ le nombre $\sum_{j,k\leq d} x^j c^{jk} y^k$.

Comme dans le chapitre I, on note $J(X)$ l'ensemble des temps de discontinuités fixes de X, et $D(X) = R_+ \setminus J(X)$.

On appelle *fonction de troncation* toute fonction $h: R^d \to R^d$ vérifiant

1.2 $$\exists a \in]0,\infty[\quad \text{avec } |h(x)| \leq a, \quad |x| \leq \tfrac{a}{2} \Longrightarrow h(x) = x,$$
$$|h^j(x)| \leq |x^j|, \quad \text{et } |x| \geq a \Longrightarrow h(x) = 0 .$$

Si h est une fonction de centrage, on pose

1.3 $$X^h_t = X_t - \sum_{s\leq t} \{\Delta X_s - h(\Delta X_s)\},$$

ce qui définit un nouveau processus càdlàg X^h (car X n'a qu'un nombre fini de sauts $|\Delta X_s| > a/2$ sur tout intervalle fini), et $\Delta X^h_t = h(\Delta X_t)$.

§a - CARACTERISTIQUES DES PAI SANS DISCONTINUITES FIXES.

THEOREME 1.4: *Soit* h *une fonction de troncation.*

a) Soit X *un PAI sans discontinuités fixes, vérifiant 1.1. Il existe un triplet* (B^h,C,ν) *et un seul, constitué de:*

1.5 $B^h = (B^{h,j})_{j\leq d}$ une fonction continue: $R_+ \to R^d$ avec $B^h_0 = 0$ (le *drift*);

1.6 $C = (C^{jk})_{j,k\leq d}$ une fonction continue: $R_+ \to R^d \otimes R^d$ avec $C_0 = 0$, telle que pour $s\leq t$ la matrice $C_t - C_s$ soit symétrique nonnégative (*variance de la partie gaussienne*);

1.7 ν une mesure positive sur $R_+ \times R^d$ vérifiant $\nu(R_+\times\{0\}) = 0$, $\nu(\{t\}\times R^d) = 0$ et $\int_{R^d} |x|^2\wedge 1\ \nu([0,t]\times dx) < \infty$ pour tout t (*mesure de Lévy*);

et tel que pour tous $s<t$, $u\in R^d$ *on ait:*

1.8 $$E(e^{iu.(X_t-X_s)}) = \exp\{iu.(B^h_t-B^h_s) - \tfrac{1}{2}u.(C_t-C_s).u + \int_s^t\int_{R^d} (e^{iu.x}-1-iu.h(x))\nu(dr\times dx)\}.$$

b) Inversement, si (B^h,C,ν) *vérifie 1.5, 1.6, 1.7, il existe un PAI* X *satisfaisant 1.1 et 1.8, et* $\mathcal{L}(X)$ *est entièrement déterminée par le triplet* (B^h,C,ν)*, et en outre* X *n'a pas de discontinuités fixes.*

Un PAI qui vérifie 1.1 étant "centré" au sens de Lévy, ce théorème est intégralement contenu dans le livre [10] de Doob (pp. 417-419), à ceci près qu'on a choisi une version différente de la formule de Lévy-Khintchine 1.8.

Comme les notations le suggèrent, C et ν ne dépendent pas de la fonction de troncation h , mais B^h en dépend. Avant de donner la formule reliant B^h et $B^{h'}$ pour deux fonctions de troncation h et h', introduisons des notations supplémentaires:

1.9 $$f*\nu_t = \int_0^t\int_{R^d} f(x)\ \nu(ds\times dx) \quad \text{si cette intégrale existe,}$$

1.10 $$\tilde{C}^h = (\tilde{C}^{h,jk})_{j,k\leq d}: \quad \tilde{C}^{h,jk} = C^{jk} + (h^j h^k)*\nu .$$

La fonction $\tilde{C}^h$ vérifie les mêmes conditions que la fonction C ($\tilde{C}^h$ est bien définie, car $|h|^2 \leq c^{te}(|x|^2\wedge 1)$ et on a 1.7).

L'unicité dans 1.8 donne alors immédiatement la relation suivante:

1.11 $$B^{h'} = B^h + (h' - h)*\nu .$$

Il existe une autre caractérisation du triplet (B^h,C,ν) en termes de martingales, qui sera plus utile pour nous:

THEOREME 1.12: *Soit* X *une PAI sans discontinuités fixes, vérifiant 1.1, sur l'espace* $(\Omega,\underline{\underline{F}},(\underline{\underline{F}}_t),P)$. *Soit* (B^h,C,ν) *le triplet défini ci-dessus, et* X^h *et* $\tilde{C}^h$ *définis par 1.3 et 1.10; alors*

(i) $\tilde{X}^h = X^h - B^h$ *est une martingale* (d-dimensionnelle);

(ii) pour tous $j,k\leq d$, $\tilde{X}^{h,j}\,\tilde{X}^{h,k} - \tilde{C}^{h,jk}$ *est une martingale;*

(iii) pour toute fonction f *borélienne bornée nulle sur un voisinage de* 0 *, le processus* $\sum_{s\leq t} f(\Delta X_s) - f*\nu_t$ *est une martingale.*

De plus, (B^h,C,ν) *est l'unique triplet vérifiant 1.5-1.7 et ayant ces propriétés.*

Preuve. D'abord, il est clair que X^h, $\tilde{X}^h$ et $Y^f_t = \sum_{s\leq t} f(\Delta X_s)$ (où f est comme dans (iii)) sont des PAI sur $(\Omega,\underline{\underline{F}},(\underline{\underline{F}}_t),P)$.

Ensuite, d'après la discussion des pp. 421-424 de [10], on a $E(Y^f_t) = f*\nu_t$ et

$$E(\exp iu.X^h_t) = \exp\{iu.B^h_t - \tfrac{1}{2}u.C_t.u + (e^{iu.h} - 1 - iu.h)*\nu_t\}.$$

Il est facile de vérifier que $\phi_t(u) = E(\exp iu.X^h_t)$ est deux fois dérivable, avec $\partial\phi_t/\partial u^j(0) = iB^{h,j}_t$ et $\partial^2\phi_t/\partial u^j\partial u^k(0) = -\{B^{h,j}_t\,B^{h,k}_t + C^{jk}_t + (h^j h^k)*\nu_t\}$. Donc

$$E(\tilde{X}^{h,j}_t) = 0 , \qquad E(\tilde{X}^{h,j}_t\,\tilde{X}^{h,k}_t) = \tilde{C}^{h,jk}_t.$$

Mais alors, (i) (ii) et (iii) proviennent des remarques évidentes suivantes: si Y est un PAI tel que $y(t) = E(Y_t)$ existe pour tout t, alors $Y_t - y(t)$ est une martingale; si de plus $y(t) = 0$ et $z^{jk}(t) = E(Y^j_t Y^k_t)$ existe pour tout t, alors $Y^j_t Y^k_t - z^{jk}(t)$ est aussi une martingale.

Enfin l'unicité vient de ce qu'une martingale déterministe nulle en 0 est identiquement nulle. ∎

§b - CARACTERISTIQUES DES PAI QUELCONQUES. Commençons par énoncer la généralisation du théorème 1.4.

THEOREME 1.13: *Soit* h *une fonction de troncation.*

a) Si X *est un PAI vérifiant 1.1, il existe un triplet* (B^h,C,ν) *et un seul constitué de:*

1.14 $B^h = (B^{h,j})_{j\leq d}$ une fonction càdlàg: $R_+ \to R^d$ avec $B^h_0 = 0$;

1.15 $C = (C^{jk})_{j,k\leq d}$ une fonction continue: $R_+ \to R^d \times R^d$ avec $C_0 = 0$, telle que pour $s \leq t$ la matrice $C_t - C_s$ soit symétrique nonnégative;

1.16 ν une mesure positive sur $R_+ \times R^d$ vérifiant pour tout $t>0$:

(i) $\nu(\{0\}\times R^d) = 0, \quad \nu(R_+\times\{0\}) = 0, \quad \nu([0,t]\times\{x:|x|>\varepsilon\}) < \infty \quad \forall\varepsilon>0$

(ii) $\nu(\{t\}\times R^d) \leq 1$; on pose alors $\delta^h_t = \int \nu(\{t\}\times dx)\, h(x)$;

(iii) $\int_0^t \int_{R^d} |h(x) - \delta^h_s|^2\, \nu(ds\times dx) + \sum_{s\leq t} \{1 - \nu(\{s\}\times R^d)\}|\delta^h_s|^2 < \infty$

(iv) $\sum_{s\leq t} |\int h(x-\delta^h_s)\, \nu(\{s\}\times dx)| + \{1 - \nu(\{s\}\times R^d)\}|h(-\delta^h_s)| < \infty$

et vérifiant en outre:

1.17 $$\Delta B^h_t = \delta^h_t ,$$

tel que si $D_o = \{t>0: \nu(\{t\}\times R^d) = 0\}$ *on ait pour tous* $s<t$, $u \in R^d$:

1.18
$$E(\exp iu.(X_t - X_s)) = \prod_{s<r\leq t}\{\{ 1 + \int \nu(\{r\}\times dx)(e^{iu.x} - 1)\}\, e^{-iu.\Delta B^h_r}\}$$
$$\times \exp\{iu.(B^h_t - B^h_s) - \tfrac{1}{2}u.(C_t - C_s).u + \int_s^t \int_{R^d} (e^{iu.x} - 1 - iu.h(x))1_{D_o}(r)\, \nu(dr\times dx)\}$$

De plus on a alors:

1.19 $$J(X) = \{t: \nu(\{t\}\times R^d) > 0\}$$

1.20 $$P(\Delta X_t \in A) = \nu(\{t\}\times A) \quad \text{pour tout borélien } A \text{ avec } 0 \notin A.$$

b) Inversement, si (B^h,C,ν) *vérifient 1.14, 1.15, 1.16, 1.17, alors le produit infini et la dernière intégrale figurant dans 1.18 sont absolument convergents; il existe un PAI* X *qui vérifie 1.1 et 1.18, et sa loi* $\mathcal{L}(X)$ *est entièrement déterminée par le triplet* (B^h,C,ν).

Exactement comme au §a, C et ν ne dépendent pas de h, mais B^h en dépend: si h' est une autre fonction de troncation, en utilisant 1.17 et l'unicité dans 1.18 on voit qu'on a encore la relation 1.11 (on utilisera toujours la notation 1.9; comme $h-h'$ est bornée et nulle sur un voisinage de 0, 1.16-(i) montre que $(h-h')*\nu$ est bien défini).

Avant de démontrer ce théorème, on va énoncer une série de lemmes techniques (et fastidieux!: les démonstrations ne sont pas à lire) dans le but, notamment, de prouver que la condition 1.16 ne dépend pas de la fonction de troncation h .

LEMME 1.21: *Soit 1.16 et 1.17.*

a) Si B^h *est à variation finie, on a:*

$\underline{1.22}$ $$\sum_{s\leq t} |\int h(x)\, \nu(\{s\}\times dx)| < \infty \quad \forall t>0$$

b) Si on a 1.22, on a aussi:

$\underline{1.23}$ $$(|x|^2\wedge 1)*\nu_t < \infty \qquad \forall t>0.$$

Preuve. a) est immédiat d'après 1.17 et la définition de δ^h. Supposons qu'on ait 1.22; il vient

$$|h(x)|^2*\nu_t \leq 2\int_0^t\int_{R^d} (|h(x) - \delta_s^h|^2 + |\delta_s^h|^2)\, \nu(ds\times dx)$$

$$\leq 2\int_0^t\int_{R^d} |h(x) - \delta_s^h|^2\, \nu(ds\times dx) + 2\sum_{s\leq t} |\delta_s^h|^2$$

Le premier terme ci-dessus est fini d'après 1.16-(iii), le second est fini car $\sum_{s\leq t} |\delta_s^h|<\infty$ par hypothèse. Donc $|h(x)|^2*\nu_t<\infty$. Etant donné 1.16-(i), on en déduit 1.23.∎

LEMME 1.24: *a) Si* ν *vérifie 1.16-(i),(ii), l'ensemble* $\{s: s\leq t, |\delta_s^h|>b\}$ *est fini pour tous* $t>0$, $b>0$.

b) Si ν *vérifie 1.16 pour* h *, alors* ν *vérifie 1.16 pour toute autre fonction de troncation* h'.

Preuve. Soit $a>0$ associé par 1.2 à h. Si $b\leq a/2$ on a $|\delta_s^h| \leq a\, \nu(\{s\}\times\{|x|>b\}) + b$. On déduit alors l'assertion (a) de 1.16-(i). Pour montrer (b), on pose d'abord:

$\underline{1.25}$ $$\alpha_t^h = \int_0^t\int_{R^d} |h(x) - \delta_s^h|^2\, \nu(ds\times dx) + \sum_{s\leq t}\{1 - \nu(\{s\}\times R^d)\}|\delta_s^h|^2$$

$\underline{1.26}$ $$\gamma_t^h = \int_{R^d} h(x-\delta_t^h)\, \nu(\{t\}\times dx) + \{1 - \nu(\{t\}\times R^d)\}\, h(-\delta_t^h) .$$

Comme $h-h'$ est borné et nul sur un voisinage de 0, 1.6-(i) implique

$\underline{1.27}$ $$\sum_{s\leq t} |\delta_s^h - \delta_s^{h'}| \leq |h - h'|*\nu_t < \infty$$

et a-fortiori

$\underline{1.28}$ $$\sum_{s\leq t} |\delta_s^h - \delta_s^{h'}|^2 < \infty$$

On a aussi $|\delta_t^{h'}|^2 \leq 2|\delta_t^h|^2 + 2|\delta_t^h - \delta_t^{h'}|^2$ et

$$|h'(x) - \delta_t^{h'}|^2 \leq 3|h(x) - \delta_t^h|^2 + 3|\delta_t^h - \delta_t^{h'}|^2 + 3|h(x) - h'(x)|^2$$

$$\alpha_t^{h'} \leq 3\alpha_t^h + 3|h - h'|^2*\nu_t + 5\sum_{s\leq t}|\delta_s^h - \delta_s^{h'}|^2 < \infty$$

d'après 1.28 et l'hypothèse $\alpha_t^h<\infty$. Enfin, il existe $b>0$ tel que $h(x) = h'(x) = x$

si $|x|\leq b$; il existe une constante K telle que

$$|\gamma_t^h - \gamma_t^{h'}| \leq K\{1_{\{|\delta_t^h|>b/3\}} + 1_{\{|\delta_t^{h'}|>b/3\}} + \nu(\{t\}\times\{|x|>\tfrac{b}{3}\})\} + |\delta_t^h - \delta_t^{h'}|.$$

Donc d'après la partie (a), 1.16-(i) et 1.27, on a $\sum_{s\leq t}|\gamma_s^h-\gamma_s^{h'}|<\infty$. Comme $\sum_{s\leq t}|\gamma_s^h|<\infty$ on en déduit que $\sum_{s\leq t}|\gamma_s^{h'}|<\infty$, ce qui achève la démonstration. ∎

On suppose encore que ν vérifie 1.16-(i),(ii). La formule suivante définit (grâce à 1.24) une autre mesure $\bar{\nu}^h$ qui vérifie aussi 1.16-(i),(ii):

1.29 $$\bar{\nu}^h(A) = \int 1_A(s,x-\delta_s^h)\, 1_{\{x\neq\delta_s^h\}}\, \nu(ds\times dx) + \sum_s \{1 - \nu(\{s\}\times R^d)\} 1_A(s,-\delta_s^h) 1_{\{\delta_s^h\neq 0\}}.$$

LEMME 1.30: *Supposons qu'on ait 1.16-(i),(ii). Pour que ν vérifie 1.16-(iii) (resp. 1.16-(iv)) il faut et il suffit que $\bar{\nu}^h$ vérifie 1.23 (resp. 1.22).*

Preuve. On utilise les notations 1.25 et 1.26; un calcul simple montre que $\gamma_t^h = \bar{\nu}^h(\{t\}\times h)$, d'où l'équivalence de 1.16-(iv) pour ν et de 1.22 pour $\bar{\nu}^h$. Si $a>0$ vérifie 1.2 on a $|\delta^h|\leq a$, donc

$$(|x|^2\wedge 4a^2)*\bar{\nu}_t^h = \alpha_t^h + \int_0^t\int_{R^d} \{|x-\delta_s^h|^2\wedge 4a^2 - |h(x)-\delta_s^h|^2\}\, \nu(ds\times dx)$$

$$|(|x|^2\wedge 4a^2)*\bar{\nu}_t^h - \alpha_t^h| \leq 8a^2\, \nu([0,t]\times\{|x|>\tfrac{a}{2}\}) < \infty$$

d'après 1.16-(i). Comme $\bar{\nu}^h$ vérifie aussi 1.16-(i), on voit que 1.23 pour $\bar{\nu}^h$ équivaut à $(|x|^2\wedge 4a^2)*\bar{\nu}_t^h<\infty$, donc à 1.16-(iii) pour ν d'après l'inégalité ci-dessus. ∎

LEMME 1.31: *Supposons que (B^h,C,ν) vérifie 1.14-1.17; alors le produit infini et la seconde intégrale dans 1.18 sont absolument convergents; de plus (B^h,C,ν) est le seul triplet vérifiant 1.14-1.17 et 1.18.*

Preuve. Soit $\bar{\nu}^h$ donné par 1.29; un calcul simple montre que

1.32 $$\{1 + \int\nu(\{t\}\times dx)(e^{iu.x} - 1)\}e^{-iu.\delta_t^h} = 1 + \int\bar{\nu}^h(\{t\}\times dx)(e^{iu.x} - 1).$$

Soit $g_u(x) = e^{iu.x} - 1 - iu.h(x)$. Il existe $c_u>0$ tel que $|g_u(x)|\leq c_u(|x|^2\wedge 1)$. Si

1.33 $$\nu^c(dt\times dx) = 1_{D_o}(t)\, \nu(dt\times dx) = 1_{D_o}(t)\, \bar{\nu}^h(dt\times dx),$$

on déduit de 1.30 que $|g_u|*\nu_t^c<\infty$ pour tout t: donc l'intégrale figurant dans 1.18 est absolument convergente. Ensuite, si $\eta_t(u) = \int\bar{\nu}^h(\{t\}\times dx)(e^{iu.x} - 1)$, on a:

$$|\eta_t(u)| \leq |u|\, |\bar{\nu}^h(\{t\}\times h)| + \bar{\nu}^h(\{t\}\times|g_u|)$$

donc $\sum_{s\leq t}|\eta_s(u)|<\infty$ d'après 1.30 encore. Etant donné 1.32 on en déduit que le produit infini de 1.18 converge absolument.

Supposons qu'on ait 1.18; en faisant $s\uparrow t$ dans cette formule, on obtient:

$$E(\exp iu.\Delta X_t) = 1 + \int \nu(\{t\}\times dx)(e^{iu.x} - 1),$$

donc les mesures $\nu(\{t\}\times .)$ sont déterminées de manière unique, donc aussi B^h_t. Par suite la partie "exp..." de 1.18 est aussi unique, ce qui détermine B^h, C, ν^c d'après l'unicité de la représentation de Lévy-Khintchine. ∎

Passons maintenant à la preuve de 1.13. Soit X un PAI vérifiant 1.1, ce qui implique qu'il est centré (rappelons que "centré" signifie: en chaque point, il y a p.s. une limite à gauche et une limite à droite le long des suites). D'après Doob [10, pp.416-417], pour tout t la série $\sum_{s\leq t, s\in J(X)} \Delta X_s$ "converge après centrage", indépendamment de l'ordre de sommation des $s\in J(X)\cap[0,t]$. Cela veut dire qu'il existe des constantes $\rho_s \in R^d$ telles que

1.34 $\begin{cases} \text{la série } \sum_{s\leq t, s\in J(X)} (\Delta X_s - \rho_s) \text{ converge p.s., indépendamment de l'ordre} \\ \text{de sommation} \end{cases}$

(cela ne veut pas dire qu'elle converge absolument p.s., mais que pour tout ordre de sommation elle converge p.s.). Il existe donc une version càdlàg du processus

1.35 $$Z_t = \sum_{s\leq t, s\in J(X)} (\Delta X_s - \rho_s)$$

et il est clair que Z et $X' = X - Z$ sont deux PAI indépendants, càdlàg, avec $J(Z)\subset J(X)$, $J(X')\subset J(X)$.

Soit $f: R_+ \to R^d$ la fonction caractérisée par

$$\forall t \geq 0, \forall j \leq d, \qquad E\{\mathrm{Arctg}(X'^{i}_t - f^i(t))\} = 0 .$$

Comme X' est càdlàg il est facile de vérifier que f est càdlàg, et continue en tout point de $D(X')$. Si $t\in J(X')$ on a $\Delta X'_t = \rho_t$, donc

$$E\{\mathrm{Arctg}(X'^{i}_{t-} - f^i(t-))\} = E\{\mathrm{Arctg}(X'^{i}_t - \rho^i - f^i(t-))\} = 0 .$$

Donc $\Delta f(t) = \rho_t$. Par suite si $Y_t = X'_t - f(t)$, on a $J(Y) = \emptyset$. Rassemblons ces résultats:

1.36 $\begin{cases} X_t = f(t) + Y_t + Z_t, \quad Y \text{ et } Z \text{ sont des PAI indépendants;} \\ Z \text{ est donné par 1.35, } Y \text{ est sans discontinuités fixes, } \Delta f(t) = \rho_t 1_{J(X)}(t) \end{cases}$

D'après le théorème 1.4, il existe β^h vérifiant 1.5, C vérifiant 1.6 = 1.15, ν^c vérifiant 1.7, tels que:

1.37 $E(\exp iu.(Y_t-Y_s)) = \exp\{iu.(\beta_t^h-\beta_s^h) - \frac{1}{2}u.(C_t-C_s).u + g_u*\nu_t^c - g_u*\nu_s^c\}$

Etudions ensuite Z. On note μ_t la loi de la variable aléatoire $\Delta Z_t = \Delta X_t - \rho_t$. D'après le théorème des trois séries de Kolmogorov, 1.34 entraine que:

1.38 $\sum_{s\leq t, s\in J(X)} \mu_s(|x|>1) < \infty$

1.39 $\sum_{s\leq t, s\in J(X)} \mu_s(x1_{\{|x|\leq 1\}})$ converge indépendamment de l'ordre de sommation, donc converge absolument;

1.40 $\sum_{s\leq t, s\in J(X)} \{\mu_s(|x|^2 1_{\{|x|\leq 1\}}) - |\mu_s(x1_{\{|x|\leq 1\}})|^2\} < \infty$

Comme h vérifie 1.2, il est facile de déduire de ces trois propriétés que

1.41 $$\begin{cases} \sum_{s\leq t} \mu_s(|x|^2\wedge 1) < \infty, & \sum_{s\leq t} |\mu_s(h)| < \infty \\ \sum_{s\leq t} |\mu_s(x1_{\{|x|\leq b\}})| < \infty \quad \forall b>0 \end{cases}$$

(si $s\notin J(X)$ on a $\mu_s = \varepsilon_0$, masse de Dirac à l'origine; donc ci-dessus les termes indicés par $s\notin J(X)$ sont tous nuls). Etant donné 1.2 on a aussi:

$$|\int\mu_t(dx)\{h(x+\Delta f_t) - \Delta f_t\} - \mu_t(h)| \leq (2a + |\Delta f_t|)1_{\{|\Delta f_t|>a/4\}} + 3a\,\mu_t(|x|>\tfrac{a}{4})$$

et comme f est càdlàg, l'ensemble $\{s: s\leq t, |\Delta f_s|>a/4\}$ est fini. On déduit donc de 1.41 que

1.42 $$\sum_{s\leq t} |\int\mu_s(dx)\{h(x+\Delta f_s) - \Delta f_s\}| < \infty.$$

On pose alors

1.43 $$\begin{cases} B_t^h = \beta_t^h + f(t) + \sum_{s\leq t} \int\mu_s(dx)\{h(x+\Delta f_s) - \Delta f_s\} \\ \nu(A) = \nu^c(A) + \sum_{s\leq t} \int\mu_s(dx)\, 1_A(s,x+\Delta f_s)\, 1_{\{x+\Delta f_s\neq 0\}}. \end{cases}$$

On remarque que B^h vérifie 1.14, et que ν vérifie 1.16-(i),(ii) d'après les propriétés de ν^c et d'après 1.41. Par ailleurs 1.17 est évident. On a aussi $\delta_t^h = \int\mu_t(dx)h(x+\Delta f_t)$, donc 1.42 implique:

1.44 $$\sum_{s\leq t} |\delta_s^h - \Delta f_s| < \infty.$$

Avec les notations 1.25 et 1.26, on a:

1.45 $$\alpha_t^h = |h|^2*\nu_t^c + \sum_{s\leq t} \int\mu_s(dx)\,|h(x+\Delta f_s) - \delta_s^h|^2$$

$$\gamma_t^h = \int\mu_t(dx)\, h(x+\Delta f_t-\delta_t^h).$$

L'ensemble $K = \{s: |\delta_s^h| > a/8$ ou $|\Delta f_s|>a/8\}$ est localement fini (appliquer 1.24), et on a les majorations:

$$\alpha_t^h \leqq |h|^2 * \nu_t^c + \sum_{s\leqq t} \{4a^2\{1_K(s)+\mu_s(|x|>\tfrac{a}{4})\} + 2\mu_s(|x|^2 1_{\{|x|\leqq\frac{a}{4}\}}) + 2|\Delta f_s - \delta_s^h|^2\}$$

$$|\gamma_t^h| \leqq a\{1_K(t) + \mu_t(|x|>\tfrac{a}{8})\} + |\mu_t(x 1_{\{|x|\leqq a/8\}})| + |\Delta f_t - \delta_t^h| .$$

On déduit alors de 1.41 et 1.44 que $\alpha_t^h<\infty$ et $\sum_{s\leqq t}|\gamma_s^h|<\infty$, donc ν vérifie 1.16-(iii),(iv).

Etant donné 1.30 on a aussi

$$E(e^{iu.(X_t-X_s)}) = E(e^{iu.(Y_t-Y_s)})\, e^{iu.(f(t)-f(s))} \prod_{s<r\leqq t} \int\mu_r(dx)e^{iu.x} .$$

D'après 1.37 et 1.43, cette formule n'est autre que 1.18. Cela achève la preuve de la partie (a) de 1.13, car 1.19 et 1.20 découlent immédiatement de 1.18, tandis que l'unicité de (B^h,C,ν) a été prouvée en 1.31.

Démontrons maintenant la partie (b) de 1.13; la première assertion découle de 1.31. Soit ν^c définie par 1.33: il est évident que ν^c vérifie 1.7, donc d'après le théorème 1.4 il existe un PAI sans discontinuités fixes Y qui vérifie 1.37 avec $\beta^h = 0$. Ensuite, on définit $\bar{\nu}^h$ par 1.29, et on pose

$$\mu_t(dx) = \bar{\nu}^h(\{t\}\times dx) + \{1 - \bar{\nu}^h(\{t\}\times R^d)\}\varepsilon_0(dx) .$$

D'après 1.30 les probabilités μ_t vérifient 1.41, donc d'après le théorème des trois séries on peut construire un PAI Z qui vérifie 1.1, qui est indépendant de Y, et tel que

$$E(e^{iu.(Z_t-Z_s)}) = \prod_{s<r\leqq t} \int\mu_r(dx)e^{iu.x}$$

(on construit des variables indépendantes $\tilde{Z}_s$ de loi μ_s, et on pose $Z_t = \sum_{s\leqq t} \tilde{Z}_s$, qui converge p.s. indépendamment de l'ordre de sommation). Enfin, on pose $X = B^h + Y + Z$, qui est encore un PAI vérifiant 1.1, ainsi que 1.18 par construction (utiliser 1.32). Enfin, la dernière assertion de (b) est évidente.

<u>§c - PAI ET MARTINGALES</u>. Dans ce paragraphe nous allons étendre le théorème 1.12 aux PAI quelconques. Soit X un PAI, auquel on associe le triplet (B^h,C,ν) par 1.13. Les conditions 1.16-(iii) et 1.17 permettent de définir la fonction matricielle suivante $\tilde{C}^h = (\tilde{C}^{h,jk})_{j,k\leqq d}$, qui vérifie 1.15 à l'exception de la continuité (elle est seulement càdlàg):

<u>1.46</u>
$$\left\{\begin{aligned} \tilde{C}_t^{h,jk} = C_t^{jk} &+ \int_0^t\int_{R^d} \{h^j(x)-\Delta B_s^{h,j}\}\{h^k(x)-\Delta B_s^{h,k}\}\nu(ds\times dx) \\ &+ \sum_{s\leqq t} \{1 - \nu(\{s\}\times R^d)\}\Delta B_s^{h,j}\,\Delta B_s^{h,k} \end{aligned}\right.$$

Remarquer que si X n'a pas de discontinuités fixes, on a $\Delta B^h = 0$ et on retrouve la formule 1.10; lorsque 1.22 est satisfaite, donc aussi 1.23, on a:

<u>1.47</u>
$$\tilde{C}_t^{h,jk} = C_t^{jk} + (h^jh^k)*\nu_t - \sum_{s\leqq t} \Delta B_s^{h,j}\,\Delta B_s^{h,k} .$$

Enfin, dans tous les cas on a:

<u>1.48</u> $$\Delta\tilde{C}_t^{h,jk} = \nu(\{t\}\times h^j h^k) - \Delta B_t^{h,j}\,\Delta B_t^{h,k}.$$

<u>THEOREME 1.49</u>: *Soit* X *un PAI vérifiant 1.1, sur l'espace* $(\Omega,\underline{\underline{F}},(\underline{\underline{F}}_t),P)$. *Soit* (B^h,C,ν) *le triplet défini en 1.13, et* X^h *et* $\tilde{C}^h$ *définis par 1.3 et 1.46. Alors:*

(*i*) $\tilde{X}^h = X^h - B^h$ *est une martingale;*

(*ii*) *pour tous* $j,k\leq d$, $\tilde{X}^{h,j}\tilde{X}^{h,k} - \tilde{C}^{h,jk}$ *est une martingale;*

(*iii*) *pour toute fonction* g *borélienne bornée nulle sur un voisinage de* 0 , *le processus* $\sum_{s\leq t} g(\Delta X_s) - g*\nu_t$ *est une martingale.*

De plus, (B^h,C,ν) *est le seul triplet vérifiant 1.14-1.17 et ayant ces propriétés.*

<u>Preuve</u>. L'unicité se montre comme dans 1.12. On va reprendre intégralement les notations du §b, et notamment la décomposition 1.36; soit $Z' = f + Z$.

Les PAI Y et Z' n'ont pas de sauts communs donc si g est comme en (iii) on a $\sum_{s\leq t} g(\Delta X_s) = \sum_{s\leq t} g(\Delta Y_s) + \sum_{s\leq t} g(\Delta Z'_s)$. D'après la preuve de 1.12 on a:

$$E\{\textstyle\sum_{s\leq t} g(\Delta Y_s)\} = g*\nu_t^c.$$

D'autre part $\Delta Z'_s = \Delta f_s + \Delta Z_s$ et μ_s est la loi de ΔZ_s , et Z' ne saute que sur l'ensemble dénombrable $J(X)$. Donc si $g\geq 0$ on a

$$\begin{aligned} E\{\textstyle\sum_{s\leq t} g(\Delta X_s)\} &= g*\nu_t^c + \textstyle\sum_{s\leq t,\, s\in J(X)} E\{g(\Delta Z_s + \Delta f_s)\} \\ &= g*\nu_t^c + \textstyle\sum_{s\leq t,\, s\in J(X)} \int \mu_s(dx)\, g(x+\Delta f_s) = g*\nu_t \end{aligned}$$

d'après 1.43. On en déduit alors (iii) comme dans 1.12.

A nouveau, le même argument que dans 1.12 montre que pour obtenir (i) et (ii) il suffit de prouver que si $\phi_t(u) = E(\exp iu.X_t^h)$ on a:

<u>1.50</u> $$\partial\phi_t/\partial u^j(0) = i\, B_t^{h,j}, \qquad \partial^2\phi_t/\partial u^j\partial u^k(0) = -B_t^{h,j}B_t^{h,k} - \tilde{C}_t^{h,jk}.$$

Comme Y et Z' n'ont pas de sauts communs, $X^h = Y^h + Z'^h$, et Y^h et Z'^h sont indépendants. Comme les caractéristiques de Y sont (β^h,C,ν^c), d'après la preuve de 1.12 on a:

$$E(e^{iu.Y_t^h}) = \exp\{iu.\beta_t^h - \tfrac{1}{2}u.C_t.u + g_u*\nu_t^c\}.$$

Par ailleurs un calcul simple montre que $Z_t'^h = f(t) + \sum_{s\leq t} \{h(\Delta Z_s+\Delta f_s) - \Delta f_s\}$, la série du second membre convergeant p.s. indépendamment de l'ordre de sommation. Donc

$$E(e^{iu.Z_t'^h}) = e^{iu.f(t)} \prod_{s\leq t} \int \mu_s(dx)\, e^{iu.(h(x+\Delta f_s)-\Delta f_s)}.$$

Comme d'une part $X^h = Y^h + Z'^h$ avec Y^h et Z'^h indépendants, comme d'autre part

$B^h_t = \beta^h_t + f(t) + \sum_{s\leq t}(\delta^h_s - \Delta f_s)$ d'après 1.43, il vient:

$$\phi_t(u) = \{\prod_{s\leq t} \int \mu_s(dx) e^{iu.(h(x+\Delta f_s)-\delta^h_s)}\} \exp\{iu.B^h_t - \frac{1}{2}u.C_t.u + g_u * \nu^c_t\} .$$

Par ailleurs, $(|x|^2 \wedge 1) * \nu^c_t < \infty$, donc on peut dériver deux fois $g_u * \nu^c_t$ sous le signe somme; de plus si $\eta_s(u) = \int \mu_s(dx)\{\exp iu.(h(x+\Delta f_s)-\delta^h_s) - 1\}$ on a $|\eta_s(u)| \leq 1/2$ pour tout u assez petit (uniformément en s), donc on peut remplacer ci-dessus le produit infini par: $\exp \sum_{s\leq t} \mathrm{Log}(1+\eta_s(u))$. Mais on a aussi

$$\eta_s(u) = \int \mu_s(dx)\{e^{iu.(h(x+\Delta f_s)-\delta^h_s)} - 1 - iu.(h(x+\Delta f_s)-\delta^h_s)\}$$

(car $\delta^h_s = \int \mu_s(dx) h(x+\Delta f_s)$), et d'après 1.45,

$$\sum_{s\leq t} \int \mu_s(dx) \; |h(x+\Delta f_s) - \delta^h_s|^2 < \infty .$$

Donc dans $\exp \sum_{s\leq t} \mathrm{Log}(1+\eta_s(u))$ on peut dériver deux fois terme à terme. Cela donne:

$$\partial\phi_t(u)/\partial u^j = i\phi_t(u) \{\sum_{s\leq t} \{1+\eta_s(u)\}^{-1} \int \mu_s(dx)(e^{iu.(h(x+\Delta f_s)-\delta^h_s)} - 1)(h^j(x+\Delta f_s) - \delta^{h,j}_s)$$
$$+ B^{h,j}_t - \sum_{k\leq d} u^k C^{jk}_t + (e^{iu.h} - 1)h^j * \nu^c_t\} .$$

On en déduit facilement la première relation 1.50, ainsi que:

$$\partial^2\phi_t/\partial u^j \partial u^k(0) = - B^{h,j}_t B^{h,k}_t - \{\sum_{s\leq t} \int \mu_s(dx)\{h^j(x+\Delta f_s) - \delta^{h,j}_s\}\{h^k(x+\Delta f_s) - \delta^{h,k}_s\}$$
$$+ C^{jk}_t + (h^j h^k) * \nu^c_t\}$$

Il reste à utiliser 1.43 et $\delta^h = \Delta B^h$ pour identifier la relation ci-dessus avec la seconde relation 1.50.∎

<u>REMARQUE 1.51</u>: On verra dans la chapitre suivant une réciproque à ce théorème: si (B^h,C,ν) vérifient 1.14-1.17 et si X est càdlàg adapté nul en 0, les conditions (i)(ii)(iii) impliquent que X est un PAI.∎

Terminons par un résultat facile:

<u>PROPOSITION 1.52</u>: *Soit* X *un PAI de caractéristiques* (B^h,C,ν); *soit* g *une fonction bornée*: $\mathbb{R}^d \to \mathbb{R}_+$, *nulle sur un voisinage de* 0. *Alors* $X'_t = \sum_{s\leq t} g(\Delta X_s)$ *est un PAI, et*

$$E(\exp - \sum_{s\leq t} g(\Delta X_s)) = \exp\{-\int_0^t \int_{\mathbb{R}^d} (1-e^{-g(x)}) 1_{D_o}(s)\nu(ds\times dx) + \sum_{s\leq t} \mathrm{Log}\{1-\nu(\{s\}\times(1-e^{-g}))\}\}.$$

<u>Preuve</u>. Il existe bien-sûr une démonstration directe de ce résultat, mais nous allons utiliser 1.49. D'abord, comme $0 \leq 1-e^{-g} < 1$ le Log ci-dessus est bien défini. Soit $b = \mathrm{Sup}|g|$. On peut choisir une fonction de troncation h telle que $a > 2b$, donc

$|\Delta X'| \leq a/2$ et $X'^h = X'$.

Il est clair que X' est un PAI, dont on note (B'^h, C', ν') les caractéristiques. D'après 1.18, si $C'_t \neq 0$ la variable X'_t est somme d'une gaussienne non dégénérée et d'une autre variable, indépendante de la première. Comme $X'_t \geq 0$ il faut donc que $C'_t = 0$. D'après 1.49-(iii) appliqué à X, $X' - g*\nu$ est une martingale, donc 1.49-(i) appliqué à X' montre que $B'^h = g*\nu$. Enfin $\sum_{s\leq t} f(\Delta X'_s) - (f\circ g)*\nu_t = \sum_{s\leq t} f\circ g(\Delta X_s) - (f\circ g)*\nu_t$ est une martingale pour toute fonction f bornée nulle autour de 0, donc $f*\nu' = (f\circ g)*\nu$. Finalement, on a:

1.53 $$B'^h = g*\nu, \quad C' = 0, \quad \nu'(A) = \int 1_A(s,g(x))\, \nu(ds\times dx).$$

Par ailleurs, comme $X' \geq 0$ on peut utiliser la transformée de Laplace, et en particulier remplacer iu par -1 dans 1.18: compte tenu de 1.53, cela donne la formule de l'énoncé.∎

2 - CONDITION NECESSAIRE ET SUFFISANTE DE CONVERGENCE VERS UN PAI SANS DISCONTINUITES FIXES

§a - ENONCE DE LA CONDITION. Nous considérons maintenant une suite (X^n) de PAI à valeurs dans R^d, qui tous vérifient 1.1. Nous allons énoncer une condition nécessaire et suffisante pour que $X^n \xrightarrow{\mathcal{L}} X$ (dans ce cas, X est aussi un PAI), lorsque le processus X n'a pas de discontinuités fixes (par contre, et c'est important pour les applications, les X^n peuvent avoir des discontinuités fixes).

On fixe une fonction de troncation h. On appelle $(B^{h,n}, C^n, \nu^n)$ les caractéristiques de X^n, et (B^h, C, ν) celles de X. on définit les fonctions $\tilde{C}^{h,n}$ et $\tilde{C}^h$ par la formule 1.46, à partir de $(B^{h,n}, C^n, \nu^n)$ et (B^h, C, ν) respectivement. Rappelons que X n'a pas de discontinuités fixes si et seulement si (B^h, C, ν) vérifie les conditions 1.5, 1.6, 1.7.

THEOREME 2.1: *On suppose la fonction de troncation* h *continue, et le PAI* X *sans discontinuités fixes. Pour que* $X^n \xrightarrow{\mathcal{L}} X$ *il faut et il suffit qu'on ait les trois conditions suivantes:*

[Sup-β] $B^{h,n} \to B^h$ uniformément sur les compacts;

[γ] $\tilde{C}^{h,n}_t \to \tilde{C}^h_t$ pour tout t dans une partie dense A de R_+

[δ] $f*\nu^n_t \to f*\nu_t$ pour tout t dans une partie dense A de R_+ et toute fonction f continue bornée: $R^d \to R_+$ nulle sur un voisinage de 0.

Dans ce cas on peut prendre $A = R_+$ *dans* [γ] *et* [δ], *et on a aussi:*

[Sup-γ] $\tilde{C}^{h,n} \to \tilde{C}^h$ uniformément sur les compacts

[Sup-δ] $f*\nu^n \to f*\nu$ uniformément sur les compacts, pour toute fonction f continue bornée: $R^d \to R_+$ nulle sur un voisinage de 0.

Faisons quelques commentaires. D'abord, dans [γ] et [δ] les fonctions qu'on considère sont croissantes, et les limites $\tilde{C}^h$ et $f*\nu$ sont continues; l'assertion 1.21 du chapitre I montre alors que [γ] = [Sup-γ] et [δ] = [Sup-δ], et on peut prendre $A = R_+$.

Ensuite, on peut affaiblir [δ] ainsi (c'est important, car la condition [δ] porte sur une infinité non dénombrable de fonctions f).

LEMME 2.2: *Soit* $\{f_p\}$ *une suite de fonctions, dense pour la convergence uniforme dans l'ensemble des fonctions positives bornée sur* R^d, *nulles sur un voisinage de* 0 *et uniformément continues. Alors* [δ] *équivaut à:*

[δ'] $f_d*\nu^n_t \to f_p*\nu_t$ pour tout $p \geq 1$ et tout t dans une partie dense A de R_+.

Preuve. Pour la même raison que ci-dessus, on peut supposer que $A = R_+$ dans [δ']. Mais si $f_p*\nu^n_t \to f_p*\nu_t$ pour un t et pour tout p, un raisonnement classique sur la convergence étroite montre que $f*\nu^n_t \to f*\nu_t$ pour toute fonction continue bornée nulle sur un voisinage de 0. ∎

Supposons de plus que chaque processus X^n soit sans discontinuités fixes. Alors

$$E(\exp iu.X^n_t) = \exp\{iu.B^{h,n}_t - \frac{1}{2}u.C^n_t.u + (e^{iu.x} - 1 - iu.h(x))*\nu^n_t\}$$

et $\tilde{C}^{h,n}_t = C^n_t + (h\otimes h)*\nu^n_t$, et de même pour X. D'après les résultats classiques de convergence des lois indéfiniment divisibles (voir par exemple Gnedenko et Kolmogorov [15]) on a

2.3 $$X^n_t \xrightarrow{\mathcal{L}} X_t \iff \begin{cases} B^{h,n}_t \to B^h_t, \quad \tilde{C}^{h,n}_t \to \tilde{C}^h_t, \quad f*\nu^n_t \to f*\nu_t \text{ pour toute } f \\ \text{continue bornée nulle autour de } 0, \end{cases}$$

à condition que h soit continue: on ne peut donc pas espérer avoir 2.1 lorsque h n'est pas continue, en général (plus précisément on a 2.1 pour h discontinue, à condition que $\nu([0,t]\times dx)$ ne charge pas l'ensemble des points de discontinuité de h, pour aucune valeur de t).

COROLLAIRE 2.4: *Supposons que les PAI* X^n *n'aient pas de discontinuités fixes, et que* $X^n_t \xrightarrow{\mathcal{L}} X_t$ *pour tout* t. *Pour que* $X^n \xrightarrow{\mathcal{L}} X$ *il faut et il suffit qu'on ait* [Sup-β].

En particulier si X^n et X sont des PAI homogènes, on a $B_t^{h,n} = b^n t$ et $B_t^h = bt$; $\tilde{C}_t^{h,n} = \tilde{c}^n t$ et $\tilde{C}_t^h = \tilde{c}t$; $\nu^n(dt\times dx) = dt\times F^n(dx)$ et $\nu(dt\times dx) = dt\times F(dx)$. On retrouve alors le résultat bien connu suivant:

COROLLAIRE 2.5: *Si les* X^n *et* X *sont des PAI homogènes, pour que* $X^n \xrightarrow{\mathcal{L}} X$ *il faut et il suffit que* $X_1^n \xrightarrow{\mathcal{L}} X_1$.

Pour terminer, signalons que le théorème 2.1 s'étend au cas où le processus X admet des discontinuités fixes: voir [28]. Voici l'énoncé dans le cas général:

THEOREME: *Suppons* h *continue. Pour que* $X^n \xrightarrow{\mathcal{L}} X$ *il faut et il suffit qu'on ait* $[\gamma]$ *et*

[Sk-β] $\quad B^{h,n} \to B^h$ pour la topologie de Skorokhod

[Sk-δ] $\quad f*\nu^n \to f*\nu$ pour la topologie de Skorokhod, pour toute fonction f continue bornée: $R^d \to R_+$ nulle sur un voisinage de 0.

Dans ce cas, on a aussi:

[Sk-γ] $\quad \tilde{C}^{h,n} \to \tilde{C}^h$ pour la topologie de Skorokhod.

(Noter que si X n'a pas de discontinuités fixes, on a trivialement [Sk-β]=[Sup-β] et [Sk-δ]=[δ]).

§b - UN LEMME FONDAMENTAL POUR LA CONDITION NECESSAIRE. On se place dans le même cadre qu'au §a: les X^n et X sont des PAI, et X n'a pas de discontinuités fixes. Nous ne répèterons pas cette hypothèse. Le point clé pour la démonstration de la condition nécessaire est la

PROPOSITION 2.6: *Supposons que* $X^n \xrightarrow{\mathcal{L}} X$. *Supposons que pour chaque* $t>0$ *la suite* $\{\sup_{s\leq t} |X_s^n|\}_{n\geq 1}$ *de variables aléatoires soit uniformément intégrable. Si* $\alpha_n(t) = E(X_t^n)$ *et* $\alpha(t) = E(X_t)$, *alors* $\alpha_n \to \alpha$ *uniformément sur les compacts.*

Cette proposition fonctionne de la manière suivante, pour prouver la condition nécessaire; supposons par exemple qu'on veuille montrer [Sup-β]; si $X^n \xrightarrow{\mathcal{L}} X$ et si h est continue, il est facile de voir que $X^{h,n} \xrightarrow{\mathcal{L}} X^h$; comme $B_t^{h,n} = E(X_t^{h,n})$ et $B_t^h = E(X_t^h)$ d'après 1.49, on obtient [Sup-β] si on parvient à montrer l'uniforme intégrabilité de la suite $\{\sup_{s\leq t} |X_s^{h,n}|\}_{n\geq 1}$.

Signalons que le fait que chaque X^n soit un PAI est essentiel pour cette proposition.

Commençons par une série de lemmes.

LEMME 2.7: *Si* $X^n \xrightarrow{\mathcal{L}} X$ *on a* $\sup_n f*\nu^n_t<\infty$ *pour tout* t *et toute fonction* f *positive bornée, nulle sur un voisinage de* 0.

Preuve. Il suffit de montrer le résultat lorsque f est continue et vérifie $0\leq f\leq 1$. L'application $\alpha \rightsquigarrow \sum_{s\leq .} f(\Delta\alpha(s))$ étant alors continue de D^d dans D^1 pour la topologie de Skorokhod (cela découle de l'assertion I-1.19 appliquée avec un $u>0$ tel que $f(x) = 0$ pour $|x|\leq u$), et comme $D(X) = R_+$, on a

2.8
$$\sum_{s\leq t} f(\Delta X^n_s) \xrightarrow{\mathcal{L}} \sum_{s\leq t} f(\Delta X_s) .$$

Posons

$$b_n = \int_0^t\int_{R^d} 1_{D(X^n)}(s)(1 - e^{-f(x)})\nu^n(ds\times dx) - \sum_{s\leq t}\mathrm{Log}\{1 - \nu^n(\{s\}\times(1-e^{-f}))\}$$

et $b = (1-e^{-f})*\nu_t$. La proposition 1.52 et 2.8 entrainent alors que $b_n \to b$ (rappelons que $\nu(\{s\}\times R^d) = 0$ pour tout s).

Par ailleurs il existe un nombre $\gamma>1$ tel que

$$0 \leq x \leq 1 \Longrightarrow x \leq \gamma(1-e^{-x}), \qquad 0 \leq y \leq 1-\frac{1}{e} \Longrightarrow y \leq -\gamma\,\mathrm{Log}(1-y),$$

donc

$$\begin{aligned} f*\nu^n_t &\leq \gamma(1 - e^{-f})*\nu^n_t \\ &\leq \gamma\{\int_0^t\int_{R^d} 1_{D(X^n)}(s)(1 - e^{-f(x)})\nu^n(ds\times dx) - \gamma\sum_{s\leq t}\mathrm{Log}\{1 - \nu^n(\{s\}\times(1-e^{-f}))\}\} \\ &\leq \gamma^2 b_n . \end{aligned}$$

Comme $b_n \to b$, on a le résultat. ∎

LEMME 2.9: *Si* $X^n \xrightarrow{\mathcal{L}} X$ *on a* [δ].

Preuve. Il suffit de montrer que $f*\nu^n_t \to f*\nu_t$ pour f continue, nulle sur un voisinage de 0 , et vérifiant $0\leq f\leq a/2$, où a vérifie 1.2. Soit

$$X'^n_t = \sum_{s\leq t} f(\Delta X^n_s) , \qquad X'_t = \sum_{s\leq t} f(\Delta X_s) .$$

D'après 1.53, les caractéristiques $(B'^{h,n},C'^n,\nu'^n)$ du PAI réel X'^n sont

$$B'^{h,n} = f*\nu^n , \quad C'^n = 0 , \quad \nu'^n(A) = \int 1_A(s,f(x))\, \nu^n(ds\times dx) .$$

En particulier ν'^n vérifie 1.22 et 1.23 et $\tilde{C}'^{h,n}$ est donné par 1.47, donc vérifie $\tilde{C}'^{h,n}_t \leq f^2*\nu^n_t$. Par suite $M^n = X'^n - f*\nu^n$ vérifie d'après 1.49:

$$E\{(M^n_t)^2\} = \tilde{C}'^{h,n}_t \leq f^2*\nu^n_t$$

D'après le lemme 2.7 on a $\mathrm{Sup}_n f^2*\nu^n_t<\infty$, donc l'inégalité précédente montre que la suite $(M^n_t)_{n\geq 1}$ est uniformément intégrable. Comme on a aussi $\sup_n f*\nu^n_t<\infty$, et comme $X'^n = M^n + f*\nu^n$, la suite $(X'^n_t)_{n\geq 1}$ est également uniformément intégrable. D'après 2.8 on a $X'^n_t \xrightarrow{\mathcal{L}} X'_t$, donc $f*\nu^n_t = E(X'^n_t) \to E(X'_t) = f*\nu_t$. ∎

Preuve de la proposition 2.6. Comme $D(X) = R_+$ on a $X_t^n \xrightarrow{\mathcal{L}} X_t$ pour tout t, donc $\alpha_n(t) \to \alpha(t)$ pour tout t à cause de l'hypothèse d'uniforme intégrabilité. Il suffit donc de montrer que la suite (α_n) est relativement compacte pour la topologie de Skorokhod de D^d.

On utilise les modules de continuité w_N et w'_N du chapitre I. On va montrer:

2.10 $$\forall N \in \mathbb{N}^*, \qquad \lim_{\delta \downarrow 0} \limsup_n w_N(\alpha_n, \delta) = 0 .$$

Comme $\sup_{t \leq N} |\alpha_n(t)| \leq |\alpha_n(0)| + k\, w_N(\alpha_n, N/k)$ et comme $\alpha_n(0) = 0$, 2.10 implique que la suite (α_n) vérifie la condition I-1.16-(i). Comme $w'_N(\alpha_n,\delta) \leq w_N(\alpha_n, 2\delta)$, 2.10 implique aussi que la suite (α_n) vérifie la condition I 1.16 (ii), si bien qu'on aura le résultat d'après le théorème I-1.16.

Il reste à montrer 2.10; soit $N \in \mathbb{N}^*$ et $\varepsilon > 0$. Soit $M^n = \sup_{s \leq N} |X_s^n|$. D'après l'hypothèse, il existe $\theta > 0$ tel que

2.11 $$\sup_n E^n(M^n 1_{\{M^n > \theta\}}) \leq \varepsilon .$$

D'après I-2.7 il existe $\delta_o > 0$ et $n_o \in \mathbb{N}^*$ tels que

2.12 $$n > n_o \implies P^n(B^n) \leq \varepsilon/\theta , \quad \text{où } B^n = \{w'_N(X^n, \delta_o) > \varepsilon\}.$$

Soit g une fonction continue sur R^d, nulle autour de 0, vérifiant $0 \leq g \leq 1$ et $g(x) = 1$ si $|x| \geq \varepsilon$. D'après 2.9, $g*\nu^n \to g*\nu$ uniformément sur les compacts, et $g*\nu$ est continue, donc uniformément continue sur les compacts. Il existe donc $n_1 \geq n_o$ et $\delta_1 \in]0, \delta_o]$ tels que

2.13 $$n > n_1 \implies \sup_{s \leq N} (g*\nu^n_{s+\delta_1} - g*\nu^n_s) \leq \varepsilon/\theta .$$

Soit $C_s^n = \{\sup_{s < r \leq s+\delta_1} |\Delta X_r^n| > \varepsilon\}$. Il vient

$$P^n(C_s^n) \leq E^n(\textstyle\sum_{s<r\leq s+\delta_1} g(\Delta X_r^n)) \leq g*\nu^n_{s+\delta_1} - g*\nu^n_s ,$$

d'où d'après 2.13:

2.14 $$n > n_1 , \quad s \leq N \implies P^n(C_s^n) \leq \varepsilon/\theta .$$

Soit alors $s \leq t \leq s+\delta_1 \leq N$. Si $\omega \in (C_s^n \cup B^n)^c$ il est facile de vérifier que $|X_t^n - X_s^n| \leq 3\varepsilon$. Par suite

$$|X_t^n - X_s^n| \leq 3\varepsilon + 2M^n 1_{C_s^n \cup B^n} \leq 3\varepsilon + 2\theta(1_{C_s^n} + 1_{B^n}) + 2M^n 1_{\{M^n > \theta\}}$$

et 2.11, 2.12, 2.14 entrainent pour $n > n_1$:

$$|\alpha_n(t) - \alpha_n(s)| \leq 3\varepsilon + 2\theta(\frac{\varepsilon}{\theta} + \frac{\varepsilon}{\theta}) + 2\varepsilon = 9\varepsilon .$$

On en déduit que $w(\alpha_n,]s, s+\delta_1]) \leq 9\varepsilon$ si $n > n_1$ et $s+\delta_1 \leq N$; donc $w_N(\alpha_n, \delta_1) \leq 9\varepsilon$ si $n > n_1$, d'où finalement 2.10.∎

§c - DEMONSTRATION DE LA CONDITION NECESSAIRE. Commençons par un lemme plus ou moins classique sur les fonctions caractéristiques (voir Gnedenko et Kolmogorov [15] ou Petrov [45]).

LEMME 2.15: *Pour tous* $a>0$, $\theta>0$ *il existe une constante* $c(a,\theta)$ *ayant la propriété suivante: si* G *est une probabilité sur* R^d *vérifiant* $G(|x|>a) = 0$, *si* $\delta=\int xG(dx)$ *et si* ϕ_G *est la fonction caractéristique de* G , *on a*

$$\int G(dx)\ |x-\delta|^2 \leqq c(a,\theta) \int_{|u|\leqq\theta} \{1 - |\phi_G(u)|^2\}du$$

Preuve. Soit $\tilde{G}$ la symétrisée de G:

$$\int\tilde{G}(f) = \int\int G(dx)\ G(dy)\ f(x-y)\ .$$

D'après la définition de δ, on a $\int G(dx)|x-\delta|^2 \leqq \int G(dx)|x-y|^2$ pour tout $y\in R^d$. Donc

2.16 $$\int\tilde{G}(dx)\ |x|^2 = \int G(dy) \int G(dx)\ |x-y|^2 \geqq \int G(dx)\ |x-\delta|^2.$$

Par ailleurs $\phi_{\tilde{G}}(u) = \int\tilde{G}(dx)\cos(u.x) = |\phi_G(u)|^2$, donc

2.17 $$\int_{|u|\leqq\theta} \{1 - |\phi_G(u)|^2\}du = \int_{|x|\leqq 2a} \tilde{G}(dx) \int_{|u|\leqq\theta} (1 - \cos(u.x))du$$

car $\tilde{G}(|x|>2a) = 0$. Pour chaque $x\in R^d\setminus\{0\}$ on note $C(x,\theta)$ un hypercube de R^d inscrit dans la sphère $\{u: |u|\leqq\theta\}$ et dont l'un des cotés est parallèle au vecteur x . Les cotés de $C(x,\theta)$ ont pour longueur $b = 2\theta\sqrt{d}$, et si $x \neq 0$ on a

$$\begin{aligned}\int_{|u|\leqq\theta}(1-\cos(u.x))du &\geqq \int_{C(x,\theta)}(1-\cos(u.x))du \\ &\geqq b^{d-1}\int_{-b/2}^{b/2}(1-\cos s|x|)ds = b^d(1-\tfrac{2}{b|x|}\sin\tfrac{b|x|}{2}).\end{aligned}$$

Il existe $b'>0$ tel que

$$0 < |x| \leqq 2a \quad\Longrightarrow\quad b^d(1-\tfrac{2}{b|x|}\sin\tfrac{b|x|}{2}) \geqq b'|x|^2.$$

Si on compare 2.16 et 2.17, on voit que la formule de l'énoncé est vraie avec $c(a,\theta) = 1/b'$. ∎

LEMME 2.18: *Si* $X^n \xrightarrow{\mathcal{L}} X$, *on a*

(*i*) $X^{h,n} \xrightarrow{\mathcal{L}} X^h$;

(*ii*) *pour tout* $t>0$ *et tout* $j\leqq d$, *on a* $\sup_n \tilde{C}_t^{h,n,jj} < \infty$.

Preuve. Comme h est continue et vérifie 1.2, une modification évidente de la preuve de I-1.19 montre que l'application: $\alpha \rightsquigarrow \alpha^h(t) = \alpha(t) - \sum_{s\leqq t}\{\Delta\alpha(s)-h(\Delta\alpha(s))\}$ est continue de D^d dans D^d pour la topologie de Skorokhod, d'où (i).

Posons $\nu^{nc}(ds\times dx) = \nu^n(ds\times dx)1_{D(X^n)}(s)$ et notons μ_s^n la loi de la variable ΔX_s^n . Etant donnés 1.19 et 1.20, la formule 1.18 s'écrit:

$$E^n(e^{iu.X_t^n}) = (\prod_{s\leqq t} e^{-iu.\Delta B_s^{h,n}} \int\mu_s^n(dx)e^{iu.x})\exp\{iu.B_t^{h,n} - \tfrac{1}{2}u.C_t^n.u + g_u * \nu_t^{nc}\}$$

Soit $t>0$ fixé. Il existe $\theta>0$ tel que $\operatorname{Inf}_{|u|\leq\theta}|E(\exp iu.X_t)| \geq 3/4$. Comme $X_t^n \xrightarrow{\mathcal{L}} X_t$, il existe n_o tel que: $\operatorname{Inf}_{|u|\leq\theta, n>n_o} |E(\exp iu.X_t^n)| \geq 1/2$. Donc

$$\text{2.19} \qquad n>n_o,\ |u|\leq\theta \Longrightarrow |\textstyle\prod_{s\leq t}\int\mu_s^n(dx)e^{iu.x}| \quad \exp\{-\frac{1}{2}u.C_t^n.u - (1-\cos(u.x))*\nu_t^{nc}\} \geq \frac{1}{2}.$$

En particulier, chacun des facteurs du second membre ci-dessus est minoré par 1/2.

Appliquons d'abord ceci à l'exponentielle dans 2.19:

$$n>n_o,\ |u|\leq\theta \Longrightarrow \frac{1}{2}u.C_t^n.u + (1-\cos(u.x))*\nu_t^{nc} \leq \operatorname{Log} 2.$$

Soit a le nombre figurant dans 1.2; soit $u\in R^d$ de composantes $u^j=\theta\wedge\frac{\pi}{a}$ et $u^k=0$ pour $k\neq j$. On a $y^2 \leq (\pi^2/2)(1-\cos y)$ si $|y|\leq\pi$, donc si $\lambda=\theta\wedge\frac{\pi}{a}$ il vient:

$$\begin{aligned}1-\cos(u.x) &\geq (1-\cos(u.x))1_{\{|x|\leq a\}} \geq \frac{2}{\pi^2}|u.x|^2\, 1_{\{|x|\leq a\}} \\ &= \frac{2\lambda^2}{\pi^2}|x^j|^2\, 1_{\{|x|\leq a\}} \geq \frac{2\lambda^2}{\pi^2}|h^j(x)|^2\end{aligned}$$

et donc

$$\text{2.20} \qquad n>n_o \Longrightarrow C_t^{n,jj} + |h^j|^2*\nu_t^{nc} \leq \frac{\pi^2}{2\lambda^2}\operatorname{Log} 2 .$$

Passons maintenant à l'étude du produit infini de 2.19. Pour tout $y>0$ on a $1-y^2 \leq -2\operatorname{Log} y$, donc 2.19 implique

$$\text{2.21} \qquad n>n_o,\ |u|\leq\theta \Longrightarrow \textstyle\sum_{s\leq t}\{1-|\int\mu_s^n(dx)\, e^{iu.x}|^2\} \leq 2\operatorname{Log} 2.$$

Soit par ailleurs $\overline{\mu}_s^n$ la loi de la variable $h(\Delta X_s^n)$, donc $\overline{\mu}_s^n(f)=\mu_s^n(f\circ h)$ et

$$|\textstyle\int\mu_s^n(dx)e^{iu.x} - \int\overline{\mu}_s^n(dx)e^{iu.x}| = |\int\mu_s^n(dx)(e^{iu.h(x)}-e^{iu.x})| \leq 2\mu_s^n(|x|>\frac{a}{2}).$$

Le lemme 2.7 appliqué à $f(x)=1_{\{|x|>a/2\}}$ entraine que $K := \sup_n \sum_{s\leq t}\mu_s^n(|x|>\frac{a}{2})$ est fini (on rappelle que $\mu_s^n(A)=\nu^n(\{s\}\times A)$ si $0\notin A$), donc

$$\textstyle\sum_{s\leq t}||\int\mu_s^n(dx)e^{iu.x}|^2 - |\int\overline{\mu}_s^n(dx)e^{iu.x}|^2| \leq 2\sum_{s\leq t}|\int\mu_s^n(dx)e^{iu.x} - \int\overline{\mu}_s^n(dx)e^{iu.x}| \leq 4K.$$

Donc 2.21 implique:

$$n>n_o,\ |u|\leq\theta \Longrightarrow \textstyle\sum_{s\leq t}\{1-|\int\overline{\mu}_s^n(dx)\, e^{iu.x}|^2\} \leq 4K + 2\operatorname{Log} 2.$$

On applique alors le lemme 2.15 aux probabilités $\overline{\mu}_s^n$: on a $\overline{\mu}_s^n(|x|>a)=0$ par construction, et $\Delta B_s^{h,n} = \int\overline{\mu}_s^n(dx)\,x$ d'après 1.17, donc si $b(\theta)$ désigne le volume de la sphère de R^d de rayon θ il vient:

$$\text{2.22} \qquad n>n_o \Longrightarrow \textstyle\sum_{s\leq t}\int\overline{\mu}_s^n(dx)|\,x-\Delta B_s^{h,n}|^2 \leq b(\theta)c(a,\theta)(4K+2\operatorname{Log} 2).$$

En utilisant 1.46, un calcul simple montre que:

$$\tilde{C}_t^{n,h,jj} = C_t^{n,jj} + |h^j|^2 * \nu_t^{nc} + \sum_{s\leq t} \int \mu_s^n(dx)\, |h^j(x) - \Delta B_s^{h,n,j}|^2$$
$$= C_t^{n,jj} + |h^j|^2 * \nu_t^{nc} + \sum_{s\leq t} \int \overline{\mu}_s^n(dx)\, |x^j - \Delta B_s^{h,nj}|^2 .$$

Il suffit alors d'ajouter 2.20 et 2.22 pour obtenir que $\tilde{C}_t^{h,njj} \leq (\pi^2/2\lambda^2)\,\mathrm{Log}\,2 + b(\theta)c(a,\theta)(4K + 2\,\mathrm{Log}\,2)$ si $n>n_o$, d'où le résultat. ∎

Poursuivons par un lemme général sur les martingales. La "variation quadratique prévisible" $\langle M,M\rangle$ d'une martingale localement de carré intégrable M a été définie au §I-3-b.

LEMME 2.23: *Soit* M *une martingale localement de carré intégrable (réelle), vérifiant* $|\Delta M|\leq b$ *identiquement. Il existe deux constantes* γ_1 *et* γ_2 *telles que:*

$$E(\sup_{s\leq t} |M_s|^4) \leq b^2 \gamma_1 E(\langle M,M\rangle_t) + \gamma_2 E(\langle M,M\rangle_t^2)$$

(γ_1 et γ_2 ne dépendent ni de M, ni de b).

Preuve. On note $[M,M]$ le processus "variation quadratique" de M. D'une part on sait que $N = [M,M] - \langle M,M\rangle$ est une martingale locale, qui vérifie $|\Delta N|\leq 2b^2$, d'autre part d'après les inégalités de Davis-Burkhölder-Gundy (voir par exemple [9]) il existe des constantes universelles α et β telles que pour toute martingale locale X on ait

2.24 $$E(\sup_{s\leq t} |X_s|^2) \leq \alpha\, E([X,X]_t), \quad E(\sup_{s\leq t} |X_s|^4) \leq \beta E([X,X]_t^2).$$

La martingale locale N est à variation finie sur les compacts, et son processus variation est majoré par le processus croissant $[M,M] + \langle M,M\rangle$; donc $[N,N] \leq 2b^2([M,M] + \langle M,M\rangle)$ (car $|\Delta N|\leq 2b^2$ et $[N,N]_t = \sum_{s\leq t} \Delta N_s^2$). Par suite 2.24 entraine

$$E(\sup_{s\leq t} N_s^2) \leq \alpha E([N,N]_t) \leq 2b^2\alpha E([M,M]_t + \langle M,M\rangle_t) = 4b^2\alpha E(\langle M,M\rangle_t).$$

Par ailleurs $[M,M]^2 \leq 2N^2 + 2\langle M,M\rangle^2$. Une nouvelle application de 2.24 donne

$$E(\sup_{s\leq t} M_s^4) \leq \beta 8b^2\alpha E(\langle M,M\rangle_t) + 2\beta E(\langle M,M\rangle_t^2) .$$ ∎

Démonstration de la condition nécessaire de 2.1. On suppose que $X^n \xrightarrow{\mathcal{L}} X$. D'après le lemme 2.9 on a $[\delta]$. D'après 1.49 on a

$$B_t^{h,n} = E(X_t^{h,n}), \qquad B_t^h = E(X_t^h)$$

et on sait que $X^{h,n} \xrightarrow{\mathcal{L}} X^h$ d'après 2.18, et les $X^{h,n}$ sont des PAI. La condition $[\text{Sup-}\beta]$ découlera de 2.6 si on prouve que:

2.25 la suite $\{(X^{h,n*})_t\}_{n\geq 1}$ est uniformément intégrable, où $(X^{h,n*})_t = \sup_{s\leq t} |X_s^{h,n}|$.

Soit $\tilde{X}^{h,n} = X^{h,n} - B^{h,n}$ et $\tilde{X}^{h} = X^{h} - B^{h}$. Soit $(\tilde{X}^{h,n})^{*}_{t} = \sup_{s\leq t}|\tilde{X}^{h,n}_{s}|$ et $(B^{h,n})^{*}_{t} = \sup_{s\leq t}|B^{h,n}_{s}|$. D'après 1.49, on a $<\tilde{X}^{h,n,j},\tilde{X}^{h,n,j}> = \tilde{C}^{h,n,jj}$. Par construction on a aussi $|\Delta\tilde{X}^{h,nj}| \leq 2a$. Donc 2.18 et 2.23 entrainent que

$$\sup_{n} E^{n}\{(\tilde{X}^{h,n})^{*4}_{t}\} < \infty.$$

Donc

2.26 la suite $\{(\tilde{X}^{h,n})^{*p}_{t}\}_{n\geq 1}$ est uniformément intégrable pour $p<4$.

Cette propriété entraine:

2.27 $$\lim_{b\uparrow\infty} \sup_{n} P^{n}\{(\tilde{X}^{h,n})^{*}_{t}>b\} = 0.$$

Comme $X^{h,n} \xrightarrow{\mathcal{L}} X^{h}$, les variables $(X^{h,n})^{*}_{t}$ vérifient aussi 2.27 d'après le théorème I-2.7. Comme $(B^{h,n})^{*}_{t} \leq (X^{h,n})^{*}_{t} + (\tilde{X}^{h,n})^{*}_{t}$, les $(B^{h,n})^{*}_{t}$ vérifient aussi 2.27; mais les "variables aléatoires" $(B^{h,n})^{*}_{t}$ sont déterministes, donc $\sup_{n}(B^{h,n})^{*}_{t}<\infty$. Finalement, comme $(X^{h,n})^{*}_{t} \leq (B^{h,n})^{*}_{t} + (\tilde{X}^{h,n})^{*}_{t}$, 2.25 découle de 2.26 et on a donc [Sup-β].

Enfin, on a $B^{h,n}_{t} \to B^{h}_{t}$ d'après [Sup-β], donc $\tilde{X}^{h,n}_{t} \xrightarrow{\mathcal{L}} \tilde{X}^{h}_{t}$; comme $\tilde{C}^{h,n,jk}_{t} = E(\tilde{X}^{h,n,j}_{t}\tilde{X}^{h,n,k}_{t})$ on déduit [γ] de 2.26.

§d - DEMONSTRATION DE LA CONDITION SUFFISANTE DU THEOREME 2.1. On va d'abord démontrer que les conditions [Sup-β], [γ], [δ] = [Sup-δ] entrainent que la suite $\{\mathcal{L}(X^{n})\}$ est tendue. Pour cela, on va utiliser le lemme I-2.17 et le théorème I-3.24.

Pour tout $b>0$ on définit h_{b} par $h_{b}(x) = bh(x/b)$: la fonction h_{b} est encore une fonction de troncation continue, avec a remplacé par ab . Pour chaque $q\in\mathbb{N}^{*}$ on pose $\tilde{X}^{h_{q},n} = X^{h_{q},n} - B^{h_{q},n}$.

LEMME 2.28: *Soit [γ] et [δ]. Pour chaque $q\in\mathbb{N}^{*}$, la suite $\{\mathcal{L}(\tilde{X}^{h_{q},n})\}_{n\geq 1}$ est tendue.*

Preuve. D'après 1.49, $\tilde{X}^{h_{q},n}$ est une martingale localement de carré intégrable, et

$$\begin{aligned}<\tilde{X}^{h_{q},n,j},\tilde{X}^{h_{q},n,j}>_{t} &= \tilde{C}^{h_{q},n,jj}_{t}\\ &= \tilde{C}^{h,n,jj}_{t} + \{(h^{j}_{q})^{2} - (h^{j})^{2}\}*\nu^{n}_{t} + \sum_{s\leq t}\{\nu^{n}(\{s\}\times h^{j})^{2} - \nu^{n}(\{s\}\times h^{j}_{q})^{2}\}\end{aligned}$$

d'après 1.46. Donc si $A^{n} = \sum_{j\leq d} <\tilde{X}^{h_{q},n,j},\tilde{X}^{h_{q},n,j}>$ on a, avec la notation $\prec$ signifiant la domination forte pour les processus croissants (cf. I-2.19):

$$\begin{aligned}A^{n} &\prec \sum_{j\leq d}\{\tilde{C}^{h,n,jj} + |(h^{j}_{q})^{2} - (h^{j})^{2}|*\nu^{n} + \sum_{s\leq .} \nu^{n}(\{s\}\times|h^{j}_{q} + h^{j}|)\ \nu^{n}(\{s\}\times|h^{j}_{q} - h^{j}|)\}\\ &\prec G^{nq} := \sum_{j\leq d} \tilde{C}^{h,n,jj} + \hat{h}_{q}*\nu^{n} ,\end{aligned}$$

où

$$\hat{h}_q = \sum_{j\leq d} \{|(h_q^j)^2 - (h^j)^2| + a(1+q)|h_q^j - h^j|\}$$

car $|h|\leq a$ et $|h_q|\leq aq$. Comme $\hat{h}_q$ est une fonction continue bornée sur R^d, nulle sur un voisinage de 0, on voit d'après [γ] et [δ] que les fonctions croissantes G^{nq} convergent simplement (quand $n\uparrow\infty$), donc uniformément sur les compacts, vers la fonction croissante continue $G^q = \sum_{j\leq d} C^{h,jj} + \hat{h}_q * \nu$. Il suffit alors d'appliquer la proposition I-2.20 et le théorème I-3.24 pour obtenir le résultat. ∎

<u>LEMME 2.29</u>: *Soit* [Sup-β], [γ], [δ]. *La suite* $\{\mathcal{L}(X^n)\}$ *est alors tendue.*

<u>Preuve</u>. Pour chaque $q \in \mathbb{N}^*$ on pose

$$U^{nq} = \tilde{X}^{h_q,n} + (h_q - h_{1/q}) * \nu^n, \quad V^{nq} = B^{h,n} + (h_{1/q} - h) * \nu^n, \quad W^{nq} = \sum_{s\leq .} \{\Delta X_s^n - h_q(\Delta X_s^n)\}$$

de sorte que d'après 1.11 et la définition de $\tilde{X}^{h_q,n}$, on a $X^n = U^{nq} + V^{nq} + W^{nq}$. Il reste donc à vérifier que ces processus vérifient les conditions du lemme I-2.17.

D'après [δ] = [Sup-δ] on a $(h_q - h_{1/q}) * \nu^n \to (h_q - h_{1/q}) * \nu$ uniformément sur les compacts, donc la suite $\{(h_q - h_{1/q}) * \nu^n\}_{n\geq 1}$ de "processus" déterministes est C-tendue et la condition I-2.17-(i) découle alors du lemme 2.28 et du corollaire I-2.28.

De même, d'après [Sup-β] et [Sup-δ] on a $V^{nq} \to B^h + (h_{1/q} - h) * \nu$ uniformément sur les compacts; comme $|\Delta V^{nq}| \leq 1/q$ par construction, on a I-2.17-(ii).

Enfin, soit g_q une fonction continue sur R^d, vérifiant $0\leq g_q \leq 1$ et $g_q(x) = 1$ si $|x|>aq/2$, et $g_q(x) = 0$ si $|x|\leq aq/4$. Soit $N \in \mathbb{N}^*$, $\varepsilon>0$. Il existe $q \in \mathbb{N}^*$ tel que $g_q * \nu_N \leq \varepsilon$; il existe n_o tel que $g_q * \nu_N^n \leq 2\varepsilon$ si $n>n_o$. De plus,

$$\begin{aligned} P^n(\sup_{s\leq N} |W_s^{nq}|>0) &\leq P^n(\textstyle\sum_{s\leq N} g_q(\Delta X_s^n)>0) \\ &\leq E^n\{\textstyle\sum_{s\leq N} g_q(\Delta X_s^n)\} = g_q * \nu_N^n \leq 2\varepsilon \qquad \text{si } n>n_o, \end{aligned}$$

donc on a bien la condition I-2.17-(iii). ∎

<u>Démonstration de la condition suffisante</u>. On suppose qu'on a [Sup-β], [γ], [δ]. On vient de voir que la suite $\{\mathcal{L}(X^n)\}$ est tendue. Pour obtenir le résultat il suffit donc de montrer que si $\mathcal{L}(X')$ est une loi limite de la suite $\{\mathcal{L}(X^n)\}$, alors $\mathcal{L}(X') = \mathcal{L}(X)$. Quitte à prendre une sous-suite, on peut supposer que $X^n \xrightarrow{\mathcal{L}} X'$.

Il est évident que X' est un PAI. Soit t un temps de discontinuité fixe de X', et $\varepsilon>0$. Soit g une fonction continue sur R^d, vérifiant $0\leq g \leq 1$ et $g(x)=1$ si $|x|>\varepsilon$, et $g(x) = 0$ si $|x|<\varepsilon/2$. Il existe $s,s' \in D(X')$ avec $s<t<s'$ et $g*\nu_{s'} - g*\nu_s \leq \varepsilon$. D'après [δ] il existe donc n_o tel que $g*\nu_{s'}^n - g*\nu_s^n \leq 2\varepsilon$ si $n > n_o$, donc

<u>2.30</u> $$P^n(\sup_{s<r\leq s'} |\Delta X_r^n| \geq \varepsilon) \leq E^n\{\textstyle\sum_{s<r\leq s'} g(\Delta X_r^n)\} = g*\nu_{s'}^n - g*\nu_s^n \leq 2\varepsilon \quad \text{si } n>n_o.$$

Par ailleurs l'application: $\alpha \rightsquigarrow \sup_{s<r\leq s'} |\Delta\alpha(r)|$ est continue pour la topologie de Skorokhod aux points α tels que $\Delta\alpha(s) = \Delta\alpha(s') = 0$ d'après I-1.19; comme $X^n \xrightarrow{\mathcal{L}} X'$ et comme $s,s' \in D(X')$, on en déduit que: $\sup_{s<r\leq s'} |\Delta X^n_r| \xrightarrow{\mathcal{L}} \sup_{s<r\leq s'} |\Delta X'_r|$. D'après 2.30 il vient alors

$$P(|\Delta X'_t| \geq \varepsilon) \leq P(\sup_{s<r\leq s'} |\Delta X'_r| \geq \varepsilon) \leq \lim\sup_n P^n(\sup_{s<r\leq s'} |\Delta X^n_r| \geq \varepsilon) \leq 2\varepsilon.$$

Ceci étant vrai pour tout $\varepsilon>0$, on a $P(\Delta X'_t \neq 0) = 0$.

Autrement dit, X' est un PAI sans discontinuités fixes. Notons (B'^h, C', ν') ses caractéristiques. D'après la condition nécessaire, on a

$$\begin{cases} B^{h,n}_t \to B'^h_t, \quad \tilde{C}^{h,n}_t \to \tilde{C}'^h_t, \quad f*\nu^n_t \to f*\nu'_t \text{ pour } f \text{ continue bornée} \\ \text{nulle dans un voisinage de } 0. \end{cases}$$

On en déduit que $B'^h = B^h$, puis que $\nu' = \nu$, puis que $\tilde{C}'^h = \tilde{C}^h$, donc $C' = C$. L'unicité dans 1.4-(b) entraine alors que $\mathcal{L}(X') = \mathcal{L}(X)$, ce qui achève la démonstration.

§e - <u>AUTRES CONDITIONS SUFFISANTES</u>. Les conditions du théorème 2.1 sont certes optimales, mais parfois difficiles à vérifier à cause de la présence de la fonction de troncation h. C'est pourquoi il peut être utile de rappeler d'autres conditions suffisantes, non nécessaires, mais plus faciles à vérifier.

Dans ce § on se place toujours sous les hypothèses du théorème 2.1: en particulier, h est <u>continue</u> et X <u>n'a pas de discontinuités fixes</u>.

Le premier résultat est un corollaire immédiat de 2.1, compte tenu de la formule 1.47:

<u>PROPOSITION 2.31</u>: *On suppose que chaque* ν^n *vérifie 1.22* (et donc 1.23; noter que par hypothèse, ν vérifie toujours ces conditions). *Pour que* $X^n \xrightarrow{\mathcal{L}} X$ *il suffit qu'on ait* [Sup-β], [δ] *et les deux conditions:*

(i) $\sum_{s\leq t} |\Delta B^{h,n}_s|^2 \to 0 \quad \forall t > 0$;

(ii) $C^{n,jk}_t + (h^j h^k)*\nu^n_t \to C^{jk}_t + (h^j h^k)*\nu_t \quad \forall t > 0$.

Pour le second résultat, on va supposer que:

<u>2.32</u> $$|x|^2*\nu^n_t < \infty, \qquad |x|^2*\nu_t < \infty \qquad \forall t > 0.$$

Cela entraine 1.22 et 1.23. De plus les fonctions $(x-h(x))*\nu^n$ et $(x-h(x))*\nu$ sont bien définies, et on peut poser:

<u>2.33</u> $$B^n = B^{h,n} + (x-h(x))*\nu^n, \qquad B = B^h + (x-h(x))*\nu.$$

Formellement, on a $B^n = B^{h',n}$ avec $h'(x) = x$: cela revient à dire que sous 2.32 il n'y a pas besoin de "tronquer" les sauts de X^n ou de X. Noter que

2.34 $$\Delta B_t^n = \int \nu^n(\{t\}\times dx)\, x , \qquad \Delta B_t = 0.$$

PROPOSITION 2.35: *On suppose que* ν *et que chaque* ν^n *vérifient 2.32. On suppose aussi que*

(*i*) $\lim_{b\uparrow\infty} \limsup_n \ |x|^2 1_{\{|x|>b\}} * \nu_t^n = 0 \quad \forall t > 0.$

Alors, pour que $X^n \xrightarrow{\mathcal{L}} X$ *il faut et il suffit qu'on ait* [δ] *et*

[Sup-β'] $\quad B^n \to B$ uniformément sur les compacts;

[γ'] $\quad C_t^{n,jk} + (x^j x^k)*\nu_t^n - \sum_{s\leq t} \Delta B_s^{n,j} \Delta B_s^{n,k} \to C_t^{jk} + (x^j x^k)*\nu_t \qquad \forall t>0,\ \forall j,k\leq d.$

Preuve. Il suffit de montrer que, sous (i) et [δ], on a: [Sup-β] $\Longleftrightarrow$ [Sup-β'] et [γ] $\Longleftrightarrow$ [γ']. On remarque que (i) et [δ] entrainent immédiatement:

[δ'] $f*\nu^n \to f*\nu$ uniformément sur les compacts, pour toute fonction f continue, nulle autour de 0, telle que $f(x)/|x|^2$ soit bornée.

(En effet, pour tout $b>0$ il existe une fonction f' continue bornée, telle que $f'(x) = f(x)$ si $|x|\leq b$). Les composantes de la fonction $x-h(x)$ vérifient les conditions de [δ'], donc $(x-h(x))*\nu^n$ converge uniformément sur les compacts vers $(x-h(x))*\nu$. On déduit alors l'équivalence [Sup-β] $\Longleftrightarrow$ [Sup-β'] des relations 2.33. De même, la fonction $x^j x^k - h^j(x)h^k(x)$ vérifie les conditions de [δ'], donc

$$h^j h^k * \nu^n - x^j x^k * \nu^n \to h^j h^k * \nu - x^j x^k * \nu \qquad \text{uniformément sur les compacts.}$$

Etant donné 1.47, il nous reste donc à montrer que

2.36 $$\Big|\sum_{s\leq t} \{\Delta B_s^{h,n,j}\, \Delta B_s^{h,n,k} - \Delta B_s^{n,j}\, \Delta B_s^{n,k}\}\Big| \to 0 .$$

Le premier membre de 2.36 est majoré par

$$\frac{1}{4} \sum_{s\leq t} \big| (\Delta B_s^{h,n,j} + \Delta B_s^{h,n,k})^2 - (\Delta B_s^{h,n,j} - \Delta B_s^{h,n,k})^2 - (\Delta B_s^{n,j} + \Delta B_s^{n,k})^2 + (\Delta B_s^{n,j} - \Delta B_s^{n,k})^2 \big|$$

$$\leq \frac{1}{2}\Big\{\sum_{s\leq t} \{|\Delta B_s^{h,n,j} - \Delta B_s^{n,j}| + |\Delta B_s^{h,n,k} - \Delta B_s^{n,k}|\}\Big\} \sup_{s\leq t} \{|\Delta B_s^{h,n,j}| + |\Delta B_s^{n,j}| + |\Delta B_s^{h,n,k}| + |\Delta B_s^{n,k}|\}$$

$$\leq \frac{1}{2}\{\{|h^j(x)-x^j| + |h^k(x)-x^k|\}*\nu_t^n\} \ \sup_{s\leq t} \{|\Delta B_s^{h,n,j}| + |\Delta B_s^{n,j}| + |\Delta B_s^{h,n,k}| + |\Delta B_s^{n,k}|\} .$$

Soit $\varepsilon>0$ et g une fonction continue telle que $0\leq g(x)\leq |x|$ et $g(x) = |x|$ si $|x|>\varepsilon$, et $g(x) = 0$ si $|x|<\varepsilon/2$. D'après [δ'] et le fait que $g*\nu$ est continue, on voit que:

$$\sup_{s\leq t} \{|\Delta B_s^{h,n,j}| + |\Delta B_s^{n,j}| + |\Delta B_s^{h,n,k}| + |\Delta B_s^{n,k}|\} \leq \sup_{s\leq t} \Delta(g*\nu^n)_s \to 0 .$$

Par ailleurs, toujours d'après [δ'], on a

$$\{|h^j(x) - x^j| + |h^k(x) - x^k|\}*\nu_t^n \to \{|h^j(x)-x^j| + |h^k(x)-x^k|\}*\nu_t$$

et on en déduit qu'on a 2.36, d'où le résultat.∎

COROLLAIRE 2.37: *On suppose que chaque* ν^n *vérifie 2.32. On suppose que* X *est un PAI continu, de caractéristiques* $(B^h,C,0)$. *On suppose enfin qu'on a la condition de Lindeberg:*

(L) $\forall \varepsilon>0, \forall t>0, \qquad |x|^2 1_{\{|x|>\varepsilon\}}*\nu_t^n \to 0.$

Alors, pour que $X^n \xrightarrow{\mathcal{L}} X$ *il faut et il suffit qu'on ait:*

[Sup-β'] $B^n \to B = B^h$ uniformément sur les compacts;

[γ'] $C_t^{n,jk} + (x^jx^k)*\nu_t^n - \sum_{s\leq t} \Delta B_s^{n,j}\, \Delta B_s^{n,k} \to C_t^{jk}$.

(lorsque $B^h = 0$ et $C_t^{jk} = \delta_{jk}t$, X est un mouvement brownien d-dimensionnel standard).

Preuve. Il suffit de remarquer que (L) entraine [δ] et 2.35-(i).∎

3 - APPLICATION AUX SOMMES DE VARIABLES INDEPENDANTES

Historiquement, les premiers résultats de convergence en loi ont concerné les sommes - normalisées - de variables indépendantes. La situation naturelle consiste à considérer des tableaux triangulaires (on devrait plutôt dire "rectangulaires"!)

Considérons donc une double suite $(Y_m^n)_{n\geq 1, m\geq 1}$ de variables aléatoires à valeurs dans R^d. On note ρ_m^n la loi de Y_m^n.

L'hypothèse fondamentale consiste à supposer que *dans chaque ligne* n, *les variables* $(Y_m^n)_{m\geq 1}$ *sont indépendantes*. Pour chaque n, on considère aussi une fonction $\gamma^n: R_+ \to \mathbb{N}$ croissante, càdlàg, n'ayant que des sauts unité. Soit

3.1 $$X_t^n = \sum_{1\leq m\leq \gamma^n(t)} Y_m^n$$

Il est évident que X^n est un PAI. Ce processus est "de saut pur", constant entre les instants de saut de la fonction γ^n, et ces instants de saut sont précisément les temps de discontinuités fixes de X^n. Ainsi, les caractéristiques $(B^{h,n},C^n,\nu^n)$ sont (cela découle immédiatement de 1.18 et de 3.1):

$$\begin{cases} B_t^{h,n} = \sum_{1\leq m\leq \gamma^n(t)} \int \rho_m^n(dx)h(x), & C^n = 0 \\ \nu^n([0,t]\times A) = \sum_{1\leq m\leq \gamma^n(t)} \rho_m^n(A) & \text{si } 0\notin A \end{cases}$$

Le théorème 2.1 s'énonce alors ainsi:

THEOREME 3.2: *Soit* X^n *défini par 3.1. Soit* X *un PAI sans discontinuités fixes, de caractéristiques* (B^h, C, ν) . *Pour que* $X^n \xrightarrow{\mathcal{L}} X$ *il faut et il suffit qu'on ait les trois conditions suivantes:*

[Sup-β] $\sup_{s\leq t} | \sum_{1\leq m\leq \gamma^n(s)} \int \rho_m^n(dx)h(x) - B_s^h| \to 0.$

[γ] $\sum_{1\leq m\leq \gamma^n(t)} \{\int \rho_m^n(dx)h^j(x)h^k(x) - \int \rho_m^n(dx)h^j(x)\int \rho_m^n(dx)h^k(x)\} \to C_t^{jk} + h^j h^k * \nu_t$

[δ] $\sum_{1\leq m\leq \gamma^n(t)} \int \rho_m^n(dx)f(x) \to f*\nu_t$ pour toute fonction continue bornée f nulle sur un voisinage de 0.

Un cas très important consiste à partir d'une seule suite de variables indépendantes $(Z_n)_{n\geq 1}$, de même loi ρ , et des fonctions $\gamma^n(t) = [nt]$ (= partie entière de nt). Soit alors

3.3 $$X_t^n = c_n \sum_{1\leq m\leq [nt]} Z_m$$

où les c_n sont des constantes de normalisation. Dans la terminologie précédente, cela revient à poser $Y_m^n = c_n Z_m$ et $\rho_m^n(f) = \int \rho(dx)f(c_n x)$.

Supposons par exemple que $d=1$ et que $\int \rho(dx)x = 0$ et $\int \rho(dx)x^2 = 1$. Les mesures ν^n vérifient 2.32 et, avec la notation 2.33 on a $B^n = 0$. On a aussi, pour $c_n = 1/\sqrt{n}$:

$$x^2 1_{\{|x|>\varepsilon\}} * \nu_t^n = \frac{[nt]}{n} \rho(|x| > \varepsilon\sqrt{n}) \to 0 \qquad \forall \varepsilon>0, \forall t>0$$

$$C_t^n + x^2 * \nu_t^n - \sum_{s\leq t} (\Delta B_s^n)^2 = \frac{[nt]}{n} \to t \qquad \forall t.$$

Par suite, on déduit de 2.37 le théorème de Donsker:

THEOREME 3.4: *Soit* (Z_m) *des variables indépendantes, de même loi, centrées et de variance* 1. *La suite de processus* $X_t^n = \frac{1}{\sqrt{n}} \sum_{1\leq m\leq [nt]} Z_m$ *converge en loi vers un mouvement brownien standard.*

Plus généralement, il n'est pas difficile de voir, en utilisant cette fois-ci le théorème 2.1, plus les conditions nécessaires et suffisantes de convergence de tableaux triangulaires indépendants par ligne et qui sont à termes asymptotiquement négligeables (voir le livre [15] de Gnedenko et Kolmogorov) que:

THEOREME 3.5: *Soit* (Z_m) *des variables indépendantes, de même loi appartenant au domaine d'attraction normale d'une loi stable* μ ; *soit* c_n *des constantes de normalisation telles que* $c_n \sum_{1\leq m\leq n} Z_m$ *converge en loi vers* μ . *Alors la suite de processus donnée par 3.3 converge en loi vers un PAI homogène caractérisé par* $\mathcal{L}(X_1) = \mu$.

III - CONVERGENCE DE SEMIMARTINGALES VERS UN PROCESSUS A ACCROISSEMENTS INDEPENDANTS

1 - SEMIMARTINGALES ET CARACTERISTIQUES LOCALES

Le paragraphe 1 est entièrement consacré à des rappels: beaucoup sont donnés sans démonstration. Pour la théorie générale des semimartingales, nous renvoyons à [9] ou [43], ou aussi aux deux livres récents [12] et [42]; pour les caractéristiques locales, nous renvoyons à [23]. Nous supposerons connues certaines choses de base, comme les notions de temps d'arrêt (prévisibles) et de tribu optionnelle et tribu prévisible.

§a - LES SEMIMARTINGALES. Soit $(\Omega,\underline{\underline{F}},(\underline{\underline{F}}_t),P)$ un espace probabilisé filtré, fixé jusqu'à la fin du §1. On suppose la filtration $(\underline{\underline{F}}_t)$ continue à droite.

Une semimartingale (réelle) est un processus càdlàg adapté X qui s'écrit comme $X = M + A$, où M est une martingale locale et A est un processus càdlàg à variation finie ("variation finie" signifie ici: variation finie sur les compacts). On note $Var(A)$ le processus variation de A. La décomposition $X = M + A$ n'est évidemment pas unique.

Une semimartingale vectorielle (d-dimensionnelle) est un processus $X = (X^i)_{i\leq d}$ à valeurs dans R^d, dont chaque composante X^i est une semimartingale.

L'espace des semimartingales possède de nombreuses propriétés attrayantes. C'est d'abord un espace vectoriel; il est stable pour toute une série de transformations: par exemple par arrêt (i.e.: si X est une semimartingale et T un temps d'arrêt, le processus arrêté $X^T_t = X_{T\wedge t}$ est encore une semimartingale), ou par changement équivalent de probabilité. Mais surtout, les semimartingales sont les processus les plus généraux par rapport auxquels on peut intégrer (intégrales stochastiques) tous les processus prévisibles bornés ou localement bornés: c'est le théorème de Bichteler-Dellacherie-Mokobodzki [9].

Voici quelques notions et résultats utiles:

1.1 La semimartingale X est dite spéciale si elle admet une décomposition $X = X_0 + M + A$ avec $M_0 = A_0 = 0$, M martingale locale, A processus à variation localement intégrable (i.e. il existe une suite (T_n) de temps d'arrêt croissant vers ∞, telle que $E\{Var(A)_{T_n}\} < \infty$ pour tout n).

Dans ce cas, X admet une décomposition $X = X_0 + M + A$ du type précédent avec A prévisible: cette décomposition est unique, et est appelée <u>décomposition canonique</u> de X.

<u>1.2</u> (i) Toute martingale locale est une semimartingale spéciale.

(ii) Toute semimartingale à sauts bornés est spéciale.

(iii) Tout processus prévisible càdlàg à variation finie est à variation localement intégrable, et est une semimartingale spéciale.

<u>1.3</u> Si A est un processus càdlàg adapté à variation intégrable, avec $A_0 = 0$, il s'écrit de manière unique, d'après 1.1, comme $A = M + B$ avec $M_0 = B_0 = 0$, M martingale locale, B prévisible à variation finie. Le processus B s'appelle la <u>projection prévisible duale</u> (ou compensateur) de A. De plus:

(i) si A est croissant, alors B est croissant,

(ii) si $|\Delta A| \leq a$ identiquement, alors $|\Delta B| \leq a$.

<u>1.4</u> Toute martingale locale M s'écrit de manière unique comme $M = M_0 + M^c + M^d$ où $M_0^c = M_0^d = 0$, M^c est une martingale locale <u>continue</u>, M^d est une martingale locale <u>somme compensée de sauts</u>, i.e. $M^d N$ est encore une martingale locale pour toute martingale locale continue N.

<u>1.5</u> Soit X une semimartingale de décomposition $X = M + A$. La martingale locale continue M^c ne dépend pas de la décomposition choisie: on la note X^c et on l'appelle <u>partie martingale continue</u> de X.

Soit M une martingale locale, localement de carré intégrable (toute martingale locale à sauts bornés, a-fortiori si elle est continue, est localement de carré intégrable). On a déjà introduit au chapitre I sa <u>variation quadratique prévisible</u> $\langle M,M\rangle$, qui est un processus croissant. Noter que M^2 est une semimartingale spéciale, et $\langle M,M\rangle$ est la partie "prévisible à variation finie" de sa décomposition canonique. Si N est une autre martingale localement de carré intégrable, on definit par polarisation:

<u>1.6</u> $$\langle M,N\rangle = \frac{1}{4}(\langle M+N,M+N\rangle - \langle M-N,M-N\rangle)$$

et $\langle M,N\rangle$ est la partie prévisible à variation finie de la décomposition canonique de la semimartingale spéciale MN.

Si X est une semimartingale, sa <u>variation quadratique</u> est le processus càdlàg croissant suivant:

<u>1.7</u> $$[X,X]_t = \langle X^c,X^c\rangle_t + \sum_{s\leq t} (\Delta X_s)^2$$

Ce processus est p.s. fini (on verra plus loin une autre définition de la variation quadratique). Lorsque X est une martingale localement de carré intégrable, $<X,X>$ est la projection prévisible duale de $[X,X]$.

Si X et Y sont deux semimartingales, on pose aussi

1.8 $$[X,Y] = \frac{1}{4}([X+Y,X+Y] - [X-Y,X-Y])$$

et on a:

1.9 $$[X,Y]_t = <X^c,Y^c>_t + \sum_{s\leq t} \Delta X_s \Delta Y_s .$$

Soit X une semimartingale et H un processus prévisible localement borné: cela signifie qu'il existe une suite (T_n) de temps d'arrêt croissant vers $+\infty$, telle que $\sup_{s\leq T_n(\omega),\omega\in\Omega} |H_s(\omega)|$ est fini pour tout n (par exemple, si Y est un processus càdlàg réel adapté, le processus Y_- est localement borné). On sait qu'on peut définir l'intégrale stochastique de H par rapport à X; on utilisera indifféremment les trois notations suivantes:

$$H\bullet X_t \equiv \int_0^t H_s\, dX_s \equiv \int_{]0,t]} H_s\, dX_s .$$

Lorsque X est à variation finie, cette intégrale coïncide avec l'intégrale de Stieltjes, par trajectoires.

Le processus $H\bullet X$ est une semimartingale et vérifie

1.10 $$[H\bullet X,H\bullet X] = H^2\bullet[X,X] \quad , \quad (H\bullet X)^c = H\bullet X^c .$$

1.11 Formule d'Ito. Soit $X = (X^i)_{i\leq d}$ une semimartingale d-dimensionnelle, et f une fonction deux fois différentiable sur $\mathbb{R}^d$, à valeurs dans $\mathbb{R}$ (resp. dans $\mathbb{C}$). Le processus $Y = f(X)$ est alors une semimartingale réelle (resp. complexe, ce qui signifie que partie réelle et partie imaginaire sont des semimartingales), et on a:

$$f(X_t) = f(X_0) + \sum_{i\leq d}\int_0^t \frac{\partial f}{\partial x^i}(X_{s-})dX_s^i + \frac{1}{2}\sum_{j,k\leq d}\int_0^t \frac{\partial^2 f}{\partial x^j\partial x^k}(X_{s-})d<X^{j,c},X^{k,c}>_s$$
$$+ \sum_{s\leq t} \{f(X_s) - f(X_{s-}) - \sum_{i\leq d} \frac{\partial f}{\partial x^i}(X_{s-})\Delta X_s^i\}$$

En particulier, si X et Y sont des semimartingales réelles, la formule précédente appliquée à $f(x,y) = xy$ montre que

1.12 $$XY = X_0Y_0 + X_-\bullet Y + Y_-\bullet X + [X,Y].$$

1.13 L'exponentielle de Doléans-Dade. Soit X une semimartingale réelle. Il existe une semimartingale Y et une seule qui vérifie l'équation $Y = 1 + Y_-\bullet X$. Cette semimartingale Y est notée $\mathcal{E}(X)$ et appelée exponentielle de Doléans-Dade de X, et elle admet l'expression explicite suivante:

$$\mathcal{E}(X)_t = \prod_{s\leq t}\{(1+\Delta X_s)e^{-\Delta X_s}\} \exp(X_t - X_0 - \frac{1}{2}\langle X^c,X^c\rangle_t),$$

le produit infini ci-dessus étant absolument convergent (ce dernier point découle facilement de ce que $\sum_{s\leq t}(\Delta X_s)^2<\infty$).

§b - DEFINITION DES CARACTERISTIQUES LOCALES. Dans le chapitre II nous avons défini les caractéristiques (B^h,C,ν) d'un PAI de deux manières différentes: par la formule II-1.18, ou par la caractérisation du théorème II-1.49. C'est cette dernière caractérisation qui se généralise aisément aux semimartingales. Nous verrons plus loin comment interprèter II-1.18.

Commençons par quelques notations concernant les mesures aléatoires. D'abord, on appelle mesure aléatoire (positive) sur $R_+\times R^d$ une collection $\nu = (\nu(\omega;dt\times dx):\omega\in\Omega)$ de mesures positives sur $R_+\times R^d$. Pour toute fonction W sur $\Omega\times R_+\times R^d$ on pose

1.14 $$W*\nu_t(\omega) = \int_0^t\int_{R^d} \nu(\omega;ds\times dx)\, W(\omega,s,x)$$

(et $W*\nu_t = +\infty$ si cette intégrale n'a pas de sens): on définit ainsi un processus $W*\nu$ (c'est exactement la même notation que II-1.9).

On note $\underline{\underline{P}}$ la tribu prévisible sur $\Omega\times R_+$. On dit que la mesure aléatoire ν est prévisible si le processus $W*\nu$ est prévisible pour toute fonction positive $\underline{\underline{P}}\otimes\underline{\underline{R}}^d$-mesurable W sur $\Omega\times R_+\times R^d$.

Soit $X = (X^j)_{j\leq d}$ une semimartingale d-dimensionnelle. Soit h une fonction de troncation (cf. II-1.2). Par analogie avec II-1.3 on pose

1.15 $$X_t^h = X_t - \sum_{s\leq t}\{\Delta X_s - h(\Delta X_s)\},$$

ce qui définit une nouvelle semimartingale $X^h = (X^{h,j})_{j\leq d}$, qui est à sauts bornés par a.

THEOREME 1.16: *Soit* X *une semimartingale* d-*dimensionnelle et* h *une fonction de troncation. Il existe un triplet et un seul* (B^h,C,ν) *constitué de:*

1.17 $B^h = (B^{h,j})_{j\leq d}$, un processus prévisible càdlàg à variation finie, $B_0^h = 0$

1.18 $C = (C^{jk})_{j,k\leq d}$, un processus continu adapté avec $C_0 = 0$, tel que pour $s\leq t$ la matrice $C_t(\omega) - C_s(\omega)$ soit symétrique nonnégative.

1.19 ν, une mesure aléatoire positive sur $R_+\times R^d$, prévisible, vérifiant:

(i) $\nu(\omega;\{0\}\times R^d) = 0$, $\nu(\omega;R_+\times\{0\}) = 0$

(ii) $\nu(\omega;\{t\}\times R^d) \leq 1$

(iii) $\int_0^t\int_{R^d} (|x|^2\wedge 1)\ \nu(\omega;ds\times dx) < \infty$

(iv) $\sum_{s\leq t} |\int_{R^d} h(x)\ \nu(\omega;\{s\}\times dx)| < \infty$

et vérifiant en outre

1.20 $$\Delta B_t^h(\omega) = \int_{R^d} h(x)\ \nu(\omega;\{t\}\times dx)$$

tel qu'on ait les trois propriétés suivantes:

(i) $\tilde{X}^h = X^h - B^h$ *est une martingale locale*

(ii) pour tous $j,k\leq d$, $\tilde{X}^{h,j}\ \tilde{X}^{h,k} - \tilde{C}^{h,jk}$ *est une martingale locale, où*

1.21 $$\tilde{C}_t^{h,jk} = C_t^{jk} + (h^j h^k)*\nu_t - \sum_{s\leq t} \Delta B_s^{h,j}\ \Delta B_s^{h,k}$$

(iii) pour toute fonction borélienne bornée g *sur* R^d, *nulle sur un voisinage de* 0, *le processus* $\sum_{s\leq t} g(\Delta X_s) - g*\nu_t$ *est une martingale locale.*

De plus, on a (iii) pour toute fonction borélienne g *telle que* $g(x)/(|x|^2\wedge 1)$ *soit borné, et*

1.22 $$\nu(\{T\}\times A) = P(\Delta X_T\in A|\underline{\underline{F}}_{T-}) \text{ sur } \{T<\infty\} \text{ si } T \text{ est un temps prévisible et si } 0\notin A$$

1.23 $$C^{jk} = \langle\tilde{X}^{h,c,j},\tilde{X}^{h,c,k}\rangle$$

où $\tilde{X}^{h,c,j}$ *désigne la "partie martingale continue" de* $\tilde{X}^{h,j}$.

REMARQUES 1.24: 1) L'unicité s'entend ainsi: si (B'^h,C',ν') est un autre triplet vérifiant les mêmes conditions, il existe un ensemble négligeable N tel que pour $\omega\notin N$ on ait $B^h_\cdot(\omega) = B'^h_\cdot(\omega)$, $C_\cdot(\omega) = C'_\cdot(\omega)$ et $\nu(\omega;.) = \nu'(\omega;.)$.

2) D'après 1.19-(iii) et 1.17 le processus $\tilde{C}^h$ donné par 1.21 est bien défini; d'après 1.18 et 1.19-(ii) et 1.20, il vérifie 1.18 à l'exception de la continuité.∎

Preuve. a) D'après 1.1, un processus prévisible nul en 0, à variation finie, et qui est aussi une martingale locale, est identiquement nul (p.s.). On en déduit l'unicité de (B^h,C,ν) ainsi: d'abord, celle de B^h et de $\tilde{C}^h$, puis celle de $g*\nu$ pour chaque fonction g; puis, comme les mesures $\nu(\omega;[0,t]\times.)$ sont déterminées par leurs intégrales sur une suite bien choisie de fonctions boréliennes bornées g nulles autour de 0, l'unicité de ν ; enfin, d'après 1.21, on en déduit enfin l'unicité de C.

b) La semimartingale X^h est nulle en 0 et vérifie $|\Delta X_t^h| \leq a$, donc elle est spéciale; il existe donc B^h vérifiant 1.17 et (i).

c) On définit C par 1.23. Pour tout $u \in R^d$, $u.C.u = \langle N,N \rangle$, où $N = \sum_{j \leq d} u^j \tilde{X}^{h,c,j}$; donc $u.C.u$ est croissant et continu. Il est alors facile de trouver une version de C qui vérifie 1.18.

d) Soit $\underline{\underline{G}}$ l'espace des fonctions boréliennes g telles que $g(x)/|x|^2 \wedge 1$ soit borné. Si $g \in \underline{\underline{G}}$, $g \geq 0$, soit $N_t^g = \sum_{s \leq t} g(\Delta X_s)$. Ce processus croissant est majoré par une constante multipliée par $\sum_{j \leq d} \langle X^j, X^j \rangle$, dont il est à valeurs finies, et aussi à sauts bornés. D'après 1.3 il admet une projection prévisible duale A^g. De plus on a $A_t^g + A_t^{g'} = A_t^{g+g'}$ et $A_t^{\lambda g} = \lambda A_t^g$ p.s.: il n'est alors pas difficile de construire une mesure aléatoire positive prévisible ν vérifiant 1.19-(i), telle que $A_t^g = g * \nu_t$ p.s. (pour plus de détails, voir [23]), et on a (iii).

En appliquant ceci à $g(x) = |x|^2 \wedge 1$, on voit qu'on a 1.19-(iii). Rappelons le résultat classique suivant: si M est une martingale uniformément intégrable et si T est un temps prévisible, on a:

1.25 $$E(\Delta M_T | \underline{\underline{F}}_{T-}) = 0 \quad \text{sur } \{T<\infty\}$$

et ceci s'étend à toute martingale locale pour laquelle $\Delta M_T \in L^1$. En appliquant ceci à $g = 1_A$ où A est un borélien de R^d situé à une distance strictement positive de l'origine (donc $g \in \underline{\underline{G}}$) et à $M = N^g - A^g$, on voit qu'on a 1.22. Il est alors clair que cette relation s'étend à tout borélien A tel que $0 \notin A$. En particulier $\nu(\{T\} \times R^d) \leq 1$ (en utilisant 1.19-(i)). Comme l'ensemble $\{(\omega,t): \nu(\omega;\{t\} \times R^d) > 1\}$ est prévisible, une application du théorème de section prévisible montre qu'on peut modifier ν sur un ensemble négligeable, de sorte qu'on ait 1.19-(ii) identiquement.

De même 1.25 appliqué à $\tilde{X}^h$ et 1.22 entrainent que

1.26 $$B_T^h = E\{h(\Delta X_T) | \underline{\underline{F}}_{T-}\} = \int \nu(\{T\} \times dx)\, h(x) \qquad \text{sur } \{T<\infty\}$$

pour tout temps prévisible T. Là encore, le même raisonnement que ci-dessus montre qu'on peut modifier ν de sorte qu'on ait 1.20 identiquement. Enfin 1.19-(iv) découle de 1.17 et 1.20.

e) Il reste à montrer (ii), ce qui revient à montrer que $\tilde{C}^{h,jk} = \langle \tilde{X}^{h,j}, \tilde{X}^{h,k} \rangle$. On a d'après 1.9 et la définition de $\tilde{X}^h$:

$$[\tilde{X}^{h,j}, \tilde{X}^{h,k}]_t = C_t^{jk} + \sum_{s \leq t} \{h^j(\Delta X_s) - \Delta B_s^{h,j}\}\{h^k(\Delta X_s) - \Delta B_s^{h,k}\}$$

1.27 $$= C_t^{jk} + \sum_{s \leq t} (h^j h^k)(\Delta X_s) + \sum_{s \leq t} \Delta B_s^{h,j} \Delta B_s^{h,k} - \sum_{s \leq t} \{h^j(\Delta X_s) \Delta B_s^{h,k} + h^k(\Delta X_s) \Delta B_s^{h,j}\}$$

De plus $<\tilde{X}^{h,j},\tilde{X}^{h,k}>$ est la projection prévisible duale du processus ci-dessus. C^{jk} et $\sum_{s\leq t}\Delta B_s^{h,j}\Delta B_s^{h,k}$ sont leurs propres projections prévisibles duales; comme $h^j h^k \in \underline{\underline{G}}$, celle de $\sum_{s\leq t}(h^j h^k)(\Delta X_s)$ est $(h^j h^k)*\nu$ d'après (d). Soit enfin F le dernier processus de 1.26. Comme les sauts de B^h sont prévisibles, il existe des temps d'arrêt prévisibles T_n à graphes deux-à-deux disjoints, tels que $F_t = \sum_n \Delta F_{T_n} 1_{\{T_n\leq t\}}$. D'après un résultat bien connu, la projection prévisible duale de F est alors:

$$\begin{aligned}\tilde{F}_t &= \sum_n 1_{\{T_n\leq t\}}\, E(\Delta F_{T_n}|\underline{\underline{F}}_{T_n-})\\ &= \sum_n 1_{\{T_n\leq t\}}\, \{\Delta B^{h,k}_{T_n}\, E\{h^j(\Delta X_{T_n})|\underline{\underline{F}}_{T_n-}\} + \Delta B^{h,j}_{T_n}\, E\{h^k(\Delta X_{T_n})|\underline{\underline{F}}_{T_n-}\}\}\\ &= 2\sum_n 1_{\{T_n\leq t\}}\, \Delta B^{h,j}_{T_n}\Delta B^{h,k}_{T_n} \;=\; 2\sum_{s\leq t}\Delta B^{h,j}_s\, \Delta B^{h,k}_s\end{aligned}$$

(utiliser 1.26). En rassemblant ces résultats, on voit que $<\tilde{X}^{h,j},\tilde{X}^{h,k}> = \tilde{C}^{h,jk}$.∎

Le triplet (B^h,C,ν) s'appelle le triplet des <u>caractéristiques locales</u> (associé à la fonction de troncation h) de la semimartingale X. Soit h' une autre fonction de troncation. D'après (iii) la mesure ν n'est pas modifiée. On a

$$X_t^{h'} = X_t^h + \sum_{s\leq t}\{h'(\Delta X_s) - h(\Delta X_s)\}$$

de sorte que d'après (i) et (iii), il vient

<u>1.28</u> $$B^{h'} = B^h + (h'-h)*\nu.$$

Enfin $\tilde{X}^{h'} - \tilde{X}^h$ est à variation finie, donc $\tilde{X}^{h',c} = \tilde{X}^{h,c}$, et C, qui est donné par 1.23, ne dépend pas de la fonction de troncation.

En comparant au théorème II-1.49, on obtient le résultat suivant:

<u>THEOREME 1.29</u>: *Soit* X *un PAI sur* $(\Omega,\underline{\underline{F}},(\underline{\underline{F}}_t),P)$ *auquel on associe les caractéristiques* (déterministes) (B^h,C,ν) *par II-1.13. Pour que* X *soit une semimartingale, il faut et il suffit que la fonction* B^h *soit à variation finie sur les compacts, et dans ce cas* (B^h,C,ν) *est aussi le triplet des caractéristiques locales de* X.

(remarquer que d'après le lemme II-1.21, si ν et B^h vérifient II-(1.14,1.16, 1.17) et si B^h est à variation finie, alors ν vérifie aussi III-1.19).

<u>Preuve</u>. Soit X un PAI; le processus $X - X^h$ est à variation finie, et X^h est une martingale. Donc X est une semimartingale si et seulement si le processus (déterministe) B^h en est une, ce qui revient à dire (cf. [23] par exemple) que B^h est à variation finie sur les compacts. Enfin, la dernière assertion est évidente si on compare II-1.49 et 1.16.∎

§c - UNE DEFINITION EQUIVALENTE DES CARACTERISTIQUES LOCALES. Soit (B^h,C,ν) un triplet vérifiant 1.17-1.20. Pour tout $u\in R^d$ il existe une constante c_u telle que $|g_u(x)| \le c_u(|x|^2\wedge 1)$, où $g_u(x) = e^{iu.x} - 1 - iu.h(x)$. On définit donc un processus prévisible à variation finie (à valeurs complexes) $A(u)$ en posant:

1.30 $$A(u) = iu.B^h - \frac{1}{2}\,u.C.u + g_u*\nu$$

(si h' est une autre fonction de troncation, et si $B^{h'}$ est donné par 1.28, on obtient évidemment le même processus $A(u)$ en utilisant h' et $(B^{h'},C,\nu)$).

THEOREME 1.31: *Soit* X *un processus* d-*dimensionnel càdlàg. Soit* (B^h,C,ν) *vérifiant 1.17-1.20, et* $A(u)$ *donné par 1.30. Il y a équivalence entre:*

a) X *est une semimartingale, admettant* (B^h,C,ν) *pour caractéristiques locales.*

b) Pour tout $u\in R^d$, *le processus* $e^{iu.X} - (e^{iu.X_-})\bullet A(u)$ *est une martingale locale.*

c) pour toute fonction bornée deux fois différentiable f *sur* R^d *le processus*

$$A_t^f = f(X_t) - f(X_0) - \sum_j \frac{\partial f}{\partial x^j}(X_-)\bullet B_t^{h,j} - \frac{1}{2}\sum_{j,k}\frac{\partial^2 f}{\partial x^j\partial x^k}(X_-)\bullet C_t^{jk}$$
$$- \left[f(X_-+x) - f(X_-) - \sum_j \frac{\partial f}{\partial x^j}(X_-)\,h^j(x)\right]*\nu_t$$

est une martingale locale.

Preuve. (a) $\Longrightarrow$ (c): Soit f bornée de classe C^2. On a

$$X_t = B_t^h + \tilde{X}_t^h + \sum_{s\le t}\{\Delta X_s - h(\Delta X_s)\},$$

de sorte que la formule d'Ito appliquée à f donne, grâce à 1.23:

$$f(X_t) - f(X_0) = \sum_j \frac{\partial f}{\partial x^j}(X_-)\bullet \tilde{X}_t^{h,j} + \sum_j \frac{\partial f}{\partial x^j}(X_-)\bullet B_t^{h,j} + \sum_j \sum_{s\le t}\frac{\partial f}{\partial x^j}(X_{s-})(\Delta X_s^j - h^j(\Delta X_s))$$
$$+ \frac{1}{2}\sum_{j,k}\frac{\partial^2 f}{\partial x^j\partial x^k}(X_-)\bullet C_t^{jk} + \sum_{s\le t}\{f(X_{s-}+\Delta X_s) - f(X_{s-}) - \sum_j \frac{\partial f}{\partial x^j}(X_{s-})\Delta X_s^j\}$$

1.32 $$= \sum_j \frac{\partial f}{\partial x^j}(X_-)\bullet \tilde{X}_t^{h,j} + \sum_j \frac{\partial f}{\partial x^j}(X_-)\bullet B_t^{h,j} + \frac{1}{2}\sum_{j,k}\frac{\partial^2 f}{\partial x^j\partial x^k}(X_-)\bullet C_t^{jk} + \sum_{s\le t} W(s,\Delta X_s)$$

où $W(\omega,t,x) = f(X_{t-}(\omega)+x) - f(X_{t-}(\omega)) - \sum_j \frac{\partial f}{\partial x^j}(X_{t-}(\omega))h^j(x)$.

Considérons 1.32; la première somme est une martingale locale. La seconde et la troisième sont des processus prévisibles à variation finie. La quatrième est un processus optionnel à variation finie, dont les sauts sont localement bornés: il est donc à variation localement intégrable. Il suffit donc de montrer que sa projection prévisible duale est $W*\nu$ pour obtenir (c). Mais W est $\underline{\underline{P}}\otimes\underline{\underline{R}}^d$-mesurable: étant données des propriétés classiques des mesures aléatoires [23], ce résultat découle de 1.16-(iii).

(c) $\Longrightarrow$ (b): Il suffit de remarquer que si $f(x) = e^{iu.x}$, on a $A^f = e^{iu.X} - e^{iu.X_0} - (e^{iu.X_-})\bullet A(u)$.

(b) $\Longrightarrow$ (a): Par hypothèse $e^{iu.X}$ est une semimartingale complexe pour tout $u \in R^d$. Donc $\sin(bX^j)$ et $\cos(bX^j)$ sont des semimartingales réelles pour tous $b \in R$, $j \leq d$. Il existe une fonction g de classe C^2 sur R^2, telle que $g(\sin x, \cos x) = x$ pour $|x| \leq 1/2$. Donc si $T_n = \inf(t: |X_t^j| \geq 2n)$, le processus X^j coincide avec la semimartingale $n\, g(\sin(X^j/n), \cos(X^j/n))$ sur l'intervalle $[0,T_n[$. Comme $T_n \uparrow \infty$ cela suffit à prouver que X^j est une semimartingale.

Soit alors (B'^h, C', ν') les caractéristiques locales de X, auxquelles on associe $A'(u)$ par 1.30. D'après les implications (a) $\Longrightarrow$ (c) $\Longrightarrow$ (b), le processus $e^{iu.X} - (e^{iu.X_-})\bullet A'(u)$ est une martingale locale, donc aussi le processus $B(u) = (e^{iu.X_-})\bullet A(u) - (e^{iu.X_-})\bullet A'(u)$, donc aussi le processus $A(u) - A'(u) = (e^{-iu.X_-})\bullet B(u)$. Mais $A(u) - A'(u)$ est également prévisible à variation finie et nul en 0, donc $A(u)_t = A'(u)_t$ p.s. Comme $A(u)_t$ et $A'(u)_t$ sont continus en u et càdlàg en t, on a $A(u)_t(\omega) = A'(u)_t(\omega)$ pour tous $u \in R^d, t>0$, si ω n'appartient pas à un ensemble négligeable N. L'unicité dans la formule de Lévy-Khintchine entraine alors que si $\omega \notin N$ on a $B^h_\cdot(\omega) = B'^h_\cdot(\omega)$, $C_\cdot(\omega) = C'_\cdot(\omega)$ et $\nu(\omega;.) = \nu'(\omega;.)$, d'où le résultat. ∎

Soit maintenant le processus prévisible

1.33 $$G(u)_t = e^{A(u)_t} \prod_{s \leq t} \{(1 + \Delta A(u)_s)\, e^{-\Delta A(u)_s}\}.$$

Comme $A(u)_0 = 0$ et comme la partie martingale continue de $A(u)$ est nulle (car $A(u)$ est à variation finie), on voit que $G(u) = \mathcal{E}\{A(u)\}$ est l'exponentielle de Doléans-Dade de $A(u)$. Bien que ces processus soient complexes, et non réels, on vérifie aisément que $G(u)$ est l'unique solution de l'équation:

1.34 $$G(u) = 1 + G(u)_- \bullet A(u)$$

et $G(u)$ est à variation finie (cela peut se vérifier directement sur 1.33, bien-sûr).

THEOREME 1.35: *Soit* X *une semimartingale de caractéristiques locales* (B^h, C, ν); *soit* $A(u)$ *et* $G(u)$ *définis par 1.30 et 1.33. Soit* $T(u) = \inf(t: \Delta A(u)_t = -1)$.

a) $T(u)$ *est un temps d'arrêt prévisible, tel que* $T(u) = \inf(t: G(u)_t = 0)$. *On a* $G(u)_t = 0$ *si* $t \geq T(u)$ *et le processus* $(e^{iu.X}/G(u))\, 1_{[0,T(u)[}$ *est une martingale locale sur l'intervalle stochastique* $[0,T(u)[$.

b) $G(u)$ *est "l'unique" processus prévisible à variation finie ayant ces propriétés, au sens suivant: si* G' *est un processus prévisible à variation finie, avec*

$G'_0 = 1$, *tel que, si* $T' = \inf(t: G'_t = 0)$, *le processus* $(e^{iu.X}/G')1_{[0,T'[}$ *soit une martingale locale sur* $[0,T'[$, *alors* $T' \leq T(u)$ *et* $G' = G(u)$ *sur* $[0,T'[$.

Dans cet énoncé, "martingale locale sur $[0,T(u)[$" signifie la chose suivante: soit T un temps d'arrêt prévisible; on sait qu'il existe une suite (T_n) de temps d'arrêt croissant vers T, avec $T_n<T$ p.s. si $T>0$. Un processus M est appelé <u>martingale locale sur</u> $[0,T[$ si chaque processus arrêté M^{T_n} est une martingale locale. Il est facile de voir que cette notion ne dépend pas de la suite (T_n) annonçant T.

<u>Preuve</u>. Pour simplifier, on pose $A = A(u)$, $T = T(u)$, $G = G(u)$, $Y = e^{iu.X}$ et $Z = \frac{Y}{G} 1_{[0,T[}$.

a) Que T soit prévisible provient de la prévisibilité de A; que $T = \inf(t: G_t = 0)$ découle de 1.33, ainsi que la propriété $G_t = 0$ si $t \geq T$. De même, G étant prévisible, chaque temps d'arrêt $R_n = \inf(t: |G_t| \leq 1/n)$ est prévisible. Comme $R_n > 0$ pour $n \geq 2$, il existe un temps d'arrêt S_n tel que $S_n<R_n$ et $P(S_n<R_n-1/n) \leq 2^{-n}$. Si alors $T_n = \sup_{m\leq n} S_m$ on a $\lim_n \uparrow T_n = T$ p.s., et $T_n < T$, et $|G_t| \geq 1/n$ si $t\leq T_n$.

Il reste alors à prouver que chaque processus arrêté Z^{T_n} est une martingale locale. Comme à l'évidence A^{T_n} est associé à X^{T_n} comme A à X, et comme $G^{T_n} = \mathcal{E}(A^{T_n})$, en remplaçant X par X^{T_n} il reste à prouver que Z est une martingale locale, sachant que $|G| \geq 1/n$ identiquement. En particulier $Z = \frac{Y}{G}$.

Appliquons la formule d'Ito à une fonction f de classe C^2 sur R^4, qui vérifie $f(x,y,z,u) = \frac{x+iy}{z+iu}$ lorsque $|z+iu| \geq 1/n$. Il vient

$$Z = 1 + \frac{1}{G_-}\bullet Y - \left(\frac{Z}{G}\right)_-\bullet G + \sum_{s\leq .}\left(\Delta Z_s - \frac{\Delta Y_s}{G_{s-}} + \frac{Z_{s-}}{G_{s-}}\Delta G_s\right).$$

D'après 1.31, $N = Y - Y_-\bullet A$ est une martingale locale. D'après 1.34, $G = 1+G_-\bullet A$. Un calcul simple montre alors que $\Delta Z - (\Delta Y/G_-) + Z_-(\Delta G/G_-) = -\frac{\Delta N \Delta A}{1+\Delta A}$, d'où

<u>1.36</u> $$Z = 1 + \frac{1}{G_-}\bullet N - \frac{1}{1+\Delta A}\cdot\left(\sum_{s\leq .}\Delta N_s\, \Delta A_s\right)$$

Mais N est une martingale locale, et A est prévisible à variation finie, donc le processus $\sum_{s\leq .} \Delta N_s\, \Delta A_s$ est aussi une martingale locale (d'après le lemme de Yoeurp [9]). D'après 1.36, on en déduit que Z est une martingale locale.

b) Soit G' et T' vérifiant les propriétés de l'énoncé, et $Z' = \frac{Y}{G'} 1_{[0,T'[}$. Exactement comme en (a), on peut trouver une suite (T'_n) de temps d'arrêt croissant vers T', telle que $T'_n < T'$ et $|G'_t| \geq 1/n$ si $t \leq T'_n$. Soit $A'(n) = \frac{1}{G'_-}\bullet G'^{T'_n}$. Comme $G'_0 = 1$ on a par inversion: $G'^{T'_n} = 1 + (G'^{T'_n})_-\bullet A'(n)$, de sorte que $G'^{T'_n} = \mathcal{E}(A'(n))$.

On a aussi $Y^{T'_n} = Z'^{T'_n} G'^{T'_n}$ et $Z'^{T'_n}$ est une martingale locale par hypothèse, donc la formule d'Ito donne:

$$Y^{T'_n} = (Z'^{T'_n})_- \bullet G'^{T'_n} + (G'^{T'_n})_- \bullet Z'^{T'_n} + \sum_{s\leq .} \Delta G'^{T'_n}_s \Delta Z'^{T'_n}_s$$

$$= (Z'^{T'_n})_- \bullet G'^{T'_n} + (G'^{T'_n})_- \bullet Z'^{T'_n}$$

$$= Y_- \bullet A'(n) + (G'^{T'_n})_- \bullet Z'^{T'_n} ;$$

donc $Y^{T'_n} - Y_- \bullet A'(n)$ est une martingale locale. D'après 1.31, $Y^{T'_n} - Y_- \bullet A^{T'_n}$ est aussi une martingale locale, donc le processus $Y_- \bullet A'(n) - Y_- \bullet A^{T'_n}$ est une martingale locale prévisible à variation finie, nulle en 0, donc le processus $A'(n) - A^{T'_n}$ vérifie les mêmes propriétés: par suite il est p.s. nul. De $A'(n) = A^{T'_n}$ p.s. on déduit $G'^{T'_n} = G^{T'_n}$ p.s.: par suite $G' = G$ sur $[0,T'_n]$ pour chaque n, donc $G' = G$ sur $[0,T'[$. D'après la définition de T et T', on en déduit également que $T' \leq T$. ∎

Supposons que X soit un PAI de caractéristiques (B^h,C,ν) et que B^h soit à variation finie. Le processus $A(u)$ donné par 1.30 est alors déterministe, et si on compare à 1.33 et II-1.18, on obtient:

<u>1.37</u>
$$G(u)_t = E(e^{iu.X_t}).$$

La propriété d'accroissements indépendants implique par ailleurs immédiatement que $e^{iu.X}/G(u)$ soit une martingale sur $[0,T(u)[$ (ici, $T(u)$ est aussi déterministe). Ainsi, le théorème 1.35 est la version "naturelle" de la formule II-1.18 pour les semimartingales, du moins lorsque $s = 0$.

Plus précisément, remarquons que 1.35 ne dit rien sur ce qui se passe après le temps $T(u)$. Si on veut avoir une transposition complète de II-1.18, il faut utiliser le résultat suivant, qui se démontre exactement comme 1.35:

<u>1.38</u> Pour tout $s\geq 0$, soit $T(u,s) = \inf(t>s: \Delta A(u)_t = -1)$ et

$$G(u,s)_t = \begin{cases} 1 & \text{si } t<s \\ e^{A(u)_t - A(u)_s} \prod_{s<r\leq t} \{(1+\Delta A(u)_r)e^{-\Delta A(u)_r}\} & \text{si } t\geq s . \end{cases}$$

Alors $\{e^{iu.(X_t-X_s)}/G(u,s)_t\} 1_{\{t<T(u,s)\}}$ est une martingale locale (en t) sur $[0,T(u,s)[$.

<u>COROLLAIRE 1.39</u>: *Pour qu'une semimartingale* X *soit aussi un PAI* (relativement à la filtration $(\underline{F}_t)$) *il faut et il suffit que ses caractéristiques locales soient déterministes.*

Preuve. La condition nécessaire a été montrée en 1.29. Inversement, supposons les caractéristiques locales (B^h,C,ν) déterministes. Utilisons les notations 1.38. Si $s<t<T(u,s)$ on a

$$E(e^{iu.(X_t-X_s)}|\underline{\underline{F}}_s) = G(u,s)_t .$$

Si $t = T(u,s)$, il vient d'après 1.22

$$E(e^{iu.(X_t-X_s)}|\underline{\underline{F}}_s) = E\{e^{iu.(X_{t-}-X_s)} E(e^{iu.X_t}|\underline{\underline{F}}_{t-})|\underline{\underline{F}}_s\} = G(u,s)_{t-}(\Delta A(u)_t+1) = 0.$$

Enfin si $T(u,s) < t$ il existe s',s'' avec $s\leq s'$, $s'' = T(u,s')$ et $s''\leq t<T(u,s'')$.

$$E(e^{iu.(X_t-X_s)}|\underline{\underline{F}}_s) = E\{e^{iu.(X_{s''}-X_s)} E(e^{iu.(X_t-X_{s''})}|\underline{\underline{F}}_{s''})|\underline{\underline{F}}_s\}$$

$$= G(u,s'')_t\ E\{e^{iu.(X_{s'}-X_s)} E(e^{iu.(X_{s''}-X_{s'})}|\underline{\underline{F}}_{s'})|\underline{\underline{F}}_s\} = 0.$$

Donc $E(e^{iu.(X_t-X_s)}|\underline{\underline{F}}_s) = G(u,s)_t$ dans tous les cas. On en déduit alors le résultat. ∎

§d - EXEMPLES.

1) Diffusions et diffusions avec sauts. La condition (c) du théorème 1.31 permet de voir immédiatement que de nombreux processus de Markov à valeurs dans R^d sont des semimartingales à valeurs dans R^d, dont les caractéristiques locales sont liées de manière évidente au generateur infinitésimal.

Supposons en effet qu'on ait un processus de Markov fort càdlàg $(\Omega,\underline{\underline{F}},\underline{\underline{F}}_t,X,P^x)$ à valeurs dans R^d, normal (i.e. $P^x(X_0=x) = 1$). Supposons que son générateur infinitésimal (faible) soit de la forme suivante: (A,D_A) avec

1.40
$$\begin{cases} \text{Si } f \text{ est bornée de classe } C^2 \text{ sur } R^d, \text{ on a } f\in D_A \text{ et} \\ Af(x) = \sum_j b^j(x)\dfrac{\partial f}{\partial x^j}(x) + \dfrac{1}{2}\sum_{j,k} c^{jk}(x)\dfrac{\partial^2 f}{\partial x^j\partial x^k}(x) \\ \qquad + \int_{R^d} N(x,dy)\{f(x+y) - f(y) - \sum_j \dfrac{\partial f}{\partial x^j}(x)\, h^j(y)\} \end{cases}$$

où $b = (b^j)_{j\leq d}$ est une fonction localement bornée: $R^d \to R^d$, $c = (c^{jk})_{j,k\leq d}$ est une fonction localement bornée sur R^d à valeurs dans l'espace des matrices $d\times d$ symétriques nonnégatives, et N est un noyau de R^d dans R^d tel que la fonction $\int N(.,dy)(|y|^2\wedge 1)$ soit aussi localement bornée. D'après la formule de Dynkin, le processus $f(X_t) - f(X_0) - \int_0^t Af(X_s)ds$ est alors une P^x-martingale locale pour tout x lorsque f est comme ci-dessus. On en déduit que:

1.41 Pour chaque propbabilité P^x, le processus X est une semimartingale de caractéristiques locales: $B^h_t = \int_0^t b(X_s)ds$, $C_t = \int_0^t c(X_s)ds$, $\nu(dt\times dx) = dt\times N(X_{t-},dx)$.

On dit aussi que P^x résoud le problème de martingales associé à X, aux caractéristiques locales (B^h,C,ν) définies en 1.41, et à la condition initiale $X_0 = x$. Remarquer qu'on retrouve les diffusions de Stroock et Varadhan ([53],[55]) lorsque $N = 0$, et les diffusions "avec sauts" de Stroock [54] si $N \neq 0$.

2) Processus ponctuels. On appelle processus ponctuel un processus de comptage N, c'est-à-dire un processus à valeurs dans $\mathbb{N}$, croissant, nul en 0, n'ayant que des sauts unité. Un tel processus est localement intégrable, et on note A sa projection prévisible duale. Si on choisit une fonction de troncation h qui vérifie $h(x) = x$ pour $|x| \leq 1$, les caractéristiques locales de N sont:

1.42 $$B^h = 0, \qquad C = 0, \qquad \nu(dt\times dx) = dA_t \times \varepsilon_1(dx).$$

3) Processus discrets. Soit $(V_n)_{n\geq 1}$ une suite de variables aléatoires à valeurs dans R^d, définies sur un espace $(\Omega,\underline{\underline{F}},P)$. Soit $(\underline{\underline{G}}_n)_{n\geq 0}$ une suite croissante de sous-tribus de $\underline{\underline{F}}$, telle que chaque V_n soit $\underline{\underline{G}}_n$-mesurable.

1.43 On appelle changement de temps une suite $(\sigma_t)_{t\geq 0}$ de variables aléatoires telle que:
(i) $\sigma_0 = 0$; chaque σ_t est à valeurs dans $\mathbb{N}$ et est un $(\underline{\underline{G}}_n)$-temps d'arrêt;
(ii) chaque trajectoire $t \rightsquigarrow \sigma_t$ est croissante, càdlàg, à sauts unité.

On considère alors le processus

1.44 $$X_t = \sum_{1\leq n\leq \sigma_t} V_n \quad (= 0 \text{ si } \sigma_t = 0), \quad \text{adapté à } \underline{\underline{F}}_t = \underline{\underline{G}}_{\sigma_t}$$

et les temps
$$\tau_k = \inf(t: \sigma_t = k) \quad \text{pour } k\in\mathbb{N}.$$

PROPOSITION 1.45: *a) La filtration* $(\underline{\underline{F}}_t)$ *est continue à droite; les* τ_k *sont des temps prévisibles pour cette filtration; si* $k\geq 1$ *on a* $\underline{\underline{F}}_{\tau_k} \cap \{\tau_k<\infty\} = \underline{\underline{G}}_k \cap \{\tau_k<\infty\}$ *et* $\underline{\underline{F}}_{\tau_k-} \cap \{\tau_k<\infty\} = \underline{\underline{G}}_{k-1} \cap \{\tau_k<\infty\}$.

b) X *est une semimartingale relativement à* $(\underline{\underline{F}}_t)$, *de caractéristiques locales*

1.46 $$\begin{cases} B^h_t = \sum_{1\leq n\leq \sigma_t} E(h(V_n)|\underline{\underline{G}}_{n-1}) \\ C = 0 \\ \nu([0,t]\times A) = \sum_{1\leq n\leq \sigma_t} P(V_n\in A,\ V_n\neq 0|\underline{\underline{G}}_{n-1}) \end{cases}$$

Noter qu'on pourrait admettre que σ_t prenne la valeur $+\infty$: dans ce cas, bien-sûr, il faudrait ajouter une condition de façon à ce que la série donnée par 1.44 converge; mais sous cette condition supplémentaire, la proposition 1.45 resterait valide.

Preuve. a) $\underline{\underline{F}}_t$ est l'ensemble des $A \in \underline{\underline{F}}$ tels que $A \cap \{\sigma_t = n\} \in \underline{\underline{G}}_n$ pour tout n. Si $A \in \underline{\underline{F}}_{t+}$ on a $A \cap \{\sigma_s = n\} \in \underline{\underline{G}}_n$ pour tout $s>t$; d'après 1.43-(ii) on a $A \cap \{\sigma_t = n\} = \lim_{s \downarrow t, s>t} A \cap \{\sigma_s = n\}$ appartient à $\underline{\underline{G}}_n$, donc $A \in \underline{\underline{F}}_t$, donc $\underline{\underline{F}}_{t+} = \underline{\underline{F}}_t$.

Soit $k \geq 1$. Si $A \in \underline{\underline{G}}_k$ on a

$$A \cap \{\tau_k \leq t\} \cap \{\sigma_t = n\} = \begin{cases} \emptyset & \text{si } n<k \\ A \cap \{\sigma_t = n\} & \text{si } n \geq k, \end{cases}$$

et cet ensemble appartient à $\underline{\underline{G}}_n$. Par suite $A \cap \{\tau_k \leq t\} \in \underline{\underline{F}}_t$; en prenant $A = \Omega$ on en déduit que τ_k est un $(\underline{\underline{F}}_t)$-temps d'arrêt; on en déduit aussi que $\underline{\underline{G}}_k \subset \underline{\underline{F}}_{\tau_k}$.

Soit ensuite $A \in \underline{\underline{F}}_{\tau_k}$. Alors $A_t := A \cap \{\tau_k \leq t < \tau_{k+1}\}$ est dans $\underline{\underline{F}}_t$, donc $A_t \cap \{\sigma_t = k\} = A_t$ est dans $\underline{\underline{G}}_k$. Comme $A \cap \{\tau_k < \infty\} = \bigcup_{t \in \mathbb{Q}_+} A_t$ il s'ensuit que $A \cap \{\tau_k < \infty\} \in \underline{\underline{G}}_k$. D'après ce qui précède, il vient $\underline{\underline{F}}_{\tau_k} \cap \{\tau_k < \infty\} = \underline{\underline{G}}_k \cap \{\tau_k < \infty\}$.

On a $\{\tau_k > t\} = \{\sigma_t \leq k-1\} \in \underline{\underline{G}}_{k-1} \subset \underline{\underline{F}}_{\tau_{k-1}}$. Donc τ_k est $\underline{\underline{G}}_{k-1}$-mesurable et $\tau_k > \tau_{k-1}$ si $\tau_{k-1} < \infty$. On en déduit que τ_k est un temps d'arrêt prévisible pour $(\underline{\underline{F}}_t)$, car il est annoncé par les $(\underline{\underline{F}}_t)$-temps d'arrêt $\{\tau_{k-1} + (\tau_k - \tau_{k-1} - 1/n)^+\} \wedge n$. Si $A \in \underline{\underline{G}}_{k-1}$ on a $A \cap \{\tau_k < \infty\} \in \underline{\underline{G}}_{k-1} \subset \underline{\underline{F}}_{\tau_{k-1}} \subset \underline{\underline{F}}_{(\tau_k)-}$. Si $A \in \underline{\underline{F}}_t$, alors $A \cap \{t < \tau_k\} = A \cap \{\sigma_t \leq k-1\} \in \underline{\underline{G}}_{k-1}$. Comme $\underline{\underline{F}}_{(\tau_k)-}$ est engendré par $\underline{\underline{F}}_0 = \underline{\underline{G}}_0$ et par les ensembles $A \cap \{t < \tau_k\}$ pour $t>0$, $A \in \underline{\underline{F}}_t$, on en déduit que $\underline{\underline{F}}_{(\tau_k)-} \subset \underline{\underline{G}}_{k-1}$. On a donc terminé la preuve de (a).

b) X est un processus à variation finie, adapté, donc c'est une semimartingale dont la seconde caractéristique C est nulle d'après 1.23. Comme X s'écrit aussi $X = \sum_k \Delta X_{\tau_k} 1_{\{\tau_k \leq t\}}$ et comme les τ_k sont des temps prévisibles, on a:

$$X_t^h = \sum_k h(\Delta X_{\tau_k}) 1_{\{\tau_k \leq t\}}, \qquad \sum_{s \leq t} g(\Delta X_s) = \sum_k g(\Delta X_{\tau_k}) 1_{\{\tau_k \leq t\}}$$

et d'après 1.16 et un résultat classique sur les projections prévisibles duales de processus à variation finie purement discontinus et à sauts prévisibles cela implique que les caractéristiques B^h et ν vérifient:

$$B_t^h = \sum_k 1_{\{\tau_k \leq t\}} E\{h(\Delta X_{\tau_k}) | \underline{\underline{F}}_{(\tau_k)-}\}, \qquad g*\nu_t = \sum_k 1_{\{\tau_k \leq t\}} E\{g(\Delta X_{\tau_k}) | \underline{\underline{F}}_{(\tau_k)-}\}$$

si g est nul autour de 0 et borné. Comme $\Delta X_{\tau_k} = V_k$, il suffit d'utiliser (a) pour obtenir 1.46. ∎

§e - SEMIMARTINGALES LOCALEMENT DE CARRE INTEGRABLE. Ce § a un intérêt secondaire, bien que la plupart des semimartingales qu'on rencontre soient du type suivant: on dit qu'une semimartingale X est localement de carré intégrable s'il existe une suite (T_n) de temps d'arrêt croissant vers l'infini, telle que pour chaque n:

1.47 $$E(\sup_{s \leq T_n} |X_t - X_0|^2) < \infty$$

PROPOSITION 1.48: *Soit* X *une semimartingale* d-*dimensionnelle de caractéristiques locales* (B^h,C,ν). *Pour qu'elle soit localement de carré intégrable, il faut et il suffit que*

1.49 $$|x|^2*\nu_t < \infty \qquad \text{p.s.} \quad \forall t>0$$

Dans ce cas, X *est une semimartingale spéciale de décomposition canonique* $X = X_0 + M + B$ *et*

(*i*) $B = B^h + (x-h(x))*\nu$;

(*ii*) *les crochets* $\langle M^j,M^k\rangle$ *sont les processus*

1.50 $$\tilde{C}_t^{jk} = C_t^{jk} + (x^jx^k)*\nu_t - \sum_{s\leq t} \Delta B_s^j\, \Delta B_s^k$$

(*iii*) $\sum_{s\leq t} g(\Delta X_s) - g*\nu_t$ *est une martingale locale pour toute fonction borélienne* g *telle que* $g(x)/|x|^2$ *soit bornée.*

En d'autres termes, si X est localement de carré intégrable on peut prendre $h(x) = x$ pour "fonction de troncation". Remarquer que (i) et 1.20 entrainent:

1.51 $$\Delta B_t = \int \nu(\{t\}\times dx)\ x$$

Preuve. Supposons d'abord qu'on ait 1.49 . Soit g une fonction borélienne positive telle que $g(x)/|x|^2$ soit borné. Soit (f_p) une suite de fonctions positives bornées nulles autour de 0, nulles pour $|x|>p$, et croissant vers 1. D'après 1.16-(iii) on a pour tout temps d'arrêt T:

$$E\{\sum_{s\leq T} g(\Delta X_s) f_p(\Delta X_s)\} = E\{(gf_p)*\nu_T\}.$$

En passant à la limite en p, on en déduit que $E\{\sum_{s\leq T} g(\Delta X_s)\} = E(g*\nu_T)$. Comme le processus croissant $g*\nu$ est à valeurs finies et est prévisible, il est localement intégrable et on déduit facilement (iii) de la relation précédente (si g n'est pas positive, on l'écrit comme différence de deux fonctions positives ayant la même propriété de bornitude).

Soit $B = B^h + (x-h(x))*\nu$, qui est bien défini à cause de 1.49 . D'après (iii), $N_t = \sum_{s\leq t} \{\Delta X_s - h(\Delta X_s)\} - (x-h(x))*\nu_t$ est une martingale locale, donc $M = X^h + N - B^h$ égalcment. On a $X = X_0 + M + B$, donc X est une semimartingale spéciale. La démonstration de (ii) se fait exactement comme celle de 1.16-(ii), en remplaçant la fonction $h(x)$ par la fonction x . En particulier, comme $[M^j,M^j]$ est localement intégrable pour chaque j, la martingale locale M est localement de carré intégrable, donc vérifie 1.47. D'après 1.51 , on a aussi $\sum_{s\leq t} |\Delta B_s|^2 \leq |x|^2*\nu_t$, donc B est à variation localement de carré intégrable, donc vérifie 1.47. On en déduit que X également vérifie 1.47.

Supposons inversement que 1.47 soit satisfait par X. Il existe une suite de temps d'arrêt croissant vers l'infini, avec

$$E(\sum_{s\leq T_n} |\Delta X_s|^2) \leq E(\sum_{j\leq d} [X^j,X^j]_{T_n} + 4 \sup_{s\leq T_n} |X_s - X_0|^4) < \infty.$$

En utilisant 1.51 pour $T = T_n$ et $g(x) = |x|^2$ et en passant à la limite en p, on voit que $E(|x|^2 * \nu_{T_n}) < \infty$, d'où 1.49.∎

2 - CONVERGENCE DE SEMIMARTINGALES VERS UN PAI

§a - ENONCE DU RESULTAT PRINCIPAL - PRINCIPE DE LA METHODE. Dans ce § nous considérons une suite de semimartingales d-dimensionnelles X^n, nulles en 0 ; chaque X^n est définie sur l'espace filtré $(\Omega^n,\underline{\underline{F}}^n,(\underline{\underline{F}}^n_t),P^n)$ et admet les caractéristiques locales $(B^{h,n},C^n,\nu^n)$. Par ailleurs, soit X un PAI càdlàg d-dimensionnel sans discontinuités fixes, défini sur $(\Omega,\underline{\underline{F}},(\underline{\underline{F}}_t),P)$, et admettant les caractéristiques (déterministes) (B^h,C,ν).

A chaque $(B^{h,n},C^n,\nu^n)$ on associe le processus $\tilde{C}^{h,n}$ défini par 1.21; de même à (B^h,C,ν) on associe la fonction $\tilde{C}^h$ définie par II-1.46.

THEOREME 2.1: *On suppose la fonction de troncation* h *continue, et le PAI* X *sans discontinuités fixes. Pour que* $X^n \xrightarrow{\mathcal{L}} X$ *il suffit qu'on ait les trois conditions:*

[Sup-β] $\sup_{s\leq t} |B^{h,n}_s - B^h_s| \xrightarrow{\mathcal{L}} 0$ pour tout $t>0$;

[γ] $\tilde{C}^{h,n}_t \xrightarrow{\mathcal{L}} \tilde{C}^h_t$ pour tout t dans une partie dense A de R_+ ;

[δ] $f*\nu^n_t \xrightarrow{\mathcal{L}} f*\nu_t$ pour tout t dans une partie dense A de R_+ et toute fonction continue bornée $f: R^d \to R_+$, nulle sur un voisinage de 0.

Dans ce cas, on a [γ] *et* [δ] *avec* $A=R_+$ *et on a aussi:*

[Sup-γ] $\sup_{s\leq t} |\tilde{C}^{h,n}_s - \tilde{C}^h_s| \xrightarrow{\mathcal{L}} 0$ pour tout $t>0$;

[Sup-δ] $\sup_{s\leq t} |f*\nu^n_s - f*\nu_s| \xrightarrow{\mathcal{L}} 0$ pour tout $t>0$ et toute f comme en [δ].

Nous donnons à ces conditions le même nom que dans le théorème II-2.1, car elles sont rigoureusement identiques; ici, bien-sûr, $B^{h,n}$, $\tilde{C}^{h,n}$ et $f*\nu^n$ sont aléatoires (mais pas B^h, $\tilde{C}^h$, $f*\nu$); c'est pourquoi il y a lieu de faire intervenir la convergence en loi (ou de manière équivalente, la convergence en probabilité).

La différence, essentielle, avec le théorème II-2.1, est que ces conditions ne sont plus nécessaires pour que $X^n \xrightarrow{\mathcal{L}} X$.

Nous démontrerons ce théorème au §c. Cependant, indiquons tout de suite pourquoi on a les équivalences $[\gamma] \Longleftrightarrow [\text{Sup-}\gamma]$ et $[\delta] \Longleftrightarrow [\text{Sup-}\delta]$.

LEMME 2.2: *a) On a* $[\gamma] \Longleftrightarrow [\text{Sup-}\gamma]$ *et* $[\delta] \Longleftrightarrow [\text{Sup-}\delta]$.

b) Soit (f_p) *une suite de fonctions, dense pour la convergence uniforme dans l'ensemble des fonctions positives bornées sur* R^d *qui sont nulles sur un voisinage de* 0 *et qui sont uniformément continues. Alors* $[\delta]$ *équivaut à:*

$[\delta']$ $\quad f_p * \nu^n_t \xrightarrow{\mathcal{L}} f_p * \nu_t$ pour tout $p \geqq 1$ et tout t dans une partie dense $A \subset R_+$.

Preuve. Il est clair que $[\delta] \Longrightarrow [\delta']$. On va montrer que $[\delta'] \Longrightarrow [\text{Sup-}\delta]$. On montrerait de même l'équivalence $[\text{Sup-}\gamma] \Longleftrightarrow [\gamma]$ (c'est même un peu plus simple).

Supposons $[\delta']$. Quitte à prendre le produit de tous les espaces de probabilité, on peut supposer que tous les processus sont définis sur le même espace. Pour montrer $[\text{Sup-}\delta]$, il suffit de montrer que de toute sous-suite infinie (n_k) on peut extraire une sous-sous-suite (n_{k_q}) telle qu'en dehors d'un ensemble négligeable N on ait $\sup_{s \leqq t} |f * \nu_s^{n_{k_q}}(\omega) - f * \nu_s| \to 0$ quand $q \uparrow \infty$. Mais par un procédé diagonal, on peut toujours extraire de (n_k) une sous-sous-suite (n_{k_q}) telle qu'en dehors d'un négligeable N on ait $f_p * \nu_t^{n_{k_q}}(\omega) \to f_p * \nu_t$ pour tout $p \geqq 1$ et tout $t \in A$. Il suffit alors d'appliquer le lemme II-2.2 en chaque point $\omega \notin N$ pour obtenir le résultat.∎

Pour montrer que $X^n \xrightarrow{\mathcal{L}} X$, et en suivant les considérations du chapitre I, nous allons d'abord montrer que la suite $\{\mathcal{L}(X^n)\}$ est tendue, puis que la loi $\mathcal{L}(X)$ est la seule loi limite possible pour la suite $\{\mathcal{L}(X^n)\}$. A cet effet, nous verrons deux méthodes:

1) l'une, basée sur la caractérisation II-1.12 du PAI X en termes de martingales, sera expliquée au chapitre IV;

2) l'autre, expliquée ci-dessous utilise le fait que $X^n \xrightarrow{\mathcal{L}(R_+)} X$. Cette seconde méthode contrairement à la première, utilise explicitement le fait que la limite est un PAT.

§b - RELATIVE COMPACITE DE LA SUITE $\{\mathcal{L}(X^n)\}$. La démonstration est essentiellement la même qu'au chapitre II pour les PAI. Pour tout $b>0$ on définit h_b par $h_b(x) = bh(x/b)$; on pose $\tilde{X}^{h_q,n} = X^{h_q,n} - B^{h_q,n}$ pour $q \in \mathbb{N}^*$.

LEMME 2.3: *Soit* $[\gamma]$ *et* $[\delta]$. *Pour chaque* $q \in \mathbb{N}^*$, *la suite* $\{\mathcal{L}(\tilde{X}^{h_q,n})\}_{n \geqq 1}$ *est tendue.*

Preuve. Il suffit de recopier la preuve de II-2.28. La seule différence est que A^n et G^{nq} sont aléatoires. D'après $[\text{Sup-}\gamma]$ et $[\text{Sup-}\delta]$, on a $\sup_{s \leqq t} |G^{nq}_s - G^q_s| \xrightarrow{\mathcal{L}} 0$.

Ceci entraine à l'évidence que $G^{nq} \xrightarrow{\mathcal{L}} G^q$, donc la suite $\{\mathcal{L}(G^{nq})\}_{n\geq 1}$ est C-tendue, donc aussi la suite $\{\mathcal{L}(A^n)\}_{n\geq 1}$ d'après I-2.20 et on conclut par le théorème I-3.24.∎

LEMME 2.4: *Soit* [Sup-β], [γ], [δ]. *La suite* $\{\mathcal{L}(X^n)\}$ *est tendue.*

Preuve. Là encore, on recopie la preuve de II-2.29. Les seules différences sont, d'une part que $(h_q - h_{1/q})*\nu^n$ et V^{nq} sont des processus et non des fonctions (ce qui ne change pas les arguments), d'autre part que W^{nq} n'est pas traité tout-à-fait de la même manière. En effet, on a

$$P^n(\sup_{s\leq N} |W_s^{nq}|>0) \leq P^n\{\textstyle\sum_{s\leq N} g_q(\Delta X_s^n)>\frac{1}{2}\} .$$

Le processus $F_t^{nq} = \sum_{s\leq t} g_q(\Delta X_s^n)$ est dominé au sens de Lenglart (cf. I-3.16) par $g_q*\nu^n$, car $F^{nq} - g_q*\nu^n$ est une martingale locale. Donc d'après I-3.18 on a

$$P^n(\sup_{s\leq N} |W_s^{nq}|>0) \leq 2\varepsilon + P^n(g_q*\nu_N^n>2\varepsilon).$$

Comme dans II-2.29, il existe q tel que $g_q*\nu_N\leq\varepsilon$; d'après [δ] il existe n_o tel que $P^n(g_q*\nu_N^n>2\varepsilon) \leq \varepsilon$ si $n>n_o$, ce qui implique $P^n(\sup_{s\leq N} |W_s^{nq}|>0) \leq 3\varepsilon$. On a donc bien la condition I-2.17-(iii).∎

§c - CONVERGENCE FINI-DIMENSIONNELLE DES X^n. Pour chaque n on associe aux caractéristiques de X^n les processus prévisibles $A^n(u)$ et $G^n(u)$ par 1.30 et 1.33. De même, $A(u)$ et $G(u)$ sont associés à X. D'après 1.37, on a $G(u)_t = E(e^{iu.X_t})$.

PROPOSITION 2.5: *Supposons que* $G^n(u)_t \xrightarrow{\mathcal{L}} G(u)_t$ *pour tout* $t\in A$ *et tout* $u\in R^d$. *Alors* $X^n \xrightarrow{\mathcal{L}(A)} X$.

Nous verrons dans la preuve que la propriété de X de ne pas avoir de discontinuités fixes n'est pas pleinement utilisée; seul est utilisé le fait que la fonction $u \rightsquigarrow G(u)_t$ ne s'annule pour aucun $t\in A$ (c'est évidemment vrai lorsque X n'a pas de discontinuités fixes).

Lorsque les X^n sont eux-mêmes des PAI, le résultat est trivial car on a aussi $G^n(u)_t = E(\exp iu.X_t^n)$

Preuve. Soit $0 = t_o<\ldots<t_p$ avec $t_j\in A$ (on peut évidemment supposer que $0\in A$). On va montrer par récurrence sur p que $(X_{t_o}^n,\ldots,X_{t_p}^n) \xrightarrow{\mathcal{L}} (X_{t_o},\ldots,X_{t_p})$. On suppose cette assertion vraie pour $p-1$ (elle est évidente pour $p=0$ car $X_0^n = X_0 = 0$). Il faut donc montrer que pour tous u_j, $u\in R^d$ on a

$$E^n\{\exp i\{\textstyle\sum_{1\leq j\leq p-1} u_j.X_{t_j}^n + u.(X_{t_p}^n - X_{t_{p-1}}^n)\}\} \to E\{\exp i\{\textstyle\sum_{1\leq j\leq p-1} u_j.X_{t_j} + u.(X_{t_p} - X_{t_{p-1}})\}\}$$

Soit $\zeta^n = \exp i \sum_{1\leq j\leq p-1} u_j.X^n_{t_j}$ et $\zeta = \exp i \sum_{1\leq j\leq p-1} u_j.X_{t_j}$. L'hypothèse de récurrence implique:

2.6 $$E^n(\zeta^n) \to E(\zeta) ,$$

et il faut montrer que

$$v_n := E^n\{\zeta^n \exp iu.(X^n_{t_p} - X^n_{t_{p-1}})\} \to v := E\{\zeta \exp iu.(X_{t_p} - X_{t_{p-1}})\} = E(\zeta)\frac{G(u)_{t_p}}{G(u)_{t_{p-1}}}.$$

(on a utilisé la propriété de PAI de X pour la dernière égalité).

Soit $T^n(u) = \inf(t: G^n(u)_t = 0)$. Soit $b = |G(u)_{t_p}|$. On a $b>0$ et $R^n = \inf(t: |G^n(u)_t| \leq b/2)$ est un temps prévisible (car $G^n(u)$ est prévisible). De plus, $|G^n(u)_t|$ est un processus décroissant en t (cela se voit immédiatement sur 1.33) et par hypothèse $|G^n(u)_{t_p}| \xrightarrow{\mathcal{L}} b$: donc $P^n(R^n \leq t_p) \to 0$.

R^n étant un temps prévisible, il existe un $(\underline{\underline{F}}^n_t)$-temps d'arrêt S^n tel que $S^n < R^n$ et $P^n(S^n < R^n - 1/n) \leq 1/n$. Par suite

2.7 $$P^n(S^n \leq t_p) \to 0.$$

Comme $S^n < T^n(u)$, le processus $M^n_t = \{\exp iu.X^n_{t\wedge S^n}\}/G^n(u)_{t\wedge S^n}$ est une martingale locale d'après le théorème 1.35. De plus $|G^n(u)_t| \geq b/2$ si $t \leq S^n$, donc $|M^n| \leq 2/b$ et ce processus M^n est même une martingale. Par suite

2.8 $$E^n(\beta^n | \underline{\underline{F}}^n_{t_{p-1}}) = 1 , \quad \text{où} \quad \beta^n = M^n_{t_p} / M^n_{t_{p-1}} .$$

Soit aussi $\gamma^n = G^n(u)_{t_p} / G^n(u)_{t_{p-1}}$ (avec $a/0 = 0$ par convention) et $\gamma = G(u)_{t_p} / G(u)_{t_{p-1}}$. D'après 2.8, on a:

$$v_n = E^n\{\zeta^n 1_{\{S^n \leq t_p\}} \exp iu.(X^n_{t_p} - X^n_{t_{p-1}})\} + E^n\{\zeta^n \beta^n (\gamma^n 1_{\{S^n > t_p\}} - \gamma)\} + E^n(\zeta^n)\gamma ,$$

d'où

2.9 $$|v_n - v| \leq P^n(S^n \leq t_p) + E^n(|\beta^n| . |\gamma^n 1_{\{S^n > t_p\}} - \gamma|) + |\gamma| . |E^n(\zeta^n) - E(\zeta)|$$

D'après 2.6 et 2.7, les premier et troisième termes ci-dessus tendent vers 0. Par ailleurs $|\beta^n| \leq 2/b$, $|\gamma^n 1_{\{S^n > t_p\}}| \leq 2/b$, et l'hypothèse plus 2.7 entrainent:

$$\gamma^n 1_{\{S^n > t_p\}} \xrightarrow{P} \gamma$$

Donc le second terme du second membre de 2.9 tend également vers 0. Par suite on a $v_n \to v$, d'où le résultat.∎

Preuve du théorème 2.1. On suppose [Sup-β], [γ], [δ], donc aussi [Sup-γ] et [Sup-δ]. Etant donnés 2.4 et 2.5, il suffit de montrer que si $u \in \mathbb{R}^d$, $t>0$, on a: $G^n(u)_t \xrightarrow{\mathcal{L}} G(u)_t$. Pour cela, il suffit que, (n_k) étant une suite infinie de $\mathbb{N}$, on puisse en extraire une sous-suite (n_{k_q}) telle que $G^{n_{k_q}}(u)_t \xrightarrow{\mathcal{L}} G(u)_t$.

Soit les fonctions f_p du lemme 2.2. Par un procédé diagonal, on peut extraire une sous-suite (n_{k_q}) de (n_k) telle qu'en dehors d'un ensemble négligeable N, on ait (rappelons que, quitte à prendre le produit de tous les espaces, on peut toujours supposer que tous les processus sont définis sur un même espace de probabilité):

$$B^{h,n_{k_q}}(\omega) \longrightarrow B^h \quad \text{uniformément sur les compacts;}$$

$$\tilde{C}^{h,n_{k_q}}(\omega) \longrightarrow \tilde{C}^h \quad \text{uniformément sur les compacts;}$$

$$f_p * \nu_t^{n_{k_q}}(\omega) \longrightarrow f_p * \nu_t \quad \text{pour tous } p \in \mathbb{N},\ t \in \mathbb{Q}_+.$$

D'après le théorème 2.1 et le lemme 2.2 du chapitre II, pour tout $\omega \notin N$ les lois des PAI admettant les caractéristiques $(B^{h,n_{k_q}}(\omega), C^{n_{k_q}}(\omega), \nu^{n_{k_q}}(\omega))$ convergent vers $\mathcal{L}(X)$. Comme $G^{n_{k_q}}(u)_t(\omega)$ est l'espérance de $\exp iu.Y_t$ lorsque Y est le PAI ci-dessus, on en déduit que:

$$\omega \notin N \Longrightarrow G^{n_{k_q}}(u)_t(\omega) \to G(u)_t \qquad \forall u \in R^d, \forall t>0,$$

d'où le résultat. ∎

La preuve ci-dessus est très courte, car elle s'appuie sur la condition suffisante du théorème II-2.1, elle-même basée sur la condition nécessaire du même théorème. Il existe bien-sûr une démonstration directe de la condition suffisante de II-2.1 et, partant, du théorème III-2.1.

Plus précisément, on peut montrer directement que [Sup-β]+[γ]+[δ] entrainent que $G^n(u)_t \xrightarrow{\mathcal{L}} G(u)_t$. De même, on peut montrer directement (nous ne le ferons pas ici: il suffit de suivre Gnedenko et Kolmogorov [15]) que si pour <u>une</u> valeur de t on a

$[\beta_t]$ $\quad B_t^{h,n} \xrightarrow{\mathcal{L}} B_t^h$

$[\gamma_t]$ $\quad \tilde{C}_t^{h,n} \xrightarrow{\mathcal{L}} \tilde{C}_t^h$

$[\delta_t]$ $\quad f*\nu_t^n \xrightarrow{\mathcal{L}} f*\nu_t$ pour toute f continue bornée positive nulle autour de 0

$[UP_t]$ $\quad \mathrm{Sup}_{s\leq t}\ \nu^n(\{s\}\times\{|x|>\varepsilon\}) \xrightarrow{\mathcal{L}} 0 \qquad \forall \varepsilon>0$

alors $G^n(u)_t \xrightarrow{\mathcal{L}} G(u)_t$ pour tout $u \in R^d$ (dans [15] ce résultat est montré lorsque $B^{h,n}, C^n, \nu^n$ sont déterministes, donc les X^n des PAI; on passe au cas aléatoire exactement comme ci-dessus). Etant donné 2.5, on en déduit:

<u>THEOREME 2.10</u>: *Sous* $[\beta_t]$, $[\gamma_t]$, $[\delta_t]$, $[UP_t]$, *on a* $X_t^n \xrightarrow{\mathcal{L}} X_t$.

<u>REMARQUES 2.11</u>: 1) Si on a [δ], il est facile de voir que $[UP_t]$ est satisfait pour tout t, car $\nu(\{t\}\times R^d) = 0$ par hypothèse.

2) Supposons qu'on ait $[\beta_t]$ pour tout t, et [γ] et [δ] (donc $[\gamma_t]$,

$[\delta_t]$, $[UP_t]$ pour tout t). D'après 2.5 on a aussi $X^n \xrightarrow{\mathcal{L}(R_+)} X$. Cependant, il n'y a pas nécessairement convergence en loi $X^n \xrightarrow{\mathcal{L}} X$ pour la topologie de Skorokhod (c'est la même situation qu'en II-2.4).

3) Le théorème 2.10 ne fait pas intervenir X en tant que processus; seules interviennent les caractéristiques de Lévy-Khintchine B_t^h, C_t, $\nu([0,t]\times .)$ de la loi (indéfiniment divisible) de X_t; par contre, on utilise pleinement les propriétés des processus X^n (jusqu'à l'instant t), et pas seulement les lois $\mathcal{L}(X_t^n)$. ∎

§d - APPLICATION AUX SEMIMARTINGALES LOCALEMENT DE CARRE INTEGRABLE. Dans ce paragraphe on suppose que les X^n sont des semimartingales nulles en 0, qui sont localement de carré intégrable, ce qui d'après 1.48 équivaut à:

2.12 $$|x|^2 * \nu_t^n < \infty \qquad \forall t > 0.$$

On suppose aussi que

2.13 $$|x|^2 * \nu_t < \infty \qquad \forall t > 0,$$

et on pose

2.14 $$B^n = B^{h,n} + (x - h(x)) * \nu^n , \qquad B = B^h + (x - h(x)) * \nu.$$

PROPOSITION 2.15: *On suppose 2.12, 2.13, et le PAI* X *sans discontinuités fixes. Pour que* $X^n \xrightarrow{\mathcal{L}} X$ *il suffit qu'on ait* $[\delta]$ *et*

(i) $\lim_{b\uparrow\infty} \limsup_n P^n(|x|^2 1_{\{|x|>b\}} * \nu_t^n > \varepsilon) = 0 \qquad \forall \varepsilon>0, \forall t>0;$

[Sup-β'] $\sup_{s\leq t} |B_s^n - B_s| \xrightarrow{\mathcal{L}} 0 \qquad \forall t>0;$

[γ'] $C_t^{n,jk} + (x^j x^k) * \nu_t^n - \sum_{s\leq t} \Delta B_s^{n,j} \Delta B_s^{n,k} \xrightarrow{\mathcal{L}} C_t^{jk} + (x^j x^k) * \nu_t \qquad \forall t>0, \forall j,k\leq d.$

Preuve. Comme pour les preuves du lemme 2.2 ou du théorème 2.1, on va appliquer le "principe des sous-suites". Soit (n_k) une suite infinie de $\mathbb{N}$. Par le lemme de Borel-Cantelli et en utilisant un procédé diagonal, il est facile d'en extraire une sous-suite (n_{k_q}) telle que si ω n'appartient pas à un ensemble négligeable N, les triplets "déterministes" $(B^{h,n_{k_q}}(\omega), C^{n_{k_q}}(\omega), \nu^{n_{k_q}}(\omega))$ vérifient $[\delta']$, [Sup-β'] [γ'] et la condition (i) de II-2.35. D'après la proposition II-2.35, pour chaque $\omega \notin N$ les triplets déterministes ci-dessus vérifient [Sup-β], [γ], [δ] et on en déduit que les triplets "aléatoires" $(B^{h,n}, C^n, \nu^n)$ vérifient également ces trois conditions, d'où le résultat. ∎

Exactement comme au §II-2-e on en déduit le:

COROLLAIRE 2.16: *On suppose 2.12; on suppose que* X *est un PAI continu, de caractéristiques* $(B^h,C,0)$. *Pour que* $X^n \xrightarrow{\mathcal{L}} X$ *il suffit que:*

$$[L] \quad |x|^2\, 1_{\{|x|>\varepsilon\}}*\nu_t^n \xrightarrow{\mathcal{L}} 0 \qquad \forall t>0, \forall\varepsilon>0$$

$$[\text{Sup-}\beta'] \quad \text{Sup}_{s\leq t}\, |B_s^n - B_s| \xrightarrow{\mathcal{L}} 0 \qquad \forall t>0 \quad (\text{ici, } B = B^h)$$

$$[\gamma'] \quad C_t^{n,jk} + (x^jx^k)*\nu_t^n - \sum_{s\leq t} \Delta B_s^{n,j}\Delta B_s^{n,k} \xrightarrow{\mathcal{L}} C_t^{jk} \qquad \forall t>0, \forall j,k\leq d.$$

COROLLAIRE 2.17: *On suppose que les* X^n *sont des martingales localement de carré intégrable réelles, et que* X *est un PAI continu de caractéristiques* $(0,C,0)$ (i.e., X est une martingale continue gaussienne, et $C_t = E(X_t^2)$). *Pour que* $X^n \xrightarrow{\mathcal{L}} X$, *il suffit qu'on ait* [L] *et*

$$\langle X^n,X^n\rangle_t \xrightarrow{\mathcal{L}} C_t \qquad \forall t>0.$$

Preuve. Il suffit d'appliquer 2.16 en remarquant que $B^n = 0$ et que le premier membre de $[\gamma']$ égale $\langle X^n,X^n\rangle_t$. ∎

Voici une application simple de ce corollaire. Soit $(M^n)_{n\geq 1}$ une suite de martingales indépendantes et de même loi, telles que

$$\text{2.18} \qquad C_t = E\{(M_t^n)^2\} < \infty \qquad \forall t>0.$$

et qui vérifient $M_0^n = 0$. On suppose en outre que la fonction C est continue; alors:

PROPOSITION 2.19: *Sous les hypothèses précédentes, les processus* $X^n = \frac{1}{\sqrt{n}}(M^1+..+M^n)$ *convergent en loi vers un PAI de caractéristiques* $(0,C,0)$ (martingale gaussienne continue).

Preuve. Les X^n sont des martingales de carré intégrable. Comme les M^n sont indépendantes, on a

$$\langle X^n,X^n\rangle = \frac{1}{n}\sum_{1\leq p\leq n} \langle M^p,M^p\rangle$$

et les variables $\langle M^p,M^p\rangle_t$ sont indépendantes, de même loi, de moyenne C_t . D'après la loi forte des grands nombres, on a donc $\langle X^n,X^n\rangle_t \to C_t$ p.s.

Par ailleurs, C étant continu, les martingales M^n n'ont pas de discontinuités fixes. Il s'ensuit que les M^n n'ont p.s. pas de sauts communs. Si ν'^p désigne la troisième caractéristique locale de M^p, et ν^n celle de X^n, on a donc:

$$f*\nu_t^n = \sum_{1\leq p\leq n} \int \nu'^p([0,t]\times dx)\, f(x/\sqrt{n})$$

D'où

$$x^2 1_{\{|x|>\varepsilon\}}*\nu_t^n = \frac{1}{n}\sum_{1\leq p\leq n} x^2 1_{\{|x|>\varepsilon\sqrt{n}\}}*\nu_t'^p ,$$

$$E(x^2 1_{\{|x|>\varepsilon\}} * \nu_t^n) = E(x^2 1_{\{|x|>\varepsilon\sqrt{n}\}} * \nu_t^{\prime P}),$$

qui tend vers 0 quand $n\uparrow\infty$, pour tout $\varepsilon>0$. On a donc [L], d'où le résultat.∎

§e - APPLICATION AUX TABLEAUX TRIANGULAIRES. Rappelons le cadre introduit au §II-3. Pour chaque n on considère un espace $(\Omega^n, \underline{\underline{F}}^n, P^n)$ muni d'une suite $(Y_m^n)_{m\geq 1}$ de variables à valeurs dans R^d, adaptés à une filtration discrète $(\underline{\underline{G}}_m^n)_{m\geq 0}$. On considère un "changement de temps" $(\sigma_t^n)_{t\geq 0}$ au sens de 1.43, et le processus

2.20 $$X_t^n = \sum_{1\leq m\leq \sigma_t^n} Y_m^n \qquad (= 0 \text{ si } \sigma_t^n = 0),$$

qui est adapté à la filtration $\underline{\underline{F}}_t^n = \underline{\underline{G}}_{\sigma_t^n}^n$.

THEOREME 2.21: *Soit* X *un PAI sans discontinuités fixes, de caractéristiques* (B^h, C, ν) *associées à une fonction de troncation* h *continue. Pour que* $X^n \xrightarrow{\mathcal{L}} X$ *il suffit qu'on ait:*

[Sup-β] $\sup_{s\leq t} |\sum_{1\leq m\leq \sigma_s^n} E^n\{h(Y_m^n)|\underline{\underline{G}}_{m-1}^n\} - B_s^h| \xrightarrow{\mathcal{L}} 0 \quad \forall t>0;$

[γ] $\sum_{1\leq m\leq \sigma_t^n} \{E^n\{(h^j h^k)(Y_m^n)|\underline{\underline{G}}_{m-1}^n\} - E^n\{h^j(Y_m^n)|\underline{\underline{G}}_{m-1}^n\}\; E^n\{h^k(Y_m^n)|\underline{\underline{G}}_{m-1}^n\}\}$

$$\xrightarrow{\mathcal{L}} C_t^{jk} + (h^j h^k) * \nu_t \qquad \forall t>0,\ \forall j,k\leq d;$$

[δ] $\sum_{1\leq m\leq \sigma_t^n} E^n\{f(Y_m^n)|\underline{\underline{G}}_{m-1}^n\} \xrightarrow{\mathcal{L}} f*\nu_t$ pour tout t et toute fonction f continue bornée sur R^d, nulle sur un voisinage de l'origine.

Preuve. Il suffit d'appliquer le théorème 2.1 aux semimartingales X^n, en utilisant la forme 1.45 de leurs caractéristiques locales.∎

De même, l'application du corollaire 2.16 donne:

COROLLAIRE 2.22: *Soit* X *un PAI continu de caractéristiques* $(B^h = B, C, 0)$. *Pour que* $X^n \xrightarrow{\mathcal{L}} X$ *il suffit, dans le cas où* $Y_m^n \in L^2$ *pour tous* m,n, *que*

[L] $\sum_{1\leq m\leq \sigma_t^n} E^n\{|Y_m^n|^2 1_{\{|Y_m^n|>\varepsilon\}}|\underline{\underline{G}}_{m-1}^n\} \xrightarrow{\mathcal{L}} 0 \qquad \forall t>0,\ \forall\varepsilon>0$

[Sup-β'] $\sup_{s\leq t} |\sum_{1\leq m\leq \sigma_s^n} E^n(Y_m^n|\underline{\underline{G}}_{m-1}^n) - B_s| \xrightarrow{\mathcal{L}} 0 \qquad \forall t>0$

[γ'] $\sum_{1\leq m\leq \sigma_t^n} \{E^n(Y_m^{n,j} Y_m^{n,k}|\underline{\underline{G}}_{m-1}^n) - E^n(Y_m^{n,j}|\underline{\underline{G}}_{m-1}^n)\, E^n(Y_m^{n,k}|\underline{\underline{G}}_{m-1}^n)\} \xrightarrow{\mathcal{L}} C_t^{jk} \qquad \forall t>0.$

Un cas très important est la généralisation du théorème de Donsker (c'est le "théorème limite central fonctionnel" classique pour les martingales discrètes). On part d'une martingale discrète $(U_n)_{n\geq 0}$ nulle en 0 sur l'espace $(\Omega, \underline{\underline{F}}, (\underline{\underline{G}}_m), P)$ qui est supposée de carré intégrable (i.e. chaque U_n est dans L^2). Soit aussi

$$C_n = \sum_{1\leq p\leq n} E\{(U_p - U_{p-1})^2 | \underline{\underline{G}}_{p-1}\}.$$

THEOREME 2.23. *Supposons que:*

(*i*) $\frac{1}{n} \sum_{1\leq m\leq [nt]} E\{(U_p - U_{p-1})^2 1_{\{|U_p - U_{p-1}|>\varepsilon\sqrt{n}\}} | \underline{\underline{G}}_{p-1}\} \xrightarrow{\mathcal{L}} 0 \qquad \forall t>0,\ \forall \varepsilon>0;$

(*ii*) $\frac{1}{n} C_{[nt]} \xrightarrow{\mathcal{L}} t \qquad \forall t>0.$

Alors les processus $X_t^n = \frac{1}{\sqrt{n}} U_{[nt]}$ *convergent en loi vers un mouvement brownien standard.*

Preuve. On applique le corollaire précédent aux variables $Y_m^n = \frac{1}{\sqrt{n}} (U_m - U_{m-1})$ et aux changements de temps $\sigma^n(t) = [nt]$.∎

§f - COMMENTAIRES BIBLIOGRAPHIQUES. Les premiers théorèmes limite ont concerné, bien entendu, les tableaux triangulaires. En ce qui concerne le genre de méthodes utilisé ici, signalons les articles fondamentaux de B. Brown [5] et de B. Brown et Eagleson [6] sur la convergence (non fonctionnelle) d'accroissements de martingales. Les résultats de convergence fonctionnelle, dans des cas assez particuliers, sont plus anciens: voir par exemple Rosen [49], et Billingsley [3]. Ensuite, les résultats cités de B. Brown ont été généralisé au cadre fonctionnel par McLeish [40] qui a démontré le théorème 2.22, puis Durrett et Resnick [11] (pour des limites qui sont des PAI non continus) et [26].

Les résultats de convergence de processus continus indicés par R_+ sont également d'origine assez ancienne, notamment pour les processus de Markov, et sous des hypothèses assez restrictives; ces résultats sont habituellement basés sur la méthode des "problèmes de martingales" que nous exposerons dans le chapitre IV. L'étude de la convergence de processus discontinus, et encore plus la convergence vers un processus limite discontinu, sont des choses plus récentes: les premiers résultats généraux sont dûs à T. Brown ([7] pour les processus ponctuels, [8] pour les processus ponctuels multivariés) et à Rebolledo ([46],[48]: convergence de martingales vers un brownien). La méthode utilisée dans ce chapitre est basée sur l'article [29] de Kabanov, Liptcer et Shiryaev; des formulations plus générales ont ensuite été proposées par Liptcer et Shiryaev [35] et aussi dans [24] et [26].

3 - DEUX EXEMPLES

Le théorème 2.1 peut sembler séduisant au lecteur et, au vu du théorème II-2.1, optimal. Les deux exemples ci-dessous visent à tempérer cet enthousiasme! le premier montre que, même dans un cas simple, les conditions de ce théorème 2.1 peuvent se révéler très difficiles à vérifier. Le second montre qu'en outre elles ne sont pas nécessaires (on examinera plus à fond la nécessité de ces conditions au chapitre V).

§a-SOMMES NORMALISEES DE SEMIMARTINGALES INDEPENDANTES DE MEME LOI. On va généraliser ci-dessous la situation de la proposition 2.19, en considérant une suite de copies indépendantes de la même semimartingale, nulle en 0, uni-dimensionnelle pour simplifier: soit $Y(p)$ la $p^{\text{ième}}$ copie, définie sur l'espace $(\Omega(p),\underline{\underline{F}}(p),(\underline{\underline{F}}_t(p)),P(p))$ (noté aussi $\mathcal{B}(p)$ pour simplifier), et de caractéristiques locales $(B^h(p),C(p),\nu(p))$; on lui associe aussi le processus $A(p)(u)$ défini par 1.30.

Soit $(\Omega,\underline{\underline{F}},(\underline{\underline{F}}_t),P)$ le produit des espaces $\mathcal{B}(p)$; Soit β_n des constantes de normalisation. On considère les processus

3.1
$$X^n = \beta_n \sum_{1\leq p\leq n} Y(p).$$

Il est évident que X^n est une martingale, dont on note $(B^{h,n},C^n,\nu^n)$ les caractéristiques locales, et $A^n(u)$ le processus associé par 1.30.

LEMME 3.2: *Supposons que les* $Y(p)$ *n'aient pas de temps fixes de discontinuité. On a alors*

3.3
$$B^{h,n} = \sum_{1\leq p\leq n} \{\beta_n B^h(p) + \{h(\beta_n x) - \beta_n h(x)\}*\nu(p)\}$$

3.4
$$C^n = \sum_{1\leq p\leq n} \beta_n^2\, C(p)$$

3.5
$$g*\nu^n = \sum_{1\leq p\leq n} g(\beta_n x)*\nu(p)$$

3.6
$$A^n(u) = \sum_{1\leq p\leq n} A(p)(\beta_n u)$$

Preuve. On a $X^{n,c} = \sum_{1\leq p\leq n} \beta_n Y(p)^c$, et comme les $Y(p)$ sont indépendantes, on a $\langle Y(p)^c,Y(q)^c\rangle = 0$ si $p \neq q$. La formule 3.4 découle alors de 1.23. L'hypothèse entraine que les $Y(p)$ n'ont p.s. pas de sauts communs. On a donc

$$\sum_{s\leq t} g(\Delta X^n_s) = \sum_{1\leq p\leq n} \sum_{s\leq t} g(\beta_n \Delta Y(p)_s)$$

et 3.5 découle de la caractérisation 1.16-(iii). Toujours pour la même raison, on a:

$$X^{h,n}_t = X^n_t - \sum_{s\leq t} \{\Delta X^n_s - h(\Delta X^n_s)\}$$

$$= \sum_{1\leq p\leq n}\{\beta_n\{Y(p)_t^h + \sum_{s\leq t}\{\Delta Y(p)_s - h(\Delta Y(p)_s)\}\} - \sum_{s\leq t}\{\beta_n \Delta Y(p)_s - h(\beta_n \Delta Y(p)_s)\}\}$$

$$= \sum_{1\leq p\leq n}\{\beta_n Y(p)_t^h + \sum_{s\leq t}\{h(\beta_n \Delta Y(p)_s) - \beta_n h(\Delta Y(p)_s)\}\}.$$

Donc 3.3 découle de la caractérisation 1.16-(i). Enfin un calcul simple permet d'obtenir 3.6 à partir des trois formules précédentes.∎

D'après la forme de ν^n donnée par 3.5, il est clair que la vérification des conditions [Sup-β], [γ], [δ] risque de se révéler très difficile.

Afin toutefois d'écrire un résultat positif, nous allons particulariser la situation. Nous supposons que $Y(p)$ s'écrit $Y(p) = H(p)\bullet Z(p)$, intégrale stochastique d'un processus prévisible $H(p)$ sur $\mathcal{B}(p)$, qu'on suppose <u>borné</u> pour simplifier, par rapport à une semimartingale $Z(p)$ qui est un <u>PAI sans discontinuités fixes</u>. Bien entendu, les couples $(Z(p),H(p))$ sont des copies d'un <u>même couple</u> (Z,H).

On note $\hat{A}(u)$ la fonction associée aux caractéristiques du PAI Z par 1.30. Comme ce PAI n'a pas de discontinuités fixes, on a:

<u>3.7</u> $$E(e^{iu.Z_t}) = \exp \hat{A}(u)_t, \qquad \text{et} \quad \hat{A}(u)_t = \int_0^t a_s(u)\, d\gamma_s$$

où γ est une fonction croissante continue, et pour chaque s la fonction $a_s(u)$ est encore du type de Lévy-Khintchine (i.e., donnée par une formule du type 1.30).

Nous allons faire les deux hypothèses suivantes:

<u>3.8</u> Z est <u>symétrique</u> (i.e. Z et $-Z$ ont même loi). Cela revient à dire que $\hat{A}(u) = \hat{A}(-u)$, ou encore qu'on peut choisir a_s de sorte que $a_s(u)=a_s(-u)$.

<u>3.9</u> Il existe $\alpha\in]0,2]$ tel que pour tout s on ait

$$n\, a_s(u/n^{1/\alpha}) \to -|u|^\alpha \qquad \forall u\in R$$

(si $a_s(u) = -|u|^\alpha$, ce qui revient à dire que Z est le processus stable homogène symétrique d'indice α, cette hypothèse est évidemment satisfaite).

Le théorème suivant est un cas particulier d'un résultat dû à Giné et Marcus [14].

<u>THEOREME 3.10</u>: *Sous les hypothèses 4.8 et 4.9, les processus*

$$X^n = \frac{1}{n^{1/\alpha}} \sum_{1\leq p\leq n} H(p)\bullet Z(p)$$

convergent en loi vers un PAI X *stable symétrique d'indice* α, *caractérisé par*

<u>3.11</u> $$E(e^{iu.X_t}) = \exp -|u|^\alpha \int_0^t \delta(s)\, d\gamma_s,$$

où $\delta(s) = E(|H_s|^\alpha)$ $(= E(|H(p)_s|^\alpha)$ pour tout p).

Preuve. a) Nous allons d'abord calculer les termes $(B^h(p),C(p),\nu(p))$ et $A(p)(u)$ associés à $Y(p) = H(p)\bullet Z(p)$. Soit $(\hat{B}^h,\hat{C},\hat{\nu})$ les caractéristiques de Z. Comme $Y(p)^c = H(p)\bullet Z(p)^c$, on a

$$C(p) = H(p)^2\bullet\hat{C}.$$

On a $\Delta Y(p) = H(p)\Delta Z(p)$, donc $\sum_{s\leq t} g(\Delta Y(p)_s) = \sum_{s\leq t} g\{H(p)_s\Delta Z(p)_s\}$. D'après la propriété 1.16-(iii) généralisée aux fonctions "prévisibles" sur $\Omega\times R_+\times R$ (cf. la preuve de l'implication (a) $\Longrightarrow$ (c) dans 1.31), on en déduit que

$$g*\nu(p) = g\{H(p)x\}*\hat{\nu}.$$

Enfin, un calcul analogue à celui de la preuve de 3.2, reliant $Y(p)^h$ et $Z(p)^h$, permet d'obtenir

3.12 $$B^h(p) = H(p)\bullet\hat{B}^h + \{h(H(p)x) - H(p)h(x)\}*\hat{\nu}.$$

Etant donnée la forme 3.7 de $\hat{A}(u)$, un calcul élémentaire permet de conclure que

$$A(p)(u)_t = \int_0^t a_s(H(p)_s u)\, d\gamma_s$$

et la formule 3.6 appliquée avec $\beta_n = 1/n^{1/\alpha}$ donne

3.13 $$A^n(u)_t = \int_0^t \sum_{1\leq p\leq n} a_s(H(p)_s \frac{u}{n^{1/\alpha}})\, d\gamma_s.$$

b) Passons à la démonstration proprement dite. D'après 3.13 on a

$$A^n(u)_t = \int_0^t d\gamma_s \frac{1}{n}\sum_{1\leq p\leq n}\{n\, a_s(H(p)_s \frac{u}{n^{1/\alpha}}) + |H(p)_s u|^\alpha\}$$
$$- |u|^\alpha \frac{1}{n}\sum_{1\leq p\leq n}\int_0^t |H(p)_s|^\alpha\, d\gamma_s.$$

Noter que la convergence dans 3.9 est uniforme en u sur les compacts; comme $H(p)$ est borné (uniformément en p) on en déduit que le premier terme du second membre ci-dessus converge vers 0 pour tout ω. Quant au second terme, d'après la loi des grands nombres, il converge vers $-|u|^\alpha\int_0^t \delta(s)\, d\gamma_s$ en dehors d'un ensemble négligeable N_t indépendant de u. Si alors (B^h,C,ν) sont les caractéristiques du PAI X décrit par 3.11, la fonction $A(u)$ associée à ces caractéristiques par 1.30 est précisément $A(u)_t = -|u|^\alpha \int_0^t \delta(s)d\gamma_s$. D'après les résultats classiques de convergence des lois indéfiniment divisibles, on en déduit que

$$\omega\notin N \Longrightarrow \begin{cases} B_t^{h,n}(\omega) \to B_t^h, \quad C_t^n(\omega) + h^2*\nu_t^n(\omega) \to C_t + h^2*\nu_t \\ f*\nu_t^n(\omega) \to f*\nu_t \quad \text{pour } f \text{ continue bornée nulle autour de } 0. \end{cases}$$

En particulier, la suite (X^n) vérifie, relativement au PAI X, les conditions $[\gamma]$ et $[\delta]$. Ce qui précède ne suffit pas à obtenir [Sup-β], et c'est là qu'on va utiliser l'hypothèse de symétrie 3.8: cette hypothèse implique que $\hat{\nu}$ est symétrique (i.e., $\hat{\nu}([0,t]\times.)$ est symétrique sur R), et que $\hat{B}^h = 0$ si on choisit une fonction de troncation h impaire. D'après 3.12 on a alors $B^h(p) = 0$, donc aussi $B^{h,n} = 0$; par ailleurs X est aussi un processus symétrique, donc on a $B^h = 0$ (toujours avec h impaire): on a donc automatiquement [Sup-β]. Le résultat découle alors du théorème 1.2. ∎

Terminons par quelques commentaires sur le résultat obtenu par Giné et Marcus. Ce résultat est plus général que 3.10 sous trois aspects. En premier lieu, ils admettent des processus H(p) qui ne sont pas bornés, ce qui est une généralisation mineure. En second lieu, ils admettent des processus Z(p) avec discontinuités fixes: cela amène de légères complications dans les calculs, mais dans ce cours nous nous sommes restreints au cas où le PAI limite n'admet pas de discontinuités fixes, ce qui n'est possible ici que si les Z(p) n'ont pas non plus de discontinuités fixes.

En troisième lieu, ils utilisent une hypothèse notablement plus faible que 3.9. Plus précisément, ils supposent que le processus générique Z est dans le domaine d'attraction normale d'un PAI stable symétrique d'indice α, ce qui veut dire que les PAI $n^{-1/\alpha}\sum_{1\leq p\leq n} Z(p)$ convergent vers un PAI stable symétrique d'indice α. Sous l'hypothèse de symétrie 3.8 cela revient à dire que

3.14 $$n\, A_s(u\, n^{-1/\alpha}) \to -\gamma_s |u|^\alpha \qquad \forall u\in R,\ \forall s>0,$$

une condition évidemment plus faible que 3.9.

Lorsque le processus générique H est de la forme

3.15 $$H_s = \sum_{q\geq 0} V_q\, 1_{\{t_q<s\leq t_{q+1}\}}:\quad 0=t_0<t_1<\ldots<t_q<\ldots,\ \lim_q\uparrow t_q = \infty,$$

où les V_q sont $\underline{F}_{t_q}$-mesurables et bornés, on peut reprendre la preuve précédente et obtenir le même résultat sous 3.14 au lieu de 3.9, sans aucune complication supplémentaire. Par contre si H est prévisible borné quelconque, il faut l'approcher en un sens convenable par des processus du type 3.15, ce qui implique la nécessité de compléter 3.14 par une hypothèse supplémentaire, à savoir

3.16 $$\forall t>0,\quad \mathrm{Sup}_{K \text{ prévisible}, |K|\leq 1}\ \mathrm{Sup}_{\lambda>0}\, \lambda^\alpha\, P(|K\bullet Z_t|>\lambda) < \infty$$

(on peut montrer que 3.9 entraine 3.16). On a alors le résultat 3.10, sous les hypothèses 3.8, 3.14 et 3.16; la démonstration nécessite des majorations délicates sur les intégrales stochastiques, qui sortent du cadre de ce cours.

§b - FONCTIONNELLES DE PROCESSUS DE MARKOV STATIONNAIRES. Nous avons vu en 2.23 un théorème central limite pour des variables U_p qui sont des accroissements de martingale; il existe aussi un théorème analogue lorsque les U_p forment une suite stationnaire mélangeante, ou sont des "fonctionnelles" d'une telle suite: voir par exemple le livre [3] de Billingsley. On en déduit des théorèmes limite pour les fonctionnelles de chaînes ou processus de Markov stationnaires (voir par exemple Maigret [39]).

A titre d'exemple, nous allons exposer ci-dessous un résultat récent de Touati [56], sans chercher le maximum de généralité.

Soit $(\Omega,\underline{\underline{F}},\underline{\underline{F}}_t,\theta_t,Y_t,P_x)$ un processus de Markov fort, continu à droite, à valeurs dans un espace topologique E. On fait l'hypothèse:

3.17 Ce processus de Markov admet une probabilité invariante μ, et la tribu des évènements invariants par le semi-groupe $(\theta_t)_{t\geq 0}$ est P_μ-triviale.

On note aussi (A,D_A) le générateur infinitésimal faible du processus, dans l'espace des fonctions boréliennes bornées sur E.

THEOREME 3.18: *Soit 3.17; soit* f *une fonction borélienne bornée qui s'écrit* $f = Ag$, *où* g *et* g^2 *appartiennent à* D_A. *Alors les processus*

3.19 $$X^n_t = \frac{1}{\sqrt{n}} \int_0^{nt} f(Y_s)\, ds \quad , \text{ pour la loi } P_\mu$$

convergent en loi vers $\sqrt{\beta}W$, *où* W *est un mouvement brownien standard et où*

3.20 $$\beta = -2\int g(x)\, Ag(x)\, \mu(dx).$$

Ce résultat est un contre-exemple à la nécessité des conditions [Sup-β], [γ], [δ] dans 2.1. En effet X^n, relativement à la filtration $\underline{\underline{F}}^n_t = \underline{\underline{F}}_{nt}$, est une semi-martingale continue à variation finie, donc ses caractéristiques locales sont

$$B^{h,n} = X^n, \quad C^n = 0, \quad \nu^n = 0.$$

Par ailleurs, les caractéristiques du PAI $\sqrt{\beta}W$ sont $(0,\beta t,0)$; On a donc [δ], mais [γ] et [Sup-β] ne sont pas vérifiées. Noter que β, défini par 3.20, est positif.

Preuve. On a déjà rappelé (cf. §1-d) que le processus

$$M_t = g(Y_t) - g(Y_0) - \int_0^t Ag(Y_s)\, ds$$

est une martingale locale pour toute loi initiale, donc en particulier pour P_μ. De plus il est bien connu (et facile à démontrer en utilisant la formule de Dynkin pour g^2 et la formule d'Ito) que, si $\Gamma(g,g) = Ag^2 - 2gAg$, alors

$$\langle M,M\rangle_t = \int_0^t \Gamma(g,g)(Y_s)\, ds.$$

Soit alors $M_t^n = M_{nt}/\sqrt{n}$, qui est une martingale pour $\underline{\underline{F}}_t^n = \underline{\underline{F}}_{nt}$; pour cette filtration, le crochet de M^n est clairement

<u>3.21</u>
$$<M^n,M^n>_t = \frac{1}{n}\int_0^{nt} \Gamma(g,g)(Y_s)\, ds.$$

On va alors appliquer le corolaire 2.17 aux martingales M^n, avec pour X le PAI $X = \sqrt{\beta}W$ de caractéristiques $(0,\beta t,0)$. Il existe une constante K qui majore $|g|$, donc $|\Delta M^n| \leq 2K/\sqrt{n}$ et la troisième caractéristique locale de M^n ne charge donc que l'ensemble $R_+\times\{x: |x| \leq 2K/\sqrt{n}\}$. Il est alors évident qu'on a la condition [L]. Il reste à montrer que sous P_μ on a

<u>3.22</u>
$$<M^n,M^n>_t \xrightarrow{\mathcal{L}} \beta t.$$

Etant donné 3.17, on peut appliquer la version continue du théorème ergodique ponctuel, à savoir que pour toute variable bornée V sur $(\Omega,\underline{\underline{F}})$ on a

$$\frac{1}{t}\int_0^t (V\circ\theta_s)\, ds \rightarrow E_\mu(V) \quad P_\mu\text{-p.s.}$$

quand $t\uparrow\infty$. Etant donné 3.21, il vient

$$<M^n,M^n>_t \rightarrow t\, E_\mu\{\Gamma(g,g)(Y_0)\} = t\int\mu(dx)\, \Gamma(g,g)(x) \qquad P_\mu\text{-p.s.}$$

Mais alors si β est défini par 3.20, on a bien 3.22, une fois remarqué que $\int\mu(dx)\, Ag^2(x) = 0$ puisque μ est une mesure invariante.

On a donc démontré que $M^n \xrightarrow{\mathcal{L}} \sqrt{\beta}W$. Pour conclure il suffit de remarquer que $X_t^n = M_t^n - \frac{1}{\sqrt{n}}\{g(Y_{nt}) - g(Y_0)\}$, tandis que g est bornée. ∎

IV - CONVERGENCE VERS UNE SEMIMARTINGALE

1 - UN THEOREME GENERAL DE CONVERGENCE

§a - ENONCE DES RESULTATS. Nous allons maintenant étudier la convergence d'une suite (X^n) de semimartingales vers une semimartingale X qui n'est pas nécessairement un PAI. Comme dans les chapitres précédents, on veut exprimer les conditions en fonctions des caractéristiques locales $(B^{h,n},C^n,\nu^n)$ et (B^h,C,ν).

Toutefois, dans la condition [Sup-β] par exemple, il faut effectuer la différence $B^{h,n} - B^h$, alors qu'ici les processus X^n et X sont a-priori définis sur des espaces différents (ce problème ne se pose évidemment pas lorsque B^h est déterministe). On tourne la difficulté en faisant l'hypothèse suivante sur X:

1.1 | X est le processus canonique sur l'espace de Skorokhod D^d: $X_t(\alpha) = \alpha(t)$, et P est une probabilité sur $(D^d,\underline{\underline{D}}^d)$ telle que:

(i) $P(X_0 = 0) = 1$;

(ii) pour P, et relativement à la filtration $(\underline{\underline{D}}^d_{t+})$ définie en I-1.20, le processus X est une semimartingale, dont on note (B^h,C,ν) les caractéristiques locales.

Ci-dessous, et dans tout le chapitre, h est une fonction de troncation fixée. On notera que 1.1 n'est pas une restriction sérieuse car, si Y est une semimartingale quelconque le processus canonique X sur D^d vérifie 1.1 pour $P = \mathcal{L}(Y)$ dès que $Y_0 = 0$ p.s.

On fera aussi l'hypothèse simplificatrice que X est quasi-continu à gauche sur $(D^d,\underline{\underline{D}}^d,(\underline{\underline{D}}^d_{t+}),P)$: voir la définition I-3.12; d'après III-1.22 cela équivaut à dire qu'on peut choisir une version de ν qui vérifie identiquement $\nu(\{t\}\times R^d)=0$, et dans ce cas, B^h est aussi continu. Ainsi, lorsque X est un PAI, la quasi-continuité à gauche équivaut à l'absence de discontinuités fixes, une hypothèse qu'on a toujours faite dans les chapitres précédents.

Par ailleurs, on suppose comme au chapitre III que:

1.2 | Pour chaque n, X^n est une semimartingale d-dimensionnelle sur $(\Omega^n,\underline{\underline{F}}^n,(\underline{\underline{F}}^n_t),P^n)$ vérifiant $X^n_0 = 0$ et de caractéristiques locales $(B^{h,n},C^n,\nu^n)$.

A ces triplets, on associe les processus $\tilde{C}^h$ et $\tilde{C}^{h,n}$ par III-1.21.

Chaque X^n est une application: $\Omega^n \to D^d$. Par composition avec X^n, on peut donc associer à toute variable (tout processus) sur D^d une variable (processus) sur Ω^n. Ainsi on a un triplet $(B^h\circ X^n, C\circ X^n, \nu\circ X^n)$ sur Ω^n, qui vérifie clairement les mêmes propriétés III-(1.17-1.20) que $(B^{h,n}, C^n, \nu^n)$. On peut donc comparer $B^{h,n}$ et B^h en calculant la différence $B^{h,n} - B^h\circ X^n$.

Introduisons maintenant la version adéquate des conditions [Sup-β], [γ], [δ].

[β] $B_t^{h,n} - B_t^h\circ X^n \xrightarrow{\mathcal{L}} 0$ pour tout t dans une partie dense $A\subset R_+$

[γ] $\tilde{C}_t^{h,n} - \tilde{C}_t^h\circ X^n \xrightarrow{\mathcal{L}} 0$ pour tout t dans une partie dense $A\subset R_+$

[δ] $f*\nu_t^n - (f*\nu_t)\circ X^n \xrightarrow{\mathcal{L}} 0$ pour tout t dans une partie dense $A\subset R_+$ et toute fonction f continue bornée: $R^d \to R_+$, nulle sur un voisinage de 0.

[Sup-β] $\sup_{s\leq t}|B_s^{h,n} - B_s^h\circ X^n| \xrightarrow{\mathcal{L}} 0$ pour tout $t>0$

[Sup-γ] $\sup_{s\leq t}|\tilde{C}_s^{h,n} - \tilde{C}_s^h\circ X^n| \xrightarrow{\mathcal{L}} 0$ pour tout $t>0$

[Sup-δ] $\sup_{s\leq t}|f*\nu_s^n - (f*\nu_s)\circ X^n| \xrightarrow{\mathcal{L}} 0$ pour tout $t>0$ et toute f comme dans [δ].

Ces conditions se réduisent évidemment aux conditions de même nom, du chapitre III, lorsque X est un PAI. On a évidemment [Sup-β] $\Longrightarrow$ [β], [Sup-γ] $\Longrightarrow$ [γ] et [Sup-δ] $\Longrightarrow$ [δ], avec $A = R_+$.

LEMME 1.3:

Soit (f_p) une suite de fonctions, dense pour la convergence uniforme dans l'ensemble des fonctions positives bornées sur R^d qui sont nulles sur un voisinage de 0 et qui sont uniformément continues. Alors [δ] équivaut à:

[δ'] $f_p*\nu_t^n - (f_p*\nu_t)\circ X^n \xrightarrow{\mathcal{L}} 0$ pour tout $p\in \mathbb{N}^*$ et tout t dans une partie dense $A\subset R_+$.

Preuve. Il suffit de reprendre mot pour mot la preuve du lemme III-2.2, en remplaçant partout $f_p*\nu_t$ par $(f_p*\nu_t)\circ X^n$ et en supprimant "$\sup_{s\leq t}$". ■

Voici maintenant trois autres conditions qui vont jouer un rôle essentiel.

1.4 Condition d'unicité. P est l'unique probabilité sur $(D^d, \underline{\underline{D}}^d)$ qui vérifie les conditions (i) et (ii) de 1.1. ■

1.5 Condition de majoration. Pour chaque $t\geq 0$ les fonctions $\alpha \rightsquigarrow C_t(\alpha)$ et $\alpha \rightsquigarrow (|x|^2\wedge 1)*\nu_t(\alpha)$ sont bornées sur D^d (donc aussi $\alpha \rightsquigarrow \tilde{C}_t^h(\alpha)$). ■

1.6 Condition de continuité. Pour chaque $t>0$ et chaque fonction f sur R^d, continue bornée et nulle sur un voisinage de 0, les fonctions

$$\alpha \rightsquigarrow B_t^h(\alpha) , \qquad \alpha \rightsquigarrow \tilde{C}_t^h(\alpha) , \qquad \alpha \rightsquigarrow f*\nu_t(\alpha)$$

sont continues pour la topologie de Skorokhod sur D^d. ∎

Remarquer que si X est un PAI, ces trois conditions sont satisfaites: c'est évident pour 1.5 et 1.6; pour 1.4, cela découle de III-1.39 et de l'unicité dans II-1.4-(b).

La condition d'unicité 1.4 est la plus difficile à vérifier; elle est vraie pour les processus de diffusion, avec ou sans sauts, sous des conditions assez générales de continuité des coefficients: [54],[55]. On verra plus loin qu'on peut notablement affaiblir 1.5.

THEOREME 1.7: *On suppose la fonction de troncation* h *continue, et la semimartingale* X *quasi-continue à gauche; on suppose les condition 1.4, 1.5 et 1.6. Pour que* $X^n \xrightarrow{\mathcal{L}} X$ *il suffit alors que:*

(i) la suite $\{\mathcal{L}(X^n)\}$ *soit tendue;*

(ii) on ait $[\beta],[\gamma],[\delta]$.

Sous cette forme, ce théorème est dû à Grigelionis et Mikulevicius [17], et partiellement à Rebolledo [47]. Mais c'est une extension simple de résultats déjà anciens: voir par exemple le livre [55] de Stroock et Varadhan (et même l'article [53]), et aussi de résultats présentés sous une forme un peu différente: par exemple le théorème de Kurtz [30] que nous rappellerons plus loin.

Ce théorème n'est évidemment pas très satisfaisant, à cause de la condition (i) qui semble difficile à vérifier dans les applications. Voici cependant un critère assurant la validité de (i), et qui est une extension d'un résultat de Liptcer et Shiryaev [38].

1.8 Condition de majoration forte. a) Il existe une fonction croissante continue F sur R_+, nulle en 0, telle que pour tout $\alpha \in D^d$ et tout $j \leq d$ on ait

$$\mathrm{Var}\{B^h(\alpha)^j\} \prec F , \qquad C^{jj}(\alpha) \prec F , \qquad (|x|^2\wedge 1)*\nu(\alpha) \prec F$$

(où $\prec$ désigne la domination forte des processus croissants: cf. I-2.19)

b) pour tout $t>0$ on a:

$$\lim_{b\uparrow\infty} \sup_{\alpha\in D^d} \nu(\alpha;[0,t]\times\{|x|>b\}) = 0. \blacksquare$$

Bien entendu, on a $\lim_{b\uparrow\infty} \nu(\alpha;[0,t]\times\{|x|>b\}) = 0$ pour tout α , mais la condition 1.8-(b) implique une uniformité. Il est évident que 1.8 $\Longrightarrow$ 1.5; remarquer aussi que sous 1.8 on a: $\mathrm{Var}\{C^{jk}(\alpha)\} \prec 4F$.

THEOREME 1.9: *On suppose la fonction de troncation* h *continue. Sous les conditions 1.8,* [Sup-β] , [γ], [δ] , *la suite* $\{\mathcal{L}(X^n)\}$ *est tendue.*

On obtient finalement le corollaire suivant, qui admet le théorème III-2.1 comme cas particulier lorsque X est un PAI-semimartingale.

THEOREME 1.10: *On suppose la fonction de troncation* h *continue. On suppose qu'on a les conditions 1.4, 1.6, et 1.8. Pour que* $X^n \xrightarrow{\mathcal{L}} X$, *il suffit qu'on ait* [Sup-β], [γ], [δ].

§b - DEMONSTRATION DU THEOREME DE COMPACITE 1.9. Commençons par un lemme: dans son énoncé, F est une fonction croissante continue sur R_+, nulle en 0 (celle de 1.8), et les G^n sont des processus càdlàg nuls en 0 (d'après notre convention, G^n est défini sur Ω^n).

LEMME 1.11: *La suite* $\{\mathcal{L}(G^n)\}$ *est C-tendue dès qu'on a l'une des deux conditions suivantes:*

a) $G^n_t - G_t\circ X^n \xrightarrow{\mathcal{L}} 0$ $\forall t>0$, *où* G *est un processus croissant sur* D^d *qui vérifie* $G(\alpha)\prec F$ *pour tout* $\alpha\in D^d$, *et chaque* G^n *est croissant.*

b) $\sup_{s\leq t} |G^n_s - G_s\circ X^n| \xrightarrow{\mathcal{L}} 0$ $\forall t>0$, *où* G *est un processus à variation finie sur* D^d *qui vérifie* $\mathrm{Var}(G(\alpha))\prec F$ *pour tout* $\alpha\in D^d$.

Preuve. On va utiliser le module de continuité w_N du chapitre I. Plus précisément, on va montrer que si $N\in\mathbb{N}^*$, $\varepsilon>0$, $\eta>0$ sont fixés, il existe n_o et $\theta>0$ tels que

1.12 $$n > n_o \longrightarrow P^n(w_N(G^n,\theta)>\eta) \leq \varepsilon .$$

Comme $G^n_0 = 0$, on a $\sup_{s\leq N}|G^n_s| \leq (N/\theta + 1)w_N(G^n,\theta)$, donc 1.12 et la proposition I-2.11 permettent de conclure.

Comme F est continue, il existe une subdivision $0=t_o<..<t_p=N$ telle que $F_{t_{j+1}} - F_{t_j} \leq \eta/6$. Soit $\theta = \inf_j (t_{j+1}-t_j)$. Sous la condition (a) il existe n_o avec

1.13 $$n > n_o \Longrightarrow P^n(D^n) \geq 1 - \varepsilon \quad \text{où } D^n = \{\sup_{j\leq p}|G^n_{t_j} - G_{t_j}\circ X^n| < \frac{\eta}{6}\}.$$

Comme les G^n sont croissants, et comme $G\prec F$, on a:

$$w_N(G^n,\theta) \leqq 2\sup_{j\leqq p-1} (G^n_{t_{j+1}} - G^n_{t_j})$$
$$\leqq 2\sup_{j\leqq p-1} (G_{t_{j+1}}\circ X^n - G_{t_j}\circ X^n) + 4\sup_{j\leqq p} |G^n_{t_j} - G_{t_j}\circ X^n|$$
$$\leqq 2\sup_{j\leqq p-1} (F_{t_{j+1}} - F_{t_j}) + 4\sup_{j\leqq p} |G^n_{t_j} - G_{t_j}\circ X^n|$$

qui est $\leqq\eta$ sur D^n. Par suite 1.12 découle de 1.13.

Sous la condition (b), il existe n_o tel que

<u>1.14</u> $\quad n > n_o \Longrightarrow P^n(D^n) \geq 1-\varepsilon\ ,\quad$ où $D^n = \{\sup_{s\leqq N} |G^n_s - G_s\circ X^n| < \frac{\eta}{4}\}$.

Comme $\mathrm{Var}(G) \prec F$, il vient

$$w_N(G^n,\theta) \leqq w_N(G\circ X^n,\theta) + 2\sup_{s\leqq N} |G^n_s - G_s\circ X^n|$$
$$\leqq 2\sup_{j\leqq p-1}\{\mathrm{Var}(G\circ X^n)_{t_{j+1}} - \mathrm{Var}(G\circ X^n)_{t_j}\} + 2\sup_{s\leqq N} |G^n_s - G_s\circ X^n|$$
$$\leqq 2\sup_{j\leqq p-1} (F_{t_{j+1}} - F_{t_j}) + 2\sup_{s\leqq N} |G^n_s - G_s\circ X^n|,$$

qui est $\leqq\eta$ sur D^n. Par suite 1.12 découle de 1.14.∎

Pour obtenir le théorème 1.9, on recopie une seconde fois la preuve du §II-2-d. Pour tout b, soit $h_b(x) = bh(x/b)$, qui est une nouvelle fonction de troncation continue. Pour chaque $q\in\mathbb{N}^*$, soit $\tilde{X}^{h_q,n} = X^{h_q,n} - B^{h_q,n}$, qui est une martingale localement de carré intégrable, d-dimensionnelle. Avec les notations de la preuve de II-2.28, on a

$$A^n := \sum_{j\leqq d} \langle\tilde{X}^{h_q,n,j},\tilde{X}^{h_q,n,j}\rangle \prec G^{nq} := \sum_{j\leqq d} \tilde{C}^{h,n,jj} + \hat{h}_q * \nu^n ,$$

où $\hat{h}_q$ est une fonction continue bornée nulle autour de 0. Ainsi chaque C^{nq} est croissant, et d'après $[\gamma]$ et $[\delta]$ on a $G^{nq}_t - G^q_t\circ X^n \xrightarrow{\mathcal{L}} 0$, où

$$G^q = \sum_{j\leqq d} C^{jj} + \{\hat{h}_q + |h|^2\}*\nu.$$

Il existe donc d'après 1.8 une constante β telle que $G^q \prec \beta F$. Le lemme 1.11 entraine alors que la suite $\{\mathcal{L}(G^{nq})\}_{n\geqq 1}$ est C-tendue, donc la suite $\{\mathcal{L}(A^n)\}$ également d'après I-2.20. Le théorème I-3.24 entraine alors:

<u>1.15</u> $\quad$ pour chaque $q\in\mathbb{N}^*$, la suite $\{\mathcal{L}(\tilde{X}^{h_q,n})\}_{n\geqq 1}$ est tendue.

On pose alors

$$U^{nq} = \tilde{X}^{h_q,n} + (h_q - h_{1/q})*\nu^n\ ,\quad V^{nq} = B^{h,n} + (h_{1/q} - h)*\nu^n,\quad W^{nq} = \sum_{s\leqq .}\{\Delta X^n_s - h_q(\Delta X^n_s)\}$$

de sorte que $X^n = U^{nq} + V^{nq} + W^{nq}$.

D'après $[\delta]$, on a $(h_q - h_{1/q})*\nu^n_t - (h_q - h_{1/q})*\nu\circ X^n_t \xrightarrow{\mathcal{L}} 0$ et il existe une constante β_q telle que $\mathrm{Var}\{(h_q - h_{1/q}) * \nu\} \prec |h_q - h_{1/q}| * \nu \prec \beta_q F$. Donc d'après le lemme 1.11, la suite $\{\mathcal{L}((h_q - h_{1/q}) * \nu^n)\}_{n \geq 1}$ est C-tendue. Mais alors 1.15 et I-2.18 entrainent que la suite $\{\mathcal{L}(U^{nq})\}_{n \geq 1}$ est tendue.

Le même argument montre que la suite $\{\mathcal{L}((h_{1/q} - h) * \nu^n)\}_{n \geq 1}$ est C-tendue. D'après [Sup-β] et 1.8, le lemme 1.11 entraine aussi que la suite $\{\mathcal{L}(B^{h,n})\}_{n \geq 1}$ est C-tendue, donc la suite $\{\mathcal{L}(V^{nq})\}_{n \geq 1}$ également d'après I-2.18. Par ailleurs, on a $|\Delta V^{nq}| \leq a/q$ par construction.

Soit enfin g_q une fonction continue sur R^d avec $0 \leq g_q \leq 1$, $g_q(x) = 1$ si $|x| > aq/2$ et $g_q(x) = 0$ si $|x| < aq/4$. Soit $N \in \mathbb{N}^*$, $\varepsilon > 0$. D'après 1.8,b il existe $q_o \in \mathbb{N}^*$ tel que $\sup_{q > q_o, \alpha \in D^d} g_q * \nu_N(\alpha) \leq \varepsilon$. D'après [Sup-δ] il existe $n_o(q) \in \mathbb{N}^*$ tel que

$$q > q_o, \quad n > n_o(q) \quad \Longrightarrow \quad P^n(g_q * \nu_N^n > 2\varepsilon) \leq \varepsilon.$$

Par ailleurs le processus $\sum_{s \leq .} g_q(\Delta X_s^n)$ est dominé au sens de Lenglart par $g_q * \nu^n$. Comme $\sup_{s \leq N} |W_s^{nq}| > 0$ implique que $\sum_{s \leq N} g_q(\Delta X_s^n) \geq 1$, I-3.16 implique

$$q > q_o, \quad n > n_o(q) \quad \Longrightarrow \quad P^n(\sup_{s \leq N} |W_s^{nq}| > 0) \leq 2\varepsilon + P^n(g_q * \nu_N^n > 2\varepsilon) \leq 3\varepsilon .$$

Ainsi, on a démontré que les décompositions $X^n = U^{nq} + V^{nq} + W^{nq}$ vérifient les conditions du lemme I-2.17, donc $\{\mathcal{L}(X^n)\}$ est tendue.

<u>§c -DEMONSTRATION DU THEOREME DE CONVERGENCE 1.8</u>. La démonstration est basée sur les lemmes suivants, où $\tilde{P}$ désigne une probabilité sur $(D^d, \underline{\underline{D}}^d)$ et $\tilde{E}$ est l'espérance relative à $\tilde{P}$.

<u>LEMME 1.16</u>: *On suppose que* $\mathcal{L}(X^n) \longrightarrow \tilde{P}$. *Soit* $(Z_i)_{i \in I}$ *une famille de fonctions sur* D^d *qui sont* $\tilde{P}$-*p.s. continue pour la topologie de Skorokhod. Soit* $(Z_i^n)_{i \in I}$ *des variables aléatoires définies sur* Ω^n *et vérifiant:*

(i) la famille $(Z_i^n)_{i \in I, n \geq 1}$ *est uniformément intégrable;*

(ii) $Z_i^n - Z_i \circ X^n \xrightarrow{\mathcal{L}} 0$ *pour tout* $i \in I$.

Alors, la famille $(Z_i)_{i \in I}$ *est* $\tilde{P}$-*uniformément intégrable, et* $E^n(Z_i^n) \to \tilde{E}(Z_i)$.

<u>Preuve</u>. a) Soit d'abord des variables Z^n et Z telles que $|Z^n| \leq N$, $|Z| \leq N$ pour une constante N, telles que $Z^n - Z \circ X^n \xrightarrow{\mathcal{L}} 0$ et telles que Z soit $\tilde{P}$-p.s. continue. On a

$$|E^n(Z^n) - \tilde{E}(Z)| \leq E^n(|Z^n - Z \circ X^n|) + |E^n(Z \circ X^n) - \tilde{E}(Z)| .$$

Les hypothèses impliquent $E^n(Z \circ X^n) \to \tilde{E}(Z)$. On a aussi $E^n(|Z^n - Z \circ X^n|) \leq \varepsilon + 2NP^n(|Z^n - Z \circ X^n| > \varepsilon)$ pour tout $\varepsilon > 0$, donc $E^n(|Z^n - Z \circ X^n|) \to 0$, et finalement on a $E^n(Z^n) \to \tilde{E}(Z)$.

b) Passons à la démonstration proprement dite. Les parties positives $(Z_i^{n,+}, Z_i^+)$ et négatives $(Z_i^{n,-}, Z_i^-)$ vérifiant encore les hypothèses, ce n'est pas une restriction que de supposer $Z_i^n \geq 0$, $Z_i \geq 0$. Si $N \in \mathbb{N}^*$, soit g_N la fonction continue:

$$g_N(x) = \begin{cases} x & \text{si } 0 \leq x \leq N \\ 2N - x & \text{si } N < x < 2N \\ 0 & \text{si } 2N \leq x. \end{cases}$$

D'après (a), on a $E^n(g_N(Z_i^n)) \to \tilde{E}(g_N(Z_i))$. Comme $\tilde{E}(Z_i) = \lim_N \uparrow \tilde{E}(g_N(Z_i))$ on a $\tilde{E}(Z_i) \leq \sup_{n \geq 1, N \geq 1} E^n(g_N(Z_i^n)) \leq \sup_{n \geq 1} E^n(Z_i^n) < \infty$ à cause de (i).

En fait, la condition (i) équivaut à: si $a_N = \sup_{i \in I, n \geq 1} E^n\{Z_i^n - g_N(Z_i^n)\}$, on a $a_N \to 0$ quand $N \uparrow \infty$. Soit alors $\varepsilon > 0$; il existe N tel que $a_N \leq \varepsilon$ et que $\tilde{E}\{Z_i - g_N(Z_i)\} \leq \varepsilon$ (a-priori, N dépend de i); il existe n_o tel que pour $n > n_o$ on ait $|E^n(g_N(Z_i^n)) - \tilde{E}(g_N(Z_i))| \leq \varepsilon$: donc $|E^n(Z_i^n) - \tilde{E}(Z_i)| \leq 3\varepsilon$. Par suite on en déduit que $E^n(Z_i^n) \to \tilde{E}(Z_i)$.

Enfin, ce qui précède montre que $E^n(Z_i^n - g_N(Z_i^n)) \to \tilde{E}(Z_i - g_N(Z_i))$ pour tous $i \in I$, $N \geq 1$. Donc

$$\sup_{i \in I} \tilde{E}(Z_i - g_N(Z_i)) \leq a_N \to 0 \quad \text{si } N \uparrow \infty,$$

ce qui implique la $\tilde{P}$-uniforme intégrabilité de la famille $(Z_i)_{i \in I}$. ■

<u>LEMME 1.17</u>: *On suppose que* $\mathcal{L}(X^n) \longrightarrow \tilde{P}$. *Soit* M *un processus càdlàg nul en* 0 *sur* D^d, *tel que pour tout* t *appartenant à une partie dense* A *de* R_+ *la fonction* $\alpha \rightsquigarrow M_t(\alpha)$ *soit* P-*p.s. continue. Pour chaque* $n \geq 1$, *soit* M^n *une martingale sur* $(\Omega^n, \underline{\underline{F}}^n, (\underline{\underline{F}}_t^n), P^n)$, *et supposons que*

(i) la famille $(M_t^n)_{n \geq 1, t \geq 0}$ *est uniformément intégrable,*

(ii) $M_t^n - M_t \circ X^n \xrightarrow{\mathcal{L}} 0$ *pour tout* $t \in A$.

Alors, si M *est adapté à* $(\underline{\underline{D}}_{t+}^d)$, *c'est une martingale (uniformément intégrable) pour* $\tilde{P}$.

<u>Preuve</u>. Soit $s, t \in A$ avec $s < t$; soit Z une fonction continue bornée sur D^d, qui soit $\underline{\underline{D}}_{s-}^d$-mesurable. 1.16 implique que $E^n(Z \circ X^n(M_t^n - M_s^n)) \to \tilde{E}(Z(M_t - M_s))$, et par ailleurs $Z \circ X^n$ est $\underline{\underline{F}}_{s-}^n$-mesurable, donc $E^n(Z \circ X^n(M_t^n - M_s^n)) = 0$. On en déduit que $\tilde{E}(Z(M_t - M_s)) = 0$, et d'après I-1.20 et un argument de classe monotone, cette égalité reste vraie pour toute fonction Z bornée et $\underline{\underline{D}}_{s-}^d$-mesurable. Par suite

<u>1.18</u> $$s < t, \ s, t \in A \Longrightarrow \tilde{E}(M_t - M_s | \underline{\underline{D}}_{s-}^d) = 0 .$$

Soit alors $s < t$ dans R_+. Il existe des suites (s_n) et (t_n) dans A, décroissant strictement vers s et t respectivement. 1.18 entraine que

$\tilde{E}(M_{t_n} - M_{s_n} | \underline{\underline{D}}^d_{s+}) = 0$. Par hypothèse $M_{t_n} - M_{s_n} \to M_t - M_s$, et d'après 1.16 encore les variables $(M_{t_n} - M_{s_n})$ sont $\tilde{P}$-uniformément intégrables. Donc $\tilde{E}(M_t - M_s | \underline{\underline{D}}^d_{s+}) = 0$ en passant à la limite sous l'espérance conditionnelle, et on a le résultat.■

On va maintenant démontrer le théorème 1.7, dont on suppose satisfaites les hypothèses. Comme la suite $\{\mathcal{L}(X^n)\}$ est tendue, la seule chose à montrer est que P est l'unique point limite de cette suite. Soit donc $\tilde{P}$ un point limite de la suite $\{\mathcal{L}(X^n)\}$ (il en existe au moins un). Quitte à prendre une sous-suite, on peut supposer que $\mathcal{L}(X^n) \longrightarrow \tilde{P}$.

Considérons les processus suivants, avec les notations du chapitre III:

$$\underline{1.19}\quad \begin{cases} \tilde{X}^h = X^h - B^h \\ Z = (Z^{jk})_{j,k\leq d} \quad \text{avec} \quad Z^{jk} = \tilde{X}^{h,j}\,\tilde{X}^{h,k} - \tilde{C}^{h,jk} \\ N^g_t = \sum_{s\leq t} g(\Delta X_s) - g*\nu_t \,, \quad \text{pour } g \text{ continue bornée positive nulle autour de } 0. \end{cases}$$

On va démontrer que ces processus sont des martingales sur $(D^d, \underline{\underline{D}}^d, (\underline{\underline{D}}^d_{t+}), \tilde{P})$. Comme $X - \tilde{X}^h = B^h + \sum_{s\leq .} (\Delta X_s - h(\Delta X_s))$ est à variation finie, cela démontrera que pour $\tilde{P}$, le processus X est une semimartingale qui, d'après la caractérisation du théorème III-1.16, admet (B^h, C, ν) pour caractéristiques locales. Comme par ailleurs $X^n_0 = 0$ il est évident que $\tilde{P}(X_0 = 0) = 1$. La condition d'unicité 1.4 entrainera alors $\tilde{P} = P$, et le théorème sera démontré.

On a vu au chapitre II (cf. 2.8 et 2.18) que les fonctions

$$\underline{1.20} \qquad \alpha \rightsquigarrow X^h_t(\alpha) \;, \qquad \alpha \rightsquigarrow \sum_{s\leq t} g(\Delta X_s(\alpha)) \quad \text{pour } g \text{ comme en } 1.19$$

sont continues sur D^d en tout point α tel que $\Delta\alpha(t) = 0$. Etant donné 1.6, on en déduit:

<u>1.21</u> Si $A = \{t: \tilde{P}(\Delta X_t \neq 0) = 0\}$, les fonctions $\alpha \rightsquigarrow \tilde{X}^h_t(\alpha)$, $\alpha \rightsquigarrow Z_t(\alpha)$ et $\alpha \rightsquigarrow N^g_t(\alpha)$ sont $\tilde{P}$-p.s. continues pour tout $t \in A$.

Soit alors $t \in A$ fixé. Il reste alors à trouver des martingales M^n sur $(\Omega^n, \underline{\underline{F}}^n, (\underline{\underline{F}}^n_t), P^n)$ telles qu'on ait les conditions (i) et (ii) de 1.17, pour M de la forme suivante:

$$M_s = \tilde{X}^h_{s\wedge t} \,, \qquad M_s = Z^{jk}_{s\wedge t} \,, \qquad M_s = N^g_{s\wedge t}.$$

a) <u>Le cas de</u> $M_s = N^g_{s\wedge t}$. Rappelons que g est continue, positive, nulle autour de 0, et bornée par une constante K. Soit $N^{g,n}_s = \sum_{r\leq s} g(\Delta X^n_r) - g*\nu^n_s$: c'est une martingale localement de carré intégrable, et un calcul analogue à celui de la partie (e) de la preuve de III-1.16 montre que

$$<N^{g,n},N^{g,n}>_s = g^2 * \nu^n_s - \sum_{r\leq s} \nu^n(\{r\}\times g)^2 \leq g^2 * \nu^n_s .$$

D'après 1.5 il existe une constante K' telle que $g^2 * \nu_t(\alpha) \leq K'$ pour tout $\alpha \in D^d$. Soit alors le temps d'arrêt $T^n = \inf(s: g^2 * \nu^n_s \geq 2K')$, et M^n le processus arrêté: $M^n_s = N^{g,n}_{s\wedge T^n\wedge t}$, qui est une martingale locale. D'après l'inégalité de Doob,

$$E^n(\sup_s |M^n_s|^2) \leq 4\, E^n(<M^n,M^n>_\infty) \leq 4\, E^n(g^2 * \nu^n_{T^n}) \leq 4(2K' + K^2).$$

On en déduit d'abord que M^n est une martingale, ensuite que la condition 1.17-(i) est satisfaite, car $\sup_{n,s} E^n(|M^n_s|^2) < \infty$.

Par ailleurs $N^{g,n}_s - N^g_s \circ X^n = g * \nu^n_s - (g * \nu_s)\circ X^n$. Donc

$$P^n(|M^n_s - M_s\circ X^n| > \varepsilon) \leq P^n(|g * \nu^n_{s\wedge t} - (g * \nu_{s\wedge t})\circ X^n| > \varepsilon) + P^n(T^n < t).$$

D'après [δ], $g^2 * \nu^n_t - (g^2 * \nu_t)\circ X^n \xrightarrow{\mathcal{L}} 0$, tandis que $(g^2 * \nu_t)\circ X^n \leq K'$: d'après la définition de T^n, on a donc $P^n(T^n < t) \to 0$. Une nouvelle application de [δ] montre qu'on a 1.17-(ii).

b) <u>Le cas de</u> $M_s = \tilde{X}^h_{s\wedge t}$. D'après 1.5 il existe une constante K telle que $\sum_{j\leq d} \tilde{C}^{h,jj}_t(\alpha) \leq K$. Soit $T^n = \inf(s: \sum_{j\leq d} \tilde{C}^{h,n,jj}_s \geq 2K)$ et $M^s_s = \tilde{X}^{h,n}_{s\wedge t\wedge T^n}$, qui est une martingale locale. Comme en (a), on a:

<u>1.22</u> $$E^n(\sup_s |M^n_s|^2) \leq 4 \sum_{j\leq d} E^n(\tilde{C}^{h,n,jj}_{T^n\wedge t}) \leq 4(2K + 4a^2)$$

(car $|\Delta\tilde{C}^{h,n}| \leq 4a^2$). Donc chaque M^n est une martingale, et on a 1.17-(i). Par ailleurs $\tilde{X}^{h,n}_s - \tilde{X}^h_s\circ X^n = B^{h,n}_s - B^h_s\circ X^n$, donc

$$P^n(|M^n_s - M_s\circ X^n| > \varepsilon) \leq P^n(|B^{h,n}_{s\wedge t} - B^h_{s\wedge t}\circ X^n| > \varepsilon) + P^n(T^n < t).$$

D'après [γ] on a $P^n(T^n < t) \to 0$, donc on obtient 1.17-(ii) grâce à [β].

c) <u>Le cas de</u> $M_s = Z^{jk}_{s\wedge t}$. Soit K et T^n comme en (b). Soit $M^n_s = \tilde{X}^{h,n,j}_{s\wedge t\wedge T^n} \tilde{X}^{h,n,k}_{s\wedge t\wedge T^n} - \tilde{C}^{h,n,jk}_{s\wedge t\wedge T^n}$, qui est une martingale locale. On a

$$P^n(|M^n_s - M_s\circ X^n| > \varepsilon) \leq P^n(|\tilde{X}^{h,n,j}_{s\wedge t} \tilde{X}^{h,n,k}_{s\wedge t} - (\tilde{X}^{h,j}_{s\wedge t} \tilde{X}^{h,k}_{s\wedge t})\circ X^n| > \tfrac{\varepsilon}{2})$$
$$+ P^n(|\tilde{C}^{h,n,jk}_{s\wedge t} - \tilde{C}^{h,jk}_{s\wedge t}\circ X^n| > \tfrac{\varepsilon}{2}) + P^n(T^n < t).$$

On a vu en (b) que $P^n(T^n < t) \to 0$, et que $\tilde{X}^{h,n}_{s\wedge t} - \tilde{X}^h_{s\wedge t}\circ X^n \xrightarrow{\mathcal{L}} 0$. En utilisant [γ], on voit donc que l'expression précédente tend vers 0, d'où 1.17-(ii).

Il reste à montrer que la famille $(M^n_s)_{s\geq 0, n\in \mathbb{N}^*}$ est uniformément intégrable. Comme $|\tilde{C}^{h,n,jk}_{s\wedge t\wedge T^n}| \leq 2(2K + 4a^2)$, il suffit de montrer que la famille $(|\tilde{X}^{h,n}_{s\wedge T^n}|^2)_{s,n}$ est uniformément intégrable. Soit $b > 4a^2$ et

$$R_b^n = \inf(s: |\tilde{X}^{h,n}_{s\wedge T^n}|^2 \geq b-4a^2).$$

Comme $|\Delta\tilde{X}^{h,n}| \leq 2a$, on a

1.23 $$R_b^n < T^n \Longrightarrow b - 4a^2 \leq |\tilde{X}^{h,n}_{R_b^n\wedge T^n}|^2 \leq b.$$

On déduit d'abord de 1.22 et 1.23 que

1.24 $$P^n(R_b^n < T^n) \leq \frac{1}{b-4a^2} E^n(|\tilde{X}^{h,n}_{R_b^n\wedge T^n}|^2) \leq \frac{4(2K+4a^2)}{b-4a^2}.$$

Ensuite si $|\tilde{X}^{h,n}_{s\wedge T^n}|^2 > b$ on a $R_b^n < T^n$, donc d'après 1.23 encore,

$$\begin{aligned} E^n\{(|\tilde{X}^{h,n}_{s\wedge T^n}|^2 - b)^+\} &\leq E^n(|\tilde{X}^{h,n}_{s\wedge T^n}|^2 - |\tilde{X}^{h,n}_{s\wedge T^n\wedge R_b^n}|^2) \\ &\leq E^n(\textstyle\sum_{j\leq d}(\tilde{C}^{h,n,jj}_{s\wedge T^n} - \tilde{C}^{h,n,jj}_{s\wedge T^n\wedge R_b^n})) \\ &\leq E^n(\textstyle\sum_{j\leq d} \tilde{C}^{h,n,jj}_{T^n} 1_{\{R_b^n<T^n\}}) \leq \frac{\{4(2K+4a^2)\}^2}{b-4a^2}, \end{aligned}$$

en utilisant 1.24 et la définition de T^n pour obtenir la dernière inégalité. Si maintenant $b\geq 4$, on a $|x|1_{\{|x|>b\}} \leq 2(|x|-\sqrt{b})^+$ pour tout x. Donc

$$\sup_{s\geq 0, n\in\mathbb{N}^*} E^n(|\tilde{X}^{h,n}_{s\wedge T^n}|^2 1_{\{|\tilde{X}^{h,n}_{s\wedge T^n}|^2>b\}}) \leq 2\frac{\{4(2K+4a^2)\}^2}{\sqrt{b}-4a^2},$$

qui tend vers 0 quand $b\uparrow\infty$. On a donc l'uniforme intégrabilité cherchée.

§d - APPLICATION AUX SEMIMARTINGALES LOCALEMENT DE CARRE INTEGRABLE. Donnons maintenant une version "simplifiée" des théorèmes 1.7 et 1.9, lorsque les semimartingales X^n sont localement de carré intégrable, ce qui équivaut à:

1.25 $$|x|^2 * \nu_t^n < \infty \quad \forall t > 0,$$

et on suppose aussi que $|x|^2*\nu_t(\alpha) < \infty$ $\forall t>0, \forall\alpha\in D^d$, ce qui permet de poser:

1.26 $$B^n = B^{h,n} + (x-h(x))*\nu^n, \quad B = B^h + (x-h(x))*\nu.$$

Il convient aussi de modifier les conditions 1.5 et 1.6:

1.27 Condition de majoration. Pour chaque $t>0$, les fonctions $\alpha \rightsquigarrow C_t(\alpha)$ et $\alpha \rightsquigarrow |x|^2*\nu_t(\alpha)$ sont bornées sur D^d.■

1.28 Condition de continuité. Pour chaque $t>0$ et chaque fonction continue bornée f sur R^d, nulle sur un voisinage de 0, les fonctions

$$\alpha \rightsquigarrow B_t(\alpha), \qquad \alpha \rightsquigarrow C_t^{jk}(\alpha) + (x^j x^k)*\nu_t(\alpha), \qquad \alpha \rightsquigarrow f*\nu_t(\alpha)$$

sont continues pour la topologie de Skorokhod sur R^d.

PROPOSITION 1.29: *Supposons 1.25, que* X *soit quasi-continu à gauche, et qu'on ait 1.4, 1.27 et 1.28. Pour que* $X^n \xrightarrow{\mathcal{L}} X$ *il suffit que la suite* $\{\mathcal{L}(X^n)\}$ *soit tendue, et qu'on ait les conditions:*

[β'] $B_t^n - B_t \circ X^n \xrightarrow{\mathcal{L}} 0$ pour tout t dans une partie dense $A \subset R_+$

[γ'] $C_t^{n,jk} + (x^j x^k) * \nu_t^n - \sum_{s \leq t} \Delta B_s^{n,j} \Delta B_s^{n,k} - (C_t^{jk} + (x^j x^k) * \nu_t) \circ X^n \xrightarrow{\mathcal{L}} 0$ pour tout t dans une partie dense $A \subset R_+$

[δ'] $f * \nu_t^n - (f * \nu_t) \circ X^n \xrightarrow{\mathcal{L}} 0$ pour tout t dans une partie dense $A \subset R_+$ et toute fonction f continue sur R^d, nulle sur un voisinage de 0, et telle que $f(x)/|x|^2$ soit bornée.

Preuve. Il suffit de montrer que les hypothèses de 1.7 sont satisfaites. On a 1.27 ⟹ 1.5 et il n'est pas difficile de montrer que 1.27 et 1.28 entrainent 1.6.

Avec des notations évidentes, on a [δ'] ⟹ [Sup-δ'] (cf. Lemme 1.3). On déduit aisément [β] de [β'] et de [Sup-δ']; on a [δ'] ⟹ [δ]. Enfin, exactement comme en II-2.35, on déduit [γ] de [γ'] et de la propriété:

1.30
$$|\sum_{s \leq t} (\Delta B_s^{h,n,j} \Delta B_s^{h,n,k} - \Delta B_s^{n,j} \Delta B_s^{n,k}| \xrightarrow{\mathcal{L}} 0.$$

Mais en II-2.35 on a vu que le premier membre de 1.30 est majoré par $(f * \nu_t^n) \sup_{s \leq t} |\Delta(g * \nu^n)_s|$, où f et g sont des fonctions du type de [δ']. De la continuité de $g * \nu_.$ et de [Sup-δ'] on déduit que: $\sup_{s \leq t} |\Delta(g * \nu^n)_s| \xrightarrow{\mathcal{L}} 0$. On déduit alors 1.30 de [Sup-δ'].∎

1.31 Condition de majoration forte. Il existe une fonction croissante continue F sur R_+, nulle en 0, telle que pour tous $\alpha \in D^d$, $j \leq d$, on ait

$$\mathrm{Var}(B(\alpha)^j) \prec F, \qquad C^{jj}(\alpha) \prec F, \qquad |x|^2 * \nu(\alpha) \prec F.$$∎

PROPOSITION 1.32: *Supposons 1.25. Pour que la suite* $\{\mathcal{L}(X^n)\}$ *soit tendue, il suffit qu'on ait 1.31,* [γ'], [δ'] *et*

[Sup-β'] $\sup_{s \leq t} |B_s^n - B_s \circ X^n| \xrightarrow{\mathcal{L}} 0 \qquad \forall t > 0$

Preuve. Il est évident que 1.31 ⟹ 1.8. D'après la preuve de 1.29 on a [γ], [δ], et aussi [Sup-δ'], donc clairement [Sup-β'] ⟹ [Sup-β]. Il suffit alors d'appliquer 1.9.∎

COROLLAIRE 1.33: *Supposons 1.25. Pour que* $X^n \xrightarrow{\mathcal{L}} X$, *il suffit qu'on ait 1.4, 1.28, 1.31,* [Sup-β'], [γ'], [δ'].

2 - THEOREME DE CONVERGENCE: UNE CONDITION PLUS FAIBLE

Les conditions de majoration 1.5 et 1.8 sont très fortes: dans le cas où X est une diffusion continue par exemple, elles reviennent à peu près à supposer les coefficients bornés; or, habituellement, on considère des diffusions à coefficients continus, donc localement bornés seulement. Nous allons donc donner ci-dessous une version du théorème 1.10 où les conditions "globales" de majoration ou de convergence sont remplacées par des conditions "locales", ceci au prix d'un léger renforcement de la condition d'unicité 1.4.

On suppose qu'on a 1.1, et pour $\rho>0$ on pose

$$S_\rho(\alpha) = \inf(t: |\alpha(t)| \geq \rho).$$

On rappelle que si Y est un processus et T un temps d'arrêt, on note Y^T le processus arrêté $Y_t^T = Y_{t\wedge T}$; de même ν^T est la mesure aléatoire "arrêtée": $\nu^T = 1_{[0,T]\times\mathbb{R}^d}\cdot\nu$ (ou encore; $f*\nu^T = (f*\nu)^T$).

2.1 Condition d'unicité. Si Q est une probabilité sur $(D^d,\underline{\underline{D}}^d)$ vérifiant

(i) $Q(X_0 = 0) = 1$,

(ii) X^{S_ρ} est une semimartingale pour Q, de caractéristiques locales $((B^h)^{S_\rho},C^{S_\rho},\nu^{S_\rho})$,

alors les probabilités P et Q coincident sur la tribu $\underline{\underline{D}}^d_{(S_\rho)-}$. ∎

Cette condition, plus forte que 1.4, est une sorte "d'unicité locale". Assez fréquemment, et en tous cas dans le cadre markovien examiné au §3, on a: 1.4 $\Longrightarrow$ 2.1 (voir un théorème général dans [23], §12-4-b).

2.2 Condition de majoration forte: a) Pour tout $\rho>0$ il existe une fonction croissante continue $F(\rho)$ sur R_+, nulle en 0, telle qu'on ait identiquement:

$$\mathrm{Var}(B^h(\alpha)^j)^{S_\rho(\alpha)} \prec F(\rho), \qquad (C^{jj})^{S_\rho(\alpha)} \prec F(\rho), \qquad (|x|^2\wedge 1)*\nu^{S_\rho(\alpha)} \prec F(\rho).$$

b) pour tous $\rho>0$, $t>0$, on a

$$\lim_{b\uparrow\infty} \sup_{\alpha\in D^d} \nu(\alpha;[0,t\wedge S_\rho(\alpha)]\times\{|x|>b\}) = 0. \quad ∎$$

Si Y est un processus prévisible sur D^d, et si $\alpha(s) = \alpha'(s)$ pour tout $s<t$, alors $Y_t(\alpha) = Y_t(\alpha')$ (voir [9]); la condition 2.2 équivaut alors à: pour tout $\rho>0$ les conditions 1.8-(a),(b) sont satisfaites par tout $\alpha\in D^d$ qui vérifie

$\sup_t|\alpha(t)| \leq \rho$, avec $F(\rho)$ au lieu de F dans 1.8-(a).

Par ailleurs, on suppose 1.2, et on pose

$$S^n_\rho = \inf(t: |X^n_t| \geq \rho) = S_\rho \circ X^n,$$

et on remplace [Sup-β], [γ], [δ] par

[Sup-β,loc] $\sup_{s\leq t\wedge S^n_\rho} |B^{h,n}_s - B^h_s \circ X^n| \xrightarrow{\mathcal{L}} 0$ $\forall\rho>0, \forall t>0$;

[γ,loc] $\tilde{C}^{h,n}_{t\wedge S^n_\rho} - \tilde{C}^h_{t\wedge S_\rho} \circ X^n \xrightarrow{\mathcal{L}} 0$ $\forall\rho>0$, $\forall t$ dans un ensemble dense $A\subset R_+$;

[δ,loc] $f*\nu^n_{t\wedge S^n_\rho} - (f*\nu_{t\wedge S_\rho})\circ X^n \xrightarrow{\mathcal{L}} 0$ $\forall\rho>0$, $\forall t$ dans une partie dense $A\subset R_+$, $\forall f$ continue bornée sur R^d, nulle sur un voisinage de 0.

De même qu'en 1.3, on montre que [γ,loc] et [δ,loc] équivalent respectivement à

[Sup-γ,loc] $\sup_{s\leq t\wedge S^n_\rho} |\tilde{C}^{h,n}_s - \tilde{C}^h_s \circ X^n| \xrightarrow{\mathcal{L}} 0$ $\forall\rho>0, \forall t>0$

[Sup-δ,loc] $\sup_{s\leq t\wedge S^n_\rho} |f*\nu^n_s - (f*\nu_s)\circ X^n| \xrightarrow{\mathcal{L}} 0$ $\forall\rho>0,\forall t>0$, $\forall f$ comme en [δ].

THEOREME 2.3: *On suppose la fonction de troncation* h *continue. On suppose qu'on a les conditions 1.6, 2.1, 2.2,* [Sup-β,loc], [γ,loc], [δ,loc]. *Alors* $X^n \xrightarrow{\mathcal{L}} X$.

A l'exception de 2.1, les conditions de ce théorème sont plus faibles que les conditions correspondantes du théorème 1.10.

Preuve. a) Pour simplifier l'écriture, on pose $B^h(\rho) = (B^h)^{S_\rho}$, $C(\rho) = C^{S_\rho}$, $\nu(\rho) = \nu^{S_\rho}$ et $\tilde{C}^h(\rho) = (\tilde{C}^h)^{S_\rho}$. Soit aussi $X^n(\rho) = (X^n)^{S^n_\rho}$, qui est une semimartingale de caractéristiques locales $B^{h,n}(\rho) = (B^{h,n})^{S^n_\rho}$, $C^n(\rho) = (C^n)^{S^n_\rho}$, $\nu^n(\rho) = (\nu^n)^{S^n_\rho}$. Soit enfin $\tilde{C}^{h,n}(\rho) = (\tilde{C}^{h,n})^{S^n_\rho}$.

Fixons $\rho>0$. Comme $B^h_t(\alpha) = B^h_t(\alpha')$ si $\alpha(s) = \alpha'(s)$ pour tout $s<t$, on a pour $t \geq S^n_\rho(\omega)$:

$$\{B^h(\rho)_t \circ X^n(\rho)\}(\omega) = \{B^h_{t\wedge S_\rho} \circ X^n(\rho)\}(\omega) = B^h_{t\wedge S^n_\rho(\omega)}\{X^n(\rho)(\omega)\} = B^h_{t\wedge S^n_\rho(\omega)}\{X^n(\omega)\},$$

d'où

$$\sup_{s\leq t} |B^{h,n}(\rho)_s - B^h(\rho)_s \circ X^n(\rho)| = \sup_{s\leq t\wedge S^n_\rho} |B^{h,n}_s - B^h_s \circ X^n| .$$

Ainsi, [Sup-β,loc] entraine que les processus $(X^n(\rho))_{n\geq 1}$ vérifient [Sup-β] relativement au triplet $(B^h(\rho),C(\rho),\nu(\rho))$. On montre de même qu'ils vérifient [Sup-γ] et [Sup-δ]. Par ailleurs, 2.2 entraine que ce triplet vérifie 1.8. D'après le théorème 1.9 on en déduit que la suite $\{\mathcal{L}(X^n(\rho))\}_{n\geq 1}$ est tendue.

b) Soit (n_k) une suite infinie de $\mathbb{N}$. Par un procédé diagonal on peut en extraire une sous-suite (n_{k_q}) telle que pour chaque $p\in\mathbb{N}^*$ on ait

2.4 $$\mathcal{L}(X^{n_{k_q}}(p)) \xrightarrow[q\uparrow\infty]{} Q_p$$

où Q_p est une probabilité sur $(D^d,\underline{\underline{D}}^d)$. On va montrer que $\mathcal{L}(X^{n_{k_q}}) \to P$, ce qui d'après le principe des sous-suites entrainera le résultat. Par suite, sans restreindre la généralité, on peut remplacer 2.4 par:

2.5 $$\mathcal{L}(X^n(p)) \xrightarrow[n\uparrow\infty]{} Q_p \quad \text{pour tout } p\in\mathbb{N}^*.$$

Pour tout $\rho>0$ on pose $S_{\rho+} = \lim_{\varepsilon\downarrow 0,\varepsilon>0} S_{\rho+\varepsilon}$ et $S_{\rho-} = \lim_{\varepsilon\downarrow 0,\varepsilon>0} S_{\rho-\varepsilon}$. Comme $\rho \rightsquigarrow S_\rho(\alpha)$ est croissante, il existe clairement une partie dénombrable $\tilde{A}\subset R_+$ telle que

2.6 $$\rho\notin\tilde{A} \Longrightarrow Q_p(S_{\rho-} = S_{\rho+}) = 1 \qquad \forall p\in\mathbb{N}^*.$$

Par ailleurs il est facile de déduire de la caractérisation I-1.6 de la convergence au sens de Skorokhod que

2.7 $$\alpha_n\to\alpha\ ,\ S_{\rho-}(\alpha) = S_{\rho+}(\alpha) \Longrightarrow \begin{cases} S_\rho(\alpha_n)\to S_\rho(\alpha) \\ \phi_\rho(\alpha_n)\to\phi_\rho(\alpha) \text{ dans } D^d, \end{cases}$$

où $\phi_\rho: D^d\to D^d$ est l'opérateur d'arrêt en S_ρ, défini par $\phi_\rho(\alpha)(t) = \alpha(t\wedge S_\rho(\alpha))$. D'après 2.2 les familles d'applications $(t\rightsquigarrow B^h_t(\alpha))_{\alpha\in D^d}$, $(t\rightsquigarrow \tilde{C}^h_t(\alpha))_{\alpha\in D^d}$ et $(t\rightsquigarrow g*\nu_t(\alpha))_{\alpha\in D^d}$ sont équicontinues en tout point $t\leq S_\rho(\alpha)$. On déduit alors immédiatement de 1.6 et de 2.7 que

2.8 $\begin{cases}\text{les applications: } \alpha\rightsquigarrow B^h(\rho)_t(\alpha),\ \alpha\rightsquigarrow\tilde{C}^h(\rho)_t(\alpha),\ \alpha\rightsquigarrow g*\nu(\rho)_t(\alpha) \\ \text{(pour } g \text{ continue bornée nulle sur une voisinage de } 0) \text{ sont continues au} \\ \text{point } \alpha \text{ si } S_{\rho-}(\alpha) = S_{\rho+}(\alpha).\end{cases}$

Soit maintenant p fixé dans $\mathbb{N}^*$. Soit $\rho\in]0,p[\cap\tilde{A}^c$. D'après 2.6 et 2.7, l'application ϕ_ρ est Q_p-p.s. continue. Comme $\rho\leq p$, on a $X^n(\rho) = \phi_\rho\circ X^n(p)$. Par suite si $Q'_\rho = Q_p\circ\phi_\rho^{-1}$ (= loi de X^{S_ρ} sous Q_p), on déduit de 2.5:

$$\mathcal{L}(X^n(\rho)) \to Q'_\rho .$$

On peut alors reprendre la preuve du théorème 1.7 en remplaçant $\tilde{P}$ par Q'_ρ, X^n par $X^n(\rho)$, (B^h,C,ν) par $(B^h(\rho),C(\rho),\nu(\rho))$ et $(B^{h,n},C^{h,n},\nu^n)$ par $(B^{h,n}(\rho),C^n(\rho),\nu^n(\rho))$: on a vu que ces termes vérifient 1.5 (et même 1.8), [Sup-β], [γ] et [δ]; d'après 2.8, le triplet $(B^h(\rho),C(\rho),\nu(\rho))$ vérifie la condition 1.6 de continuité en Q_p-presque tout point α (utiliser 2.6), donc aussi en Q'_ρ-presque tout point α, car $B^h(\rho)(\alpha) = B^h(\rho)(\phi_\rho(\alpha))$, et de même pour C et ν). On en déduit que:

$$\begin{cases} Q'_\rho(X_0 = 0) = 1 \\ X \text{ est une } Q'_\rho\text{-semimartingale de caractéristiques locales } (B^h(\rho), C(\rho), \nu(\rho)). \end{cases}$$

Le dernier point ci-dessus entraine a-fortiori que $X = X^{S_\rho}$ Q'_ρ-p.s., et la condition d'unicité 2.1 implique alors que $Q'_\rho = P$ en restriction à $\underline{\underline{D}}^d_{(S_\rho)-}$. Comme $\rho \leq p$ on a aussi $Q'_\rho = Q_p$ sur $\underline{\underline{D}}^d_{(S_\rho)-}$, donc $Q_p = P$ sur $\underline{\underline{D}}^d_{(S_\rho)-}$. Ceci étant vrai pour tout $\rho \in]0,p[\bigcap \tilde{A}^c$, on en déduit:

2.9 $$Q_p = P \quad \text{en restriction à } \underline{\underline{D}}^d_{(S_p)-}.$$

c) On va enfin utiliser 2.5 et 2.6 pour montrer que $\mathcal{L}(X^n) \to P$. Commençons par montrer que la suite $\{\mathcal{L}(X^n)\}$ est tendue. Pour cela on utilise le théorème I-2.7. Soit $N \in \mathbb{N}^*$, $\varepsilon>0$, $\eta>0$. Il existe $p \in \mathbb{N}^*$ tel que $P(S_{p-1} \leq N) \leq \varepsilon/3$. Comme S_{p-1} est $\underline{\underline{D}}^d_{(S_p)-}$-mesurable, 2.9 entraine que $Q_p(S_{p-1} \leq N) \leq \varepsilon/3$. Soit F la fermeture dans l'espace D^d de l'ensemble $\{\alpha: S_p(\alpha)<N\}$. Il est facile de vérifier que $F \subset \{S_{p-1} \leq N\}$, donc 2.5 entraine:

$$\limsup_n P^n(X^n(p) \in F) \leq Q_p(F) \leq Q_p(S_{p-1} \leq N) \leq \frac{\varepsilon}{3}.$$

Il existe donc n_o tel que

2.10 $$n > n_o \implies P^n(S^n_p < N) \leq P^n(X^n(p) \in F) \leq \frac{\varepsilon}{2}$$

(car $S^n_p = S_p \circ X^n = S_p \circ X^n(p)$). La suite $\{\mathcal{L}(X^n(p))\}_{n \geq 1}$ étant tendue, il existe $\delta>0$ et $K>0$ tels que pour tout n on ait

2.11 $$P^n(\sup_{s \leq N} |X^n(p)_s| > K) \leq \frac{\varepsilon}{2}, \qquad P^n(w'_N(X^n(p),\delta) > \eta) \leq \frac{\varepsilon}{2}.$$

Enfin si $S^n_p \geq N$ on a $X^n_s = X^n(p)_s$ pour tout $s \leq N$, donc 2.10 et 2.11 entrainent

$$n > n_o \implies P^n(\sup_{s \leq N} |X^n_s| > K) \leq \varepsilon, \qquad P^n(w'_N(X^n,\delta) > \eta) \leq \varepsilon.$$

Donc, d'après le théorème I-2.7, la suite $\{\mathcal{L}(X^n)\}$ est tendue.

Il reste à montrer que P est l'unique point limite de cette suite. Soit $t_1 < \ldots < t_q$ et f une fonction continue sur $(\mathbb{R}^d)^q$, bornée par 1, et soit $\psi(\alpha) = f(\alpha(t_1),\ldots,\alpha(t_q))$. Soit $N \in \mathbb{N}^*$, $\varepsilon>0$, et $p \in \mathbb{N}^*$ comme ci-dessus. D'après 2.5 on a: $E^n\{\psi(X^n(p))\} \to E^{Q_p}\{\psi(X)\}$. D'après 2.10 on a:

$$n > n_o \implies |E^n\{\psi(X^n(p))\} - E^n\{\psi(X^n)\}| \leq \frac{\varepsilon}{2}.$$

On a aussi $|E^{Q_p}\{\psi(X)\} - E^P\{\psi(X)\}| \leq \varepsilon/3$, car $P(S_p \leq N) \leq \varepsilon/3$, et à cause de 2.9. Par suite

$$\limsup_n |E^n\{\psi(X^n)\} - E^P\{\psi(X)\}| \leq \varepsilon .$$

Comme $\varepsilon>0$ est arbitraire, on a $E^n\{\psi(X^n)\} \to E^P\{\psi(X)\}$. Autrement dit, on a $X^n \xrightarrow{\mathcal{L}(R_+)} X$ (sous la loi P), et d'après I-2.6 on obtient $X^n \xrightarrow{\mathcal{L}} X$. ■

Le corollaire suivant se montre comme 1.29 et 1.32, dont on utilise les notations.

COROLLAIRE 2.12: *Supposons 1.25. Pour que* $X^n \xrightarrow{\mathcal{L}} X$ *il suffit qu'on ait 2.1, 1.28 et:*

2.13 Pour tout $\rho>0$ il existe une fonction croissante continue $F(\rho)$ sur R_+, nulle en 0, telle qu'on ait identiquement:

$$\mathrm{Var}(B(\alpha)^j)^{S_\rho(\alpha)} \prec F(\rho), \qquad (C^{jj})^{S_\rho(\alpha)} \prec F(\rho), \qquad |x|^2 * \nu^{S_\rho(\alpha)} \prec F(\rho).$$

[Sup-β',loc] $\quad \sup_{s\leq t\wedge S^n_\rho} |B^n_s - B_s \circ X^n| \xrightarrow{\mathcal{L}} 0 \qquad \forall t>0, \forall \rho>0$

[γ',loc] $\quad C^{n,jk}_{t\wedge S^n_\rho} + (x^j x^k) * \nu^n_{t\wedge S^n_\rho} - \sum_{s\leq t\wedge S^n_\rho} \Delta B^{n,j}_s \, \Delta B^{n,k}_s - (C^{jk}_{t\wedge S_\rho} + (x^j x^k) * \nu_{t\wedge S_\rho}) \circ X^n$

$\xrightarrow{\mathcal{L}} 0$, $\forall \rho>0$, $\forall t$ dans une partie dense $A \subset R_+$;

[δ',loc] $\quad f*\nu^n_{t\wedge S^n_\rho} - (f*\nu_{t\wedge S_\rho}) \circ X^n \xrightarrow{\mathcal{L}} 0 \quad \forall \rho>0, \forall t$ dans une partie dense $A \subset R_+$, $\forall f$ continue sur R^d, nulle sur un voisinage de 0, avec $f(x)/|x|^2$ bornée.

3 - CONVERGENCE DE PROCESSUS DE MARKOV

§a - RESULTATS GENERAUX. Les conditions [β], [γ], [δ] du §1 peuvent sembler un peu bizarres à première vue. En les appliquant aux processus de Markov, nous allons voir qu'au contraire elles sont très naturelles. Les résultats ci-dessous sont essentiellement de même nature que ceux du livre [55] de Stroock et Varadhan (mais dans [55] les processus limite sont continus). On comparera aussi à l'article [31] de Kurtz, qui donne des résultats intermédiaires entre le théorème 2.3 et les théorèmes ci-dessous (la limite X est markovienne, mais pas les processus X^n).

Pour chaque $n \in \mathbb{N}^*$ on considère un processus de Markov fort, normal, à valeurs dans R^d: $(\Omega^n, \underline{\underline{F}}^n, \underline{\underline{F}}^n_t, \theta^n_t, X^n_t, P^n_x)$, de générateur infinitésimal étendu (A^n, D_{A^n}) de la forme suivante:

$$\textbf{3.1}\quad \begin{cases} \text{si } f \text{ est bornée de classe } C^2, \text{on a } f\in D_{A_n} \text{ et} \\ A^n f(x) = \sum_j b_n^{h,j}(x)\frac{\partial f}{\partial x^j}(x) + \frac{1}{2}\sum_{j,k} c_n^{jk}(x)\frac{\partial^2 f}{\partial x^j \partial x^k}(x) \\ \qquad + \int_{R^d} N_n(x,dy)(f(x+y)-f(x)-\sum_j \frac{\partial f}{\partial x^j}(x)h^j(y)) \end{cases}$$

avec b_n^h, c_n, N_n vérifiant les conditions du §III-1-d. X^n est alors une semimartingale pour chaque P_x^n, de caractéristiques locales:

$$\textbf{3.2}\quad \begin{cases} B_t^{h,n} = \int_0^t b_n^h(X_s)\,ds \\ C_t^n = \int_0^t c_n(X_s)\,ds \\ f*\nu_t^n = \int_0^t N_n(X_s,f)\,ds. \end{cases}$$

Soit aussi un processus de Markov fort, normal, $(\Omega,\underline{\underline{F}},\underline{\underline{F}}_t,\theta_t,X_t,P_x)$ de générateur infinitésimal étendu (A,D_A) donné par 3.1 avec (b^h,c,N). On peut toujours supposer que $\Omega = D^d$, $\underline{\underline{F}} = \underline{\underline{D}}^d$; $\underline{\underline{F}}_t = \underline{\underline{D}}_{t+}^d$ et que X est le processus canonique sur D^d.

<u>3.3</u> <u>Condition d'unicité</u>. Pour chaque $x\in R^d$, P_x est l'unique probabilité sur $(D^d,\underline{\underline{D}}^d)$ telle que:

(i) $P_x(X_0 = x) = 1$

(ii) X est une P_x-semimartingale de caractéristiques locales $B_t^h = \int_0^t b^h(X_s)ds$, $C_t = \int_0^t c(X_s)ds$ et ν donné par $f*\nu_t = \int_0^t N(X_s,f)ds$. ∎

On pose

$$\textbf{3.4}\quad \tilde{c}_n^{h,jk}(x) = c_n^{jk}(x) + \int N_n(x,dy)h^j(y)h^k(y),\quad \tilde{c}^{h,jk}(x) = c^{jk}(x)+\int N(x,dy)h^j(y)h^k(y).$$

<u>3.5</u> <u>Condition de majoration</u>. a) les fonctions b^h, c, $\int N(.,dy)(|y|^2\wedge 1)$ sont localement bornées sur R^d;

b) $\lim_{\gamma\uparrow\infty} \sup_{|x|<\delta} N(x,\{|y|>\gamma\}) = 0 \quad \forall\,\delta>0$. ∎

<u>3.6</u> <u>Condition de continuité</u>. Les fonctions b^h, $\tilde{c}^h$, $N(.,f)$ (pour f continue bornée nulle sur un voisinage de 0) sont continues sur R^d. ∎

Les conditions 3.5 et 3.6 entraînent "presque" l'unicité 3.3: d'après [55] elles l'entraînent si, de plus, la matrice $c(x)$ n'est dégénérée pour aucun $x\in R^d$. Noter que 3.5 entraîne 3.4-(a).

<u>THEOREME 3.7</u>: *Supposons qu'on ait 3.3, 3.5, 3.6, et que la fonction de troncation* h *soit continue. Soit* $x\in R^d$. *Pour que* $\mathcal{L}(X^n/P_x^n) \to \mathcal{L}(X/P_x)$ *il suffit qu'on ait:*

$[\beta_1]$ $\quad b_n^h \to b^h$ uniformément sur les compacts;

$[\gamma_1]$ $\tilde{c}_n^h \to \tilde{c}^h$ uniformément sur les compacts;

$[\delta_1]$ $N_n(.,f) \to N(.,f)$ uniformément sur les compacts, pour f continue bornée nulle sur un voisinage de 0.

Ainsi, dans le cas des processus de Markov, les conditions $[\beta]$, $[\gamma]$, $[\delta]$ se ramènent à la "convergence"-des générateurs (A^n, D_{A^n}) vers (A, D_A), au sens où les trois conditions $[\beta_1]$, $[\gamma_1]$, $[\delta_1]$ équivalent à:

<u>3.8</u> $A^n f \to Af$ uniformément sur les compacts, pour f de classe C^3, bornée et à dérivées bornées.

Le théorème ci-dessus est donc une sorte de "théorème de Trotter-Kato" amélioré.

<u>Preuve</u>. On applique le théorème 2.3 aux semimartingales $X'^n = X^n - x$ et $X' = X - x$, qui ont mêmes caractéristiques locales, respectivement, que X^n et X (remarquer que si dans 3.5-(a) les fonctions sont bornées, et si on a 3.5-(b) avec $\delta=\infty$, il suffirait d'appliquer le théorème 1.10).

Dans notre cadre, on a 3.3 = 1.4, et cette condition implique 2.1: lorsque $N = 0$ (cas des diffusions continues) on peut se reporter à Stroock et Varadhan ([55], p. 283); dans le cas général, on peut appliquer le théorème (12.73) de [23] (en remarquant que dans ce théorème on prouve "l'unicité locale" pour les temps d'arrêt prévisibles, et aussi pour les temps d'arrêt par rapport à la filtration non continue à droite $(\underline{\underline{D}}_t^d)$, ce qui est le cas des temps S_ρ utilisés dans 2.1).

La condition 2.2 découle immédiatement de 3.5. Soit $\alpha_n \to \alpha$ dans D^d. On a $\alpha_n(s) \to \alpha(s)$, donc $b^h(\alpha_n(s)) \to b^h(\alpha(s))$ d'après 3.6, pour tout s tel que $\Delta\alpha(s)=0$; de plus, $\sup_{n\geq 1, s\leq t} |\alpha_n(s)|<\infty$, donc les fonctions $b^h(\alpha_n(s))$ sont uniformément bornées sur $[0,t]$. On en déduit que: $B_t^h(\alpha_n) \to B_t^h(\alpha)$. On montre de la même manière que $\tilde{C}_t^h(.)$ et $f{*}\nu_t(.)$ sont continues, de sorte qu'on a 1.6.

Enfin $B_t^{h,n} - B_t^h \circ X^n = \int_0^t \{b_n^h(X_s^n) - b^h(X_s^n)\}ds$. Il est alors évident, compte tenu de 1.4, que $[\beta_1] \Longrightarrow [\text{Sup-}\beta,\text{loc}]$. On montre de même les implications $[\gamma_1] \Longrightarrow [\gamma,\text{loc}]$ et $[\delta_1] \Longrightarrow [\delta,\text{loc}]$, d'où le résultat. ∎

Donnons aussi une version du corollaire 2.12 (ou 1.33) qui s'applique aux processus de Markov. On suppose que les noyaux N_n et N intègrent la fonction $|y|^2$, de sorte que (A^n, D_{A^n}) s'écrit:

<u>3.9</u>
$$A^n f(x) = \sum_j b_n^j(x) \frac{\partial f}{\partial x^j}(x) + \frac{1}{2} \sum_{j,k} c_n^{jk}(x) \frac{\partial^2 f}{\partial x^j \partial x^k}(x) + \int N_n(x,dy)\left(f(x+y) - f(x) - \sum_j \frac{\partial f}{\partial x^j}(x)\, y^j\right),$$

et de même pour (A, D_A). On a donc les relations:

$\underline{3.10}$ $\qquad b_n(x) = b_n^h(x) + \int N_n(x,dy)(y - h(y))$, $\quad b(x) = b^h(x) + \int N(x,dy)(y - h(y))$,

et on peut poser

$\underline{3.11}$ $\tilde{c}_n^{jk}(x) = c^{jk}(x) + \int N_n(x,dy)y^j y^k$, $\quad \tilde{c}^{jk}(x) = c^{jk}(x) + \int N(x,dy)y^j y^k$.

PROPOSITION 3.12: *Supposons qu'on ait 3.3 et que les noyaux* N_n *et* N *intègrent la fonction* $|y|^2$. *Pour que* $\mathcal{L}(X^n/P_x^n) \to \mathcal{L}(X/P_x)$ *il suffit qu'on ait les tions suivantes:*

3.13 les fonctions b, $\tilde{c}$, $N(.,f)$ (pour f continue nulle autour de 0 et telle que $f(x)/|x|^2$ soit bornée) sont continues sur R^d (et donc localement bornées).

$[\beta_1']$ $b_n \to b$ uniformément sur les compacts;

$[\gamma_1']$ $\tilde{c}_n \to \tilde{c}$ uniformément sur les compacts;

$[\delta_1']$ $N_n(.,f) \to N(.,f)$ uniformément sur les compacts, pour f comme dans 3.13.

§b - APPROXIMATION DE DIFFUSIONS PAR DES PROCESSUS DE SAUT PUR. Dans ce paragraphe, chaque processus X^n est un processus de Markov "de saut pur", ce qui signifie que X^n est constant par morceaux, càdlàg, et n'a qu'un nombre fini de sauts sur tout intervalle fini. On a alors

$\underline{3.14}$ $$A^n f(x) = \int N_n(x,dy)\,\{f(x+y) - f(x)\}$$

où N_n est un noyau positif intégrable. Pour simplifier, on supposera que N_n intègre la fonction $|y|^2$. On peut mettre 3.14 sous la forme 3.9, ce qui donne, avec les notations du paragraphe précédent:

$\underline{3.15}$ $\qquad b_n(x) = \int N_n(x,dy)\, y$, $\quad c_n(x) = 0$, $\quad \tilde{c}_n^{jk}(x) = \int N_n(x,dy)\, y^j y^k$.

Soit par ailleurs X le processus de Markov introduit au §1, avec les caractéristiques b,c, et $N = 0$ (donc $\tilde{c} = c$): c'est donc une diffusion (continue).

THEOREME 3.16: *Outre les hypothèses précédentes, supposons qu'on ait 3.3 et que les fonctions* b *et* c *soient continues sur* R^d. *Pour que* $\mathcal{L}(X^n/P_x^n) \to \mathcal{L}(X/P_x)$ *pour tout* $x \in R^d$ *il suffit qu'on ait:*

(i) $b_n \to b$ *uniformément sur les compacts;*

(ii) $\tilde{c}_n \to c$ *uniformément sur les compacts;*

(iii) $\sup_{|x|\leq\delta} \int N_n(x,dy)\ |y|^2\ 1_{\{|y|>\varepsilon\}} \to 0$ *pour tous* $\varepsilon>0$ *et tous* $\delta>0$.

Preuve. Il suffit d'appliquer 3.12, en remarquant que (iii) $\Longrightarrow$ $[\delta_1']$ si $N = 0$. ∎

Dans la suite, on suppose que:

3.17 $\quad b_n \to b$ uniformément sur les compacts, où b est lipschitzienne

$\tilde{c}_n \to 0$ uniformément sur les compacts.

Dans ce cas, on a aussi 3.16-(iii) et X est solution de l'équation "de diffusion" déterministe $dX_t = b(X_t)dt$. Comme b est lipschitzienne, on a donc la condition 3.3, et le théorème précédent s'applique. Plus précisément, notons $x_t(x)$ l'unique solution de l'équation différentielle (d-dimensionnelle) ordinaire:

3.18
$$dx_t(x) = b(x_t(x))\,dt, \qquad x_0(x) = x.$$

Comme la convergence de Skorokhod et la convergence uniforme sur les compacts coïncident quand la limite est continue, on a donc:

3.19
$$\sup_{s\leq t} |X^n_s - x_s(x)| \xrightarrow{\mathcal{L}(P^n_x)} 0 \qquad \forall t>0,\ \forall x\in R^d.$$

Pour évaluer la vitesse de convergence dans 3.19, on dispose d'un théorème central limite, dû à Kurtz [30]:

THEOREME 3.20: *Soit 3.17; soit* $\{\alpha_n\}$ *une suite de réels croissant vers* ∞ , *telle que:*

(i) $\alpha_n^2\ \tilde{c}_n$ *converge uniformément sur les compacts vers une fonction continue* $\hat{c} = (\hat{c}^{jk})_{j,k\leq d}$

(ii) $\lim_n \sup_{|x|\leq\delta} \alpha_n^2 \int N_n(x,dy)\ |y|^2\ 1_{\{|y|>\varepsilon/\alpha_n\}} = 0 \qquad \forall\varepsilon>0,\ \forall\delta>0.$

Soit
$$Y^n_t = \alpha_n\left(X^n_t - X^n_0 - \int_0^t b_n(X^n_s)\,ds\right)$$

Alors, pour tout $x\in R^d$, *les lois* $\mathcal{L}(Y^n/P^n_x)$ *convergent vers la loi* d'un *PAI continu de caractéristiques* $(0,C(x),0)$ (martingale gaussienne continue) *où* $C(x)_t = \int_0^t \hat{c}\{x_s(x)\}\,ds$.

Preuve. Notons $(B^{Y^n}, C^{Y^n}, \nu^{Y^n})$ les caractéristiques locales de la semimartingale localement de carré intégrable Y^n, associées à la "fonction de troncation" $h(x)=x$. Comme Y^n est en fait une martingale locale, on a $B^{Y^n} = 0$; comme Y^n est une somme compensée de sauts, on a $C^{Y^n} = 0$. Enfin $\Delta Y^n = \alpha_n \Delta X^n$, donc ν^{Y^n} égale:

$$f*\nu^{Y^n}_t = \int_0^t ds \int N_n(X^n_s,dy)\ f(\alpha_n y) .$$

On va alors appliquer le corollaire III-2.16 aux semimartingales Y^n, et au PAI Y de caractéristiques $(0,C(x),0)$ (pour un x fixé). On a [Sup-β'], et [L] vient de (ii). Il reste à montrer que si $\tilde{C}_t^{Y^n,jk} = \int_0^t \alpha_n^2 \tilde{c}_n^{jk}(X_s^n)\,ds$, on a $\tilde{C}_t^{Y^n} \xrightarrow{\mathcal{L}} C_t(x)$. Mais il vient

$$\tilde{C}_t^{Y^n} - C_t(x) = \int_0^t \{\alpha_n^2\,\tilde{c}_n(X_s^n) - \hat{c}(X_s^n)\}ds + \int_0^t \{\hat{c}(X_s^n) - \hat{c}(x_s(x))\}ds .$$

Le premier terme du second membre ci-dessus tend vers 0 en loi à cause de (i) et de 3.19; le second terme tend aussi vers 0 en loi à cause de 3.19 et de la continuité de $\hat{c}$: on a donc le résultat.∎

On a aussi $Y_t^n = \alpha_n(X_t^n - x_t(x)) + \int_0^t \{\alpha_n\{b(x_s(x)) - b_n(X_s^n)\}ds$ P_x^n-p.s.; donc si $\alpha_n(b_n - b) \to 0$ uniformément sur les compacts (ce qui est plus fort que dans 3.17), le même argument que ci-dessus montre que $\mathcal{L}(\alpha_n(X_.^n - x_.(x))/P_x^n)$ converge vers la même limite que $\mathcal{L}(Y^n/P_x^n)$, ce qui donne bien une vitesse de convergence dans 3.19.

<u>REMARQUE 3.21</u>: Nous n'avons donner ci-dessus qu'un seul exemple d'approximation de diffusion par des processus de saut pur. Il existe un très grand nombre d'autres exemples: voir la bibliographie de l'article [31] de Kurtz, notamment.

V - CONDITIONS NECESSAIRES DE CONVERGENCE

Nous avons introduit dans les chapitres précédents une série de conditions, notées [Sup-β], [γ], [δ], qui impliquent la convergence en loi $X^n \xrightarrow{\mathcal{L}} X$ pour des semimartingales, modulo quelques restrictions sur le processus limite X (par exemple les conditions 1.4, 1.6 et 1.8 du chapitre IV). Il est naturel de se demander dans quelle mesure ces conditions sont nécessaires.

A cet égard, les résultats du chapitre II (ces conditions sont nécessaires lorsque chaque X^n est un PAI) pouvaient sembler encourageantes, mais nous avons déjà donné un contre-exemple dans le §III-3-b. Voici un autre contre-exemple, qui fait mieux comprendre pourquoi ces conditions ne sont pas nécessaires: soit X un processus de Poisson standard sur $(\Omega,\underline{\underline{F}},(\underline{\underline{F}}_t),P)$; il s'écrit $X_t = \sum_{q\geq 1} 1_{\{S_q\leq t\}}$, où (S_q) est une suite de temps d'arrêt strictement croissante. Soit alors $(\Omega^n,\underline{\underline{F}}^n,(\underline{\underline{F}}^n_t),P^n) = (\Omega,\underline{\underline{F}},(\underline{\underline{F}}_t),P)$, et $X^n_t = \sum_{q\geq 1} 1_{\{S_q+1/n\leq t\}}$. Pour chaque ω, on a $X^n_.(\omega) \to X_.(\omega)$ pour la topologie de Skorokhod, donc a-fortiori $X^n \xrightarrow{\mathcal{L}} X$; de plus chaque X^n est un processus croissant càdlàg adapté, donc c'est une semimartingale. Cependant, aucune des conditions [Sup-β], [γ], [δ] n'est satisfaite.

Cela provient de ce que la convergence $X^n \xrightarrow{\mathcal{L}} X$ ne fait en aucune manière intervenir les filtrations; à l'extrême, le sens du temps n'a pas d'importance. Au contraire, les conditions [Sup-β], [γ], [δ] font intervenir de manière essentielle les propriétés "de type martingale", et donc les filtrations: dans l'exemple ci-dessus, les trajectoires de X^n et de X sont très proches (et leurs lois aussi, car X^n est un processus de Poisson standard, décalé de $1/n$ vers la droite); mais, du point de vue des filtrations, ces processus sont très différents: $X_{t+1/n} - X_t$ est indépendant de $\underline{\underline{F}}_t$, tandis que $X^n_{t+1/n} - X^n_t$ est $\underline{\underline{F}}^n_t$-mesurable.

Pour pallier cette difficulté, Aldous [2] et Helland [21,22] ont introduit un mode de convergence plus fort que la convergence en loi (mais de même type), pour lequel les conditions [Sup-β], [γ], [δ] sont essentiellement équivalentes à la convergence de X^n vers X (à condition bien-sûr d'avoir quelques conditions du genre IV-(1.4,1.6,1.8) satisfaites par X). Voir aussi [19].

Ci-dessous notre objectif est plus modeste. Pour l'essentiel, nous allons montrer que si X est une martingale locale <u>continue</u> et si les X^n ne sont pas très loin d'être des martingales locales, alors les conditions ci-dessus sont nécessaires.

1 - CONVERGENCE ET VARIATION QUADRATIQUE

§a - LES RESULTATS. Pour chaque entier n on considère une semimartingale d-dimensionnelle X^n sur $(\Omega^n,\underline{\underline{F}}^n,(\underline{\underline{F}}^n_t),P^n)$, nulle en 0 pour simplifier. Soit aussi X une semimartingale d-dimensionnelle nulle en 0 sur $(\Omega,\underline{\underline{F}},(\underline{\underline{F}}_t),P)$. h étant une fonction de troncation, on note $(B^{h,n},C^n,\nu^n)$ et (B^h,C,ν) leurs caractéristiques locales respectives.

On note aussi $[X^n,X^n]$ le processus à valeurs dans $R^d\otimes R^d$, dont les composantes sont $[X^n,X^n]^{jk} = [X^{n,j},X^{n,k}]$ (voir §III-1-a), et on définit $[X,X]$ de la même manière. Voici alors le résultat principal:

THEOREME 1.1: *On considère les conditions:*

(*i*) $X^n \xrightarrow{\mathcal{L}} X$;

(*ii*-h) $\lim_{b\uparrow\infty} \sup_n P^n\{\mathrm{Var}(B^{h,n,j})_t > b\} = 0 \quad \forall t>0, \forall j\leq d$.

Alors (*a*) *Sous* (*i*), *les conditions* (*ii*-h) *sont équivalentes entre elles, lorsque* h *parcoure l'ensemble des fonctions de troncation.*

(*b*) *Sous* (*i*) *et* (*ii*-h) *on a* $[X^n,X^n] \xrightarrow{\mathcal{L}} [X,X]$.

Ce résultat est basé sur la construction bien connue suivante de la variation quadratique ([9],[23]):

Soit $t>0$ et $\tau = \{0=t_0<\ldots<t_m=t\}$ une subdivision de $[0,t]$. Pour tout processus càdlàg Y à valeurs dans R^d on définit la variable $S_\tau(Y)$ à valeurs dans $R^d\otimes R^d$ en posant:

1.2
$$S_\tau(Y)^{jk} = \sum_{1\leq q\leq m} (Y^j_{t_q} - Y^j_{t_{q-1}})(Y^k_{t_q} - Y^k_{t_{q-1}}).$$

On rappelle alors que:

1.3 $\left\{\begin{array}{l}\text{Si } Y \text{ est une semimartingale, pour tout } t>0 \text{ on a } S_\tau(Y) \to [X,X]_t \text{ en pro-}\\ \text{babilité lorsque le pas } |\tau| \text{ de la subdivision } \tau \text{ de } [0,t] \text{ tend vers } 0.\end{array}\right.$

Le théorème 1.1 n'est donc pas surprenant. En effet si les points t_i de la subdivision τ ne sont pas des temps fixes de discontinuité de X, l'hypothèse $X^n \xrightarrow{\mathcal{L}} X$ implique que $S_\tau(X^n) \xrightarrow{\mathcal{L}} S_\tau(X)$. Toutefois la condition 1.1-(i) seule ne suffit pas à assurer la convergence $[X^n,X^n] \xrightarrow{\mathcal{L}} [X,X]$, comme le montre l'exemple suivant.

Exemple 1.4: il s'agit d'un exemple où les "processus" sont déterministes. Soit

$$X^n_t = \sum_{1\leq k\leq [n^2t]} (-1)^k \frac{1}{n}.$$

On a $|X^n| \leq 1/n$, donc $X^n \xrightarrow{\mathcal{L}} X$ où $X = 0$. On a aussi

$$[X^n,X^n]_t = \sum_{1\leq k\leq [n^2t]} (\tfrac{1}{n})^2 = \frac{[n^2t]}{n^2},$$

qui converge vers t, alors que $[X,X] = 0$. La condition 1.1-(ii-h) n'est pas satisfaite ici, car $Var(B^{h,n})_t = [n^2t]/n$ pour tout n assez grand, et cette quantité tend vers $+\infty$. ∎

COROLLAIRE 1.5: *Si les* X^n *sont des martingales locales, s'il existe une constante* c *telle que* $|\Delta X^n_t(\omega)| \leq c$ *identiquement, et si* $X^n \xrightarrow{\mathcal{L}} X$ *, alors* $[X^n,X^n] \xrightarrow{\mathcal{L}} [X,X]$.

Preuve. Il suffit de choisir la fonction de troncation h de sorte que $h(x) = x$ si $|x| \leq c$: on a alors $B^{h,n} = 0$. ∎

REMARQUES 1.6: 1) On peut renforcer la conclusion de 1.1-b ainsi: on a $(X^n,[X^n,X^n]) \xrightarrow{\mathcal{L}} (X,[X,X])$, en tant que processus à valeurs dans $R^d\times(R^d\otimes R^d)$.

2) Le théorème 1.1 reste vrai lorsque X n'est pas une semimartingale (voir [25]). Il faut alors lire la conclusion ainsi: le processus limite X admet une variation quadratique (au sens où il existe un processus càdlàg $[X,X]$ vérifiant 1.3), et $[X^n,X^n] \xrightarrow{\mathcal{L}} [X,X]$. ∎

§b - LES DEMONSTRATIONS. Pour simplifier l'écriture, on considère la condition suivante, qui s'applique à une suite (Z^n) de processus, chaque Z^n étant défini sur Ω^n:

1.7
$$\lim_{b\uparrow\infty} \sup_n P^n(\sup_{s\leq t}|Z^n_s|>b) = 0, \quad \forall t>0.$$

Ainsi, 1.1-(ii-h) se dit aussi: la suite $\{Var(B^{h,n})\}$ vérifie 1.7, où $Var(B^{h,n}) = \sum_{j\leq d} Var(B^{h,n,j})$.

LEMME 1.8: *Considérons les conditions:*

(*i*) $\lim_{b\uparrow\infty} \sup_n P^n(\sup_{s\leq t} |\Delta X^n_s|>b) = 0 \quad \forall t>0$ (i.e., la suite (ΔX^n) vérifie la condition 1.7).

(*ii*) $\lim_{b\uparrow\infty} \sup_n P^n(1_{\{|x|>b\}}*\nu^n_t>\varepsilon) = 0 \quad \forall t>0, \forall \varepsilon>0$.

Alors: (a) *On a l'équivalence:* (*i*) $\iff$ (*ii*);

(b) *si* $X^n \xrightarrow{\mathcal{L}} X$, *on a* (*i*) *et* (*ii*).

Preuve. a) Les processus $\sum_{s\leq t} 1_{\{|\Delta X^n_s|>b\}}$ et $1_{\{|x|>b\}}*\nu^n$ sont dominés l'un par l'autre au sens de Lenglart (cf. §I-3), et le dernier est prévisible. D'après I-3.16 on a donc

$$P^n(\sup_{s\leqq t} |\Delta X^n_s|>b) \leqq P^n(\textstyle\sum_{s\leqq t} 1_{\{|\Delta X^n_s|>b\}} \geqq 1)$$

$$\leqq \eta + P^n(1_{\{|x|>b\}}*\nu^n_t \geqq \eta),$$

pour tout $\eta>0$: on en déduit l'implication (ii) $\Longrightarrow$ (i). D'après I-3.16-(b) on a aussi

$$P^n(1_{\{|x|>b\}}*\nu^n_t \geqq \varepsilon) \leqq \frac{\eta}{\varepsilon} + \frac{1}{\varepsilon} E^n(\sup_{s\leqq t} 1_{\{|\Delta X^n_s|>b\}}) + P^n(\textstyle\sum_{s\leqq t} 1_{\{|\Delta X^n_s|>b\}} \geqq \eta)$$

$$\leqq \frac{\eta}{\varepsilon} + (\frac{1}{\varepsilon}+1)\ P^n(\sup_{s\leqq t} |\Delta X^n_s| > b)$$

pour tous $\varepsilon>0$, $\eta>0$: on en déduit l'implication (i) $\Longrightarrow$ (ii).

b) Si $X^n \xrightarrow{\mathcal{L}} X$, le théorème I-2.7 entraine que la suite (X^n) vérifie 1.7; comme $|\Delta X^n_t| \leqq 2 \sup_{s\leqq t} |X^n_s|$ on en déduit qu'on a (i).∎

LEMME 1.9: *Si* $X^n \xrightarrow{\mathcal{L}} X$, *la condition 1.1-(ii-h) ne dépend pas de la fonction de troncation* h *choisie.*

Preuve. Soit (ii-h), et h' une autre fonction de troncation. On a vu que $B^{h',n} = B^{h,n} + (h'-h)*\nu^n$, donc $\mathrm{Var}(B^{h',n}) \leqq \mathrm{Var}(B^{h,n}) + |h'-h|*\nu^n$. Il suffit donc de montrer que la suite de processus $\{|h'-h|*\nu^n\}$ vérifie 1.7.

Il existe une fonction continue à support compact $\hat{h}$ sur R^d, nulle autour de 0 et majorant $|h-h'|$. Soit $Y^n_t = \sum_{s\leqq t} \hat{h}(\Delta X^n_s)$ et $Y_t = \sum_{s\leqq t} \hat{h}(\Delta X_s)$. D'après l'hypothèse $X^n \xrightarrow{\mathcal{L}} X$ et I-1.19 on a $Y^n \xrightarrow{\mathcal{L}} Y$; donc 1.8 implique que la suite (Y^n) vérifie 1.7. Par ailleurs si K est la borne supérieure de $\hat{h}$, en utilisant le fait que Y^n domine au sens de Lenglart le processus $\hat{h}*\nu^n$, donc aussi le processus $|h-h'|*\nu^n$, il vient d'après I-3.16-(b):

$$P^n(|h-h'|*\nu^n_t \geqq b) \leqq \frac{\eta}{b} + \frac{K}{b} + P^n(Y^n_t \geqq \eta) \qquad \forall b>0,\ \forall \eta>0 ,$$

et on en déduit facilement le résultat.∎

Jusqu'à la fin du paragraphe, on fixe une fonction de troncation h continue, vérifiant $h(x) = x$ si $|x| \leqq 1/2$ et $h(x) = 0$ si $|x| \geqq 1$. Pour $a>0$ soit $h_a(x) = a\,h(x/a)$, qui est aussi une fonction de troncation continue. On utilise les notations $X^{h_a,n}$ et $\tilde{X}^{h_a,n}$ du théorème III-1.16.

Si $\alpha \in D^d$ et si $u>0$, on définit $t^p(\alpha,u)$ comme en I-1.19:

$$t^o(\alpha,u) = 0\ , \ \ldots\ ,\ t^{p+1}(\alpha,u) = \inf(t > t^p(\alpha,u):\ |\Delta\alpha(t)| > u)\ .$$

On note $\underline{S}(t)$ l'ensemble des subdivisions de $[0,t]$. Si $\tau = \{0=t_o<\ldots<t_m=t\} \in \underline{S}(t)$, on note $|\tau|$ son pas, et si $\alpha \in D^d$ on note $S_\tau(\alpha)$ la matrice de composantes

$$S_\tau(\alpha)^{jk} = \textstyle\sum_{1\leqq p\leqq m} \{\alpha^j(t_p) - \alpha^j(t_{p-1})\}\ \{\alpha^k(t_p) - \alpha^k(t_{p-1})\}$$

Si $\tau \in \underline{\underline{S}}(t)$, $u>0$, $\alpha \in D^d$ on note $\tau(\alpha,u)$ la subdivision de $[0,t]$ constituée:

$$\text{1.10} \qquad \begin{cases} \bullet \text{ des points de } \tau \\ \bullet \text{ des points } t^p(\alpha,u) \text{ qui vérifient } t^p(\alpha,u) \leq t. \end{cases}$$

On écrit aussi $S_{\tau(u)}(\alpha) = S_{\tau(\alpha,u)}(\alpha)$.

LEMME 1.11: *Soit* $t>0$, $\varepsilon>0$, $\eta>0$. *Sous les hypothèses de 1.1 il existe* $\rho>0$, $\delta>0$ *tels que pour tout* $u \in]0,\rho]$ *et toute subdivision* $\tau \in \underline{\underline{S}}(t)$ *vérifiant* $|\tau| \leq \delta$, *on ait*

$$\text{1.12} \qquad \sup_n P^n(|S_{\tau(u)}(X^n) - [X^n,X^n]_t| \geq \varepsilon) \leq \eta$$

(ci-dessus, $|.|$ est la norme euclidienne sur R^{d^2}).

Preuve. a) Si $\sup_{s\leq t} |\Delta X^n_s| \leq a/2$, on a $X^n = X^{h_a,n}$ sur $[0,t]$. Etant donné 1.8 il existe donc $a>0$ tel que

$$\text{1.13} \qquad \inf_n P^n(A^n) \geq 1 - \frac{\eta}{4}, \quad \text{où } A^n = \{X^n_s = X^{h_a,n}_s \text{ pour } s \leq t\}.$$

Dans la suite, ce nombre a est fixé. Soit F^n le processus croissant

$$F^n_s = \text{Var}(B^{h_a,n})_s + \sup_{r\leq s} |\tilde{X}^{h_a,n}_r|.$$

D'après 1.9, la suite $\{\text{Var}(B^{h_a,n})\}$ vérifie 1.7; comme h_a est continue, l'hypothèse $X^n \xrightarrow{\mathcal{L}} X$ entraine que $X^{h_a,n} \xrightarrow{\mathcal{L}} X^{h_a}$, donc d'après I-2.7 la suite $(X^{h_a,n})$ vérifie aussi 1.7, donc également la suite (F^n). Il existe donc $b>0$ avec

$$\text{1.14} \qquad \sup_n P^n(F^n_t > b) \leq \frac{\eta}{8}.$$

Soit aussi

$$\text{1.15} \qquad \rho = \frac{\theta}{3} \quad \text{avec } \theta = \frac{\varepsilon}{4d^2(b+3a)} \wedge \frac{\varepsilon\sqrt{\eta}}{\{8(b+2(b+3a)^2)\}^{1/2}}.$$

D'après I-2.7 encore, il existe $\delta>0$ et un entier $N \geq t$ tels que

$$\text{1.16} \qquad \sup_n P^n(w'_N(X^n,\delta) \geq \rho) \leq \frac{\eta}{8}.$$

b) Soit $\tau \in \underline{\underline{S}}(t)$ avec $|\tau| \leq \delta$, et $u \in]0,\rho]$. On va montrer qu'on a 1.12. On note $0=R^n_0<\ldots<R^n_{q_n}$ les points (aléatoires) de la subdivision $\tau(X^n,u)$; q_n est aussi une variable aléatoire, et on pose $R^n_j = t$ pour $j > q_n$. Ainsi, les R^n_p sont des temps d'arrêt. En appliquant la formule d'Ito au produit $X^{n,j} X^{n,k}$ entre les instants R^n_p et R^n_{p+1}, et en sommant sur p, on obtient:

$$\text{1.17} \qquad S_{\tau(u)}(X^n)^{jk} = [X^n,X^n]^{jk}_t + H^{n,j} \bullet X^{n,k}_t + H^{n,k} \bullet X^{n,j}_t,$$

où H^n est le processus d-dimensionnel, prévisible, continu à gauche, nul sur $]t,\infty[$, donné par

1.18 $$H^n_s = \sum_{p\geq 0} (X^n_{s-} - X^n_{R^n_p}) 1_{\{R^n_p<s\leq R^n_{p+1}\}}.$$

Posons aussi

$$\begin{cases} T^n = \inf(s: |H^n_s| > \theta \text{ ou } F^n_s > b) \\ G^{n,jk} = (H^{n,j} 1_{[0,T^n]})\bullet B^{h_a,n,k} + (H^{n,k} 1_{[0,T^n]})\bullet B^{h_a,n,j} \\ L^{n,jk} = (H^{n,j} 1_{[0,T^n]})\bullet \tilde{X}^{h_a,n,k} + (H^{n,k} 1_{[0,T^n]})\bullet \tilde{X}^{h_a,n,j}. \end{cases}$$

D'après 1.17 il vient

1.19 $$S_{\tau(u)}(X^n) - [X^n,X^n]_t = G^n_t + L^n_t \quad \text{sur} \quad A^n\bigcap\{T^n \geq t\}.$$

D'après 1.18, la définition des temps R^n_p, et la définition de w'_N, il est facile de voir que $|H^n_s| \leq 2w'_N(X^n,|\tau|) + u$ si $s \leq t$; comme $u\leq\rho$ et $|\tau|\leq\delta$ et $\theta = \rho + 2\rho$, on déduit immédiatement de 1.14 et 1.16 que

1.20 $$\sup_n P^n(T^n < t) \leq \frac{\eta}{4}.$$

c) Rappelons que si Y est une martingale de carré intégrable réelle, nulle en 0, on a $E(Y^2_T) = E([Y,Y]_T)$ pour tout temps d'arrêt T; donc les processus croissants $\sup_{s\leq .} Y^2_s$ et $[Y,Y]$ sont mutuellement dominés l'un par l'autre au sens de Lenglart. Par localisation, cette dernière propriété reste vraie si Y est seulement localement de carré intégrable. Etant donné la définition de T^n, on en déduit que le processus $(L^n_s)^2 = \sum_{j,k} (L^{n,jk}_s)^2$ est dominé au sens de Lenglart par le processus $2\theta^2 \sum_{j\leq d}[\tilde{X}^{h_a,n,j},\tilde{X}^{h_a,n,k}]_{s\wedge T^n}$, qui est lui-même dominé au sens de Lenglart par le processus $2\theta^2 (F^n_{s\wedge T^n})^2$. Comme $\Delta F^n \leq 3a$, les sauts de $2\theta^2 (F^n_{s\wedge T^n})^2$ sont majorés par $4\theta^2(b+3a)^2$ (toujours d'après la définition de T^n). D'après le lemme I-3.16-(b) il vient alors

$$P^n(\sup_{s\leq t}|L^n_s|^2 > \frac{\varepsilon^2}{4}) \leq \frac{4}{\varepsilon^2} \{4\theta^2(b+3a)^2 + 2\theta^2 b\} + P^n(2\theta^2(F^n_t)^2 > 2\theta^2 b)$$

1.21 $$\leq \frac{\eta}{4}$$

d'après 1.14 et 1.15.

Par ailleurs, en utilisant encore la définition de T^n, celle de F^n, et le fait que $F^n_{T^n} \leq b+3a$, on voit facilement que $|G^n| \leq 2d^2\theta(b+3a)$. D'après 1.15 on a donc $\sup_{s\leq t} |G^n_s| \leq \varepsilon/2$. En utilisant 1.13, 1.19, 1.20 et 1.21, il vient alors:

$$\begin{aligned} P^n(|S_{\tau(u)}(X^n) - [X^n,X^n]_t|\geq\varepsilon) &\leq P^n(T^n<t) + P^n((A^n)^c) + P^n(\sup_{s\leq t}|G^n_s|>\frac{\varepsilon}{2}) \\ &\quad + P^n(\sup_{s\leq t}|L^n_s|^2\geq \frac{\varepsilon^2}{4}) \\ &\leq \frac{\eta}{4} + \frac{\eta}{4} + 0 + \frac{\eta}{4} < \eta. \blacksquare \end{aligned}$$

Démonstration du théorème 1.1. On a déjà obtenu 1.1-(a) dans 1.9, et on va montrer la forme renforcée de 1.1-(b) (voir remarque 1.6-(a)), à savoir:

1.22 $$(X^n,[X^n,X^n]) \xrightarrow{\mathcal{L}} (X,[X,X]).$$

Pour simplifier, on écrit $A^n = [X^n,X^n]$ et $A = [X,X]$. Pour $\alpha \in D^d$, $u>0$, soit

$$h_t^{u,jk}(\alpha) = \sum_{0<s\leq t} \Delta\alpha^j(s)\Delta\alpha^k(s)\, 1_{\{|\Delta\alpha(s)|>u\}} = \sum_{p>0:t^p(\alpha,u)\leq t} \Delta\alpha^j(t^p(\alpha,u))\Delta\alpha^k(t^p(\alpha,u))$$

1.23 $$\begin{cases} \hat{A}^{n,u} = A^n - h^u(X^n) = A^n - \sum_{s\leq .} \Delta A_s^n\, 1_{\{|\Delta X_s^n|>u\}} \\ \hat{A}^u = A - h^u(X) = A - \sum_{s\leq .} \Delta A_s\, 1_{\{|\Delta X_s|>u\}} \end{cases}$$

(rappelons que $\Delta A^{jk} = \Delta X^j \Delta X^k$, et de même pour $\Delta A^{n,jk}$). Soit

$$D(X) = \{t>0:\ P(\Delta X_t = 0) = 1\}$$
$$U(X) = \{u>0:\ P(|\Delta X_t| \neq u \text{ pour tout } t>0) = 1\}.$$

Ces deux ensembles sont denses dans $\mathbb{R}_+$ (et même à complémentaire dénombrable).

D'après l'hypothèse $X^n \xrightarrow{\mathcal{L}} X$ et l'assertion I-1.19, on voit que pour tous $t_j \in D(X)$, $u_j \in U(X)$, $u \in U(X)$, et toutes subdivisions τ_j de $[0,t_j]$ dont les points appartiennent à $D(X)$, on a:

$$(X_{t_j}^n, S_{\tau_j(u_j)}(X^n), h_{t_j}^u(X^n))_{j\leq m} \xrightarrow{\mathcal{L}} (X_{t_j}, S_{\tau_j(u_j)}(X), h_{t_j}^u(X))_{j\leq m}.$$

A cause de la densité de $D(X)$ et de $U(X)$ dans $\mathbb{R}_+$, le lemme 1.11 permet d'en déduire que pour $t_j \in D(X)$, $u \in U(X)$,

1.24 $$(X_{t_j}^n, A_{t_j}^n, h_{t_j}^u(X^n))_{j\leq m} \xrightarrow{\mathcal{L}} (X_{t_j}, A_{t_j}, h_{t_j}^u(X))_{j\leq m}.$$

En particulier, (X^n,A^n) converge fini-dimensionnellement en loi le long de $D(X)$ vers (X,A). Pour obtenir 1.22, il reste donc à montrer que la suite $\{\mathcal{L}(X^n,A^n)\}$ est tendue.

Si $x \in \mathbb{R}^d \otimes \mathbb{R}^d$ est une matrice symétrique nonnégative, on a $|x| \leq \sum_j x^{jj}$; donc $|A^n| \leq \sum_j A^{n,jj}$. Comme chaque processus $A^{n,jj}$ est croissant, comme $A_t^{n,jj} \xrightarrow{\mathcal{L}} A_t^{jj}$ si $t \in D(X)$, et comme la suite $\{\mathcal{L}(X^n)\}$ est tendue, il est évident que la suite $\{\mathcal{L}(X^n,A^n)\}$ vérifie la condition I-2.7-(i).

D'après 1.23 et 1.24, on a

1.25 $$u \in U(X),\ t_j \in D(X) \implies (\hat{A}_{t_j}^{n,u})_{j\leq m} \xrightarrow{\mathcal{L}} (\hat{A}_{t_j}^u)_{j\leq m}$$

Soit $N \in \mathbb{N}^*$, $\varepsilon>0$, $\eta>0$. Soit $t \in D(X)$ avec $t\geq N$, et $u \in U(X)$ avec $u^2 \leq \varepsilon/10$. D'après 1.23 on a $\Delta\hat{A}^{u,jj} = (\Delta X^j)^2\, 1_{\{|\Delta X|\leq u\}}$, donc $\sum_j \Delta\hat{A}^{u,jj} \leq u^2 \leq \varepsilon/10$, donc il existe un nombre $\theta>0$ tel que

1.26 $$P(\sup_{s\leq t} \sum_j (\hat{A}^{u,jj}_{s+\theta} - \hat{A}^{u,jj}_s) \geq \frac{\varepsilon}{5}) \leq \frac{\eta}{4}.$$

Choisissons $\tau = \{t_0=0<t_1<..<t_m=t\}$ avec $t_j \in D(X)$ et $\theta/2 \leq |\tau| \leq \theta$. D'après 1.25 et 1.26 il existe n_o tel que

1.27 $$n > n_o \Longrightarrow P^n(B_n) \geq 1 - \frac{\eta}{2}, \text{ où } B_n = \{\sup_{1\leq p\leq m} \sum_j (\hat{A}^{n,u,jj}_{t_p} - \hat{A}^{n,u,jj}_{t_{p-1}}) < \frac{\varepsilon}{4}\}.$$

On a aussi: $\sup_{t_p<s\leq t_{p+1}} |\hat{A}^{n,u}_s - \hat{A}^{n,u}_{t_p}| \leq \sum_j (\hat{A}^{n,u,jj}_{t_{p+1}} - \hat{A}^{n,u,jj}_{t_p})$. Par suite

1.28 $$w_N(\hat{A}^{n,u}, \frac{\theta}{2}) \leq \frac{\varepsilon}{2} \quad \text{sur } B_n.$$

Enfin, la suite $\{\mathcal{L}(X^n)\}$ étant tendue, il existe $\delta>0$, $n_1 \geq n_o$ tels que

1.29 $$n > n_1 \longrightarrow P^n(C_n) \geq 1 - \frac{\eta}{2}, \quad \text{où} \quad C_n = \{w'_N(X^n,\delta) < u\wedge\frac{\varepsilon}{2}\}.$$

Soit maintenant les processus Y^n, $Y^{n,u}$, $\check{Y}^{n,u}$ à valeurs dans $R^d\times(R^d\otimes R^d)$, de composantes respectives (X^n,A^n), $(0,\hat{A}^{n,u})$ et $(X^n,A^n-\hat{A}^{n,u})$. On a $w_N(Y^{n,u},\frac{\theta}{2}) = w_N(\hat{A}^{n,u},\frac{\theta}{2})$. D'après 1.23, le processus $A^n - \hat{A}^{n,u}$ est constant sur les intervalles où $|\Delta X^n| \leq u$, donc d'après la définition de w'_N on a $w'_N(\check{Y}^{n,u},\delta) = w'_N(X^n,\delta)$ sur C_n. Enfin on a vu (cf. I-2.17) que

$$w'_N(\alpha+\beta,\rho) \leq w'_N(\alpha,\rho) + w_N(\beta,2\rho).$$

Comme $Y^n = Y^{n,u} + \check{Y}^{n,u}$, il vient sur $B_n \cap C_n$, d'après 1.28:

$$w'_N(Y^n,\delta\wedge\frac{\theta}{4}) \leq w'_N(\check{Y}^{n,u},\delta) + w_N(Y^{n,u},\frac{\theta}{2}) \leq \varepsilon .$$

En utilisant 1.27 et 1.29 on obtient alors

$$n > n_1 \Longrightarrow P^n(w'_N(Y^n,\delta\wedge\frac{\theta}{4}) > \varepsilon) \leq \eta,$$

ce qui montre que la suite $\{\mathcal{L}(X^n,A^n)\} = \{\mathcal{L}(Y^n)\}$ vérifie la condition I-2.7-(ii), d'où le résultat.∎

2 - CONDITIONS NECESSAIRES DE CONVERGENCE VERS UN PROCESSUS CONTINU

§a - LE CAS DES SUITES DE MARTINGALES LOCALES. Commençons par quelques lemmes, après avoir introduit une condition du même genre que 1.7. Cette condition s'applique aussi à une suite (Z^n) de processus, chaque Z^n étant défini sur Ω^n.

2.1 $$P^n(\sup_{s\leq t} |Z^n_s| > b) \to 0 \quad \forall b>0, \quad \forall t>0$$

LEMME 2.2: *Soit* (U^n) *une suite de processus croissants nuls en* 0, *vérifiant* $\Delta U^n \leq K$ *pour une constante* K. *Soit* V^n *le compensateur prévisible de* U^n. *Alors*

a) *La suite* (U^n) *vérifie 2.1 si et seulement si la suite* (V^n) *vérifie 2.1.*

b) *La suite* (U^n) *vérifie 1.7 si et seulement si la suite* (V^n) *vérifie 1.7.*

c) *Si la suite* (ΔU^n) *vérifie 2.1, alors la suite* (ΔV^n) *vérifie 2.1.*

Preuve. Les processus U^n et V^n sont dominés l'un par l'autre au sens de Lenglart. Donc d'après I-3.16 on a pour tous $\varepsilon>0$, $\eta>0$, $t>0$:

$$P^n(U^n_t \geq \varepsilon) \leq \frac{\eta}{\varepsilon} + P^n(V^n_t \geq \eta)$$
$$P^n(V^n_t \geq \varepsilon) \leq \frac{\eta+K}{\varepsilon} + P^n(U^n_t \geq \eta),$$

et il est facile d'en déduire (a) et (b).

Soit $b>0$ et $T^n = \inf(t:\ \Delta V^n_t > b)$. Alors T^n est une temps d'arrêt prévisible, donc

$$\begin{aligned} P^n(\sup_{s\leq t} \Delta V^n_s > b) = P^n(T^n \leq t) &\leq \frac{1}{b} E^n(\Delta V^n_{T^n} 1_{\{T^n\leq t\}}) \\ &= \frac{1}{b} E^n(\Delta U^n_{T^n} 1_{\{T^n\leq t\}}) \\ &\leq \frac{1}{b} \{\eta + K\, P^n(\sup_{s\leq t} \Delta U^n_s > \eta)\} \end{aligned}$$

pour tout $\eta>0$, et il est facile d'en déduire (c). ∎

COROLLAIRE 2.3: *Soit* X^n *des semimartingales.*

a) *Si la suite* $\{\mathcal{L}(X^n)\}$ *est C-tendue, la suite de processus* (ΔX^n) *vérifie 2.1.*

b) *La suite* (ΔX^n) *vérifie 2.1 si et seulement si la suite* (X^n) *vérifie la condition* $[\delta]$ *avec* $\nu = 0$ (ce qui revient à dire: $\nu^n([0,t]\times\{|x|>\varepsilon\}) \xrightarrow{\mathcal{L}} 0$).

Preuve. a) On a: $\sup_{s\leq N} |\Delta X^n_s| \leq w_N(X^n,\delta)$ pour tout $\delta>0$, donc le résultat découle de I-2.11.

b) Il suffit d'appliquer le lemme précédent (partie (a)) à $U^n_t = \sum_{s\leq t} 1_{\{|\Delta X^n_s|>b\}}$ et $V^n_t = 1_{\{|x|>b\}} * \nu^n_t$ pour tout $b>0$. ∎

PROPOSITION 2.4: *Soit* X^n *des martingales localement de carré intégrable, nulle en 0, d-dimensionnelles, vérifiant* $|\Delta X^n| \leq K$ *pour une constante* K. *On suppose que* $[X^n,X^n] \xrightarrow{\mathcal{L}} A$, *où* A *est un processus continu à valeurs dans* $R^d \otimes R^d$. *Alors*

a) *la suite* (ΔX^n) *vérifie 2.1;*

b) *on a* $\langle X^n,X^n\rangle \xrightarrow{\mathcal{L}} A$ *et* $[X^n,X^n] - \langle X^n,X^n\rangle \xrightarrow{\mathcal{L}} 0$.

($[X^n,X^n]$ a été défini au §1, et $\langle X^n,X^n\rangle$ désigne le processus de composantes $\langle X^{n,j},X^{n,k}\rangle$).

Preuve. D'après 2.3-(a), chaque suite $(\Delta[X^{n,j},X^{n,j}] = (\Delta X^{n,j})^2)$ vérifie 2.1, donc $(|\Delta X^n|^2)$ vérifie également 2.1, et on en déduit (a).

Soit $Y^n = [X^n,X^n] - \langle X^n,X^n\rangle$, qui est une martingale localement de carré intégrable, à valeurs dans $R^{d^2} = R^d \otimes R^d$, de composantes $(Y^{n,jk})_{j,k\leq d}$, et qui vérifie $|\Delta Y^n| \leq 2K^2$. Nous allons montrer que Y^n tend en loi vers le processus nul, ce qui entrainera (b). Pour cela, nous allons appliquer le théorème III-2.1, en montrant que les caractéristiques locales $(B^{h,n},C^n,\nu^n)$ de Y^n vérifient les conditions [Sup-β], [γ], [δ] avec $B^h = 0$, $C = 0$, $\nu = 0$.

En premier lieu, on peut choisir h (fonction de troncation sur R^{d^2}) telle que $h(x) = x$ pour $|x| \leq 2K^2$. Comme Y^n est une martingale locale, on a alors $B^{h,n} = 0$, d'où [Sup-β] avec $B^h = 0$.

En second lieu, on a vu que $(\Delta[X^n,X^n])_{n\geq 1}$ vérifie 2.1, donc chaque suite $(\Delta\langle X^{n,j},X^{nj}\rangle)_{n\geq 1}$ également d'après 2.2-(c); comme $|\Delta\langle X^n,X^n\rangle| \leq \sum_j \Delta\langle X^{n,j},X^{n,j}\rangle$ on en déduit que $(\Delta\langle X^n,X^n\rangle)_{n\geq 1}$, donc aussi $(\Delta Y^n)_{n\geq 1}$, vérifient 2.1. D'après 2.3 il en découle qu'on a [δ] avec $\nu = 0$.

En dernier lieu, $\tilde{C}^{h,n}$ est le compensateur prévisible de $[Y^n,Y^n]$ (à valeurs dans $R^{d^2} \otimes R^{d^2}$, de composantes $(\tilde{C}^{h,n,jkpq})_{j,k,p,q\leq d}$), et c'est un processus à valeurs matricielles $d^2 \times d^2$ symétriques nonnégatives; il reste à montrer que $\tilde{C}^{h,n}_t \xrightarrow{\mathcal{L}} 0$ pour tout $t>0$, et il suffit pour cela de montrer que $\tilde{C}^{h,n,jkjk}_t$ tend vers 0 en loi pour tout $t>0$ et tous $j,k\leq d$. D'après 2.2-(a) il suffit pour cela que

2.5 $$[Y^{n,jk},Y^{n,jk}]_t = \sum_{s\leq t} |\Delta Y^{n,jk}_s|^2 \xrightarrow{\mathcal{L}} 0 \qquad \forall t>0,\ \forall j,k\leq d$$

(l'égalité ci-dessus provient de ce que Y^n est à variation finie, donc sa partie martingale continue est nulle). Soit $j,k\leq d$ fixés. Soit

$$\alpha^n_t = \sup_{s\leq t} |\Delta Y^{n,jk}_s| , \qquad \beta^n_t = \mathrm{Var}(Y^{n,jk})_t ,$$

de sorte que

2.6 $$[Y^{n,jk},Y^{n,jk}]_t \leq \alpha^n_t\, \beta^n_t$$

L'hypothèse $[X^n,X^n] \xrightarrow{\mathcal{L}} A$ implique clairement que les suites $([X^{n,j},X^{n,j}])_{n\geq 1}$ vérifient 1.7, donc aussi les suites $(\langle X^{n,j},X^{n,j}\rangle)_{n\geq 1}$ d'après 2.2-(b), donc aussi la suite (F^n), où $F^n = \sum_{j\leq d}([X^{n,j},X^{n,j}] + \langle X^{n,j},X^{n,j}\rangle)$. Par ailleurs, il est clair que $\beta^n \leq F^n$. Si alors $\varepsilon>0$, $\eta>0$ sont donnés, il existe $b>0$ avec:

2.7 $$\sup_n P^n(\beta^n_t > b) \leq \frac{\eta}{2}$$

et comme (ΔY^n) vérifie 2.1 il existe n_o tel que

2.8 $$n > n_o \Longrightarrow P^n(\alpha^n_t > \frac{\varepsilon}{b}) \leq \frac{\eta}{2} .$$

En combinant 2.6, 2.7 et 2.8, on arrive à

$$n > n_o \implies P^n([Y^{n,jk},Y^{n,jk}]_t > \varepsilon) \leq \eta,$$

d'où 2.5. ∎

Comme première conséquence, nous en déduisons le corollaire suivant, tiré de Rebolledo [48].

THEOREME 2.9: *Soit* X^n *des martingales locales* d-*dimensionnelles nulles en* 0 , *vérifiant* $|\Delta X^n| \leq K$ *pour une constante* K. *Soit* X *une martingale gaussienne continue nulle en 0 (PAI), de caractéristiques* (0,C,0) . *Il y a équivalence entre:*

a) $X^n \xrightarrow{\mathcal{L}} X$

b) $[X^n,X^n] \xrightarrow{\mathcal{L}} C$

c) $[X^{n,j},X^{n,k}]_t \xrightarrow{\mathcal{L}} C_t^{jk}$ $\quad \forall t>0, \ \forall j,k\leq d$

d) $\langle X^{n,j},X^{n,k}\rangle_t \xrightarrow{\mathcal{L}} C_t^{jk}$ *pour tous* $t>0$, $j,k\leq d$; *et* $\sup_{s\leq t}|\Delta X_s^n| \xrightarrow{\mathcal{L}} 0$ *pour tout* $t>0$.

e) On a les conditions [γ] *et* [δ] *de III-2.1, ou de manière équivalente les conditions* [γ'] *et* [L] *de III-2.16* (noter que [Sup-β] et [Sup-β'] sont automatiquement satisfaites, car $B^n = 0$, et $B^{h,n} = 0$ dès que $h(x) = x$ pour $|x| < K$).

L'hypothèse $|\Delta X^n| \leq K$ est un peu forte: on pourrait faire mieux, à la manière de Rebolledo [48] ou de Liptcer et Shiryaev [37].

Preuve. D'après 2.3-(b), la condition $\sup_{s\leq t}|\Delta X_s^n| \xrightarrow{\mathcal{L}} 0$ équivaut à [δ] avec $\nu=0$. Comme $\nu^n([0,t]\times.)$ ne charge que la boule $\{x: |x|\leq K\}$, [δ] est elle-même équivalente à la condition de Lindeberg [L], et [γ] = [γ'] dès qu'on a choisit une fonction de troncation h vérifiant $h(x) = x$ pour $|x|<K$; enfin pour une telle fonction de troncation, on a $\tilde{C}^{h,n} = \langle X^n,X^n\rangle$. Les conditions (d) et (e) sont donc identiques, et elles entrainent (a) d'après III-2.1.

On a (a) ⟶ (b) d'après 1.5, et (b) ⟹ (c) est évident. L'implication (b) ⟶ (d) vient de la proposition 2.4. Il reste à montrer que (c) ⟶ (b). Comme $[X^n,X^n]_t - [X^n,X^n]_s$ est une matrice symétrique nonnégative pour $s\leq t$, cette implication se montre comme l'implication [γ] ⟶ [Sup-γ] dans III-2.1, par exemple. ∎

§b - CONVERGENCE VERS UNE SEMIMARTINGALE CONTINUE. On va maintenant donner une réciproque tout-à-fait partielle au théorème IV-3.2. Les X^n sont des semimartingales d-dimensionnelles de caractéristiques locales $(B^{h,n},C^n,\nu^n)$. On suppose par ailleurs qu'on a l'hypothèse IV-1.1 avec

2.10 $\qquad X$ est P-p.s. à trajectoires continues.

Les caractéristiques locales de X pour P sont $(B^h,C,0)$, et B^h ne dépend pas de la fonction de troncation h choisie.

On va supposer que $X^n \xrightarrow{\mathcal{L}} X$. Si $S^n_\rho = \inf(t: |X^n_t| \geq \rho)$ il en découle que $\lim_{\rho\uparrow\infty} \sup_n P^n(S^n_\rho < t) = 0$ pour tout t; par suite, avec les notations du chapitre IV, on a les équivalences:

2.11 $\qquad [\text{Sup-}\beta] \Leftrightarrow [\text{Sup-}\beta,\text{loc}], \quad [\gamma] \Leftrightarrow [\gamma,\text{loc}], \quad [\delta] \Leftrightarrow [\delta,\text{loc}]$

et comme $\nu = 0$ on a d'après 2.3:

2.12 $\qquad [\delta] \Longleftrightarrow \sup_{s\leq t}|\Delta X^n_s| \xrightarrow{\mathcal{L}} 0 \quad \forall t>0.$

THEOREME 2.13: *Soit 2.10. On suppose qu'on a les conditions IV-1.6 et IV-2.2. Si* $X^n \xrightarrow{\mathcal{L}} X$, *et si on a* $[\text{Sup-}\beta]$, *alors on a aussi* $[\gamma]$ *et* $[\delta]$.

Preuve. Etant donné 2.10, la suite $\{\mathcal{L}(X^n)\}$ est C-tendue, donc d'après 2.3 on a $[\delta]$ avec $\nu = 0$.

Soit (α_n) une suite de D^d convergeant vers α. D'après IV-1.6 on a $B^h_t(\alpha_n) \to B^h_t(\alpha)$ pour tout t. Si $N\in\mathbb{N}^*$ il existe $\rho>0$ tel que $S_\rho(\alpha_n) \geq N$ pour tout n; d'après IV-2.2 on a donc $w_N(B^h(\alpha_n),\delta) \leq w_N(F(\rho),\delta)$ pour tout n : on en déduit que la suite $B^h_\cdot(\alpha_n)$ est C-tendue, donc $B^h_\cdot(\alpha_n) \to B^h_\cdot(\alpha)$ uniformément sur les compacts. Autrement dit, l'application: $\alpha \rightsquigarrow B^h(\alpha)$ est continue de D^d dans D^d. On montre de la même manière que $\alpha \rightsquigarrow \tilde{C}^h(\alpha) = C(\alpha)$ est continue de D^d dans D^{d^2}, et on a déjà vu que $\alpha \rightsquigarrow X^h(\alpha)$ est continue de D^d dans D^d.

On déduit alors de l'hypothèse $X^n \xrightarrow{\mathcal{L}} X$ que

$$(X^{h,n}, B^h\circ X^n, \tilde{C}^h\circ X^n) \xrightarrow{\mathcal{L}} (X^h, B^h, C)$$

et par suite

2.14 $$(X^{h,n} - B^h\circ X^n, \tilde{C}^h\circ X^n) \xrightarrow{\mathcal{L}} (X^h - B^h, C).$$

On a $\tilde{X}^{h,n} = X^{h,n} - B^h\circ X^n + (B^h\circ X^n - B^{h,n})$. Donc $[\text{Sup-}\beta]$ et 2.14 entrainent

2.15 $$(\tilde{X}^{h,n}, \tilde{C}^h\circ X^n) \xrightarrow{\mathcal{L}} (\tilde{X}^h = X^h - B^h, C)) .$$

Les $\tilde{X}^{h,n}$ sont des martingales locales, à sauts bornés (uniformément en n). D'après le corollaire 1.5, renforcé par 1.6-(1), on déduit alors de 2.15 que

2.16 $$([\tilde{X}^{h,n},\tilde{X}^{h,n}], \tilde{C}^h\circ X^n) \xrightarrow{\mathcal{L}} ([\tilde{X}^h,\tilde{X}^h], C).$$

Enfin C est un processus continu, donc d'une part 2.9 entraine que

$[\tilde{X}^{h,n},\tilde{X}^{h,n}] - \tilde{C}^{h,n}$ converge en loi (donc en probabilité, uniformément sur les compacts) vers 0; d'autre part 2.16 entraine que $[\tilde{X}^{h,n},\tilde{X}^{h,n}] - \tilde{C}^{h}\circ X^{n}$ converge en loi vers $[\tilde{X}^{h},\tilde{X}^{h}] - C$, qui est nul, donc cette convergence est aussi en probabilité, uniforme sur les compacts. Par différence, on en déduit que $\tilde{C}^{h,n} - \tilde{C}^{h}\circ X^{n}$ converge en probabilité vers 0, uniformément sur les compacts, ce qui n'est autre que la condition [Sup-γ]. ∎

Comme corollaire, on en déduit un analogue du théorème 2.9, lorsque les X^n sont des martingales locales qui ne sont pas à sauts bornés; ce théorème est dû à Rootzen [50] et à Gänssler et Häusler [13] dans le cas des tableaux triangulaires, et à Liptcer et Shiryaev [36] dans le cas général.

THEOREME 2.17. *Soit* X^n *des martingales locales* d-*dimensionnelles nulles en* 0 ; *soit* X *une martingale gaussienne continue nulle en* 0 *(PAI) de caractéristiques* $(0,C,0)$. *Supposons que pour une constante réelle* K *on ait*

2.18 $$\sup_{s\leq t}|\, x\, 1_{\{|x|>K\}}*\nu^n_s| \xrightarrow{\mathcal{L}} 0 \qquad \forall t>0.$$

Il y a alors équivalence entre:

a) $X^n \xrightarrow{\mathcal{L}} X$;

b) les conditions [γ] *et* [δ] *de III-2.1 sont satisfaites.*

Noter que si $|\Delta X^n|\leq K$ on a automatiquement 2.18: on retrouve donc l'équivalence (a)$\Longleftrightarrow$(e) du théorème 2.9.

Preuve. Remarquons d'abord que la condition 2.18 a un sens: en effet, si X^n est une martingale locale, il est bien connu que $|x|1_{\{|x|>K\}}*\nu^n_t<\infty$ pour tous $t>0$, $K>0$ (voir [23]; la preuve est analogue à celle de III-1.48). De plus, pour toute fonction de troncation h, on a $B^{h,n} = \{h(x) - x\}*\nu^n$ (même chose qu'en III-1.48 -(i), avec $B = 0$).

Il est alors évident de vérifier que [δ] (avec $\nu = 0$) et 2.18 pour un $K>0$ entrainent 2.18 pour tout $K>0$, et [Sup-β] pour toute fonction de troncation h avec $B^h = 0$. L'implication (b)$\Longrightarrow$(a) vient alors de IV-2.1.

Supposons inversement (a) et 2.18. On a [δ] avec $\nu = 0$ d'après 2.3. Ce qui précède montre alors qu'on a [Sup-β], et le théorème 2.13 donne alors le résultat (on a IV-1.6 et IV-2.2, car C est déterministe). ∎

BIBLIOGRAPHIE

1 D. ALDOUS: Stopping times and tightness. Ann. Probab. 6, 335-340, 1978

2 D. ALDOUS: Weak convergence of stochastic processes, for processes viewed in the Strasbourg manner. Preprint, 1978

3 P. BILLINGSLEY: Convergence of Probability measures. Wiley, 1968

4 P. BILLINGSLEY: Conditional distributions and tightness. Ann. Probab. 2, 480-485, 1974

5 B.M. BROWN: Martingale central limit theorems. Ann. Math. Stat. 42, 59-66, 1971

6 B.M. BROWN, G.K.EAGLESON: Martingale convergence to infinitely divisible laws with finite variance. Trans. A.M.S. 162, 449-453, 1971

7 T. BROWN: A martingale approach to the Poisson convergence of simple point processes. Ann. Probab. 6, 615-628, 1978

8 T. BROWN: Some distributional approximations for random measures. PhD Thesis, Cambridge, 1979.

9 C. DELLACHERIE, P.A. MEYER: Probabilités et potentiel, tomes I et II, Hermann, 1976 et 1980

10 J.L. DOOB: Stochastic Processes. Wiley, 1953

11 R. DURRETT, S.I. RESNICK: Functional limit theorems for dependent variables. Ann. Probab. 6, 829-846, 1978

12 R. ELLIOTT: Stochastic calculus and applications. Springer, 1982

13 P. GANSSLER, E. HAUSLER: Remarks on the functional central limit theorem for martingales; Z. für Wahr. 50, 237-243, 1979

14 E. GINE, M.B. MARCUS: The central limit theorem for stochastic integrals with respect to Lévy processes. Ann. Probab. 11, 53-77, 1983

15 B.W. GNEDENKO, A.N. KOLMOGOROV: Limit distributions for sums of independent random variables. Addison-Wesley, 1954

16 B. CRIGELIONIS: On relatice compactness of sets of probability measures in $D_{[0,\infty[}(R)$. Litov. Math Sb. XIII, 4, 83-96, 1973

17 B. GRIGELIONIS, R. MIKULEVICIUS: On weak convergence of semimartingales. Litov. Math. Sb. XXI, 3, 9-24, 1981

18 B. GRIGELIONIS, V.A. LEBEDEV: Nouveaux critères de compacité pour des suites de probabilités. Usp. Math. 37, 6, 29-37, 1982

19 B. GRIGELIONIS, K. KUBILIUS, R. MIKULEVICIUS: Méthodes de martingales dans les théorèmes limite fonctionnels. U.S.P. Math. 37, 6, 39-51, 1982.

20 P. HALL, C. HEYDE: Martingale limit theory and its applications. Academic Press, 1980

21 I.S. HELLAND: On weak convergence to brownian motion. Z. für Wahr. 52, 251-265, 1980

22 I.S. HELLAND: Minimal conditions for weak convergence to a diffusion process on the line. Ann. Probab. 9, 429-452, 1981

23 J. JACOD: Calcul stochastique et problèmes de martingales. Springer Lect. Notes in Math. 714, 1979

24 J. JACOD, J. MEMIN: Sur la convergence des semimartingales vers un processus à accroissements indépendants. Sém. Proba. XIV, Lect. Notes 784, 227-249, 1980

25 J. JACOD: Convergence en loi de semimartingales et variation quadratique. Sém. Proba. XV, Lect. Notes 850, 547-560, 1981

26 J. JACOD, A. KŁOPOTOWSKI, J. MEMIN: Théorème de la limite centrale et convergence fonctionnelle vers un processus à accroissements indépendants, la méthode des martingales. An. Inst. H. Poincaré (B) XVIII, 1-45, 1982

27 J. JACOD, J. MEMIN, M. METIVIER: On tightness and stopping times. Stoch. Proc. and Appl. 14, 109-146, 1983

28 J. JACOD: Processus à accroissements indépendants: une condition nécessaire et suffisante de convergence. Z. für Wahr. 63, 109-136, 1983

29 Y. KABANOV, R. LIPTCER, A. SHIRYAEV: Some limit theorems for simple point processes. Stochastics, 3, 203-206, 1980

30 T.G. KURTZ: Limit theorems for sequences of jump Markov processes approximating ordinary differential equations. J. Appl. Probab. 8, 344-356, 1971

31 T.G. KURTZ: Semigroups of conditioned shifts and approximation of Markov processes. Ann. Probab. 3, 618-642, 1975

32 V.A. LEBEDEV: On the weak compactness of families of distributions of general semimartingales. Theor. Probab. Appl. XXVII, 1, 1982

33 E. LENGLART: Relation de domination entre deux processus. Ann. Inst. H. Poincaré (B) XIII, 171-179, 1977

34 T. LINDVALL: Weak convergence of probability measures and random functions in the function space $D[0,\infty[$. J. Appl. Probab. 10, 109-121, 1973

35 R. LIPTCER, A. SHIRYAEV: Théorème central limite fonctionnel pour les semimartingales. Theor. Probab. Appl. XXV, 683-703, 1980

36 R. LIPTCER, A. SHIRYAEV: On necessary and sufficient conditions in functional central limit theorem for semimartingales. Theor. Probab. AppL. XXVI, 132-137, 1981

37 R. LIPTCER, A. SHIRYAEV: On a problem of necessary and sufficient conditions in the functional central limit theorem for local martingales. Z. für Wahr. 59, 311-318, 1982

38 R. LIPTCER, A. SHIRYAEV: Convergence faible de semimartingales vers un processus de type diffusion. Math. Sb. 121, 2, 176-200, 1983

39 N. MAIGRET: Théorèmes de limite centrale fonctionnels pour une chaîne de Markov récurrente au sens de Harris et positive. Ann. Inst.H.Poincaré (B) XIV,425-440,1978

40 D.L. McLEISH: An extended martingale invariance principle. Ann. Probab. 6, 144-150, 1978

41 M. METIVIER: Une condition suffisante de compacité faible pour une suite de processus. Rapport Ecole Polytechnique, 1980

42 M. METIVIER: Semimartingales. De Gruyter, 1982

43 P.A. MEYER: Un cours sur les intégrales stochastiques. Sém. Proba. X, Lect. Notes in Math. 511, 245-400, 1976

44 K.R. PARTHASARATHY: Probability measures on metric spaces. Academic Press, 1967

45 V.V. PETROV: Sums of independent random variables. Springer, 1975

46 R. REBOLLEDO: La méthode des martingales appliquée à la convergence en loi des processus. Mémoire S.M.F., 62, 1979

47 R. REBOLLEDO: Sur l'existence de solutions à certains problèmes de martingales. Comptes Rendus Acad. Sc. (A) 290, 843-846, 1980

48 R. REBOLLEDO: Central limit theorems for local martingales. Z. für Wahr. 51, 269-286, 1980

49 B. ROSEN: On the central limit theorem for sums of dependent random variables. Z. für Wahr. 7, 48-82, 1967

50 H. ROOTZEN: On the functional central limit theorem for martingales. Z. für Wahr. 51, 79-94, 1980

51 A.V. SKOROKHOD: Limit theorems for stochastic processes. Theor. Probab. Appl. I, 261-290, 1956

52 C. STONE: Weak convergence of stochastic processes defined on a semi-finite time interval. Proc. Am. Math. Soc. 14, 694-696, 1963

53 D. STROOCK, S.VARADHAN: Diffusion processes with continuous coefficients, I, II, Comm. Pure Appl. Math. 22, 345-400, 479-530, 1969

54 D. STROOCK: Diffusion processes associated with Lévy generators. Z. für Wahr. 32, 209-244, 1975

55 D. STROOCK, S. VARADHAN: Multidimensional diffusion processes. Springer, 1979

56 A. TOUATI: Théorèmes de limite centrale fonctionnelle pour les processus de Markov. Ann. Inst. H. Poincaré (B) XIX, 43-55, 1983